Quantum Optics and Quantum Computation

An introduction

IOP Series in Advances in Optics, Photonics and Optoelectronics

SERIES EDITOR

 Professor Rajpal S Sirohi Consultant Scientist

About the Editor

Rajpal S Sirohi is currently working as a faculty member in the Department of Physics, Alabama A&M University, Huntsville, Alabama (USA). Prior to this, he was a consultant scientist at the Indian Institute of Science Bangalore, and before that he was chair professor in the Department of Physics, Tezpur University, Assam. During 2000–11, he was academic administrator, being vice chancellor to a couple of universities and the director of the Indian Institute of Technology Delhi. He is the recipient of many international and national awards and the author of more than 400 papers. Dr Sirohi is involved with research concerning optical metrology, optical instrumentation, holography, and speckle phenomenon.

About the series

Optics, photonics and optoelectronics are enabling technologies in many branches of science, engineering, medicine and agriculture. These technologies have reshaped our outlook, our way of interaction with each other and brought people closer. They help us to understand many phenomena better and provide a deeper insight in the functioning of nature. Further, these technologies themselves are evolving at a rapid rate. Their applications encompass very large spatial scales from nanometers to astronomical and a very large temporal range from picoseconds to billions of years. The series on the advances on optics, photonics and optoelectronics aims at covering topics that are of interest to both academia and industry. Some of the topics that the books in the series will cover include bio-photonics and medical imaging, devices, electromagnetics, fiber optics, information storage, instrumentation, light sources, CCD and CMOS imagers, metamaterials, optical metrology, optical networks, photovoltaics, freeform optics and its evaluation, singular optics, cryptography and sensors.

About IOP ebooks

The authors are encouraged to take advantage of the features made possible by electronic publication to enhance the reader experience through the use of colour, animation and video, and incorporating supplementary files in their work.

Do you have an idea of a book you'd like to explore?

For further information and details of submitting book proposals see iopscience.org/books or contact Ashley Gasque on Ashley.gasque@iop.org.

Quantum Optics and Quantum Computation

An introduction

Dipankar Bhattacharyya and Jyotirmoy Guha

Department of Physics, Santipur College, Santipur, Nadia, West Bengal, India

IOP Publishing, Bristol, UK

ISBN   978-0-7503-2715-2 (ebook)
ISBN   978-0-7503-2713-8 (print)
ISBN   978-0-7503-2716-9 (myPrint)
ISBN   978-0-7503-2714-5 (mobi)

DOI   10.1088/978-0-7503-2715-2

Version: 20220101

IOP ebooks

British Library Cataloguing-in-Publication Data: A catalogue record for this book is available from the British Library.

Published by IOP Publishing, wholly owned by The Institute of Physics, London

IOP Publishing, Temple Circus, Temple Way, Bristol, BS1 6HG, UK

US Office: IOP Publishing, Inc., 190 North Independence Mall West, Suite 601, Philadelphia, PA 19106, USA

Dedicated to Our teachers

Professor Pradip Narayan Ghosh

Former Vice Chancellor, Jadavpur University, West Bengal, India.

Former Professor of Physics, Calcutta University, West Bengal, India

and

Professor Padmanava Dasgupta

Former Professor of Physics, University of Kalyani, West Bengal, India

A teacher affects eternity; He can never tell where his influence stops.

Contents

Preface

The topics that the book deals with are very current and are the burning topics of the day. It is argued that information theory and quantum mechanics fit together very well. We in the book explore this through study of quantum mechanics as applied to some concepts that are getting increasingly popular with each passing day, e.g. quantum optics, quantum cryptography, teleportation, quantum computing etc.

Special effort and consideration have been given to make the approach attractive, putting greater emphasis on illustrations, problem solving and keeping the approach simple and to the point.

We thank IOP Publishing for publishing this book.

Dr Dipankar Bhattacharyya
(bh.dipankar@gmail.com)

Dr Jyotirmoy Guha
(jgsantipurcollege@gmail.com,
https://studio.youtube.com/channel/UCJ1JOEqmmVeyAv5XRY55fiw)

Santipur College, Santipur, Nadia, West Bengal, India

April 21, 2021

Acknowledgments

We are thankful to the Institute of Physics (IOP) Publising for giving us such a wonderful opportunity. Ms Ashley Gasque needs special mention for her support and for keeping in touch with us all through.

We would like to thank all our departmenal colleagues of Santipur College especially Dr Atreyi Paul and Dr Anita Gangopadhyay for their constant inspiration.

Dr Dipankar Bhattacharyya wants to thank his PhD supervisor Professor Pradip Narayan Ghosh, Department of Physics, University of Calcutta, India and his postdoctoral supervisor Professor Nir Davidson, Dept of Physics and Complex System, Weizmann Institute of Science, Israel, both of whom are excellent researchers and teachers who drew his attention to this field.

Dr Bhattacharyya gratefully acknowledges the suggestions and encouragement he received from Dr Amitava Bandyopadhyay, Department of Physics, Visva Bharati as well as his discussions with some PhD scholars with whom he worked for the last ten years like Dr Bankim Chandra Das, Dr Suman Mondal, Dr Arpita Das and Dr Khairul Islam, to mention a few.

Dr Bhattacharyya also acknowledges and thanks his co-author Dr Jyotirmoy Guha for his valuable suggestions and help.

Author biographies

Dipankar Bhattacharyya

Dr Dipankar Bhattacharyya is an Associate Professor of Physics, Department of Physics, Santipur College, Nadia, W.B., India. He did his PhD at the University of Calcutta, India on Laser Spectroscopy and later went to Weizmann Institute of Science, Israel for Postdoctoral research work with Feinberg Graduate School Fellowship. He has published about thirty publications all in international journals. He was the principal investigator of three research projects funded by Govt. of India.

Jyotirmoy Guha

Dr Jyotirmoy Guha, Associate Professor of Physics and currently Head of the Department of Physics, Santipur College, Santipur, Nadia, West Bengal, India has a brilliant academic record and did his PhD in Quantum cosmology. He is the author of books like *Quantum Mechanics* (*Theory, Problems and Solutions*) published by Books and Allied, Kolkata; *Solid State Physics* (*Theory, Problems and Solutions*) published by Books and Allied, Kolkata; *Modern Physics* (Volume I and II) published by Techno World, Kolkata and regularly contributes in his YouTube channel jg#physics (https://studio.you-tube.com/channel/UCJ1JOEqmmVeyAv5XRY55fiw).

Chapter 1

Bra ket algebra of Dirac

1.1 The bra and ket notation of Dirac

Dirac introduced a natural notation for quantum mechanics.

Dirac introduced the symbol $|>$ called ket symbol, and $|a>$ is called the ket vector. It is represented by a column matrix whose elements may be complex in general.

$$|a> = \begin{pmatrix} a_1 \\ a_2 \end{pmatrix} \text{ in 2D}, \quad |a> = \begin{pmatrix} a_1 \\ a_2 \\ a_3 \end{pmatrix} \text{ in 3D}.$$

where a_1, a_2, a_3 are elements of column matrix, in general complex.

Dirac also introduced the symbol $< |$ called the bra symbol and $< b |$ is called the bra vector. It is represented by a row matrix whose elements may be complex in general.

$<b| = (b_1 \ b_2)$ in 2D, $< b | = (b_1 \ b_2 \ b_3)$ in 3D where b_1, b_2, b_3 are elements of row matrix, in general complex.

We can extend the definitions to higher dimensions, say n dimensions.

- *Kets can be added.*
  $|a> + |b> = |c>$
  since
  $$\begin{pmatrix} a_1 \\ a_2 \end{pmatrix} + \begin{pmatrix} b_1 \\ b_2 \end{pmatrix} = \begin{pmatrix} a_1 + b_1 \\ a_2 + b_2 \end{pmatrix} = \begin{pmatrix} c_1 \\ c_2 \end{pmatrix}$$

- *We can multiply a ket.*
  $c|a> = |b>$
  since for a complex number c
  $$c\begin{pmatrix} a_1 \\ a_2 \end{pmatrix} = \begin{pmatrix} ca_1 \\ ca_2 \end{pmatrix} = \begin{pmatrix} b_1 \\ b_2 \end{pmatrix}$$

- *Bras can be added.*
  $< a | + < b | = < d |$
  since $(a_1 \ a_2) + (b_1 \ b_2) = (a_1 + b_1 \ a_2 + b_2) = (c_1 \ c_2)$

- *We can multiply a bra.*
  $c < a | = < b |$
  since for a complex number c
  $c(a_1 \ a_2) = (ca_1 \ ca_2) = (b_1 \ b_2)$

1.2 Hermitian conjugation

The Hermitian conjugate denoted by a dagger symbol †, is the transpose of complex conjugate or equivalently complex conjugate of transpose.

Ket and bra are Hermitian conjugates of each other i.e.

$(< |)^\dagger = | >$ (i.e. bra† = ket) and $(| >)^\dagger = < |$ (i.e. ket† = bra)

To each ket vector $|a>$ we can associate a bra vector $|a>^\dagger = < a|$

To each bra vector $< a |$ we can associate a ket vector $< a |^\dagger = |a>$

- *Examples:*

  ✓ If $|a> = \begin{pmatrix} 1 \\ i \end{pmatrix}$ then $< a | = |a>^\dagger = \begin{pmatrix} 1 \\ i \end{pmatrix}^\dagger = (1 \quad -i)$

  ✓ If $< a | = (1 \quad i)$ then $|a> = < a |^\dagger = (1 \quad i)^\dagger = \begin{pmatrix} 1 \\ -i \end{pmatrix}$

We provide some more *examples* for illustration purposes.

- $5^\dagger = 5$
- $(1 + i)^\dagger = 1 - i$

(Transpose of a number is itself)

- $\begin{pmatrix} 1 & 2 \\ -3i & 3+2i \end{pmatrix}^\dagger = \begin{pmatrix} 1 & 3i \\ 2 & 3-2i \end{pmatrix}$

- $(a_1 \quad a_2)^\dagger = \begin{pmatrix} a_1^* \\ a_2^* \end{pmatrix}$ • $\begin{pmatrix} a_1 \\ a_2 \end{pmatrix}^\dagger = (a_1^* \quad a_2^*)$

- $(i \quad 1)^\dagger = \begin{pmatrix} -i \\ 1 \end{pmatrix}$ • $\begin{pmatrix} i \\ 1 \end{pmatrix}^\dagger = (-i \quad 1)$

- $(1 \quad i)^\dagger = \begin{pmatrix} 1 \\ -i \end{pmatrix}$ • $\begin{pmatrix} 1 \\ 0 \end{pmatrix}^\dagger = (1 \quad 0)$

Hermitian conjugation means moving from ket to bra or moving from bra to ket.

- If $|a> = c|b>$ find $<a|$.

$<a| = |a>^\dagger = [c|b>]^\dagger = c* < b|$

- $<a| = c < b|$ find $|a>$.

$|a> = <a|^\dagger = [c < b|]^\dagger = c*|b>$

1.3 Definition of inner product (also called overlap)

Inner product is the product of *bra and ket*. We can construct a *number* (a *complex number* in general) using bra and ket.

Consider

$$|a> = \begin{pmatrix} a_1 \\ a_2 \end{pmatrix} \Rightarrow <a| = (a_1^* \quad a_2^*)$$

$$|b> = \begin{pmatrix} b_1 \\ b_2 \end{pmatrix} \Rightarrow <b| = \begin{pmatrix} b_1^* & b_2^* \end{pmatrix}$$

Inner product of $|a>$ and $|b>$ is

$$<a||b> = <a|b> = (a_1^* \quad a_2^*)\begin{pmatrix} b_1 \\ b_2 \end{pmatrix} = a_1^* b_1 + a_2^* b_2$$

$$= \text{a complex number (in general).}$$

Inner product of $|b>$ and $|a>$ is

$$<b||a> = <b|a> = (b_1^* \quad b_2^*)\begin{pmatrix} a_1 \\ a_2 \end{pmatrix} = b_1^* a_1 + b_2^* a_2$$

$$= \text{a complex number (in general).}$$

It is clear that:
- $<a|b> \neq <b|a>$ (*not commutative* in general)
- $<a|b> = <b|a>^* = a_1^* b_1 + a_2^* b_2$
- $<b|a> = <a|b>^* = b_1^* a_1 + b_2^* a_2$

1.4 Definition of outer product

Outer product is the product of *ket and bra*. We can construct a *matrix* using ket and bra. Again a matrix can be represented by an operator called *matrix operator*. We use the symbol of a cap or a hat to denote an operator or a matrix operator. (The symbol of a cap or a hat is also used to denote unit vectors. So we should be aware of the context of using such symbols.)

Let $|a> = \begin{pmatrix} a_1 \\ a_2 \end{pmatrix} \Rightarrow <a| = (a_1^* \quad a_2^*)$, $|b> = \begin{pmatrix} b_1 \\ b_2 \end{pmatrix} \Rightarrow <b| = (b_1^* \quad b_2^*)$

Outer product of $|a>$ and $|b>$ is

$$|a> <b| = \begin{pmatrix} a_1 \\ a_2 \end{pmatrix}(b_1^* \quad b_2^*) = \begin{pmatrix} a_1 b_1^* & a_1 b_2^* \\ a_2 b_1^* & a_2 b_2^* \end{pmatrix} = \text{matrix operator}$$

Outer product of $|b>$ and $|a>$ is

$$|b> <a| = \begin{pmatrix} b_1 \\ b_2 \end{pmatrix}(a_1^* \quad a_2^*) = \begin{pmatrix} b_1 a_1^* & b_1 a_2^* \\ b_2 a_1^* & b_2 a_2^* \end{pmatrix} = \text{matrix operator}$$

It is clear that : $|a> <b| \neq |b> <a|$
We give an *example* for illustration purposes.

- Let $|1> = \begin{pmatrix} 1 \\ 0 \end{pmatrix} \Rightarrow <1| = (1 \quad 0)$, $|2> = \begin{pmatrix} 0 \\ 1 \end{pmatrix} \Rightarrow <2| = (0 \quad 1)$

With $|1>$, $|2>$ we can construct the following *numbers*.

$$<1|2> = (1 \quad 0)\begin{pmatrix} 0 \\ 1 \end{pmatrix} = 0 \quad <2|1> = (0 \quad 1)\begin{pmatrix} 1 \\ 0 \end{pmatrix} = 0$$

$$<1|1> = (1 \quad 0)\begin{pmatrix} 1 \\ 0 \end{pmatrix} = 1 \quad <2|2> = (0 \quad 1)\begin{pmatrix} 0 \\ 1 \end{pmatrix} = 1$$

With $|1>$, $|2>$ we can construct the following *operators* or *matrix operators*.

$$|1><2| = \begin{pmatrix} 1 \\ 0 \end{pmatrix}(0 \quad 1) = \begin{pmatrix} 0 & 1 \\ 0 & 0 \end{pmatrix} \quad |2><1| = \begin{pmatrix} 0 \\ 1 \end{pmatrix}(1 \quad 0) = \begin{pmatrix} 0 & 0 \\ 1 & 0 \end{pmatrix}$$

$$|1><1| = \begin{pmatrix} 1 \\ 0 \end{pmatrix}(1 \quad 0) = \begin{pmatrix} 1 & 0 \\ 0 & 0 \end{pmatrix} \quad |2><2| = \begin{pmatrix} 0 \\ 1 \end{pmatrix}(0 \quad 1) = \begin{pmatrix} 0 & 0 \\ 0 & 1 \end{pmatrix}$$

(For two kets $|a>$ and $|b>$, we *cannot* write $|a>|b>$ since we *cannot* multiply two column matrices.)

1.5 Eigenvalue equation

An *operator* $\hat{A}$ (denoted by cap or hat sign) acts on a *function* to produce a *function* (same or different). If the *same function is reproduced* the *equation is called eigenvalue equation* (*eigen* means *characteristic*).

- $\dfrac{d}{dx} x^2 = 2x$ is *not* an eigenvalue equation since $\frac{d}{dx}$ *does not reproduce* x^2.
- $\dfrac{d}{dx} e^{5x} = 5\, e^{5x}$ is an eigenvalue equation since $\frac{d}{dx}$ *reproduces* e^{5x}.

$\hat{A}|n> = n|n>$ is an *eigenvalue equation* since $\hat{A}$ operates on $|n>$ to *reproduce* $|n>$ multiplied by a constant n. Similarly, $\hat{A}\Psi = a\Psi$ is an eigenvalue equation with eigen function Ψ and eigenvalue a.

1.6 Linear vector space

Vector space refers to the collection of elements called vectors $\vec{a}_n \rightarrow \vec{a}_1, \vec{a}_2, \vec{a}_3,\dots$ or ket vectors $|a_n> \rightarrow |a_1>, |a_2>, |a_3>,\dots$ obeying a *set of specific rules* that determine their behavior. The rules or definitions are

Rules/definitions	*Remarks*
1. $\|1>+\|2> = \|2>+\|1>$	Commutative law obeyed.
2. $\|1>+(\|2>+\|3>) = (\|1>+\|2>)+\|3>$	Associative law obeyed.
3. $0+\|n> = \|n>+0 = \|n>$	$0 =$ null element or null ket
4. $\|n>+\|-n> = 0 =$ null element	

5. Any *linear combination* of elements is also an element of the *same* vector space. *Hence it is called linear vector space.*
6. To form a linear combination of elements we need *scalars or numbers*. It is thus necessary to include scalars or numbers (real numbers as well as complex numbers) in the domain of linear vector space. *Accordingly, we think of real vector space and complex vector space.*

$a(\|n>+\|m>) = a\|n>+a\|m>$	Distributive law obeyed.
$a =$ complex number (in general)	
7. $c\|\psi> = \|c\psi>$	Product of a vector with a scalar yields a vector in the *same* linear vector space.

8. $(ab)|\psi> = a(b|\psi>)$
9. $(a + b)|\psi> = a|\psi>+b|\psi>$
10. $0|\psi> = 0 =$ null element or null ket

Linear vector space of *n* dimension is often denoted by C^n.

1.7 Linear independence

A set of elements (ket vectors |1 > , |2 > , |3 >...... say) are linearly independent if the relation

$$a_1|1 > +a_2|2 > +a_3|3 > + ... =0 \text{ holds for } a_n = 0 \text{ for } all \text{ } n.$$

This means that the ket vectors (|1 > , |2 > , |3 >...) *cannot be linked or connected*. In other words any one element *cannot be expressed* in terms of the other elements. So the *component or projection* of one element along any other element is *zero*.

- *Compare*

We *cannot* link or connect the unit vectors $\hat{i}, \hat{j}, \hat{k}$ (figure 1.1), since $a_1\hat{i} + a_2\hat{j} + a_3\hat{k} = 0$ is possible *only with* $a_1 = 0$, $a_2 = 0$, $a_3 = 0$ and $\hat{i}. \hat{j} = 0$ means $\hat{i}$ has no portion of it along $\hat{j}$ implying that they are *unrelated or independent*. (The caps on $\hat{i}, \hat{j}$ denote unit vectors.)

1.8 Linear dependence

A set of elements (ket vectors |1 > , |2 > , |3 >...... say) are linearly dependent if the relation

$$a_1|1 > +a_2|2 > +a_3|3 > + ... = 0 \text{ holds where } at \text{ } least \text{ } two \text{ } a_n \text{ s are not zero.}$$

This means that the ket vectors (|1 > , |2 > , |3 > ......) *can be linked or connected* as shown in the following example.

For four kets |1 > , |2 > , |3>, |4> suppose we can find a relation like $a_1|1 > +a_2|2 > +a_3|3 > +a_4|4 > =0$ for *at least* $a_1 \neq 0$, $a_2 \neq 0$ and then one element *can be expressed* in terms of other elements as

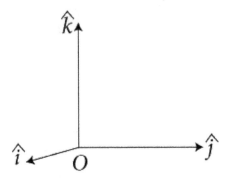

Figure 1.1. Set of independent vectors

$$|1> = -\frac{a_2}{a_1}|2> - \frac{a_3}{a_1}|3> - \frac{a_4}{a_1}|4>$$

So the component or projection of one element along other elements may not be all zero.

- *Compare*

We *can link or connect* the unit vectors $\hat{i}$, $\hat{j}$, $\hat{k}$, $\hat{r}$ (figure 1.2) and express one in terms of the others as follows:

$$\hat{r} = \frac{\vec{r}}{r} = \frac{x\hat{i} + y\hat{j} + z\hat{k}}{r} = \frac{x}{r}\hat{i} + \frac{y}{r}\hat{j} + \frac{z}{r}\hat{k}$$

$$\Rightarrow \hat{r} - \frac{x}{r}\hat{i} - \frac{y}{r}\hat{j} - \frac{z}{r}\hat{k} = 0$$

$$\Rightarrow a_1\hat{r} + a_2\hat{i} + a_3\hat{j} + a_4\hat{k} = 0 \ (a_1 = 1, a_2 = -\frac{x}{r}, a_3 = -\frac{y}{r}, a_4 = -\frac{z}{r})$$

Moving along $\hat{r}$ means, movement along $\hat{i}$, $\hat{j}$, $\hat{k}$ also. As $\hat{r} \cdot \hat{i} = \frac{x}{r} \neq 0$ this means $\hat{r}$ has a portion of it along $\hat{i}$ implying that they are dependent. ($\hat{i}$, $\hat{j}$, $\hat{k}$, $\hat{r}$ denote unit vectors.)

1.9 Span (expansion of an arbitrary ket)/expansion postulate

A set of ket vectors $|\phi_1>$, $|\phi_2>$, $|\phi_3>$, ... in the linear vector space is said to span the linear vector space if every vector (any vector), say $|\psi>$, in this space *can be expanded as a linear combination* of these vectors. So if we can write

$$|\psi> = a_1|\phi_1> + a_2|\phi_2> + a_3|\phi_3> + ...$$

then it is said that $|\phi_1>$, $|\phi_2>$, $|\phi_3>$, ... span the linear vector space.

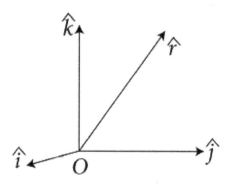

Figure 1.2. Set of dependent vectors.

In other words we can expand every vector (any vector) in this space as a *linear combination* of the set of ket basis vectors $|\phi_1>$, $|\phi_2>$, $|\phi_3>$, ... This is called expansion postulate or the *principle of linear superposition*.

NOTE

1. *Linear independence does not imply span*

    In a 3D space $\hat{i}$, $\hat{j}$ are linearly independent but they *do not* span 3D linear vector space as we *cannot* express any 3D vector completely in terms of $\hat{i}$, $\hat{j}$.

2. *Span does not imply linear independence*

Consider a vector in 3D space

$$\vec{A} = 2\hat{i} + 2\hat{j} + \hat{k} = \hat{i} + \hat{j} + \hat{k} + (\hat{i} + \hat{j}) = \hat{i} + \hat{j} + \hat{k} + \vec{e}, \text{ where } \vec{e} = \hat{i} + \hat{j}.$$

Clearly the vectors $\hat{i}, \hat{j}, \hat{k}, \vec{e}$ *span the space*. But $\hat{i}, \hat{j}, \hat{k}, \vec{e}$ are *not linearly independent* as $\vec{e} = \hat{i} + \hat{j}$ depends on $\hat{i}$ and $\hat{j}$. (Here $\hat{i}, \hat{j}, \hat{k}$ are unit vectors and $\vec{e}$ is a vector.)

1.10 Ket space, bra space, dual space

When the linear vector space is described in terms of ket vectors it is called a ket vector space. When the linear vector space is described in terms of bra vectors it is called a bra vector space or a dual linear vector space. As ket vectors and bra vectors are Hermitian conjugates of one another (i.e. are duals of each other having one-to-one correspondence), and are defined in the same linear vector space, we call it a dual space. The dual space contains both ket vectors and their dual vectors—the bra vectors.

1.11 Physical significance of inner product $<m|n>$

1. $<m|n>$ represents a number—*complex number in general.*
2. $<m|n>$ represents the *projection* of the ket $|n>$ along the ket $|m>$

- *Compare*

This is similar to the dot product $\hat{b}.\vec{a}$ between two vectors $\vec{a}$ and $\vec{b} = b\hat{b}$ which generates a scalar. Also the dot product $\hat{b}.\vec{a}$ represents the projection of vector $\vec{a}$ along vector $\vec{b}$. (In $\vec{b} = b\hat{b}$, $\vec{b}$ is the vector, b is the modulus or magnitude of the vector, $\hat{b}$ is the unit vector along the vector $\vec{b}$.)

However, scalar product in real vector space is order independent, that is, commutative, since $\hat{b}.\vec{a} = \vec{a}.\hat{b}$. But in complex linear vector space, inner product is, in general, order dependent, that is, *not commutative*, since $<m|n> \neq <n|m>$.

3. Value of $<i|j>$ decides whether ket $|i>$ and ket $|j>$ are *orthogonal or not*.

If $<i|j> = 0$ then $|i>$ and $|j>$ are *orthogonal*, while if $<i|j> \neq 0$ then they are *not orthogonal* (figure 1.3).

- *Compare*

For orthogonal vectors the dot product *vanishes*, while for non-orthogonal vectors the dot product *does not vanish*.

For instance, $\hat{i}$ and $\hat{j}$ are *orthogonal* but $\hat{r}$ and $\hat{i}$ are *not orthogonal*.

- Example:

- If $|m> = \begin{pmatrix} i \\ 1 \end{pmatrix}$, $|n> = \begin{pmatrix} 1 \\ i \end{pmatrix}$ then

$$< m|n> = (-i \quad 1)\begin{pmatrix} 1 \\ i \end{pmatrix} = -i + i = 0 \Rightarrow |m> \text{ is } orthogonal \text{ to } |n>.$$

- If $|m> = \begin{pmatrix} 0 \\ 1 \end{pmatrix}$, $|n> = \begin{pmatrix} 1 \\ i \end{pmatrix}$ then

$$< m|n> = (0 \quad 1)\begin{pmatrix} 1 \\ i \end{pmatrix} = 0 + i = i \neq 0 \Rightarrow |m> \text{ is } not \text{ orthogonal to } |n>.$$

4. Value of $<m|m>$ decides whether ket $|m>$ is *normalized to unity or not*.

If $<m|m> = 1$ then $|m>$ is *normalized* to unity, while if $<m|m> \neq 1$ then $|m>$ is *not normalized to unity* (figure 1.4).

We call $\sqrt{<m|m>} = $ *norm of ket vector* $|m>$.

If *norm* is unity then $|m>$ is normalized to unity. If *norm* is not unity then $|m>$ is not normalized to unity.

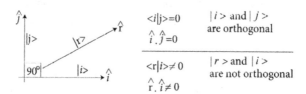

Figure 1.3. Inner product is zero for orthogonal kets and non-zero if kets are not orthogonal.

Figure 1.4. $<a|a> = 1$ means $|a>$ is normalized. $<a|a> \neq 1$ means $|a>$ is not normalized.

- *Compare*

$\vec{a} \cdot \vec{a}$ decides whether vector $\vec{a}$ has *unit norm*, i.e. unit magnitude.

For instance $\vec{a} = \hat{i}$ is *normalized to unity* since $\vec{a} \cdot \vec{a} = \hat{i} \cdot \hat{i} = 1$ but $\vec{a} = 3\hat{i}$ is *not* normalized to unity since $\vec{a} \cdot \vec{a} = 3\hat{i} \cdot 3\hat{i} = 9 \neq 1$.

- Example:

  - If $\quad |m> = \dfrac{1}{\sqrt{2}}\begin{pmatrix} i \\ 1 \end{pmatrix} \quad$ then $\quad <m|m> = \dfrac{1}{\sqrt{2}}(-i \ \ 1)\dfrac{1}{\sqrt{2}}\begin{pmatrix} i \\ 1 \end{pmatrix} = \dfrac{1}{2}(1 + 1) = 1$
    (*normalized*).
  - If $|m> = \begin{pmatrix} i \\ 1 \end{pmatrix}$ then $<m|m> = (-i \ \ 1)\begin{pmatrix} i \\ 1 \end{pmatrix} = 1 + 1 = 2 \neq 1$ (*not normalized*).

5. Inner product has an integral representation

$$<\psi|\phi> = \int_{-\infty}^{+\infty} \psi^*(x)\phi(x)\,dx$$

where $<x|\psi> = \psi(x) = $ *projection* of $|\psi>$ along $|x>$,
and $<x|\phi> = \phi(x) = $ *projection* of $|\phi>$ along $|x>$.

- *Compare*

$$\vec{a} \cdot \vec{b} = (a_1\hat{i} + a_2\hat{j} + a_3\hat{k}) \cdot (b_1\hat{i} + b_2\hat{j} + b_3\hat{k}) = a_1b_1 + a_2b_2 + a_3b_3 = \sum_i a_ib_i = a_ib_i$$

(as per *Einstein's summation convention* of summing over repeated index).

1.12 Norm and the process of normalization

The quantity $\sqrt{<a|a>} = ||a||$ (let us use this notation) is defined as *norm of the ket* $|a>$. Norm is the *magnitude* or *modulus* of the ket. In other words norm is the *length* of the ket. So norm is *non-negative*.

$$||a|| = \sqrt{<a|a>} \geqslant 0$$

If $||a|| = 0$ then $|a>$ is a null length vector or null ket: $|a> = 0$ ($\Leftarrow$ no length).

If $||a|| = \sqrt{<a|a>} = 1$ we have a *normalized ket*, i.e. we say that $|a>$ is *normalized to unity*. So a normalized ket has *unit length*.

If $||a|| = \sqrt{<a|a>} > 1$, we have a ket $|a>$, which is *unnormalized*, as its length is greater than unity.

- Compare:

$\hat{i}, \hat{j}, \hat{k}$ are unit vectors, having *unit length* or unit norm and this is evident from the dot products: $\hat{i} \cdot \hat{i} = 1$, $\hat{j} \cdot \hat{j} = 1$, $\hat{k} \cdot \hat{k} = 1$.

- *Example*

  Consider $|\phi> = \begin{pmatrix} 1 \\ -i \end{pmatrix}$, $<\phi| = (1 \ \ i) \Rightarrow <\phi|\phi> = (1 \ \ i)\begin{pmatrix} 1 \\ -i \end{pmatrix} = 1 - i^2 = 2.$

Norm of $|\phi>$ is $||\phi||=\sqrt{<\phi|\phi>}=\sqrt{2}\neq 1 \Rightarrow |\phi>$ is not *normalized to unity*.
Normalization process
Let us start with the *unnormalized ket* $|\phi>$ such that $||\phi|| = \sqrt{<\phi|\phi>} >1$.
From here we *construct another ket* $|\psi>$ that has
$||\psi|| = \sqrt{<\psi|\psi>} =1$ (i.e. $|\psi>$= normalized) as follows.

Construct $|\psi> =\dfrac{|\phi>}{||\phi||} = \dfrac{|\phi>}{\sqrt{<\phi|\phi>}}$, $<\psi|=\dfrac{<\phi|}{||\phi||} = \dfrac{<\phi|}{\sqrt{<\phi|\phi>}}$

Let us evaluate

$$<\psi|\psi> =\frac{<\phi|}{\sqrt{<\phi|\phi>}} \frac{|\phi>}{\sqrt{<\phi|\phi>}} = \frac{<\phi|\phi>}{<\phi|\phi>} = 1$$

So $|\psi>$ has *unit length*, i.e. $|\psi>$ is *normalized to unity*. This *process* is called *normalization*.

- Example:

- Convert $|\phi> =\begin{pmatrix} i \\ 2 \end{pmatrix}$ to unit norm, i.e. normalize $|\phi> =\begin{pmatrix} i \\ 2 \end{pmatrix}$.

Let us evaluate

$$<\phi|\phi> =(- i \quad 2)\begin{pmatrix} i \\ 2 \end{pmatrix} = -i^2 + 4 = 5 \Rightarrow ||\phi||=\sqrt{<\phi|\phi>}=\sqrt{5} \neq 1$$

This means that $|\phi>$ is *unnormalized* (i.e. *does not* have a unit norm or unit length).

Construct $|\psi> =\dfrac{|\phi>}{||\phi||} = \dfrac{|\phi>}{\sqrt{<\phi|\phi>}} = \dfrac{1}{\sqrt{5}}\begin{pmatrix} i \\ 2 \end{pmatrix}$

Let us evaluate

$$<\psi|\psi> =\frac{1}{\sqrt{5}}(- i \quad 2)\frac{1}{\sqrt{5}}\begin{pmatrix} i \\ 2 \end{pmatrix} = 1 \Rightarrow ||\psi||=\sqrt{<\psi|\psi>}=1$$

This means that $|\psi>$ has *unit length*. So $|\psi>$ is *normalized to unity*.
Significance of norm:
1. In quantum mechanics the ket vectors used are of *unit norm* or unit length, i.e. we use *normalized kets* only.
2. $|\psi>$ and $e^{i\delta}|\psi>$ are treated as *same kets* since they both have unit norm. If we call $|\phi> =e^{i\delta}|\psi>$ then

$$|| \phi || = \sqrt{<\phi|\phi>}=\sqrt{(e^{-i\delta} <\psi|)(e^{i\delta}|\psi>)}=\sqrt{<\psi|\psi>}=|| \psi ||$$

So in linear vector space two normalized kets can *at most* differ by a phase factor of $e^{i\delta}$ (since $||e^{i\delta}|| = 1$).
3. Normalized kets having *different directions* are treated as *different kets*.

4. In linear vector space two *parallel* kets will carry the *same information* even if they have different lengths and we customarily use that ket which has *unit length* (i.e. normalized).

1.13 Ortho-normalization (orthogonal + normalized)

- Let us give an example with a set of two kets

✓ Consider a set of kets $|\phi_1> = \frac{1}{\sqrt{2}}\begin{pmatrix}1\\i\end{pmatrix}$, $|\phi_2> = \frac{1}{\sqrt{2}}\begin{pmatrix}i\\1\end{pmatrix}$ in two-dimensional space.

$$<\phi_1|\phi_2> = \frac{1}{\sqrt{2}}(1 \quad -i)\frac{1}{\sqrt{2}}\begin{pmatrix}i\\1\end{pmatrix} = \frac{1}{2}(i - i) = 0$$

$$<\phi_2|\phi_1> = \frac{1}{\sqrt{2}}(-i \quad 1)\frac{1}{\sqrt{2}}\begin{pmatrix}1\\i\end{pmatrix} = \frac{1}{2}(-i + i) = 0$$

$$\Rightarrow <\phi_1|\phi_2> = <\phi_2|\phi_1> = 0,$$

$$|\phi_1> , \ |\phi_2> \text{ are } orthogonal.$$

$$<\phi_1|\phi_1> = \frac{1}{\sqrt{2}}(1 \quad -i)\frac{1}{\sqrt{2}}\begin{pmatrix}1\\i\end{pmatrix} = \frac{1}{2}(1 + 1) = 1$$

$$<\phi_2|\phi_2> = \frac{1}{\sqrt{2}}(-i \quad 1)\frac{1}{\sqrt{2}}\begin{pmatrix}i\\1\end{pmatrix} = \frac{1}{2}(1 + 1) = 1$$

$$\Rightarrow <\phi_1|\phi_1> = <\phi_2|\phi_2> = 1,$$

$$|\phi_1> , \ |\phi_2> \text{ are } normalized.$$

This is called the *ortho-normalization relation* of the two kets.

$$|\phi_1> = \frac{1}{\sqrt{2}}\begin{pmatrix}1\\i\end{pmatrix}, \ |\phi_2> = \frac{1}{\sqrt{2}}\begin{pmatrix}i\\1\end{pmatrix}$$

We can write the four relations

$$<\phi_1|\phi_2> = 0, \ <\phi_2|\phi_1> = 0, \ <\phi_1|\phi_1> = 1, \ <\phi_2|\phi_2> = 1$$

in a *compact form* using Kroneckar delta symbol as

$$<\phi_m|\phi_n> = \delta_{mn} = \begin{cases}0 \text{ for } m \neq n\\1 \text{ for } m = n\end{cases}$$

- Compare:

- For the set of two unit vectors $\hat{e}_1 = \hat{i}, \hat{e}_2 = \hat{j}$

$\hat{e}_1 . \hat{e}_2 = \hat{i} . \hat{j} = 0, \ \hat{e}_2 . \hat{e}_1 = \hat{j} . \hat{i} = 0 \Leftarrow \hat{e}_m . \hat{e}_n = 0 \text{ for } m \neq n \ (\hat{e}_1, \ \hat{e}_2 \text{ are orthogonal})$

$\hat{e}_1 . \hat{e}_2 = \hat{i} . \hat{i} = 1$, $\hat{e}_2 . \hat{e}_2 = \hat{j} . \hat{j} = 1 \Leftarrow \hat{e}_m . \hat{e}_n = 1$ for $m = n$ ($\hat{e}_1$, $\hat{e}_2$ are normalized)

Using Kroneckar delta symbol, we write, in *compact form*, the four relations

$\hat{e}_1 . \hat{e}_2 = 0$, $\hat{e}_2 . \hat{e}_1 = 0$, $\hat{e}_1 . \hat{e}_1 = 1$, $\hat{e}_2 . \hat{e}_2 = 1$ as $\hat{e}_m . \hat{e}_n = \delta_{mn}$

This is called the *ortho-normalization relsation* of $\hat{e}_1 = \hat{i}$, $\hat{e}_2 = \hat{j}$.
Let us give an example with a set of three kets.

- $|\phi_1 > = \begin{pmatrix} 1 \\ 0 \\ 0 \end{pmatrix}$, $|\phi_2 > = \begin{pmatrix} 0 \\ 1 \\ 0 \end{pmatrix}$, $|\phi_3 > = \begin{pmatrix} 0 \\ 0 \\ 1 \end{pmatrix}$

$<\phi_1|\phi_1 > = <\phi_2|\phi_2 > = <\phi_3|\phi_3 > = 1$ (the kets are *normalised*)

$<\phi_1|\phi_2 > = <\phi_2|\phi_1 > = 0$, $<\phi_2|\phi_3 > = <\phi_3|\phi_2 > = 0$, $<\phi_3|\phi_1 > = <\phi_1|\phi_3 > = 0$

(the kets are *orthogonal*)
The set $|\phi_1 >$, $|\phi_2 >$, $|\phi_3>$ is therefore *ortho-normalized*: $<\phi_m|\phi_n > = \delta_{mn}$

NOTE
Start with a set of kets $|\psi_1 >$, $|\psi_2 >$, $|\psi_3 >$ ... that are not ortho-normalized. We can build an ortho-normalized set of kets $|\phi_1 >$, $|\phi_2 >$, $|\phi_3 >$ ... from this given set through a method called *Graham–Schmidt ortho-normalization procedure* (section 1.29).

1.14 Orthonormal basis (orthogonal + normalized + linearly independent + span)

A 3D linear vector space is described (or built) by using three kets
$|\phi_1 > = \begin{pmatrix} 1 \\ 0 \\ 0 \end{pmatrix}$, $|\phi_2 > = \begin{pmatrix} 0 \\ 1 \\ 0 \end{pmatrix}$, $|\phi_3 > = \begin{pmatrix} 0 \\ 0 \\ 1 \end{pmatrix}$
called *orthonormal basis kets* and can be treated as the *building blocks* or *basis* of the 3D linear vector space. Their properties are
1. Orthogonal
2. Normalized $\quad <\phi_m|\phi_n > = \delta_{mn} \quad$ *ortho-normalized*
3. Linearly independent

$$a_1|\phi_1 > + a_2|\phi_2 > + a_3|\phi_3 > = 0 \ (a_1 = 0, a_2 = 0, a_3 = 0)$$

4. Spans the entire space

$|\psi > = c_1|\phi_1 > + c_2|\phi_2 > + c_3|\phi_3>$ (expansion postulate) with c_1, c_2, c_3 being called *expansion coefficients* ($|\psi>$ being an arbitrary ket).
- *Compare*:

A 3D linear vector space is built by using three unit vectors $\hat{e}_1 = \hat{i}$, $\hat{e}_2 = \hat{j}$, $\hat{e}_3 = \hat{k}$ called orthonormal basis vectors having the following properties:
$\hat{e}_m . \hat{e}_n = \delta_{mn}$ (ortho-normalized = *orthogonal + normalized*);

$$a_1\hat{i} + a_2\hat{j} + a_3\hat{k} = 0 \text{ with } a_1 = 0, a_2 = 0, a_3 = 0(linearly\ independent);$$

Spans the entire space: $\vec{A} = A_1\hat{i} + A_2\hat{j} + A_3\hat{k}$ where A_1, A_2, A_3 are coefficients called expansion coefficients ($\vec{A}$ being an arbitrary vector) (Expansion postulate).

1.15 Expansion postulate

Consider a two-dimensional space described by *ortho-normalized* basis kets $|\phi_1>$, $|\phi_2>$(where $<\phi_m|\phi_n> = \delta_{mn}$).

We can *expand an arbitrary ket* $|\psi>$ in this space as

$|\psi> = c_1|\phi_1> + c_2|\phi_2>$ (expansion postulate) with c_1, c_2 being called expansion coefficients.

The expansion coefficients are obtained as follows:

$$<\phi_1|\psi> = <\phi_1|[c_1|\phi_1> + c_2|\phi_2>] = c_1 <\phi_1|\phi_1> + c_2 <\phi_1|\phi_2> = c_1$$

(where we used $<\phi_1|\phi_1> = 1$, $<\phi_1|\phi_2> = 0$)

$$<\phi_2|\psi> = <\phi_2|[c_1|\phi_1> + c_2|\phi_2>] = c_1 <\phi_2|\phi_1> + c_2 <\phi_2|\phi_2> = c_2$$

(where we used$<\phi_2|\phi_2> = 1$, $<\phi_2|\phi_1> = 0$)

Hence we can write (as scalars c_1, c_2 can be placed anywhere)

$$|\psi> = c_1|\phi_1> + c_2|\phi_2> = |\phi_1> c_1 + |\phi_2> c_2$$

Putting the values $c_1 = <\phi_1|\psi>$, $c_2 = <\phi_2|\psi>$ we get

$$|\psi> = |\phi_1> <\phi_1|\psi> + |\phi_2> <\phi_2|\psi>$$

We can extend this result to n dimensions and rewrite the expansion postulate

$$|\psi> = c_1|\phi_1> + c_2|\phi_2> + \ldots + c_n|\phi_n>$$

$$|\psi> = \sum_{i=1}^{n} c_i|\phi_i> = \sum_{i=1}^{n}|\phi_i> c_i = \sum_{i=1}^{n}|\phi_i> <\phi_i|\psi> \text{ (with } c_i = <\phi_i|\psi>).$$

1.16 Projection operator

Projection operator is an operator *that projects out* a certain element from a combination of elements.

Let $|\phi_1>$, $|\phi_2>$, ... be the *orthonormal* basis ($<\phi_i|\phi_j> = \delta_{ij}$) in three-dimensional space and we *expand* a ket $|\psi>$ in this basis as follows

$$|\psi> = \sum_{i=1}^{n} c_i|\phi_i> = \sum_{i=1}^{n}|\phi_i> c_i$$

$$= \sum_{i=1}^{n}|\phi_i> <\phi_i|\psi> = \sum_{i=1}^{n}\hat{P_i}|\psi> \text{ (as } c_i = <\phi_i|\psi>)$$

where we define an operator $\hat{P}_i = |\phi_i> <\phi_i|$ which is called *projection operator for the state* $|\phi_i>$ (operators are denoted by placing hat or cap atop)

Let us operate $|\psi> = \sum_{j=1}^{n} c_j|\phi_j>$ by the *projection operator* $\hat{P}_i = |\phi_i> <\phi_i|$

$$\hat{P}_i|\psi> = |\phi_i> <\phi_i| \ |\psi> = |\phi_i> <\phi_i| \sum_{j=1}^{n} c_j|\phi_j>$$

$$= |\phi_i> \sum_{j=1}^{n} c_j < \phi_i|\phi_j> = |\phi_i> \sum_{j=1}^{n} c_j\delta_{ij} = |\phi_i> c_i$$

$$\Rightarrow P_i|\psi> = c_i|\phi_i> \qquad \text{since} \ \sum_{j=1}^{n} c_j\delta_{ij} = c_i$$

Clearly $\hat{P}_1|\psi> = c_1|\phi_1>$ is the *first term* in the expansion of $|\psi> = \sum_{i=1}^{n} \hat{P}_i|\psi>$

$$= \sum_{i=1}^{n} c_i|\phi_i> = c_1|\phi_1> +c_2|\phi_2> +c_3|\phi_3> + \ldots$$

This means $\hat{P}_1 = |\phi_1> <\phi_1|$ is the projection operator for $|\phi_1>$, i.e. it *projects out* $|\phi_1>$ from $|\psi>$. In other words $\hat{P}_1$ *projects out* the portion of $|\psi>$ along $|\phi_1>$.

Similarly, $\hat{P}_2|\psi> = c_2|\phi_2>$, with $\hat{P}_2 = |\phi_2> <\phi_2|$ etc.

- *Compare*:

In the orthonormal basis $\hat{e}_1 = \hat{i}$, $\hat{e}_2 = \hat{j}$, $\hat{e}_3 = \hat{k}$ ($\hat{e}_m. \hat{e}_n = \delta_{mn}$) let us expand a vector $\vec{A}$ as $\vec{A} = A_1\hat{i} + A_2\hat{j} + A_3\hat{k} = \vec{A}_1 + \vec{A}_2 + \vec{A}_3$

Projection of $\vec{A}$ along $\hat{i}$ is $\hat{i}. \vec{A} = \hat{i}. (A_1\hat{i} + A_2\hat{j} + A_3\hat{k}) = A_1$

$\Rightarrow A_1\hat{i} = \hat{i} A_1 = \hat{i}(\hat{i}. \vec{A}) = \vec{A}_1$. This represents *portion of* $\vec{A}$ along $\hat{i}$.

Similarly, $\hat{j}(\hat{j}. \vec{A}) = \vec{A}_2$, $\hat{k}(\hat{k}. \vec{A}) = \vec{A}_3$

Projection operator is an idempotent operator.

$\hat{P}_i = |\phi_i> <\phi_i|$ is the projection operator that *projects out* the state $|\phi_i>$

$$\hat{P}_i\hat{P}_j = (|\phi_i> <\phi_i|)(|\phi_j> <\phi_j|) = |\phi_i> <\phi_i|\phi_j> <\phi_j| = |\phi_i> \delta_{ij} < \phi_j|$$

$$= \delta_{ij}|\phi_i> <\phi_j| = \delta_{ij}|\phi_i> <\phi_i| = \delta_{ij}\hat{P}_i \text{ (as it survives } only \ for \ i \neq j)$$

$$\hat{P}_i^2 = \hat{P}_i\hat{P}_i = [\ |\phi_i> <\phi_i| \ ] \ [\ |\phi_i> <\phi_i| \ ] = |\phi_i> <\phi_i|\phi_i> <\phi_i| = |\phi_i> <\phi_i|$$

(using $<\phi_i|\phi_i> = 1$ as $|\phi_i>$ is *normalized*).

$$\hat{P}_i^2 = |\phi_i> <\phi_i| = \hat{P}_i \ (idempotent \ matrix \ operator)$$

Eigenvalues of projection operator

$$\hat{P}_i^2 = \hat{P}_i \Rightarrow \hat{P}_i\hat{P}_i = \hat{P}_i\,\hat{I} \Rightarrow \hat{P}_i\hat{P}_i - \hat{P}_i\,\hat{I} = 0 \Rightarrow \hat{P}_i(\hat{P}_i - \hat{I}) = 0$$

where $\hat{I}$ = unit matrix operator. This suggests eigenvalues of $\hat{P}_i$ are 0 and 1.

Projection operator leads to the expansion postulate (the span criterion) and completeness criterion

If we *join or superpose* all the projections made by the projection operators, we get back the *original*, i.e. the complete ket should be restored. This is called the *completeness condition*. This is evident from the relation

$$|\psi> = \sum_{i=1}^{n} c_i|\phi_i> = \sum_{i=1}^{n} \hat{P}_i|\psi> \Rightarrow \sum_{i=1}^{n} \hat{P}_i = \hat{I}.$$

Using $\hat{P}_i = |\phi_i><\phi_i|$ we have $\sum_{i=1}^{n}|\phi_i><\phi_i| = \hat{I}$

For n dimensions the completeness condition or closure relation is

$$\sum_{i=1}^{n}|\phi_i><\phi_i| = \sum_{i=1}^{n}\hat{P}_i = \hat{I}$$

It establishes that the basis kets are *completely* able to *describe and span* the entire linear vector space. We illustrate this for three dimensions. Consider the expansion

$$|\psi> = c_1|\phi_1> + c_2|\phi_2> + c_3|\phi_3>$$

(First part) + (Second part) + (Third part) = The complete part = $|\psi>$

$$\hat{P}_1|\psi> + \hat{P}_2|\psi> + \hat{P}_3|\psi> = |\psi> \Rightarrow (\hat{P}_1 + \hat{P}_2 + \hat{P}_3)|\psi> = |\psi>$$

$$\left(|\phi_1><\phi_1| + |\phi_2><\phi_2| + |\phi_3><\phi_3|\right)|\psi> = |\psi> \text{ (as } \hat{P}_i = |\phi_i><\phi_i|)$$

As $|\psi>$ is arbitrary and since $\hat{I}|\psi> = |\psi>$ comparison gives

$$|\phi_1><\phi_1| + |\phi_2><\phi_2| + |\phi_3><\phi_3| = \hat{I}$$

- Let us find the *form* of projection operator for the basis $|\phi_1> = \begin{pmatrix} 1 \\ 0 \end{pmatrix}$, $|\phi_2> = \begin{pmatrix} 0 \\ 1 \end{pmatrix}$ and *check* completeness condition.

$$\hat{P}_1 = |\phi_1><\phi_1| = \begin{pmatrix} 1 \\ 0 \end{pmatrix}(1 \ \ 0) = \begin{pmatrix} 1 & 0 \\ 0 & 0 \end{pmatrix}, \ \hat{P}_2 = |\phi_2><\phi_2| = \begin{pmatrix} 0 \\ 1 \end{pmatrix}(0 \ \ 1) = \begin{pmatrix} 0 & 0 \\ 0 & 1 \end{pmatrix}$$

$$\hat{P}_1 + \hat{P}_2 = |\phi_1><\phi_1| + |\phi_2><\phi_2| = \begin{pmatrix} 0 & 0 \\ 0 & 1 \end{pmatrix} + \begin{pmatrix} 1 & 0 \\ 0 & 0 \end{pmatrix} = \begin{pmatrix} 1 & 0 \\ 0 & 1 \end{pmatrix} = \hat{I}$$

1.17 Normal matrix

A matrix $\hat{A}$ is called normal matrix if it *commutes* with its Hermitian conjugate i.e. if
$[\hat{A}^{\dagger}, \hat{A}] = \hat{A}^{\dagger}\hat{A} - \hat{A}\hat{A}^{\dagger} = 0$
where [ ] refers to commutator bracket (section 1.25).

e.g. $\hat{A} = \frac{1}{2}\begin{pmatrix} 1+i & 1-i \\ 1-i & 1+i \end{pmatrix}$.

1.18 Spectral theorem

Any *normal matrix* $\hat{A}$ can be expressed as

$$\hat{A} = \sum_i \lambda_i \hat{P}_i$$

where $\hat{A}$ has eigenvalues λ_i s and eigenkets $|\phi_i\rangle$ and $\hat{P}_i = |\phi_i\rangle\langle\phi_i|$ is the *projection operator*.

- Illustration

  - Verify the relation $\hat{A} = \sum_i \lambda_i \hat{P}_i$ for the matrix $\hat{X} = \begin{pmatrix} 0 & 1 \\ 1 & 0 \end{pmatrix}$.

We first find the *eigen function* and *eigenvalue* of operator $\hat{X}$. *Eigenvalue equation* for $\hat{X}$ is

$$\hat{X}|\phi\rangle = \lambda|\phi\rangle, \quad |\phi\rangle = \begin{pmatrix} x_1 \\ x_2 \end{pmatrix} = \text{eigenfunction}, \lambda = \text{eigenvalue}, \hat{I} = \begin{pmatrix} 1 & 0 \\ 0 & 1 \end{pmatrix}$$

$$\hat{X}|\phi\rangle = \lambda\hat{I}|\phi\rangle \Rightarrow \begin{pmatrix} 0 & 1 \\ 1 & 0 \end{pmatrix}\begin{pmatrix} x_1 \\ x_2 \end{pmatrix} = \lambda\begin{pmatrix} 1 & 0 \\ 0 & 1 \end{pmatrix}\begin{pmatrix} x_1 \\ x_2 \end{pmatrix}$$

$$\Rightarrow \left[\begin{pmatrix} 0 & 1 \\ 1 & 0 \end{pmatrix} - \lambda\begin{pmatrix} 1 & 0 \\ 0 & 1 \end{pmatrix}\right]\begin{pmatrix} x_1 \\ x_2 \end{pmatrix} = \begin{pmatrix} 0 \\ 0 \end{pmatrix} \Rightarrow \begin{pmatrix} -\lambda & 1 \\ 1 & -\lambda \end{pmatrix}\begin{pmatrix} x_1 \\ x_2 \end{pmatrix} = \begin{pmatrix} 0 \\ 0 \end{pmatrix} \quad (1.1)$$

For *non-trivial* solution ($x_1 \neq 0, x_2 \neq 0$)

$$\begin{vmatrix} -\lambda & 1 \\ 1 & -\lambda \end{vmatrix} = 0 \Rightarrow \lambda^2 = 1 \quad \Rightarrow \lambda = \pm 1$$

With $\lambda = 1$, equation (1.1) yields	With $\lambda = -1$, equation (1.1) yields
$\begin{pmatrix} -1 & 1 \\ 1 & -1 \end{pmatrix}\begin{pmatrix} x_1 \\ x_2 \end{pmatrix} = \begin{pmatrix} 0 \\ 0 \end{pmatrix}$	$\begin{pmatrix} 1 & 1 \\ 1 & 1 \end{pmatrix}\begin{pmatrix} x_1 \\ x_2 \end{pmatrix} = \begin{pmatrix} 0 \\ 0 \end{pmatrix}$
$\begin{pmatrix} -x_1 + x_2 \\ x_1 - x_2 \end{pmatrix} = \begin{pmatrix} 0 \\ 0 \end{pmatrix} \Rightarrow \begin{array}{l} -x_1 + x_2 = 0 \\ x_1 - x_2 = 0 \end{array}$	$\begin{pmatrix} x_1 + x_2 \\ x_1 + x_2 \end{pmatrix} = \begin{pmatrix} 0 \\ 0 \end{pmatrix} \Rightarrow x_1 + x_2 = 0$

$$x_1 = x_2, \ x_2 = \text{arbitrary} = 1(\text{say}) \qquad\qquad x_1 = -x_2, \ x_2 = \text{arbitrary} = -1(\text{say})$$

$$|\phi_1> \ = \begin{pmatrix} x_1 \\ x_2 \end{pmatrix} = \begin{pmatrix} 1 \\ 1 \end{pmatrix} \rightarrow N\begin{pmatrix} 1 \\ 1 \end{pmatrix} \qquad\qquad |\phi_2> \ = \begin{pmatrix} x_1 \\ x_2 \end{pmatrix} = \begin{pmatrix} 1 \\ -1 \end{pmatrix} \rightarrow N\begin{pmatrix} 1 \\ -1 \end{pmatrix}$$

(N = Normalisation constant, with $<\phi_1|\phi_1> \ =1$)

(N = Normalisation constant, with $<\phi_2|\phi_2> \ =1$)

$$<\phi_1|\phi_1> \ =N{*}N(1 \ \ 1)\begin{pmatrix} 1 \\ 1 \end{pmatrix} \qquad\qquad <\phi_2|\phi_2> \ =N{*}N(1 \ \ -1)\begin{pmatrix} 1 \\ -1 \end{pmatrix}$$

$$1 = |N|^2 \ 2 \Rightarrow N = \frac{1}{\sqrt{2}} \qquad\qquad 1 = |N|^2 \ 2 \Rightarrow N = \frac{1}{\sqrt{2}}$$

$$|\phi_1> \ = \frac{1}{\sqrt{2}}\begin{pmatrix} 1 \\ 1 \end{pmatrix} = \ |+1> \ (\text{as } \lambda = 1) \qquad |\phi_2> \ = \frac{1}{\sqrt{2}}\begin{pmatrix} 1 \\ -1 \end{pmatrix} = \ |-1> \ (\text{as } \lambda = -1)$$

Construct : $\hat{P}_1 = |\phi_1> \ <\phi_1| = \frac{1}{\sqrt{2}}\begin{pmatrix} 1 \\ 1 \end{pmatrix}\frac{1}{\sqrt{2}}(1 \ \ 1) \qquad = \frac{1}{2}\begin{pmatrix} 1 & 1 \\ 1 & 1 \end{pmatrix} \qquad (\lambda = 1)$

$$\hat{P}_2 = |\phi_2> \ <\phi_2| = \frac{1}{\sqrt{2}}\begin{pmatrix} 1 \\ -1 \end{pmatrix}\frac{1}{\sqrt{2}}(1 \ \ -1) = \frac{1}{2}\begin{pmatrix} 1 & -1 \\ -1 & 1 \end{pmatrix}(\lambda = -1)$$

We see that spectral theorem $\hat{A} = \sum_i \lambda_i \hat{P}_i$ follows from here since

$$\begin{pmatrix} 0 & 1 \\ 1 & 0 \end{pmatrix} = (1)\frac{1}{2}\begin{pmatrix} 1 & 1 \\ 1 & 1 \end{pmatrix} + (-1)\frac{1}{2}\begin{pmatrix} 1 & -1 \\ -1 & 1 \end{pmatrix}$$

$$\Rightarrow \hat{X} = \lambda_1\hat{P}_1 + \lambda_2\hat{P}_2 = \sum_i \lambda_i\hat{P}_i \quad (\text{This is } \textit{spectral theorem.})$$

1.19 Elements of a matrix in Bra Ket notation

Consider 2D linear vector space where the *basis vectors* are

$$|\phi_1> \ = \begin{pmatrix} 1 \\ 0 \end{pmatrix}, \ |\phi_2> \ = \begin{pmatrix} 0 \\ 1 \end{pmatrix}.$$

The 2×2 matrix operator $\hat{A}$ can be expressed as

$$\hat{A} = \begin{pmatrix} A_{11} & A_{12} \\ A_{21} & A_{22} \end{pmatrix} = A_{11}\begin{pmatrix} 1 & 0 \\ 0 & 0 \end{pmatrix} + A_{12}\begin{pmatrix} 0 & 1 \\ 0 & 0 \end{pmatrix} + A_{21}\begin{pmatrix} 0 & 0 \\ 1 & 0 \end{pmatrix} + A_{22}\begin{pmatrix} 0 & 0 \\ 0 & 1 \end{pmatrix}$$

We note that

$$|\phi_1> \ <\phi_1| = \begin{pmatrix} 1 \\ 0 \end{pmatrix}(1 \ \ 0) = \begin{pmatrix} 1 & 0 \\ 0 & 0 \end{pmatrix} = \hat{P}_1$$

$$|\phi_1> <\phi_2| = \begin{pmatrix} 1 \\ 0 \end{pmatrix}(0 \quad 1) = \begin{pmatrix} 0 & 1 \\ 0 & 0 \end{pmatrix}$$

$$|\phi_2> <\phi_1| = \begin{pmatrix} 0 \\ 1 \end{pmatrix}(1 \quad 0) = \begin{pmatrix} 0 & 0 \\ 1 & 0 \end{pmatrix}$$

$$|\phi_2> <\phi_2| = \begin{pmatrix} 0 \\ 1 \end{pmatrix}(0 \quad 1) = \begin{pmatrix} 0 & 0 \\ 0 & 1 \end{pmatrix} = \hat{P}_2$$

$$\hat{A} = A_{11}|\phi_1> <\phi_1| + A_{12}|\phi_1> <\phi_2| + A_{21}|\phi_2> <\phi_1| + A_{22}|\phi_2> <\phi_2|$$

Clearly an *operator* can be expressed as a *linear combination of projection operators* formed by the basis vectors

$$\hat{A} = \sum_{nm} A_{nm}|\phi_n> <\phi_m|$$

($n = m$ are the *diagonal* elements, $n \neq m$ are the *off- diagonal* elements)

The *form* of the matrix elements will be evident from the following. Let us evaluate

$$<\phi_1|\hat{A}|\phi_1> = <\phi_1| \ [\hat{A}] \ |\phi_1>$$

$$= <\phi_1| \ [\ A_{11}|\phi_1> <\phi_1| + A_{12}|\phi_1> <\phi_2| + A_{21}|\phi_2> <\phi_1| + A_{22}|\phi_2> <\phi_2| \ ] \ |\phi_1>$$

$$= A_{11} < \phi_1| \ \phi_1> <\phi_1|\phi_1> + A_{12} < \phi_1| \ \phi_1> <\phi_2|\phi_1>$$

$$+ A_{21} < \phi_1| \ \phi_2> <\phi_1|\phi_1> + A_{22} < \phi_1| \ \phi_2> <\phi_2|\phi_1> = A_{11}$$

Hence $<\phi_1|\hat{A}|\phi_1> = A_{11}$ (using $<\phi_m|\phi_n> = \delta_{mn}$ (orthonormalizaion))

$$<\phi_1|\hat{A}|\phi_2> = <\phi_1| \ [\hat{A}] \ |\phi_2>$$

$$= <\phi_1| \ [\ A_{11}|\phi_1> <\phi_1| + A_{12}|\phi_1> <\phi_2| + A_{21}|\phi_2> <\phi_1| + A_{22}|\phi_2> <\phi_2| \ ] \ |\phi_2>$$

$$= A_{11} < \phi_1| \ \phi_1> <\phi_1|\phi_2> + A_{12} < \phi_1| \ \phi_1> <\phi_2|\phi_2>$$

$$+ A_{21} < \phi_1| \ \phi_2> <\phi_1|\phi_2> + A_{22} < \phi_1| \ \phi_2> <\phi_2|\phi_2> = A_{12}$$

We similarly have $A_{21} = <\phi_2|\hat{A}|\phi_1>$, $\quad A_{22} = <\phi_2|\hat{A}|\phi_2>$.

$$\hat{A} = \begin{pmatrix} <\phi_1|\hat{A}|\phi_1> & <\phi_1|\hat{A}|\phi_2> \\ <\phi_2|\hat{A}|\phi_1> & <\phi_2|\hat{A}|\phi_2> \end{pmatrix} = \begin{pmatrix} A_{11} & A_{12} \\ A_{21} & A_{22} \end{pmatrix}$$

It is clear that the matrix elements A_{11}, A_{12}, A_{21}, A_{22} have been expressed in terms of the basis $|\phi_1>$, $|\phi_2>$.

General expression for *mn* th matrix element is: $A_{mn} = <\phi_m|\hat{A}|\phi_n>$

1.20 Hermitian matrix operator

If $\hat{A} = \hat{A}^\dagger$ then $\hat{A}$ is called a Hermitian operator or Hermitian matrix.

Explanation
We recall that ket and bra are Hermitian conjugates of each other. So an operator can be made to move from ket to bra or from bra to ket by taking its Hermitian conjugate (i.e. *by putting dagger*). This is shown in the following.

- $<\psi_1|\hat{A}\psi_2> = <\hat{A}^\dagger\psi_1|\psi_2>$

  $\Leftarrow \hat{A}$ in the ket has been moved to bra at the cost of dagger sign

- $<\hat{A}\psi_1|\psi_2> = <\psi_1|\hat{A}^\dagger\psi_2>$

  $\Leftarrow \hat{A}$ in the bra has been moved to ket at the cost of dagger sign

If the operator $\hat{A}$ moves freely from ket to bra or from bra to ket then it is a Hermitian operator and we can write

$$< \psi_1|\hat{A}\psi_2> = <\hat{A}\psi_1|\psi_2>.$$
$$= <\psi_1|\hat{A}^\dagger\psi_2>$$

$$\Rightarrow <\psi_1|\hat{A}|\psi_2> = <\psi_1|\hat{A}^\dagger|\psi_2>$$

Comparison gives $\hat{A} = \hat{A}^\dagger$
Integral representation of condition of hermiticity.

$$<\psi_1|\hat{A}\psi_2> = <\hat{A}\psi_1|\psi_2> \Rightarrow \int_{-\infty}^{\infty} \psi_1^*\hat{A}\psi_2 dx = \int_{-\infty}^{\infty} (\hat{A}\psi_1)^*\psi_2 dx = \int_{-\infty}^{\infty} \psi_1^*\hat{A}^*\psi_2 dx$$

NOTE
Hermitian conjugation or adjoint operation is

$$<\psi_1|\hat{A}^\dagger\psi_2> = <\hat{A}\psi_1|\psi_2> \quad (\Leftarrow \text{ transferring } \hat{A} \text{ from bra to ket})$$

$$\int_{-\infty}^{\infty} \psi_1^*\hat{A}^\dagger\psi_2 dx = \int_{-\infty}^{\infty} (\hat{A}\psi_1)^*\psi_2 dx = \int_{-\infty}^{\infty} \psi_1^*\hat{A}^*\psi_2 dx$$

Eigen values of Hermitian operator are real numbers.
Consider the Hermitian matrix operator $\hat{H} = \hat{H}^\dagger$ having eigenket $|n>$, eigenvalue being n, and the *eigenvalue equation* is

$$\hat{H}|n> = n|n>$$

Operate by $<n|$ from left to get

$$<n|\hat{H}|n> = <n|n|n> \quad \Rightarrow \quad <n|\hat{H}|n> = n<n|n> \tag{1.2}$$

Hermitian conjugate of the eigenvalue equation $\hat{H}|n> = n|n>$ is

$$<n|H^\dagger = <n|n^*$$

Operate by $|n>$ from right to get

$$<n|\hat{H}^\dagger|n> =<n|n^*|n> \quad \Rightarrow \quad <n|\hat{H}|n> =n^* <n|n> \tag{1.3}$$

where we have used that $\hat{H}$ is *Hermitian*, and so $\hat{H}^\dagger = \hat{H}$.
From equation (1.2) and equation (1.3) we have

$$n <n|n> =n^* <n|n> \Rightarrow n - n^* = 0 \text{ as } <n|n> \neq 0$$

Let $n = a + ib \Rightarrow n^* = a - ib$

$$n - n^* = 0 \Rightarrow \quad (a + ib) - (a - ib) = 0 \quad \Rightarrow 2ib = 0 \quad \Rightarrow b = 0$$

$$\Rightarrow n = a + ib = a = \text{real. } (\textit{Hermitian operators have real eigenvalues. })$$

Eigenfunctions (or Eigenkets) corresponding to two distinct eigenvalues are orthogonal.

Consider the Hermitian matrix operator $\hat{H} = \hat{H}^\dagger$. Eigenkets are
$|n_1>$ with eigenvalue n_1 and eigenvalue equation is $\hat{H}|n_1> =n_1|n_1>$ and
$|n_2>$ with eigenvalue n_2 and eigenvalue equation is $\hat{H}|n_2> =n_2|n_2>$
Operate $\hat{H}|n_1> =n_1|n_1>$ by $<n_2|$ to get

$$<n_2|\hat{H}|n_1> =<n_2|n_1|n_1> \quad \Rightarrow \quad <n_2|\hat{H}|n_1> \ = n_1 <n_2|n_1> \tag{1.4}$$

Operate $\hat{H}|n_2> =n_2|n_2>$ by $<n_1|$ to get

$$<n_1|\hat{H}|n_2> =<n_1|n_2|n_2> \Rightarrow <n_1|\hat{H}|n_2> \ = n_2 <n_1|n_2>$$

Take Hermitian conjugate to get

$$<n_2|\hat{H}^\dagger|n_1> =n_2^* <n_2|n_1>$$

As $\hat{H}$ is *Hermitian* $\hat{H}^\dagger = \hat{H}$ and *eigenvalues are real* $n_2^* = n_2$. Hence

$$<n_2|\hat{H}|n_1> =n_2 <n_2|n_1> \tag{1.5}$$

Equations (1.4)–(1.5) gives

$$n_1 <n_2|n_1> -n_2 <n_2|n_1> =0$$

$$\Rightarrow (n_1 - n_2) <n_2|n_1> =0$$

As eigenvalues are *distinct* $n_1 \neq n_2$ and hence $<n_2|n_1> =0$
$\Rightarrow$ Eigenkets $|n_1>$ and $|n_2>$ are *orthogonal*. They are also *normalized to unity*.
So Hermitian operators with *distinct eigenvalues* have a *set* of *orthonormal eigenfunctions or eigenkets* $|n_i>$.

Hermitian operators have a *complete set* of *orthonormal eigenfunctions or eigen kets*.

Consider any arbitrary ket $|\psi>$ expanded in terms of orthonormalized eigenkets
$|n_1>, \ |n_2>$ of $\hat{A}$ as $[<n_i|n_j> =\delta_{ij}]$

$|\psi> = c_1|n_1> + c_2 \ |n_2>$ (and hence $<\psi| = c_1^* < n_1| + c_2^* \ < n_2|$)

The norm of $|\psi>$ is $<\psi|\psi>$ which can be expressed in terms of the eigenkets as follows

$$<\psi|\psi> = [\ c_1^* < n_1| + c_2^* \ < n_2| \ ] \ [\ c_1|n_1> + c_2 \ |n_2> \ ]$$

$$= c_1^*c_1 < n_1|n_1> + c_1^*c_2 \ < n_1|n_2> \ + c_2^*c_1 \ < n_2| \ n_1> \ + c_2^*c_2 \ < n_2 \ |n_2>$$

$$<\psi|\psi> = |c_1|^2 + \ |c_2|^2 \quad (\text{using} \ < n_i|n_j> = \delta_{ij}) \tag{1.6}$$

Multiplying $|\psi> = c_1|n_1> + c_2 \ |n_2>$ by $<n_1|$ we have

$$<n_1|\psi> = <n_1|[c_1|n_1> + \ c_2 \ |n_2> \ ]$$

$$= <n_1|c_1|n_1> + <n_1|c_2 \ |n_2> \ = c_1 < \ n_1|n_1> + c_2 < n_1| \ n_2> \ = c_1$$

Multiplying $|\psi> = c_1|n_1> + c_2 \ |n_2>$ by $<n_2|$ we have

$$<n_2|\psi> = <n_2|[c_1|n_1> + c_2 \ |n_2> \ ]$$

$$= <n_2|c_1|n_1> + <n_2|c_2 \ |n_2> \ = c_1 < n_2|n_1> + c_2 < n_2| \ n_2> \ = c_2$$

With this value of c_1, c_2 equation (1.6) becomes

$$<\psi|\psi> = |c_1|^2 + \ |c_2|^2 = | < n_1|\psi> |^2 + |<n_2|\psi> |^2$$

$$= <n_1|\psi>^* < n_1|\psi> + <n_2|\psi>^* < n_2|\psi>$$

$$= \ <\psi|n_1> < n_1|\psi> + <\psi|n_2> < n_2|\psi>$$

$$<\psi|\hat{I}|\psi> = <\psi| \ [|n_1> < n_1| + |n_2> < n_2|] \ \ |\psi>$$

$$\Rightarrow \hat{I} = |n_1> < n_1| + |n_2> < n_2| \ (\text{on comparison})$$

The orthonormalized eigenkets of Hermitian operator satisfy the *completeness condition*.

$(\hat{A}^\dagger)^\dagger = \hat{A}$: **Taking dagger twice gives back the original.**

Consider $<\phi|\hat{A}\psi> = <\hat{A}^\dagger\phi|\psi>$ and take *Hermitian conjugate* to get

$$<\phi|\hat{A}\psi>^\dagger = <\hat{A}^\dagger\phi|\psi>^\dagger$$

$$<\hat{A}\psi|\phi> \ = \ <\psi|\hat{A}^\dagger\phi>$$

$$= <(\hat{A}^\dagger)^\dagger\psi|\phi>$$

Comparison gives: $(\hat{A}^\dagger)^\dagger = \hat{A}$

Anti-Hermitian operator (or skew-Hermitian operator)

Definition:

If $\hat{A}^\dagger = -\hat{A}$ then $\hat{A}$ is an anti-Hermitian operator.

- Example

$$\hat{A} = \begin{pmatrix} 0 & i \\ i & 0 \end{pmatrix} \quad (\dagger = \text{dagger} = \text{complex conjugate} + \text{transpose})$$

$$\hat{A}^{\dagger} = \begin{pmatrix} 0 & i \\ i & 0 \end{pmatrix}^{*T} = \begin{pmatrix} 0 & -i \\ -i & 0 \end{pmatrix}^{T} = \begin{pmatrix} 0 & -i \\ -i & 0 \end{pmatrix} = -\begin{pmatrix} 0 & i \\ i & 0 \end{pmatrix} = -\hat{A}$$

1.21 Unitary matrix

An operator $\hat{U}$ which satisfies the condition

$\hat{U}^{\dagger} = \hat{U}^{-1} \Rightarrow \hat{U}\hat{U}^{\dagger} = \hat{U}^{\dagger}\hat{U} = \hat{I}$ is called unitary operator

- *Example*:

$$\frac{1}{\sqrt{2}}\begin{pmatrix} 1 & i \\ i & 1 \end{pmatrix}, \ \frac{1}{2}\begin{pmatrix} 1+i & 1-i \\ 1-i & 1+i \end{pmatrix}$$

Eigenvalues of unitary operator are of unit modulus

Let $\hat{U}$ = unitary matrix $\Rightarrow \hat{U}\hat{U}^{\dagger} = \hat{U}^{\dagger}\hat{U} = \hat{I}$

Eigenvalue equation for $\hat{U}$: $\hat{U}|\psi> = a|\psi>$ (a = eigenvalue) $\qquad$ (1.7)

Hermitian conjugate of equation (1.7) is

$$<\psi|\hat{U}^{\dagger} = <\psi|a^* \qquad (1.8)$$

Operate equation (1.8) from right by equation (1.7) to get

$$<\psi|\hat{U}^{\dagger}\hat{U}|\psi> = <\psi|a^*a|\psi>$$

$$\Rightarrow <\psi|\psi> = |a|^2 <\psi|\psi> \text{ as } \hat{U}^{\dagger}\hat{U} = \hat{I}$$

$$\Rightarrow |a|^2 = 1 \Rightarrow a = \pm 1$$

$$\Rightarrow |a| = 1 \Leftarrow \textit{Modulus of eigenvalue} \text{ of unitary matrix is } \textit{unity}.$$

Product of two unitary operators is a unitary operator.

Let the *unitary* operators be $\hat{U}_1$, $\hat{U}_2$

$$\hat{U}_1\hat{U}_1^{\dagger} = \hat{U}_1^{\dagger}\hat{U}_1 = \hat{I}, \ \hat{U}_2\hat{U}_2^{\dagger} = \hat{U}_2^{\dagger}\hat{U}_2 = \hat{I}$$

Consider

$$(\hat{U}_1\hat{U}_2)(\hat{U}_1\hat{U}_2) = \hat{U}_1\hat{U}_2\hat{U}_2^{\dagger}\hat{U}_1^{\dagger} = \hat{U}_1\hat{U}_1^{\dagger} = \hat{I}\left(\text{since}(\hat{A}\hat{B})^{\dagger} = \hat{B}^{\dagger}\hat{A}^{\dagger}\right)$$

$$(\hat{U}_1\hat{U}_2)^{\dagger}(\hat{U}_1\hat{U}_2) = \hat{U}_2^{\dagger}\hat{U}_1^{\dagger}\hat{U}_1\hat{U}_2 = \hat{U}_2^{\dagger}\hat{U}_2 = \hat{I}$$

Clearly

$$(\hat{U}_1\hat{U}_2)^{\dagger}(\hat{U}_1\hat{U}_2) = (\hat{U}_1\hat{U}_2)(\hat{U}_1\hat{U}_2)^{\dagger} = \hat{I} \Leftarrow \hat{U}_1 \ \hat{U}_2 \text{ is } \textit{unitary}$$

Inverse of unitary matrix is a unitary matrix

Let $\hat{U}$ = unitary operator $\Rightarrow \hat{U}\hat{U}^\dagger = \hat{U}^\dagger\hat{U} = \hat{I} \Rightarrow \hat{U}^\dagger = \hat{U}^{-1}$

Consider inverse of unitary operator $\hat{U}$ i.e. $\hat{U}^{-1}$

$$(\hat{U}^{-1})(\hat{U}^{-1})^\dagger = (\hat{U}^\dagger)(\hat{U}^\dagger)^\dagger = \hat{U}^\dagger\hat{U} = \hat{I} \quad \text{using } (\hat{U}^\dagger)^\dagger = \hat{U}^\dagger$$

$$(\hat{U}^{-1})^\dagger(\hat{U}^{-1}) = (\hat{U}^\dagger)^\dagger(\hat{U}^\dagger) = \hat{U}\hat{U}^\dagger = \hat{I}$$

$\Rightarrow \quad \hat{U}^{-1}$ is a *unitary* operator.

1.22 Diagonalization of a matrix—change of basis

Let us consider the 3×3 matrix

$$\hat{H} = \begin{pmatrix} H_{11} & H_{12} & H_{13} \\ H_{21} & H_{22} & H_{23} \\ H_{31} & H_{32} & H_{33} \end{pmatrix} = \begin{pmatrix} a & b & c \\ d & e & f \\ g & h & i \end{pmatrix}$$

Suppose the matrix elements are expressed in terms of the *ortho-normalized* basis ket vectors $|\phi_1>$, $|\phi_2>$, $|\phi_3>$ as follows $[<\phi_i|\phi_j> = \delta_{ij}]$

$$H_{11} = <\phi_1|\hat{H}|\phi_1> = \alpha \quad H_{12} = <\phi_1|\hat{H}|\phi_2> = b \quad H_{13} = <\phi_1|\hat{H}|\phi_3> = c$$

$$H_{21} = <\phi_2|\hat{H}|\phi_1> = d \quad H_{22} = <\phi_2|\hat{H}|\phi_2> = e \quad H_{23} = <\phi_2|\hat{H}|\phi_3> = f$$

$$H_{31} = <\phi_3|\hat{H}|\phi_1> = g \quad H_{32} = <\phi_3|\hat{H}|\phi_2> = h \quad H_{33} = <\phi_3|\hat{H}|\phi_3> = i$$

If the *off-diagonal elements are all not zero* then $\hat{H}$ is *not a diagonal matrix* in this representation, i.e. if $H_{ij} = <\phi_i|\hat{H}|\phi_j> =$ not all zero for $i \neq j$ then $|\phi_1>$, $|\phi_2>$, $|\phi_3>$ is a *non-diagonal basis*.

We can *change the basis* and make $\hat{H}$ *diagonal* to simplify the problem. This procedure is called *diagonalization* (or *principal axis transformation*) in which *all the off diagonal elements would vanish*. This is possible if we express $\hat{H}$ in its *eigen basis or eigen-representation*. Let the eigenvalue equation of $\hat{H}$ be

$$\hat{H}|\psi_1> = h_1|\psi_1> \quad \hat{H}|\psi_2> = h_2|\psi_2> \quad \hat{H}|\psi_3> = h_3|\psi_3>$$

$|\psi_1>$, $|\psi_2>$, $|\psi_3>$ are the *orthonormal eigen-basis* with eigenvalues h_1, h_2, h_3.

Using the eigenvalue equation we can construct the matrix elements of $\hat{H}$ as follows $(<\psi_m|\psi_n> = \delta_{mn})$:

$H_{11} = <\psi_1|\hat{H}|\psi_1>$	$H_{12} = <\psi_1|\hat{H}|\psi_2>$	$H_{13} = <\psi_1|\hat{H}|\psi_3>$
$= <\psi_1|h_1|\psi_1>$	$= <\psi_1|h_2|\psi_2>$	$= <\psi_1|h_3|\psi_3>$
$= h_1 <\psi_1|\psi_1> = h_1$	$= h_2 <\psi_1|\psi_2> = 0$	$= h_3 <\psi_1|\psi_3> = 0$

(*Continued*)

$H_{21} = <\psi_2|\hat{H}|\psi_1>$ $H_{22} = <\psi_2|\hat{H}|\psi_2>$ $H_{23} = <\psi_2|\hat{H}|\psi_3>$

$= <\psi_2|h_1|\psi_1>$ $= <\psi_2|h_2|\psi_2>$ $= <\psi_2|h_3|\psi_3>$

$= h_1 < \psi_2|\psi_1 > = 0$ $= h_2 < \psi_2|\psi_2 > = h_2$ $= h_3 < \psi_2|\psi_3 > = 0$

$H_{31} = <\psi_3|\hat{H}|\psi_1>$ $H_{32} = <\psi_3|\hat{H}|\psi_2>$ $H_{33} = <\psi_3|\hat{H}|\psi_3>$

$= <\psi_3|h_1|\psi_1>$ $= <\psi_3|h_2|\psi_2>$ $= <\psi_3|h_3|\psi_3>$

$= h_1 < \psi_3|\psi_1 > = 0$ $= h_2 < \psi_3|\psi_2 > = 0$ $= h_3 < \psi_3|\psi_3 > = h_3$

General matrix element is $H_{mn} = <\psi_m|\hat{H}|\psi_n> = h_m\delta_{mn}$

The *diagonalized* matrix in its *orthonormal eigen-basis* or *eigen representation* is

$$\hat{H}^{diagonal} \equiv \hat{H}^D = \begin{pmatrix} h_1 & 0 & 0 \\ 0 & h_2 & 0 \\ 0 & 0 & h_3 \end{pmatrix} \Rightarrow \text{Eigenvalues sit across the diagonal}$$

Matrix is diagonal in its eigen representation or eigenbasis.

The above *change of basis* (from $|\phi_1 >$, $|\phi_2 >$, $|\phi_3>$ in which $\hat{H}$ is *non-diagonal*) to $|\psi_1 >$, $|\psi_2 >$, $|\psi_3>$ (in which $\hat{H}$ becomes *diagonal*) is called *diagonalization* and can be done by the following method and on satisfaction of the following conditions.

Condition of diagonalizability:

Matrix $\hat{H}$ should have *linearly independent eigenvectors*.

$\hat{H}$ should be a *normal matrix*, i.e. should commute with its Hermitian conjugate, i.e. $[\hat{H}^{\dagger}, \hat{H}] = 0$.

Every *Hermitian* matrix and every *unitary* matrix is diagonalizable.

If a matrix has *distinct eigenvalues* it is diagonalizable.

The method of diagonalization

The *change of basis* from $|\phi_1 >$, $|\phi_2 >$, $|\phi_3>$ (in which matrix is *non-diagonal* denoted by $\hat{H}$) to $|\psi_1 >$, $|\psi_2 >$, $|\psi_3>$ (in which matrix becomes *diagonal* $\hat{H}^D$) is called *diagonalization* and is carried out by building a *non-singular* matrix $\hat{S}$ (*using the linearly independent eigenvectors of $\hat{H}$*) through the following operation called similarity transformation

$$\hat{S}^{-1}\hat{H}\hat{S} = \hat{H}^{D}(|\hat{S}| \neq 0), \quad \hat{S}^{-1} = \frac{1}{|\hat{S}|}\hat{S}^{cT},$$

where $\hat{S}^{cT}$ = transpose of the cofactor matrix, $\hat{S}^{-1}$ = inverse matrix of $\hat{S}$.

Since $\hat{H}$ and $\hat{H}^{D}$ matrices are similar matrices (only the bases are different) and since $\hat{S}$ connects them through a linear transformation $\hat{S}^{-1}\hat{H}\hat{S} = \hat{H}^{D}$ we call the operation a *similarity transformation*. It is easier to deal with $\hat{H}^{D}$ since it is diagonal. So in a problem if we have the *non-diagonal* $\hat{H}$ we can switch over to the *diagonal* matrix $\hat{H}^{D}$ through this similarity transformation.

- *Compare*:

Consider an ellipse whose symmetry axes are *not* the $\hat{X}$ and $\hat{Y}$ axes. Through a principal axes transformation (rotation) we shift to a new set of axes—the $\hat{x}$ axis and $\hat{y}$ axis, (figure 1.5) which are the *symmetry axes of ellipse*. $\Rightarrow$ Problem gets hugely simplified.

- Let the 3×3 *non-diagonal* matrix $\hat{H}$ (which is to be diagonalized) have eigenvalues h_1, h_2, h_3 corresponding to linearly independent eigenvectors

$$\begin{pmatrix} x_1 \\ x_2 \\ x_3 \end{pmatrix}, \begin{pmatrix} y_1 \\ y_2 \\ y_3 \end{pmatrix}, \begin{pmatrix} z_1 \\ z_2 \\ z_3 \end{pmatrix}.$$

We have to construct the diagonalizing matrix with these eigenvectors as

$$\hat{S} = \begin{pmatrix} x_1 & y_1 & z_1 \\ x_2 & y_2 & z_2 \\ x_3 & y_3 & z_3 \end{pmatrix}.$$

Then $\hat{S}^{-1}\hat{H}\hat{S} = \hat{H}^{D}$ is called a similarity transformation.
Such change of basis is a unitary transformation.

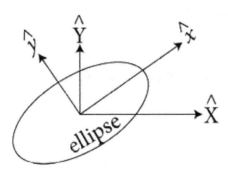

Figure 1.5. Use of symmetry axes $\hat{x}$, $\hat{y}$ simplifies the problem.

1.23 Triangle laws (inequality and equality)

In 3D linear vector space we have for three vectors $\vec{a}$, $\vec{b}$, $\vec{c}$ (figure 1.6)

$$| \vec{a} | + | \vec{b} | \;\geqslant\; | \vec{c} | \;\text{ or } a + b > c$$

If the three vectors $\vec{a}$, $\vec{b}$, $\vec{c}$ are *collinear*

$$| \vec{a} | + | \vec{b} | \;=\; | \vec{c} | \;\text{ or } a + b = c \quad (equality \text{ holds})$$

Similarly we can write for kets

$$\|\psi\| + \|\phi\| \;\geqslant\; \|\psi + \phi\|$$

$$\sqrt{<\psi|\psi>} + \sqrt{<\phi|\phi>} \;\geqslant\; \sqrt{<\psi + \phi \mid \psi + \phi>}$$

1.24 Cauchy–Schwarz laws (inequality and equality)

The scalar product in 3D linear vector space is defined as follows (figure 1.7).

$$\vec{a} . \vec{b} = ab \cos \theta \Rightarrow \frac{\vec{a} . \vec{b}}{ab} = \cos \theta \leqslant 1$$

$$\Rightarrow \vec{a} . \vec{b} \;\leqslant\; ab \;\text{ i.e. } \vec{a} . \vec{b} \;\leqslant\; | \vec{a} | \, | \vec{b} |$$

The norm is $| \vec{a} . \vec{b} | \;\leqslant\; | \vec{a} | \, | \vec{b} |$
For *collinear* vectors
$\theta = 0, \pi$ ($\Leftarrow$ vectors along the same line which means *linear dependence*)

$$\vec{a} . \vec{b} \;=\; | \vec{a} | \, | \vec{b} | \quad (equality \text{ holds})$$

Similarly we can write for a ket

$$|<\phi|\psi>| \;\leqslant\; \sqrt{<\phi|\phi>} \, \sqrt{<\psi|\psi>} \tag{1.9}$$

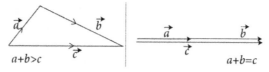

Figure 1.6. Triangle law (inequality and equality).

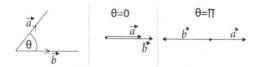

Figure 1.7. Various orientations of two vectors $\vec{a}$, $\vec{b}$.

$$|<\phi|\psi>|^2 \leqslant |<\phi|\phi>| \ |<\psi|\psi>|$$

If $|\psi> = a|\phi>$, i.e. $|\psi>$ and $|\phi>$ are *linearly dependent* then let us consider the RHS of relation (1.9)

$$\sqrt{<\phi|\phi>} \sqrt{<\psi|\psi>} = \sqrt{<\phi|\phi> <\psi|\psi>} = \sqrt{<\phi|\phi> <a\phi|\psi>}$$

$$= \sqrt{<\phi|\phi> a^\dagger <\phi|\psi>} = \sqrt{a^\dagger <\phi|\phi> <\phi|\psi>}$$

$$= \sqrt{<a\phi|\phi> <\phi|\psi>}$$

$$= \sqrt{<\psi|\phi> <\phi|\psi>} \ (\text{as} <\psi| = <a\phi|)$$

$$= \sqrt{|<\phi|\psi>|^2} = |<\phi|\psi>| = \text{LHS of relation 1.9}$$

So *equality sign* holds in relation (1.9) if kets are *linearly dependent*.

1.25 Commutator bracket

Consider two operators $\hat{A}$, $\hat{B}$. The commutator bracket of these operators is defined as

$$[\hat{A}, \ \hat{B}] = \hat{A}\hat{B} - \hat{B}\hat{A}$$

Theorem 1 : ***If two operators commute they have simultaneous eigenfunctions.***
(Proof given below)
Let $\hat{A}$, $\hat{B}$ be two *Hermitian* operators ($\hat{A} = \hat{A}^\dagger$, $\hat{B} = \hat{B}^\dagger$)
They commute $[\hat{A}, \hat{B}] = 0 \Rightarrow \hat{A}\hat{B} - \hat{B}\hat{A} = 0$
Eigenvalue equation of $\hat{A}$ is : $\hat{A}|a> = a|a>$

$$\text{Operate with } \hat{B}: \qquad \hat{B}\hat{A}|a> = \hat{B}a|a> \qquad\qquad (1.10)$$

$$= a\hat{B}|a>$$

As $\hat{A}\hat{B} - \hat{B}\hat{A} = 0 \quad \Rightarrow \quad \hat{B}\hat{A} = \hat{A}\hat{B}$ we have from equation (1.10)

$$\hat{A}\hat{B}|a> = a\hat{B}|a>$$

$$\Rightarrow \hat{A} \ [\hat{B}|a>] = a \ [\hat{B}|a>]$$

This is the eigenvalue equation for $\hat{A}$ and hence $\hat{B}|a>$ is the eigenfunction of $\hat{A}$ with eigenvalue a.

Assume that $\hat{A}$ has *non-degenerate eigenfunction* corresponding to the eigen value a.

So corresponding to the same eigenvalue a, $\hat{A}$ has unique eigenfunction. Hence the eigenfunction $\hat{B}|a>$ and the eigenfunction $|a>$ are the same or *linearly dependent*. Hence

$$\lambda_1 \hat{B}|a> + \lambda_2|a> = 0, \quad \lambda_1, \lambda_2 \rightarrow \text{constants}$$

$$\Rightarrow \hat{B}|a> = -\frac{\lambda_2}{\lambda_1}|a> \Rightarrow \hat{B}|a> = b|a> \left(b = -\frac{\lambda_2}{\lambda_1} \right)$$

So $|a>$ is eigen function of $\hat{B}$ with eigenvalue b. Clearly $|a>$ is the *common* (or *simultaneous*) eigen function of $\hat{A}$, $\hat{B}$ with eigenvalues a, b, respectively. Hence we can denote $|a>$ by $|a, b>$ and rewrite the eigenvalue equations of the operators $\hat{A}$, $\hat{B}$ as

$$\hat{A}|a, b> = a|a, b>$$

$$\hat{B}|a, b> = b|a, b>$$

Theorem 2: If *two operators have simultaneous eigenfunctions then they commute*.

Or equivalently

**If *two o*perators have simultaneous eigenfunctions then they are diagonal in the common basis.

(Proof given below)

Let $\hat{A}$, $\hat{B}$ be two *Hermitian* operators ($\hat{A} = \hat{A}^{\dagger}$, $\hat{B} = \hat{B}^{\dagger}$). $|a, b>$ is the simultaneous eigen ket. The eigenvalue equations are

$\hat{A}|a, b> = a|a, b>$	$\hat{B}|a, b> = b|a, b>$
Operate by $\hat{B}$	Operate by $\hat{A}$
$\hat{B}\hat{A}|a, b> = \hat{B}a|a, b> = a\hat{B}|a, b>$	$\hat{A}\hat{B}|a, b> = \hat{A}b|a, b> = b\hat{A}|a, b>$
$= ab|a, b>$ (1.11a)	$= ba|a, b>$ (1.11b)

Subtract: equation (1.11b)–equation (1.11a) to get

$$\hat{A}\hat{B}|a, b> - \hat{B}\hat{A}|a, b> = 0 \quad \Rightarrow (\hat{A}\hat{B} - \hat{B}\hat{A})|a, b> = 0$$

$$\Rightarrow [\hat{A}, \hat{B}]|a, b> = 0$$

$$\Rightarrow [\hat{A}, \hat{B}] = 0 \Rightarrow \hat{A}\hat{B} - \hat{B}\hat{A} = 0 \quad \Rightarrow \hat{A}\hat{B} = \hat{B}\hat{A}$$

So $\hat{A}$, $\hat{B}$ *commutes* as they have simultaneous eigenfunction.

Alternatively we can represent the problem as follows:

Let the *common eigenbasis* (or *simultaneouseigen function*) be $|1>$, $|2>$. They are *ortho-normalized*, i.e. $<i|j>=\delta_{ij}$. The eigenvalues of $\hat{A}$ are a_1, a_2 and the eigenvalues of $\hat{B}$ are b_1, b_2. The *eigenvalue equations* are

---

$\hat{A}|1>=a_1|1>$ $\hat{B}|1>=b_1|1>$

$\hat{A}|2>=a_2|2>$ $\hat{B}|2>=b_2|2>$

Matrix elements of $\hat{A}$ are   Matrix elements of $\hat{B}$ are

$<1|\hat{A}|1>=<1|a_1|1>=a_1<1|1>=a_1$ $<1|\hat{B}|1>=<1|b_1|1>=b_1<1|1>=b_1$

$<1|\hat{A}|2>=<1|a_2|2>=a_2<1|2>=0$ $<1|\hat{B}|2>=<1|b_2|2>=b_2<1|2>=0$

$<2|\hat{A}|1>=<2|a_1|1>=a_1<2|1>=0$ $<2|\hat{B}|1>=<2|b_1|1>=b_1<2|1>=0$

$<2|\hat{A}|2>=<2|a_2|2>=a_2<2|2>=a_2$ $<2|\hat{B}|2>=<2|b_2|2>=b_2<2|2>=b_2$

Thus   Thus

$$\hat{A}=\begin{pmatrix}<1|\hat{A}|1> & <1|\hat{A}|2>\\ <2|\hat{A}|1> & <2|\hat{A}|2>\end{pmatrix}$$

$$=\begin{pmatrix}a_1 & 0\\ 0 & a_2\end{pmatrix}\Rightarrow\text{diagonal in the common basis}$$

$$B=\begin{pmatrix}<1|\hat{B}|1> & <1|\hat{B}|2>\\ <2|\hat{B}|1> & <2|\hat{B}|2>\end{pmatrix}$$

$$=\begin{pmatrix}b_1 & 0\\ 0 & b_2\end{pmatrix}\Rightarrow\text{diagonal in the common basis}$$

Let us check if $\hat{A}$ and $\hat{B}$ *commutes.*

$$\hat{A}\hat{B}=\begin{pmatrix}a_1 & 0\\ 0 & a_2\end{pmatrix}\begin{pmatrix}b_1 & 0\\ 0 & b_2\end{pmatrix}=\begin{pmatrix}a_1b_1 & 0\\ 0 & a_2b_2\end{pmatrix}=\begin{pmatrix}b_1 & 0\\ 0 & b_2\end{pmatrix}\begin{pmatrix}a_1 & 0\\ 0 & a_2\end{pmatrix}=\hat{B}\hat{A}$$

$$\Rightarrow\hat{A}\hat{B}-\hat{B}\hat{A}=0\Rightarrow[\hat{A},\ \hat{B}]=0$$

$\Rightarrow\hat{A}$, $\hat{B}$ *commutes, have simultaneous eigen ket and are diagonal in the common eigen basis.*

Anti-commutator bracket

Anti-commutation relation of two operators $\hat{A}$, $\hat{B}$ is

$\{\hat{A},\ \hat{B}\}=\hat{A}\hat{B}+\hat{B}\hat{A}$, $\{\ \}\rightarrow$ anti-commutator bracket

1.26 Trace

Trace = Tr = *Sum* of diagonal elements of a matrix
- *Example*

- $Tr\begin{pmatrix} 1 & 3 \\ -2 & 1 \end{pmatrix} = 1 + 1 = 2$

- $Tr\begin{pmatrix} A_{11} & A_{12} \\ A_{21} & A_{22} \end{pmatrix} = A_{11} + A_{22} = \sum_{n=1}^{2} A_{nn} = \sum_{n} <n|\hat{A}|n>$

(since the diagonal matrix element is $A_{nn} = <n|\hat{A}|n>$)

Trace of the product of a finite number of matrices is invariant under cyclic permutation of matrices

$$Tr(\hat{A}\hat{B}) = Tr(\hat{B}\hat{A}) \quad \text{for 2 matrices}$$

$$Tr(\hat{A}\hat{B}\hat{C}) = Tr(\hat{B}\hat{C}\hat{A}) = Tr(\hat{C}\hat{A}\hat{B}) \quad \text{for 3 matrices}$$

Proof:

$$Tr(\hat{A}\hat{B}) = \sum_{n} <n|\hat{A}\hat{B}|n> = \sum_{n} <n|\hat{A}\ \hat{I}\ \hat{B}|n>$$

$$= \sum_{n} <n|\hat{A}\ \sum_{m}|m> <m|\hat{B}|n>$$

(using the completeness condition $\sum_{m}|m> <m| = \hat{I}$)

$$Tr(\hat{A}\hat{B}) = \sum_{n\,m} <n|\hat{A}|m> <m|\hat{B}|n>$$

$$Tr(\hat{A}\hat{B}) = \sum_{n\,m} A_{nm}B_{mn} = \sum_{n\,m} B_{mn}A_{nm} \quad (\text{as scalars } A_{nm}, B_{mn} \text{ commute})$$

$$Tr(\hat{A}\hat{B}) = \sum_{m\,n} <m|\hat{B}|n> <n|\hat{A}|m> \quad (\text{as order of summation over } m, n$$

$$\text{can be interchanged})$$

$$= \sum_{m} <m|\hat{B}\ \sum_{n}|n> <n|\hat{A}|m> = \sum_{m} <m|\hat{B}\ \hat{I}\ \hat{A}|m>$$

(using the completeness condition $\sum_{n}|n> <n| = \hat{I}$)

$$Tr(\hat{A}\hat{B}) = \sum_{m} <m|\hat{B}\hat{A}|m> = Tr(\hat{B}\hat{A})$$

For three matrices $Tr(\hat{A}\hat{B}\hat{C})=Tr(\hat{A}\hat{P})$ where $\hat{P} = \hat{B}\hat{C}$

$$=Tr(\hat{P}\hat{A}) = Tr(\hat{B}\hat{C}\hat{A})$$

$$=Tr(\hat{B}\hat{Q}) \text{ where } \hat{Q} = \hat{C}\hat{A}$$

$$=Tr(\hat{Q}\hat{B}) = Tr(\hat{C}\hat{A}\hat{B})$$

So $Tr(\hat{A}\hat{B}\hat{C}) = Tr(\hat{B}\hat{C}\hat{A}) = Tr(\hat{C}\hat{A}\hat{B})$

The result can be generalized to any number of matrices provided cyclic permutation is made.

The Trace of commutator bracket is always zero

$$Tr[\,\hat{A}, \;\; \hat{B}\,] = Tr(\hat{A}\hat{B} - \hat{B}\hat{A}) = Tr(\hat{A}\hat{B}) - Tr(\hat{B}\hat{A}) = Tr(\hat{A}\hat{B}) - Tr(\hat{A}\hat{B}) = 0$$

$$[\text{since } Tr(\hat{B}\hat{A}) = Tr(\hat{A}\hat{B})]$$

1.27 Pauli spin matrices

The following three 2×2 complex matrices are called Pauli spin matrices

$$\hat{\sigma}_1 = \hat{\sigma}_x = \hat{X} = \begin{pmatrix} 0 & 1 \\ 1 & 0 \end{pmatrix}, \;\; \hat{\sigma}_2 = \hat{\sigma}_y = \hat{Y} = \begin{pmatrix} 0 & -i \\ i & 0 \end{pmatrix}, \;\; \hat{\sigma}_3 = \hat{\sigma}_z = \hat{Z} = \begin{pmatrix} 1 & 0 \\ 0 & -1 \end{pmatrix}$$

The spin of electron is described using Pauli spin matrices as follows

$$\vec{S} = \frac{\hbar}{2}\vec{\sigma} \Rightarrow \;\; (S_x, S_y, S_z) = (\frac{\hbar}{2}\sigma_x, \frac{\hbar}{2}\sigma_y, \frac{\hbar}{2}\sigma_z) = \frac{\hbar}{2}(\sigma_x, \sigma_y, \sigma_z)$$

Square of Pauli matrices is unity, i.e. $\hat{\sigma}_i^2 = \hat{I}$

$\hat{\sigma}_x^2 = \begin{pmatrix} 0 & 1 \\ 1 & 0 \end{pmatrix}\begin{pmatrix} 0 & 1 \\ 1 & 0 \end{pmatrix}$	$\hat{\sigma}_y^2 = \begin{pmatrix} 0 & -i \\ i & 0 \end{pmatrix}\begin{pmatrix} 0 & -i \\ i & 0 \end{pmatrix}$	$\hat{\sigma}_z^2 = \begin{pmatrix} 1 & 0 \\ 0 & -1 \end{pmatrix}\begin{pmatrix} 1 & 0 \\ 0 & -1 \end{pmatrix}$
$=\begin{pmatrix} 1 & 0 \\ 0 & 1 \end{pmatrix} = \hat{I}$	$=\begin{pmatrix} 1 & 0 \\ 0 & 1 \end{pmatrix} = \hat{I}$	$=\begin{pmatrix} 1 & 0 \\ 0 & 1 \end{pmatrix} = \hat{I}$

Pauli spin matrices are Hermitian, i.e. $\hat{\sigma}_i^\dagger = \hat{\sigma}_i$

$\hat{\sigma}_x^\dagger = \begin{pmatrix} 0 & 1 \\ 1 & 0 \end{pmatrix}^\dagger$	$\hat{\sigma}_y^\dagger = \begin{pmatrix} 0 & -i \\ i & 0 \end{pmatrix}^\dagger$	$\hat{\sigma}_z^\dagger = \begin{pmatrix} 1 & 0 \\ 0 & -1 \end{pmatrix}^\dagger$
$=\begin{pmatrix} 0 & 1 \\ 1 & 0 \end{pmatrix}^{cT} = \begin{pmatrix} 0 & 1 \\ 1 & 0 \end{pmatrix}^T$	$=\begin{pmatrix} 0 & -i \\ i & 0 \end{pmatrix}^{cT} = \begin{pmatrix} 0 & i \\ -i & 0 \end{pmatrix}^T$	$=\begin{pmatrix} 1 & 0 \\ 0 & -1 \end{pmatrix}^{cT} = \begin{pmatrix} 1 & 0 \\ 0 & -1 \end{pmatrix}^T$
$=\begin{pmatrix} 0 & 1 \\ 1 & 0 \end{pmatrix} = \hat{\sigma}_x$	$=\begin{pmatrix} 0 & -i \\ i & 0 \end{pmatrix} = \hat{\sigma}_y$	$=\begin{pmatrix} 1 & 0 \\ 0 & -1 \end{pmatrix} = \hat{\sigma}_z$

Pauli spin matrices are unitary, i.e. $\hat{\sigma}_i^\dagger = \hat{\sigma}_i^{-1}$

Consider $\hat{\sigma}_i\hat{\sigma}_i^\dagger = \hat{\sigma}_i\hat{\sigma}_i = \hat{\sigma}_i^2 = \hat{I}$ (as $\hat{\sigma}_i$ is Hermitian and its square is unity)

$$\Rightarrow \hat{\sigma}_i^\dagger = \hat{\sigma}_i^{-1}$$

Any 2×2 ***Hermitian matrix can be written as a linear combination of Pauli matrices where the coefficients are real, i.e.*** $\hat{H} = c_0\,\hat{I} + c_z\,\hat{\sigma}_z + c_x\,\hat{\sigma}_x + c_y\,\hat{\sigma}_y$ ***where*** $\hat{H}$ ***is Hermitian.***

Proof:

$$\hat{H} = \begin{pmatrix} H_{11} & H_{12} \\ H_{21} & H_{22} \end{pmatrix}$$

$$= \begin{pmatrix} \frac{1}{2}[(H_{11} + H_{22}) + (H_{11} - H_{22})] & \frac{1}{2}[(H_{12} + H_{21}) + (H_{12} - H_{21})] \\ \frac{1}{2}[(H_{12} + H_{21}) - (H_{12} - H_{21})] & \frac{1}{2}[(H_{11} + H_{22}) - (H_{11} - H_{22})] \end{pmatrix}$$

$$= \frac{1}{2}\left[\begin{pmatrix} H_{11} + H_{22} & 0 \\ 0 & H_{11} + H_{22} \end{pmatrix} + \begin{pmatrix} H_{11} - H_{22} & 0 \\ 0 & -(H_{11} - H_{22}) \end{pmatrix} \right.$$

$$\left. + \begin{pmatrix} 0 & H_{12} + H_{21} \\ H_{12} + H_{21} & 0 \end{pmatrix} + i\begin{pmatrix} 0 & -i(H_{12} - H_{21}) \\ i(H_{12} - H_{21}) & 0 \end{pmatrix} \right]$$

$$= \frac{1}{2}\left[(H_{11} + H_{22})\begin{pmatrix} 1 & 0 \\ 0 & 1 \end{pmatrix} + (H_{11} - H_{22})\begin{pmatrix} 1 & 0 \\ 0 & -1 \end{pmatrix} + (H_{12} + H_{21})\begin{pmatrix} 0 & 1 \\ 1 & 0 \end{pmatrix} + \right.$$

$$\left. i(H_{12} - H_{21})\begin{pmatrix} 0 & -i \\ i & 0 \end{pmatrix} \right]$$

$$= c_0\begin{pmatrix} 1 & 0 \\ 0 & 1 \end{pmatrix} + c_z\begin{pmatrix} 1 & 0 \\ 0 & -1 \end{pmatrix} + c_x\begin{pmatrix} 0 & 1 \\ 1 & 0 \end{pmatrix} + c_y\begin{pmatrix} 0 & -i \\ i & 0 \end{pmatrix}$$

$$\hat{H} = c_0\,\hat{I} + c_z\,\hat{\sigma}_z + c_x\,\hat{\sigma}_x + c_y\,\hat{\sigma}_y = c_0\hat{I} + \vec{c}.\,\vec{\sigma}$$

where

$$c_0 = \frac{1}{2}(H_{11} + H_{22}),\ c_z = \frac{1}{2}(H_{11} - H_{22}),\ c_x = \frac{1}{2}(H_{12} + H_{21}),\ c_y = \frac{i}{2}(H_{12} - H_{21})$$

Pauli spin matrices and unit matrix operator form a complete set of 2×2 matrices.

Pauli spin matrices do not commute, i.e. $[\hat{\sigma}_i,\ \hat{\sigma}_j] = 2i\ \varepsilon_{ijk}\,\hat{\sigma}_k \neq 0$

Proof:

$$[\hat{\sigma}_x,\ \hat{\sigma}_y] = \hat{\sigma}_x\hat{\sigma}_y - \hat{\sigma}_y\hat{\sigma}_x$$

$$= \begin{pmatrix} 0 & 1 \\ 1 & 0 \end{pmatrix}\begin{pmatrix} 0 & -i \\ i & 0 \end{pmatrix} - \begin{pmatrix} 0 & -i \\ i & 0 \end{pmatrix}\begin{pmatrix} 0 & 1 \\ 1 & 0 \end{pmatrix} = \begin{pmatrix} i & 0 \\ 0 & -i \end{pmatrix} - \begin{pmatrix} -i & 0 \\ 0 & i \end{pmatrix}$$

$$= \begin{pmatrix} 2i & 0 \\ 0 & -2i \end{pmatrix} = 2i\begin{pmatrix} 1 & 0 \\ 0 & -1 \end{pmatrix} - 2i\hat{\sigma}_z \neq 0.$$

Also $[\hat{\sigma}_y, \ \hat{\sigma}_x] = -2i\hat{\sigma}_z$

$$[\hat{\sigma}_y, \ \hat{\sigma}_z] = \hat{\sigma}_y\hat{\sigma}_z - \hat{\sigma}_z\hat{\sigma}_y$$

$$= \begin{pmatrix} 0 & -i \\ i & 0 \end{pmatrix}\begin{pmatrix} 1 & 0 \\ 0 & -1 \end{pmatrix} - \begin{pmatrix} 1 & 0 \\ 0 & -1 \end{pmatrix}\begin{pmatrix} 0 & -i \\ i & 0 \end{pmatrix} = \begin{pmatrix} 0 & i \\ i & 0 \end{pmatrix} - \begin{pmatrix} 0 & -i \\ -i & 0 \end{pmatrix}$$

$$= \begin{pmatrix} 0 & 2i \\ 2i & 0 \end{pmatrix} = 2i\begin{pmatrix} 0 & 1 \\ 1 & 0 \end{pmatrix} = 2i\hat{\sigma}_x \neq 0$$

Also $[\hat{\sigma}_z, \ \hat{\sigma}_y] = -2i\hat{\sigma}_x$

$$[\hat{\sigma}_z, \ \hat{\sigma}_x] = \hat{\sigma}_z\hat{\sigma}_x - \hat{\sigma}_x\hat{\sigma}_z$$

$$= \begin{pmatrix} 1 & 0 \\ 0 & -1 \end{pmatrix}\begin{pmatrix} 0 & 1 \\ 1 & 0 \end{pmatrix} - \begin{pmatrix} 0 & 1 \\ 1 & 0 \end{pmatrix}\begin{pmatrix} 1 & 0 \\ 0 & -1 \end{pmatrix} = \begin{pmatrix} 0 & 1 \\ -1 & 0 \end{pmatrix} - \begin{pmatrix} 0 & -1 \\ 1 & 0 \end{pmatrix}$$

$$= \begin{pmatrix} 0 & 2 \\ -2 & 0 \end{pmatrix} = 2\begin{pmatrix} 0 & 1 \\ -1 & 0 \end{pmatrix} = 2\begin{pmatrix} 0 & -i^2 \\ i^2 & 0 \end{pmatrix} = 2i\begin{pmatrix} 0 & -i \\ i & 0 \end{pmatrix} = 2i\hat{\sigma}_y \neq 0$$

Also $[\hat{\sigma}_x, \ \hat{\sigma}_z] = -2i\hat{\sigma}_y$
Combining all the results we get

$$[\hat{\sigma}_i, \ \hat{\sigma}_j] = 2i \ \varepsilon_{ijk} \ \hat{\sigma}_k \quad (\varepsilon_{ijk} \text{ is Levi Civita symbol})$$

Pauli spin matrices anti-commute. $\{\hat{\sigma}_i, \ \hat{\sigma}_j\} = \hat{\sigma}_i\hat{\sigma}_j + \hat{\sigma}_j\hat{\sigma}_i = 0$
   Proof:

$$\{\hat{\sigma}_x, \ \hat{\sigma}_y\} = \hat{\sigma}_x\hat{\sigma}_y + \hat{\sigma}_y\hat{\sigma}_x = \begin{pmatrix} 0 & 1 \\ 1 & 0 \end{pmatrix}\begin{pmatrix} 0 & -i \\ i & 0 \end{pmatrix} + \begin{pmatrix} 0 & -i \\ i & 0 \end{pmatrix}\begin{pmatrix} 0 & 1 \\ 1 & 0 \end{pmatrix}$$

$$= \begin{pmatrix} i & 0 \\ 0 & -i \end{pmatrix} + \begin{pmatrix} -i & 0 \\ 0 & i \end{pmatrix} = \begin{pmatrix} 0 & 0 \\ 0 & 0 \end{pmatrix} = \hat{0}$$

$$\{\hat{\sigma}_y, \ \hat{\sigma}_z\} = \hat{\sigma}_y\hat{\sigma}_z + \hat{\sigma}_z\hat{\sigma}_y = \begin{pmatrix} 0 & -i \\ i & 0 \end{pmatrix}\begin{pmatrix} 1 & 0 \\ 0 & -1 \end{pmatrix} + \begin{pmatrix} 1 & 0 \\ 0 & -1 \end{pmatrix}\begin{pmatrix} 0 & -i \\ i & 0 \end{pmatrix}$$

$$= \begin{pmatrix} 0 & i \\ i & 0 \end{pmatrix} + \begin{pmatrix} 0 & -i \\ -i & 0 \end{pmatrix} = \begin{pmatrix} 0 & 0 \\ 0 & 0 \end{pmatrix} = \hat{0}$$

$$\{\hat{\sigma}_z, \ \hat{\sigma}_x\} = \hat{\sigma}_z\hat{\sigma}_x + \hat{\sigma}_x\hat{\sigma}_z = \begin{pmatrix} 1 & 0 \\ 0 & -1 \end{pmatrix}\begin{pmatrix} 0 & 1 \\ 1 & 0 \end{pmatrix} + \begin{pmatrix} 0 & 1 \\ 1 & 0 \end{pmatrix}\begin{pmatrix} 1 & 0 \\ 0 & -1 \end{pmatrix}$$

$$= \begin{pmatrix} 0 & 1 \\ -1 & 0 \end{pmatrix} + \begin{pmatrix} 0 & -1 \\ 1 & 0 \end{pmatrix} = \begin{pmatrix} 0 & 0 \\ 0 & 0 \end{pmatrix} = \hat{0}$$

Pauli spin matrices are traceless $Tr\hat{\sigma}_i = 0$, $(i \rightarrow 1, 2, 3$ i.e. $x, y, z)$

$$Tr\ \hat{\sigma}_x = Tr\begin{pmatrix} 0 & 1 \\ 1 & 0 \end{pmatrix} = 0 + 0 = 0 \qquad Tr\hat{\sigma}_y = Tr\begin{pmatrix} 0 & -i \\ i & 0 \end{pmatrix} = 0 + 0 = 0 \qquad Tr\hat{\sigma}_z = Tr\begin{pmatrix} 1 & 0 \\ 0 & -1 \end{pmatrix} = 1 - 1 = 0$$

Product of the Pauli matrices $\hat{\sigma}_x, \hat{\sigma}_y, \hat{\sigma}_z = i\hat{I}$

$$\hat{\sigma}_x\hat{\sigma}_y\hat{\sigma}_z = \begin{pmatrix} 0 & 1 \\ 1 & 0 \end{pmatrix}\begin{pmatrix} 0 & -i \\ i & 0 \end{pmatrix}\begin{pmatrix} 1 & 0 \\ 0 & -1 \end{pmatrix} = \begin{pmatrix} 0 & 1 \\ 1 & 0 \end{pmatrix}\begin{pmatrix} 0 & i \\ i & 0 \end{pmatrix} = \begin{pmatrix} i & 0 \\ 0 & i \end{pmatrix} = i\begin{pmatrix} 1 & 0 \\ 0 & 1 \end{pmatrix} = i\hat{I}$$

All $\hat{\sigma}_x, \hat{\sigma}_y, \hat{\sigma}_z$ have eigenvalues ± 1. They can be diagonalized.
We find out the eigenvalues and corresponding eigen vectors of $\hat{\sigma}_x$.
Eigenvalue equation of the operator $\hat{\sigma}_x$ is
$$\hat{\sigma}_x|\chi> = a|\chi> \Rightarrow \quad \hat{\sigma}_x|\chi> = a\hat{I}|\chi> \Rightarrow \quad (\hat{\sigma}_x - a\hat{I})|\chi> = 0$$

$$[\begin{pmatrix} 0 & 1 \\ 1 & 0 \end{pmatrix} - a\begin{pmatrix} 1 & 0 \\ 0 & 1 \end{pmatrix}]\begin{pmatrix} x_1 \\ x_2 \end{pmatrix} = \begin{pmatrix} 0 \\ 0 \end{pmatrix} \quad \Rightarrow \quad \begin{pmatrix} -a & 1 \\ 1 & -a \end{pmatrix}\begin{pmatrix} x_1 \\ x_2 \end{pmatrix} = \begin{pmatrix} 0 \\ 0 \end{pmatrix} \qquad (1.12)$$

For non-trivial solution

$$\begin{vmatrix} -a & 1 \\ 1 & -a \end{vmatrix} = 0 \quad \Rightarrow \quad a^2 - 1 = 0 \quad \Rightarrow \quad a = \pm 1$$

With $a = 1$, equation (1.12) gives

$$\begin{pmatrix} -1 & 1 \\ 1 & -1 \end{pmatrix}\begin{pmatrix} x_1 \\ x_2 \end{pmatrix} = \begin{pmatrix} 0 \\ 0 \end{pmatrix}$$

$$\begin{pmatrix} -x_1 + x_2 \\ x_1 - x_2 \end{pmatrix} = \begin{pmatrix} 0 \\ 0 \end{pmatrix}$$

$\Rightarrow x_1 = x_2, x_2 = $ arbitrary $= 1$ (say)

$$|\chi> = \begin{pmatrix} x_1 \\ x_2 \end{pmatrix} = \begin{pmatrix} 1 \\ 1 \end{pmatrix}$$

$\xrightarrow{\text{Normalize}} \frac{1}{\sqrt{2}}\begin{pmatrix} 1 \\ 1 \end{pmatrix} = |1>$

With $a = -1$, equation (1.12) gives

$$\begin{pmatrix} 1 & 1 \\ 1 & 1 \end{pmatrix}\begin{pmatrix} x_1 \\ x_2 \end{pmatrix} = \begin{pmatrix} 0 \\ 0 \end{pmatrix}$$

$$\begin{pmatrix} x_1 + x_2 \\ x_1 + x_2 \end{pmatrix} = \begin{pmatrix} 0 \\ 0 \end{pmatrix}$$

$\Rightarrow x_1 = -x_2, x_2 = $ arbitrary $= -1$ (say)

$$|\chi> = \begin{pmatrix} x_1 \\ x_2 \end{pmatrix} = \begin{pmatrix} 1 \\ -1 \end{pmatrix}$$

$\xrightarrow{\text{Normalize}} \frac{1}{\sqrt{2}}\begin{pmatrix} 1 \\ -1 \end{pmatrix} = |-1>$

Diagonalization of the matrix operator $\hat{\sigma}_x$

The eigenvectors $\begin{pmatrix} 1 \\ 1 \end{pmatrix}$ and $\begin{pmatrix} 1 \\ -1 \end{pmatrix}$ are *linearly independent*. Construct the *diagonalizing matrix* with them

$$\hat{S} = \begin{pmatrix} 1 & 1 \\ 1 & -1 \end{pmatrix} \text{ and so } |\hat{S}| = \begin{vmatrix} 1 & 1 \\ 1 & -1 \end{vmatrix} = -1 - 1 = -2$$

$$\hat{S}^{-1} = \frac{\hat{S}^{cT}}{|\hat{S}|} = \frac{1}{-2}\begin{pmatrix} 1 & 1 \\ 1 & -1 \end{pmatrix}^{cT} = -\frac{1}{2}\begin{pmatrix} -1 & -1 \\ -1 & 1 \end{pmatrix}^{T} = -\frac{1}{2}\begin{pmatrix} -1 & -1 \\ -1 & 1 \end{pmatrix} = \frac{1}{2}\begin{pmatrix} 1 & 1 \\ 1 & -1 \end{pmatrix}$$

The *diagonalized matrix* is

$$\hat{\sigma}_x^D = \hat{S}^{-1}\hat{\sigma}_x\hat{S} = \frac{1}{2}\begin{pmatrix} 1 & 1 \\ 1 & -1 \end{pmatrix}\begin{pmatrix} 0 & 1 \\ 1 & 0 \end{pmatrix}\begin{pmatrix} 1 & 1 \\ 1 & -1 \end{pmatrix} = \frac{1}{2}\begin{pmatrix} 1 & 1 \\ 1 & -1 \end{pmatrix}\begin{pmatrix} 1 & -1 \\ 1 & 1 \end{pmatrix}$$

$$= \frac{1}{2}\begin{pmatrix} 2 & 0 \\ 0 & -2 \end{pmatrix} = \begin{pmatrix} 1 & 0 \\ 0 & -1 \end{pmatrix}$$

1.28 Orthogonal matrix operator

Orthogonal matrix is a square matrix obeying

$$\hat{A}\hat{A}^T = \hat{A}^T\hat{A} = \hat{I}$$

and determinant value is either $|\hat{A}| = +1$ (called *proper* matrix) or $|\hat{A}| = -1$ (called *improper* matrix).

- *Example*

$\hat{A} = \begin{pmatrix} 0 & 1 \\ 1 & 0 \end{pmatrix}$ is an *improper* orthogonal matrix since

$$\hat{A}\hat{A}^T = \begin{pmatrix} 0 & 1 \\ 1 & 0 \end{pmatrix}\begin{pmatrix} 0 & 1 \\ 1 & 0 \end{pmatrix} = \begin{pmatrix} 1 & 0 \\ 0 & 1 \end{pmatrix} = \hat{I}, \ \hat{A}^T\hat{A} = \begin{pmatrix} 0 & 1 \\ 1 & 0 \end{pmatrix}\begin{pmatrix} 0 & 1 \\ 1 & 0 \end{pmatrix} = \begin{pmatrix} 1 & 0 \\ 0 & 1 \end{pmatrix} = \hat{I}$$

and $|\hat{A}| = \begin{vmatrix} 0 & 1 \\ 1 & 0 \end{vmatrix} = -1$

1.29 Standard method of ortho-normalization Graham–Schmidt ortho-normalization procedure

We start with a set of kets that are *not* ortho-normalized

$$|\phi_1>, |\phi_2>, |\phi_3> \dots ., \quad <\phi_i|\phi_j> \neq \delta_{ij}.$$

We will build a new set of *orthonormalized kets* from this set viz.

$$|\psi_1>, |\psi_2>, |\psi_3> \dots \text{ such that } <\psi_i|\psi_j> = \delta_{ij}.$$

We can do this by following steps as suggested by Graham–Schmidt's ortho-normalization procedure:

Step 1

$|\phi_1\rangle$ is *not* ortho-normalized since $\|\phi_1\| = \sqrt{\langle\phi_1|\phi_1\rangle} > 1$.
Construct $|\psi_1\rangle = N|\phi_1\rangle$ and demand that $\langle\psi_1|\psi_1\rangle = 1$

$$\Rightarrow \langle N\phi_1|N\phi_1\rangle = 1 \quad \Rightarrow N^*N\langle\phi_1|\phi_1\rangle = 1$$

$$\Rightarrow |N|^2 = \frac{1}{\langle\phi_1|\phi_1\rangle} \quad \Rightarrow N = \frac{1}{\sqrt{\langle\phi_1|\phi_1\rangle}} = \frac{1}{\|\phi_1\|}$$

$$|\psi_1\rangle = N|\phi_1\rangle$$

$$\Rightarrow |\psi_1\rangle = \frac{|\phi_1\rangle}{\|\phi_1\|} = \frac{|\phi_1\rangle}{\sqrt{\langle\phi_1|\phi_1\rangle}} \quad \Leftarrow \text{Normalized ket } [\because \langle\psi_1|\psi_1\rangle = 1].$$

Step 2

Consider $|\phi_2\rangle$. We wish to find the part of $|\phi_2\rangle$ *orthogonal* to $|\psi_1\rangle$.
The part of $|\phi_2\rangle$ that depends on $|\psi_1\rangle$ is $\langle\psi_1|\phi_2\rangle$ *along* $|\psi_1\rangle$
i.e. $\langle\psi_1|\phi_2\rangle|\psi_1\rangle$
Hence the part of $|\phi_2\rangle$ which is *independent* of (i.e. *orthogonal* to) $|\psi_1\rangle$ is

$$|\phi_2\rangle - \langle\psi_1|\phi_2\rangle|\psi_1\rangle$$

To check orthogonality operate by $\langle\psi_1|$ to get

$$\langle\psi_1|\ [\ |\phi_2\rangle - \langle\psi_1|\phi_2\rangle|\psi_1\rangle\ ]$$

$$= \langle\psi_1|\phi_2\rangle - \langle\psi_1|\phi_2\rangle\langle\psi_1|\psi_1\rangle = \langle\psi_1|\phi_2\rangle - \langle\psi_1|\phi_2\rangle = 0 \text{ (using } \langle\psi_1|\psi_1\rangle = 1)$$

So $|\psi_1\rangle$ is *orthogonal* to $|\phi_2\rangle - \langle\psi_1|\phi_2\rangle|\psi_1\rangle$.
Let us next *normalize* $|\phi_2\rangle - \langle\psi_1|\phi_2\rangle|\psi_1\rangle$. We construct

$$|\psi_2\rangle = \frac{|\phi_2\rangle - \langle\psi_1|\phi_2\rangle|\psi_1\rangle}{\|\ |\phi_2\rangle - \langle\psi_1|\phi_2\rangle|\psi_1\rangle\ \|} \quad \text{(normalized)}$$

- *Compare*

We wish to find the part of $\vec{a}$ that is orthogonal to $\hat{i}$ (figure 1.8). The vector $\vec{a}$ can be resolved along $\hat{i}$, $\hat{j}$ and expressed as

$$\vec{a} = \hat{i}a_x + \hat{j}a_y = \hat{i}(\hat{i}\cdot\vec{a}) + \hat{j}(\hat{j}\cdot\vec{a})$$

The $\hat{i}a_x$ part of $\vec{a}$ is that part of $\vec{a}$ which is *along* $\hat{i}$. Clearly then, the part $\vec{a} - \hat{i}a_x = \vec{a} - \hat{i}(\hat{i}\cdot\vec{a}) = \hat{j}a_y = \hat{j}(\hat{j}\cdot\vec{a})$ is the part of $\vec{a}$ which is independent of $\hat{i}$ and is *along* the normal to $\hat{i}$, or orthogonal to $\hat{i}$ i.e. along $\hat{j}$ (division by $a_y = (\hat{j}\cdot\vec{a})$ gives $\frac{\vec{a} - \hat{i}a_x}{a_y} = \frac{\hat{j}(\hat{j}\cdot\vec{a})}{a_y} = \frac{\hat{j}a_y}{a_y} = \hat{j}$ which is orthonormal to $\hat{i}$).

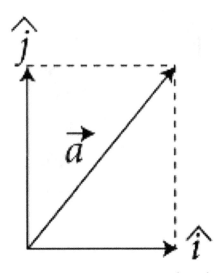

Figure 1.8. $\vec{a}$ has components along $\hat{i}$ and $\hat{j}$.

Step 3

We can continue building another ortho-normalized ket vector proceeding in the same way as follows:

Consider $|\phi_3\rangle$. We wish to find the part of $|\phi_3\rangle$ *orthogonal* to $|\psi_1\rangle$ and $|\psi_2\rangle$.

The part of $|\phi_3\rangle$ that depends on $|\psi_1\rangle$ is $\langle\psi_1|\phi_3\rangle$ *along* $|\psi_1\rangle$ (i.e. $\langle\psi_1|\phi_3\rangle |\psi_1\rangle$) and that which depends on $|\psi_2\rangle$ is $\langle\psi_2|\phi_3\rangle$ *along* $|\psi_2\rangle$ (i.e. $\langle\psi_2|\phi_3\rangle |\psi_2\rangle$).

Hence the part of $|\phi_3\rangle$ which is *independent* of (i.e. *orthogonal* to) $|\psi_1\rangle$ and $|\psi_2\rangle$ is $|\phi_3\rangle - \langle\psi_1|\phi_3\rangle |\psi_1\rangle - \langle\psi_2|\phi_3\rangle |\psi_2\rangle$.

Let us next normalize it. We construct

$$|\psi_3\rangle = \frac{|\phi_3\rangle - \langle\psi_1|\phi_3\rangle |\psi_1\rangle - \langle\psi_2|\phi_3\rangle |\psi_2\rangle}{\| \ |\phi_3\rangle - \langle\psi_1|\phi_3\rangle |\psi_1\rangle - \langle\psi_2|\phi_3\rangle |\psi_2\rangle \ \|}.$$

We can continue building more ortho-normalized ket vectors in this manner. So we are able to construct a new set of *orthonormalized kets* viz. $|\psi_1\rangle$, $|\psi_2\rangle$, $|\psi_3\rangle$ ... such that $\langle\psi_i|\psi_j\rangle = \delta_{ij}$.

- *Compare*:

We wish to find the part of $\vec{a}$ orthogonal to $\hat{i}$ and $\hat{j}$. The vector $\vec{a}$ can be resolved along $\hat{i}, \hat{j}, \hat{k}$ and expressed as (figure 1.9)

$$\vec{a} = \hat{i} a_x + \hat{j} a_y + \hat{k} a_z = \hat{i}(\hat{i}.\vec{a}) + \hat{j}(\hat{j}.\vec{a}) + \hat{k}(\hat{k}.\vec{a})$$

$$\Rightarrow \vec{a} - \hat{i}(\hat{i}.\vec{a}) - \hat{j}(\hat{i}.\vec{a}) = \vec{a} - \hat{i} a_x - \hat{j} a_y$$

$$= \hat{k} a_z = \hat{k}(\hat{k}.\vec{a}) = \text{part of } \vec{a} \text{ orthogonal to } \hat{i} \text{ and } \hat{j}.$$

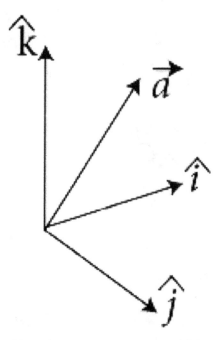

Figure 1.9. $\vec{a}$ has components along $\hat{i},\hat{j},\hat{k}$.

(Division by $a_z = (\hat{k}.\ \vec{a})$, gives $\dfrac{\vec{a} - \hat{i}a_x - \hat{j}a_y}{a_z} = \dfrac{\hat{k}(\hat{k}.\vec{a})}{a_z} = \dfrac{\hat{k}a_z}{a_z} = \hat{k}$ which is orthonormal to $\hat{i}$ and $\hat{j}$.)

1.30 Definition of average value

We define *average* of operator $\hat{A}$ in the ket state $|\psi>$ as follows

For unnormalized $\|\psi>$	For normalized $\|\psi>$ ($<\psi\|\psi> = 1$)
$<\hat{A}> = \dfrac{<\psi\|\hat{A}\|\psi>}{<\psi\|\psi>}$	$<\hat{A}> = \dfrac{<\psi\|\hat{A}\|\psi>}{<\psi\|\psi>} = <\psi\|\hat{A}\|\psi>$
$= \dfrac{\displaystyle\int_{-\infty}^{+\infty} \psi^*\hat{A}\psi\ dx}{\displaystyle\int_{-\infty}^{+\infty} \psi^*\psi\ dx}$	$= \displaystyle\int_{-\infty}^{+\infty} \psi^*\hat{A}\psi\ dx$
	(since $<\psi\|\psi> = \displaystyle\int_{-\infty}^{+\infty} \psi^*\psi\ dx = 1$)

If $|\psi> = c_1|\psi_1> + c_2|\psi_2> + c_3|\psi_3> + ... = \sum_i c_i|\psi_i>$ (*expansion postulate*)

And if $\hat{A}|\psi_i> = a_i|\psi_i>$ ($|\psi_i>$ s are *eigenkets* of the operator $\hat{A}$)

and $<\psi_i|\psi_j> = \delta_{ij}$ ($|\psi_i>$ s are *ortho-normalized*)

Then $<\psi|\hat{A}|\psi> = <\sum_j c_j\psi_j|\hat{A}|\sum_i c_i\psi_i>$

$$= \sum_j c_j^* \sum_i c_i < \psi_j|\hat{A}|\psi_i>$$

$$= \sum_j c_j^* \sum_i c_i < \psi_j|a_i|\psi_i>$$

$$= \sum_j c_j^* \sum_i c_i a_i < \psi_j|\psi_i> = \sum_j c_j^* \sum_i c_i a_i \delta_{ij}$$

$$<\psi|\hat{A}|\psi> = \sum_j c_j^* c_j a_j = \sum_j |c_j|^2 a_j$$

Also $<\psi|\psi> = <\sum_j c_j\psi_j|\sum_i c_i\psi_i> = \sum_j c_j^* \sum_i c_i < \psi_j|\psi_i>$

$$= \sum_j c_j^* \sum_i c_i \delta_{ij} = \sum_j c_j^* c_j$$

$$<\psi|\psi> = \sum_j |c_j|^2$$

For unnormalized $|\psi>$

$$<\hat{A}> = \frac{<\psi|\hat{A}|\psi>}{<\psi|\psi>} = \frac{\sum_j |c_j|^2 a_j}{\sum_j |c_j|^2}$$

For normalized $|\psi>$ ($<\psi|\psi> = 1$)

$$<\hat{A}> = <\psi|\hat{A}|\psi> = \sum_j |c_j|^2 a_j$$

1.31 Some definitions

For two ket vectors $|\phi_1>$, $|\phi_2>$ and constants c_1, c_2 we define the following:

Linear operator

An operator is said to be *linear* if

$$\hat{A}(c_1|\phi_1> + c_2|\phi_2>) = \hat{A}(c_1|\phi_1>) + \hat{A}(c_2|\phi_2>) = c_1\hat{A}|\phi_1> + c_2\hat{A}|\phi_2>$$

- *Example*

1) Differentiation operator $\frac{d}{dx}$

$$\frac{d}{dx}(c_1|\psi_1> + c_2|\psi_2>) = c_1\frac{d}{dx}|\psi_1> + c_2\frac{d}{dx}|\psi_2> .$$

So $\frac{d}{dx}$ is a linear operator.

2) Nabla square operator ∇^2

$$\nabla^2(c_1|\psi_1 > +c_2|\psi_2 >) = c_1\nabla^2|\psi_1 > +c_2\nabla^2|\psi_2 > .$$

So ∇^2 is a linear operator.

Non-linear operator
An operator is said to be *non-linear* if

$$\hat{A}(c_1|\phi_1 > +c_2|\phi_2 >) \neq c_1\hat{A}|\phi_1 > +c_2\hat{A}|\phi_2>$$

- *Example*

$$\hat{A}\psi = \psi^2 \quad \text{(squaring operator)}$$

$$\hat{A}(c_1\psi_1 + c_2\psi_2) = (c_1\psi_1 + c_2\psi_2)^2$$

$$=(c_1\psi_1)^2 + (c_2\psi_2)^2 + 2(c_1\psi_1)(c_2\psi_2)$$

$$=\hat{A}(c_1\psi_1) + \hat{A}(c_2\psi_2) + 2(c_1\psi_1)(c_2\psi_2)$$

$$\neq\hat{A}(c_1\psi_1) + \hat{A}(c_2\psi_2)$$

So squaring operator $\hat{A}$ is a *non-linear* operator due to the presence of the third term on RHS viz. $2(c_1\psi_1)(c_2\psi_2)$.

Anti-linear operator
An operator is said to be anti-linear if

$$\hat{A}(c_1|\phi_1 > +c_2|\phi_2 >) = c_1^*(\hat{A}|\phi_1 >) + c_2^*(\hat{A}|\phi_2 >)$$

- *Example*

$$\hat{A}\psi = \psi^* \text{ (complex conjugation operator)}$$

$$\hat{A}(c_1\psi_1 + c_2\psi_2) = (c_1\psi_1 + c_2\psi_2)^* = c_1^*\psi_1^* + c_2^*\psi_2^*$$

$$=c_1^*(\hat{A}\psi_1) + c_2^*(\hat{A}\psi_2) \quad [\text{as } \hat{A}\psi_1 = \psi_1^*, \ \hat{A}\psi_2 = \psi_2^*]$$

So complex conjugation operator is an *anti-linear operator*.

1.32 Kroneckar product (symbol ⊗) or direct product or tensor product

If C^n and C^m are vector spaces of dimensions n and m, respectively then their Kroneckar product generates a vector space of dimension nm and *preserves* both linearity and scalar multiplication. In other words

$$C^n \otimes C^m = C^{nm}.$$

Preservation of *linearity* means

$$|a> \otimes(|b>+|c>) = |a>\otimes|b>+|a>\otimes|c>$$

$$(|a>+|b>) \otimes |c> = |a>\otimes|c>+|b>\otimes|c>$$

Preservation of *scalar multiplication* means

$$c(|a>\otimes|b>) = |a>\otimes c|b> = c|a>\otimes|b>$$

We explain *Kroneckar product* through the following examples.

- Find Kroneckar product of $A = \begin{pmatrix} a & b \\ c & d \end{pmatrix}$ and $P = \begin{pmatrix} p & q \\ r & s \end{pmatrix}$

$$A \otimes P = \begin{pmatrix} a & b \\ c & d \end{pmatrix} \otimes P = \begin{pmatrix} aP & bP \\ cP & dP \end{pmatrix}$$

$$= \begin{pmatrix} a\begin{pmatrix} p & q \\ r & s \end{pmatrix} & b\begin{pmatrix} p & q \\ r & s \end{pmatrix} \\ c\begin{pmatrix} p & q \\ r & s \end{pmatrix} & d\begin{pmatrix} p & q \\ r & s \end{pmatrix} \end{pmatrix} = \begin{pmatrix} ap & aq & bp & bq \\ ar & as & br & bs \\ cp & cq & dp & dq \\ cr & cs & dr & ds \end{pmatrix}$$

- Find $|0>\otimes|1>$ where $|0> = \begin{pmatrix} 1 \\ 0 \end{pmatrix}$, $|1> = \begin{pmatrix} 0 \\ 1 \end{pmatrix}$

$$|0>\otimes|1> = \begin{pmatrix} 1 \\ 0 \end{pmatrix} \otimes \begin{pmatrix} 0 \\ 1 \end{pmatrix} = \begin{pmatrix} 1\begin{pmatrix} 0 \\ 1 \end{pmatrix} \\ 0\begin{pmatrix} 0 \\ 1 \end{pmatrix} \end{pmatrix} = \begin{pmatrix} 0 \\ 1 \\ 0 \\ 0 \end{pmatrix}$$

- Find Kroneckar product of Pauli matrices $\hat{X}$ and $\hat{Y}$. [$\hat{\sigma}_x \equiv \hat{X}$, $\hat{\sigma}_y = \hat{Y}$]

$$\hat{X} \otimes \hat{Y} = \begin{pmatrix} 0 & 1 \\ 1 & 0 \end{pmatrix} \otimes \begin{pmatrix} 0 & -i \\ i & 0 \end{pmatrix} = \begin{pmatrix} 0\begin{pmatrix} 0 & -i \\ i & 0 \end{pmatrix} & 1\begin{pmatrix} 0 & -i \\ i & 0 \end{pmatrix} \\ 1\begin{pmatrix} 0 & -i \\ i & 0 \end{pmatrix} & 0\begin{pmatrix} 0 & -i \\ i & 0 \end{pmatrix} \end{pmatrix} = \begin{pmatrix} 0 & 0 & 0 & -i \\ 0 & 0 & i & 0 \\ 0 & -i & 0 & 0 \\ i & 0 & 0 & 0 \end{pmatrix}$$

- If $|u> = \begin{pmatrix} p \\ q \end{pmatrix}$, $|v> = \begin{pmatrix} r \\ s \end{pmatrix}$ then show that $|u> \otimes |v> = \begin{pmatrix} pr \\ ps \\ qr \\ qs \end{pmatrix} = |uv>$

- Show that $\hat{\sigma}_x \otimes \hat{\sigma}_z = \begin{pmatrix} 0 & 0 & 1 & 0 \\ 0 & 0 & 0 & -1 \\ 1 & 0 & 0 & 0 \\ 0 & -1 & 0 & 0 \end{pmatrix}$

1.33 Further reading

[1] Guha J 2019 *Quantum Mechanics: Theory, Problems & Solutions* 3rd edn (Kolkata: Books and Allied (P) Ltd)
[2] Zettili N 2009 *Quantum Mechanics: Concepts and Applications* (New York: Wiley)
[3] Waghmare Y R 2014 *Fundamentals of Quantum Mechanics* (New Delhi: S. Chand)
[4] Guha J 2020 *Modern Physics* vols 1 and 2 (Kolkata: Techno World)

1.34 Problems

1. *Expectation value of a dynamical quantity refers to:*

   (a) *average value in repeated measurements made on identical copies of quantum state;*
   (b) *eigenvalue of operator corresponding to the observable;*
   (c) *time average value of quantum measurement;*
   (d) *most probable value in a quantum measurement.*

   Ans. (*a*).

2. *Consider a matrix* $\begin{pmatrix} 1 & i \\ -i & 1 \end{pmatrix}$ *having an eigenvalue two corresponding to eigenvector* $\begin{pmatrix} i \\ 1 \end{pmatrix}$. *What is the projection operator for the eigen space of eigen value 2?*

   $(a) \frac{1}{2} \begin{pmatrix} -1 & i \\ i & 1 \end{pmatrix} (b) \frac{1}{2} \begin{pmatrix} 1 & -i \\ i & 1 \end{pmatrix} (c) \frac{1}{2} \begin{pmatrix} 1 & i \\ -i & 1 \end{pmatrix} (d) \frac{1}{2} \begin{pmatrix} 1 & -i \\ i & -1 \end{pmatrix}$

   Ans. Projection operator for an unnormalized state $|\psi>$ is $\hat{P}_{|\psi>} = \frac{|\psi><\psi|}{<\psi|\psi>}$. Hence for

   $$|\psi> = |2> = \begin{pmatrix} i \\ 1 \end{pmatrix}, \quad \hat{P}_{|2>} = \frac{|2><2|}{<2|2>} = \frac{\begin{pmatrix} i \\ 1 \end{pmatrix}(-i \ \ 1)}{(-i \ \ 1)\begin{pmatrix} i \\ 1 \end{pmatrix}} = \frac{1}{2}\begin{pmatrix} 1 & i \\ -i & 1 \end{pmatrix}.$$

   Check: $\hat{P}_{|2>}|2> = \frac{1}{2}\begin{pmatrix} 1 & i \\ -i & 1 \end{pmatrix}\begin{pmatrix} i \\ 1 \end{pmatrix} = \frac{1}{2}\begin{pmatrix} i+i \\ -i^2+1 \end{pmatrix} = \begin{pmatrix} i \\ 1 \end{pmatrix} = |2>$

   Ans (c)

3. *An electron has ground state |0> (energy E_0) and first excited state |1> (energy E_1). What is the energy of the electron in a measurement if electron is in a state $\frac{|0> + |1>}{\sqrt{2}}$?*

    (a) $\frac{E_0 + E_1}{2}$ (b) $E_0 + E_1$

    (c) $\frac{E_0 + E_1}{\sqrt{2}}$ (d) E_0 or E_1 *with equal probability.*

    Ans. (d).

4. *An electron has ground state |0>(energy E_0) and first excited state |1> (energy E_1). What is the energy of the electron in a measurement if electron is in a state $\frac{3|0> + 4|1>}{5}$?*

    (a) $\frac{3E_0 + 4E_1}{5}$ (b) $\frac{9}{25}E_0 + \frac{16}{25}E_1$

    (c) $\frac{7}{\sqrt{5}}(E_0 + E_1)$ (d) E_0 or E_1 *with different probabilities*

    Ans. (d).

5. *Spot the matrix that is unitary (z = complex number).*

    (a) $\begin{pmatrix} \frac{1+i}{2} & \frac{1-i}{2} \\ \frac{1+i}{2} & \frac{1+i}{2} \end{pmatrix}$ (b) $\begin{pmatrix} 1 & i \\ i & 1 \end{pmatrix}$ (c) $\frac{1}{\sqrt{2}}\begin{pmatrix} 1+i & 1-i \\ 1-i & 1+i \end{pmatrix}$ (d) $\begin{pmatrix} 2 & 1-3i \\ 1+3i & 5 \end{pmatrix}$

    (e) $\frac{1}{\sqrt{2}}\begin{pmatrix} z & z^* \\ iz & -iz^* \end{pmatrix}$ (f) $\frac{1}{\sqrt{2}}\begin{pmatrix} z & z^* \\ -iz & iz^* \end{pmatrix}$ (g) $\frac{1}{\sqrt{2}}\begin{pmatrix} z & -z^* \\ iz & -iz^* \end{pmatrix}$ (h) $\frac{1}{\sqrt{2}}\begin{pmatrix} z^* & z \\ iz & -iz^* \end{pmatrix}$

    Ans. (a), (e), (f) $[\hat{U}^{\dagger}\hat{U} = \hat{I}]$

6. *If $|u_1 >$, $|u_2>$ is an orthonormalized basis, which of the following linear combinations are acceptable as properly normalized?*

    (a) $\frac{2+i}{\sqrt{11}}|u_1 > +\frac{1+3i}{\sqrt{11}}|u_2>$ (b) $\frac{2+i}{\sqrt{15}}|u_1 > +\frac{1+3i}{\sqrt{15}}|u_2>$

    (c) $\frac{2+i}{\sqrt{5}}|u_1 > +\frac{1+3i}{\sqrt{5}}|u_2>$ (d) $\frac{2+i}{\sqrt{3}}|u_1 > +\frac{1+3i}{\sqrt{3}}|u_2>$

    Ans. For $|\psi > = c_1|u_1 > +c_2|u_2>$, $| c_1 |^2 + | c_1 |^2 = 1$. Here

    $| c_1 |^2 + | c_1 |^2 = \left| \frac{2+i}{\sqrt{15}} \right|^2 + \left| \frac{1+3i}{\sqrt{15}} \right|^2 = 1$ for (b). Ans is (b).

7. *Identify the condition for the state $\begin{pmatrix} \cos\theta_1 + \cos\theta_2 \\ \sin\theta_1 + \sin\theta_2 \end{pmatrix}$ to be normalized in the orthonormalized basis $\begin{pmatrix} \cos\theta_1 \\ \sin\theta_1 \end{pmatrix}, \begin{pmatrix} \cos\theta_2 \\ \sin\theta_2 \end{pmatrix}$.*

    Ans. $|\psi > = \begin{pmatrix} \cos\theta_1 + \cos\theta_2 \\ \sin\theta_1 + \sin\theta_2 \end{pmatrix}$. Normalization condition is $<\psi|\psi > = 1$. So

    $<\psi|\psi > = (\cos\theta_1 + \cos\theta_2 \quad \sin\theta_1 + \sin\theta_2)\begin{pmatrix} \cos\theta_1 + \cos\theta_2 \\ \sin\theta_1 + \sin\theta_2 \end{pmatrix}$

    $(\cos\theta_1 + \cos\theta_2)^2 + (\sin\theta_1 + \sin\theta_2)^2 = 1$

$$\Rightarrow 1 + 1 + 2\cos(\theta_1 - \theta_2) = 1 \Rightarrow \cos(\theta_1 - \theta_2) = -\frac{1}{2} = \cos\frac{2\pi}{3}$$

$$\Rightarrow \theta_1 - \theta_2 = \frac{2\pi}{3}, \ 2n\pi \pm \frac{2\pi}{3} n = \text{integer}$$

8. *The matrix* $\frac{1}{2}\begin{pmatrix} 1+i & 1-i \\ 1-i & 1+i \end{pmatrix}$ *is*

   (a) normal(b) Hermitian(c) unitary(d) none of the these.

   Ans. $\hat{U} = \frac{1}{2}\begin{pmatrix} 1+i & 1-i \\ 1-i & 1+i \end{pmatrix}, \ \hat{U}^\dagger = \frac{1}{2}\begin{pmatrix} 1-i & 1+i \\ 1+i & 1-i \end{pmatrix}$

$$\hat{U}^\dagger\hat{U} = \frac{1}{2}\begin{pmatrix} 1-i & 1+i \\ 1+i & 1-i \end{pmatrix}\frac{1}{2}\begin{pmatrix} 1+i & 1-i \\ 1-i & 1+i \end{pmatrix} = \frac{1}{4}\begin{pmatrix} 1-i^2+1-i^2 & (1+i)^2+(1-i)^2 \\ (1+i)^2+(1-i)^2 & 1-i^2+1-i^2 \end{pmatrix}$$

$$= \frac{1}{4}\begin{pmatrix} 4 & 0 \\ 0 & 4 \end{pmatrix} = \begin{pmatrix} 1 & 0 \\ 0 & 1 \end{pmatrix} = \hat{I}$$

Also $\hat{U}\hat{U}^\dagger = \hat{I}$ (on evaluation)

$$\Rightarrow \hat{U}^\dagger\hat{U} = \hat{U}\hat{U}^\dagger = \hat{I} \Rightarrow \hat{U} \text{ is unitary}$$

$$\hat{U}^\dagger \neq \hat{U} \qquad \Rightarrow \hat{U} \text{ is not Hermitian.}$$

$[\hat{U}^\dagger, \hat{U}] = \hat{U}^\dagger\hat{U} - \hat{U}\hat{U}^\dagger = \hat{I} - \hat{I} = 0 \Rightarrow \hat{U}$ is normal matrix.
   Ans. (a), (c)

9. *What is the result of measurement of* $\hat{\sigma}_z$ *on a spin* $\frac{1}{2}$ *particle beam in a state*

   $\frac{1}{\sqrt{2}}\begin{pmatrix} 1 \\ -1 \end{pmatrix}$?

   Ans. Eigenvectors of $\hat{\sigma}_z$: $\chi_\uparrow = \begin{pmatrix} 1 \\ 0 \end{pmatrix}, \ \chi_\downarrow = \begin{pmatrix} 0 \\ 1 \end{pmatrix}$ (orthonormalized, complete set)

   Expand $\frac{1}{\sqrt{2}}\begin{pmatrix} 1 \\ -1 \end{pmatrix} = \frac{1}{\sqrt{2}}\begin{pmatrix} 1 \\ 0 \end{pmatrix} - \frac{1}{\sqrt{2}}\begin{pmatrix} 0 \\ 1 \end{pmatrix} = \frac{1}{\sqrt{2}}\chi_\uparrow - \frac{1}{\sqrt{2}}\chi_\downarrow$.

   So in a measurement the probability of finding the particle beam in spin up state is $\left|\frac{1}{\sqrt{2}}\right|^2 = \frac{1}{2}$ and in spin down state is $\left|-\frac{1}{\sqrt{2}}\right|^2 = \frac{1}{2}$. (50% in spin up state and 50% in spin down state).

10. *For a given quantum mechanical state it is given that*

    $\hat{A}|a\,b\,c> = a|a\,b\,c>, \ \hat{B}|a\,b\,c> = b|a\,b\,c>, \ \hat{C}|a\,b\,c> = c|a\,b\,c>$

    where a, b, c *are numbers. Suggest a commutation relation between* $\hat{A}, \hat{B}, \hat{C}^2$. *What is the uncertainty in simultaneous measurement of* $\hat{A}, \hat{B}, \hat{C}^2$ *in the state* $|a\,b\,c>$?

Ans. As $\hat{A}$, $\hat{B}$, $\hat{C}^2$ have common eigenstate they commute and there is no uncertainty.

$$[\hat{A}, \hat{B}] = 0, [\hat{B}, \hat{C}^2] = 0, [\hat{C}^2, \hat{A}] = 0$$

11. *Show that* $\langle r \rangle = \frac{3}{2} a_0$ *for ground state of hydrogen atom where r is the distance of electron from nucleus.* $\left(given \int_0^\infty e^{-\beta r} r^n dr = \frac{n!}{\beta^{n+1}} \right)$ a_0 *is the first Bohr radius.*

Ans. Hydrogen atom ground state wave function $\psi_{gs} = \frac{1}{\sqrt{\pi a_0^3}} e^{-r/a_0}$.

$$\langle r \rangle = \int_{all\ space} \psi_{gs}^* \hat{r}\ \psi_{gs} d\tau = \int_{all\ space} \left(\frac{1}{\sqrt{\pi a_0^3}} e^{-r/a_0} \right)^* r \left(\frac{1}{\sqrt{\pi a_0^3}} e^{-r/a_0} \right) (r^2 \sin\theta dr d\theta d\phi)$$

$$= \frac{1}{\pi a_0^3} \int_0^\infty r^3 e^{-2r/a_0} dr \int_0^\pi \sin\theta\ d\theta \int_0^{2\pi} d\phi = \frac{1}{\pi a_0^3} \frac{3!}{(2/a_0)^{3+1}} \cdot 2.\ 2\pi = \frac{3}{2} a_0$$

12. *What are the expectation values of position and momentum of a linear harmonic oscillator in the first excited state?*

Ans. First excited state of linear harmonic oscillator is

$$\psi_{ex} = \sqrt{\frac{2\alpha}{\sqrt{\pi}}} z e^{-\frac{1}{2} z^2} \text{ where } z = \alpha x, \alpha = \sqrt{\frac{mw}{\hbar}}$$

$$\langle x \rangle = \int_{-\infty}^{+\infty} \psi_{ex}^* \hat{x} \psi_{ex} dx = \int_{-\infty}^{+\infty} \left(\sqrt{\frac{2\alpha}{\sqrt{\pi}}} z e^{-\frac{1}{2} z^2} \right)^* x \left(\sqrt{\frac{2\alpha}{\sqrt{\pi}}} z e^{-\frac{1}{2} z^2} \right) dx$$

$$= \frac{2\alpha}{\sqrt{\pi}} \int_{-\infty}^{+\infty} (\alpha x)^2 e^{-(\alpha x)^2} x\ dx = 0 \text{ (odd integral)}$$

$$\langle p \rangle = \int_{-\infty}^{+\infty} \psi_{ex}^* \hat{p} \psi_{ex} dx = \int_{-\infty}^{+\infty} \left(\sqrt{\frac{2\alpha}{\sqrt{\pi}}} \alpha x e^{-\frac{(\alpha x)^2}{2}} \right)^* \left(-i\hbar \frac{\partial}{\partial x} \right) \left(\sqrt{\frac{2\alpha}{\sqrt{\pi}}} \alpha x e^{-\frac{(\alpha x)^2}{2}} \right) dx$$

$$= -i\hbar \int_{-\infty}^{+\infty} \left(\sqrt{\frac{2\alpha}{\sqrt{\pi}}} \alpha x e^{-\frac{(\alpha x)^2}{2}} \right) \left(\sqrt{\frac{2\alpha}{\sqrt{\pi}}} \alpha(1 - \alpha^2 x^2) e^{-\frac{(\alpha x)^2}{2}} \right) dx$$

$$= (-i\hbar) \frac{2\alpha^3}{\sqrt{\pi}} \int_{-\infty}^{+\infty} (x - \alpha^2 x^3)\ e^{-\alpha^2 x^2} dx = 0 \text{ (odd integral)}$$

13. *The operator $\hat{A}$ commutes with $\hat{B}$ and $\hat{C}$. Can we say that $\hat{B}$ and $\hat{C}$ are commutative?*

Ans. $\hat{B}$ and $\hat{C}$ are not commutative in general, e.g. let $\hat{A} = \hat{p}_y$, $\hat{B} = \hat{x}$, $\hat{C} = \hat{p}_x$. Though $[\hat{A}, \hat{B}] = [\hat{p}_y, \hat{x}] = 0; [\hat{A}, \hat{C}] = [\hat{p}_y, \hat{p}_x] = 0$ but $[\hat{B}, \hat{C}] = [\hat{x}, \hat{p}_x] = i\hbar \neq 0$.

Chapter 2

Postulates of quantum mechanics

Quantum physics describes nature in the small at the level of elementary particles—specifically, it describes a system whose dimension is smaller than or comparable to its de Broglie wavelength. In other words, the classical particle picture is not adequate and does not provide a faithful representation of the system for which wave property dominates.

Quantum physics is developed based on certain postulates. The postulates of quantum mechanics cannot be proved or deduced.

The postulates of quantum mechanics are abstract, weird and not visualisable—often they do not match our common sense.

The correctness of the postulates lies in the fact that the deductions from them and the results of real (or physically performed) experiments agree with each other. Quantum mechanics, to date, has withstood every experimental test.

The basic quantum postulates that lead to the idea of quantum computation will be discussed here.

2.1 First postulate: observables are replaced by operators

In quantum mechanics physical observables are replaced by *linear Hermitian operators* in an infinite dimensional linear vector space called *Hilbert space*. These operators are *identified with the act of measurement*. The *eigenvalues*, which are *real numbers*, are identified with the *results* of measurement (which are also real numbers). Let us mention the customarily chosen operators corresponding to the following observables (cap or hat atop represents operator).

doi:10.1088/978-0-7503-2715-2ch2

Observable	Corresponding operator
Position x (in 1D)	$\hat{x} = x$
Position $\vec{r}$ is (in 3D)	$\hat{\vec{r}} = \vec{r}$
Momentum p (in 1D)	$\hat{p} = -i\hbar\frac{\partial}{\partial x}$
Momentum $\vec{p}$ (in 3D)	$\hat{\vec{p}} = -i\hbar\vec{\nabla}$ since $\hat{p}_x = -i\hbar\frac{\partial}{\partial x}$, $\hat{p}_y = -i\hbar\frac{\partial}{\partial y}$, $\hat{p}_z = -i\hbar\frac{\partial}{\partial z}$
Potential energy $V(x)$	$\hat{V}(\hat{x}) = V(x)$
Hamiltonian $H = \frac{p^2}{2m} + V(x)$ (in 1D)	$\hat{H} = \frac{\hat{p}^2}{2m} + \hat{V}(\hat{x}) = -\frac{\hbar^2}{2m}\frac{\partial^2}{\partial x^2} + V(x)$
Hamiltonian $H = \frac{\vec{p}^2}{2m} + V(\vec{r})$ (in 3D)	$\hat{H} = \frac{\hat{\vec{p}}^2}{2m} + \hat{V}(\hat{\vec{r}}) = -\frac{\hbar^2}{2m}\nabla^2 + V(\vec{r})$
Energy E	$\hat{E} = i\hbar\frac{\partial}{\partial t}$
Orbital angular momentum $\vec{l} = \vec{r} \times \vec{p}$	$\hat{\vec{l}} = \hat{\vec{r}} \times \hat{\vec{p}} = \hat{\vec{r}} \times (-i\hbar\vec{\nabla})$
Component wise	Component wise
$l_x = yp_z - zp_y$	$\hat{l}_x = \hat{y}\hat{p}_z - \hat{z}\hat{p}_y = -i\hbar\left(y\frac{\partial}{\partial z} - z\frac{\partial}{\partial y}\right)$
$l_y = zp_x - xp_z$	$\hat{l}_y = \hat{z}\hat{p}_x - \hat{x}\hat{p}_z = -i\hbar\left(z\frac{\partial}{\partial x} - x\frac{\partial}{\partial z}\right)$
$l_z = xp_y - yp_x$	$\hat{l}_z = \hat{x}\hat{p}_y - \hat{y}\hat{p}_x = -i\hbar\left(x\frac{\partial}{\partial y} - y\frac{\partial}{\partial x}\right)$

Spin angular momentum:

Spin has no place in classical mechanics, i.e. *spin has no classical analogue*. We use the following spin angular momentum operator in terms of Pauli spin matrices:

$$\hat{\vec{s}} = (\hat{s}_x, \hat{s}_y, \hat{s}_z) = \frac{\hbar}{2}\hat{\vec{\sigma}} = \frac{\hbar}{2}(\hat{\sigma}_x, \hat{\sigma}_y, \hat{\sigma}_z)$$

where

$$\hat{\sigma}_x = \hat{X} = \begin{pmatrix} 0 & 1 \\ 1 & 0 \end{pmatrix}, \quad \hat{\sigma}_y = \hat{Y} = \begin{pmatrix} 0 & -i \\ i & 0 \end{pmatrix}, \quad \hat{\sigma}_z = \hat{Z} = \begin{pmatrix} 1 & 0 \\ 0 & -1 \end{pmatrix}. \tag{2.1}$$

2.2 Second postulate: state vector and wave function

The state of a physical system is characterized by a *state vector* $|\psi\rangle$ in an infinite dimensional linear vector space called *Hilbert space*. The state vector or ket vector contains *complete or full information* about the system.

The dimension of the Hilbert space has nothing to do with the physical dimension of a physical system.

We define:

$\langle x|\psi\rangle = \psi(x) \Rightarrow$ Projection of state vector $|\psi\rangle$ in the coordinate space along $|x\rangle$ and is called the *coordinate space wave function* or *configuration space wave function*

$\langle p|\psi\rangle = \psi(p) \Rightarrow$ Projection of state vector $|\psi\rangle$ in the momentum space along $|p\rangle$ and is called the *momentum space wave function*

The temporal and spatial evolution, i.e. the dynamical state of a quantum mechanical particle is described by the wave function $\psi(x, t)$ for 1D motion and by $\psi(\vec{r}, t)$ for 3D motion.

Explanation

As per de Broglie hypothesis a *moving particle* can be looked upon as a *wave*. But there is a conceptual mismatch between *localized character* of a particle and the *extended nature* of a wave. This can, however, be overcome through construction of a *wave packet containing the particle*. Such association of *wave* and *particle* is sensible since the velocity of the particle, the velocity of energy propagation and the group velocity all happen to be *equal*.

The particle can stay *anywhere* inside the wave packet, within which it is *confined*. The idea of associating a particle within a wave packet means that the *definitive* or *deterministic* way of localising a particle is *lost*. We can now only say that the particle is *somewhere* within the wave packet. We have now to speak in terms of probability. This again means that the precision or certainty of classical mechanics or classical description is lost and now we would make *inaccurate, uncertain probabilistic* statements.

In other words we would list down the *various possible states* that the particle *can assume* (consistent with the given boundary conditions and energy available to the particle) and mention the corresponding probability of their occurrence. Clearly this description of states is *imprecise* and the state of system will be a *linear combination* of these imprecise (but somewhat probable) possibilities. However, when we do a measurement then the system will *collapse to a particular state*. This *indeterminism in the micro world* is a *law of nature* and is *inescapable*—it has nothing to do with instrumental errors.

- *Example:*

    If we have a cat confined in a bag—the state of the cat is *not* definitely known (*before any real experiment, say looking at the bag, is performed*)–it *might be* alive or it *might be* dead. So the state, before any real experiment is done, will be a linear combination of *Alive + Dead*, it is a mixture of both the possibilities, alive with some probability and dead with some probability. *Only if* we do an investigation (e.g. one tries to see through the bag or open it and pester or disturb the cat) can we know whether the cat is *Alive or Dead* and then we would get a *unique or definite result*. In other words the mixture of states (*Alive + Dead*) then collapses to either *Alive state* or to *Dead state*. Clearly we can say that, without doing an *experiment* we cannot be sure whether the cat is alive or dead.

Born's probabilistic interpretation of wave function
Wave function

$$\psi = \psi(x, t)(\text{in 1D}), \psi = \psi(\vec{r}, t)(\text{in 3D})$$

Wave function ψ may be a *complex* quantity and has no direct physical significance.

In the 1D case $\psi^*\psi = |\psi|^2 = |\psi(x, t)|^2 = \rho(x, t)$ is a *real* quantity and is interpreted as the *position probability density* of finding a particle at a point x at time t while $|\psi|^2 dx$ is interpreted as the probability of finding a particle between x and $x + dx$ at time t, i.e. within length element dx located at x at time t.

In the 3D case $\psi^*\psi = |\psi|^2 = |\psi(\vec{r}, t)|^2 = \rho(\vec{r}, t)$ is a real quantity and is interpreted as the position probability density of finding particle at a point $\vec{r}$ at time t and $|\psi|^2 d\tau$ is interpreted as the probability of finding particle between $\vec{r}$ and $\vec{r} + d\vec{r}$ at time t, i.e. within volume element $d\tau$ located at $\vec{r}$ at time t.

Restrictions on the wave function and state vector

We cannot work with any and every (arbitrary) wave function. We cannot work with any and every (arbitrary) ket vector in quantum mechanics.

They have to satisfy the following properties to be physically acceptable and if that is so we call them a *well behaved* wave function.

- ψ and $\dfrac{d\psi}{dx}$ should be *finite*.

- ψ and $\dfrac{d\psi}{dx}$ should be *single valued*.

- ψ and $\dfrac{d\psi}{dx}$ should be *continuous*.

Normalization

All wave functions ψ s should be *normalized to unity* (i.e. only those ψ should be chosen that give *total probability unity*). All kets should be of *unit length. Their direction matters* (not their length). In other words two kets differing in length *possess the same information.*

Only normalized kets (state vectors) and wave functions are used in quantum mechanics. Figure 2.1 compares vectors of the *classical* world and ket vectors in the *quantum* world.

We can rephrase this postulate and say that:

States are represented by rays in Hilbert space.

The probability of finding a particle within length element dx located at x is $|\psi|^2 dx$ and it follows that the probability of finding a particle between $-\infty$ to $+\infty$ (i.e. *somewhere* on the x-axis) is unity (= the total probability) and can be expressed as

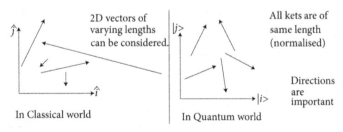

Figure 2.1. Comparison of classical vectors and quantum ket vectors

$$\int_{-\infty}^{+\infty} |\psi(x, t)|^2 \, dx = \int_{-\infty}^{+\infty} \psi^*(x, t)\psi(x, t)dx = 1 \qquad (1D)$$

$$\int_{\substack{all \\ space}} |\psi(\vec{r}, t)|^2 \, d\tau = \int_{\substack{all \\ space}} \psi^*(\vec{r}, t)\psi(\vec{r}, t)d\tau = 1 \qquad (3D)$$

Let for the ket $|\phi>$, the square of length $<\phi|\phi> \neq 1$, i.e. $|\phi>$ is not of unit length (i.e. $|\phi>$ is not normalized).

We construct a ket

$$|\psi> = N|\phi>$$

where N is called *normalization constant* and demand that $|\psi>$ is normalized. So

$$<\psi|\psi> = 1$$

$$<N\phi|N\phi> = 1 \quad \Rightarrow \quad N^*N < \phi|\phi> = 1 \quad \Rightarrow \quad N = \frac{1}{\sqrt{<\phi|\phi>}}$$

$$|\psi> = N|\phi> = \frac{|\phi>}{\sqrt{<\phi|\phi>}} = \frac{|\phi>}{||\phi||} = \text{Normalized ket.}$$

Normalization is conserved. Total probability is conserved.

If wave function is normalized it *remains normalized forever*. This is referred to as *conservation of probability*.

If ψ is normalized then

$$\int_{-\infty}^{+\infty} \psi^*\psi dx = 1.$$

Differentiating both sides w.r.t time we have

$$\frac{\partial}{\partial t}\int_{-\infty}^{+\infty} \psi^*\psi dx = \frac{\partial}{\partial t}(1) = 0.$$

So normalization $\int_{-\infty}^{+\infty} \psi^*\psi dx = 1$ does not change with time and is conserved.

Normalizability of wave functions

Wave functions for which the LHS integration $\int_{-\infty}^{+\infty} \psi^*\psi dx$ gives a finite value and hence can be normalized are called *quadratically integrable* or *square integrable*.

Wave functions for which the LHS integration $\int_{-\infty}^{+\infty} \psi^*\psi dx$ diverges (i.e. not square integrable) can be normalized through techniques of Box normalization and Dirac delta normalization.

Let the 1D wave function be plane wave type $\phi = e^{ikx}$. This wave extends from $-\infty$ to $+\infty$ and hence can be associated with a free particle (whose location can be anywhere on the x-axis as it is free). We verify that ϕ is not integrable and hence not normalizable since

$$\int_{-\infty}^{+\infty} \phi^*\phi dx = \int_{-\infty}^{+\infty} [e^{ikx}]^*[e^{ikx}]dx = \int_{-\infty}^{+\infty} e^{-ikx}e^{ikx}dx = \int_{-\infty}^{+\infty} dx = [x]_{-\infty}^{+\infty} \rightarrow \infty$$

To use the plane wave as a possible wave function let us normalize it through any of the following two techniques: (1) Box normalization (2) Dirac delta normalization.

(1) **Box normalization**

We *confine the particle* into a box of length L and the freedom which it enjoyed is taken care of through a *periodicity requirement* that the associated wave function should repeat after length L. The normalization condition is thus

$$\int_0^L \psi^*\psi dx = 1, \text{ with } \psi = N\phi \text{ (N is normalization constant)}$$

$$\Rightarrow \int_0^L [Ne^{ikx}]^*[Ne^{ikx}]dx = 1 \quad \Rightarrow |N|^2 \int_0^L e^{-ikx}e^{ikx}dx = 1$$

$$\Rightarrow |N|^2 \int_0^L dx = 1 \quad \Rightarrow |N|^2 L = 1 \quad \Rightarrow N = \frac{1}{\sqrt{L}}$$

$$\psi = N\phi = \frac{1}{\sqrt{L}}e^{ikx}$$

is the Box normalized wave function.

(2) **Dirac delta normalization**

The *integral representation of Dirac Delta function* is

$$\delta(k - k') = \frac{1}{2\pi}\int_{-\infty}^{+\infty} e^{i(k-k')x}dx$$

Let us denote or label the wave function by propagation constant, i.e. attach the k value it represents.

$$\psi_k = N\phi_k = Ne^{ikx}, \quad \psi_{k'} = N\phi_{k'} = Ne^{ik'x}$$

Impose the normalization condition

$$\int_{-\infty}^{+\infty} \psi_{k'}^* \psi_k dx = \delta(k - k') \Rightarrow \int_{-\infty}^{+\infty} [Ne^{ik'x}]^* [Ne^{ikx}]dx = \delta(k - k')$$

$$|N|^2 \int_{-\infty}^{+\infty} e^{-ik'x} e^{ikx} dx = \delta(k - k') \Rightarrow |N|^2 \int_{-\infty}^{+\infty} e^{i(k-k')x} dx = \delta(k - k')$$

$$|N|^2 2\pi\delta(k - k') = \delta(k - k') \Rightarrow |N|^2 2\pi = 1 \Rightarrow N = \frac{1}{\sqrt{2\pi}}$$

$$\psi_k = N\phi_k = \frac{1}{\sqrt{2\pi}}e^{ikx}$$

is the Dirac delta normalized wave function.

2.3 Third postulate: process of measurement

Collapse of wave function, Copenhagen interpretation

Process of measurement of a physical observable means that the operator corresponding to an observable, will act on the state vector, and carry it into its eigenvector.

Let us consider the classical observable A. Let the corresponding quantum mechanical Hermitian operator be $\hat{A}$. The eigenvalue equation of $\hat{A}$ is

$$\hat{A}|n_i> = n_i|n_i>$$

The eigenstates of $\hat{A}$ are $|n_i> = |n_1>$, $|n_2>$, $|n_3>$ ..... with respective eigenvalues being n_1, n_2, n_3,....

We choose the eigen basis of the operator $\hat{A}$ viz. $|n_i>$ as basis for description of a system. The basis $|n_i>$ satisfies the following four properties:

(1) Orthogonality $<n_i|n_j>|_{i\neq j} = 0$

(2) Normalization $<n_i|n_j>|_{i=j} = 1$

(3) Linear independence $\sum_i a_i|n_i> = 0$ $(a_i = 0$ for all $i)$

(4) Spans the linear vector space (the Hilbert space) $|\psi> = \sum_\alpha c_\alpha|n_\alpha>$

where $|\psi>$ = any ket in the same vector space

Principle of linear superposition or expansion postulate

Any arbitrary state vector $|\psi>$ can be expanded as a *linear superposition* of the *complete orthonormal set* of eigen basis states $|n_i>$ s (of the operator $\hat{A}$). Hence

$$|\psi> = \sum_i c_i|n_i > = c_1|n_1 > +c_2|n_2 > +c_3|n_3 > ...$$

Significance of the expansion coefficients c_i

1) Consider $<n_j|\psi> = <n_j|\sum_i c_i|n_i > = \sum_i c_i < n_j|n_i> = \sum_i c_i\delta_{ij} = c_j$

= *projection* of state vector $|\psi>$ along the basis $|n_j>$.

2) To be a meaningful state vector, $|\psi>$ should be normalized, i.e. $<\psi|\psi> = 1$

Using $<\psi| = \sum_i c_i^*<n_i|,$ $|\psi> = \sum_j c_j|n_j>$

$$<\psi|\psi> = \sum_i c_i^* < n_i|\sum_j c_j|n_j > = 1$$

$$\Rightarrow \sum_i \sum_j c_i^* c_j < n_i|n_j > = \sum_i\sum_j c_i^* c_j \delta_{ij} = \sum_i | c_i |^2 = 1$$

$$\Rightarrow <\psi|\psi> = | c_1 |^2 + | c_2 |^2 + | c_3 |^2 + ... = 1 \; (\textit{closure or completeness} \text{ relation})$$

As the RHS represents *total probability*, which is unity, it clearly follows that the LHS represents how the total probability is divided into probability of occurrence of the various eigenstates. For instance,

$|c_1|^2 = c_1^* c_1 =$ probability of occurrence of eigenvalue n_1, corresponding to eigenstate $|n_1>$

$|c_2|^2 = c_2^* c_2 =$ probability of occurrence of eigenvalue n_2, corresponding to eigenstate $|n_2>$

and so on. So we identify

$$c_i^* c_i = |c_i|^2 = |<n_i|\psi>|^2$$

as the probability of occurrence of eigenvalue n_i corresponding to eigenstate $|n_i>$ of $\hat{A}$ when measurement on the state vector $|\psi>$ is done. This is called the *Copenhagen interpretation*.

Before measurement of observable A on the system it is in state $|\psi>$. When we measure, the state $|\psi>$ collapses to either $|n_1>$ (the probability of such collapse being $|c_1|^2$) or $|n_2>$ (the probability of such collapse being $|c_2|^2$) and so on.

3) The relation

$$\sum_i |c_i|^2 = |c_1|^2 + |c_2|^2 + |c_3|^2 + \ldots = 1$$

ensures that we take into account *all the possibilities*, i.e. consider probability of occurrence of all eigenstates $|n_i>$ of $\hat{A}$ (none should be left out). In other words the relation should furnish a complete picture regarding the distribution of probabilities *before* actual measurement is made. Hence this relation is called *completeness* condition or *closure* relation. Put differently, the completeness condition says that the total probability should be normalized to *unity*. The set of *orthonormalized basis* (which are the building blocks of the linear vector space) is said to be *complete*.

Clearly, any arbitrary vector in the Hilbert space can be expanded in terms of the complete set of orthonormalized eigenstates of an operator corresponding to an observable.

The process of measurement is a *discontinuous* process during which the state of system $|\psi>$ *collapses* to one of the eigenstates $|n_i>$. So the state goes from $|\psi>$ to $\hat{P}_i|\psi>$ ($\hat{P}_i$ is the projection operator for the state $|n_i>$).

Let us *normalize* the post-measurement state $\hat{P}_i|\psi>$

$$\frac{\hat{P}_i|\psi>}{\sqrt{<\psi|\hat{P}_i\hat{P}_i|\psi>}} = \frac{\hat{P}_i|\psi>}{\sqrt{<\psi|\hat{P}_i^2|\psi>}} = \frac{\hat{P}_i|\psi>}{\sqrt{<\psi|\hat{P}_i|\psi>}} \text{ (since } \hat{P}_i^2 = \hat{P}_i)$$

- *Illustration:*

  Consider the operator

$$\hat{X} = \begin{pmatrix} 0 & 1 \\ 1 & 0 \end{pmatrix}$$

(section 1.27) having *eigenvector* $\frac{1}{\sqrt{2}}\begin{pmatrix} 1 \\ 1 \end{pmatrix}$ =|1> with *eigenvalue* $\lambda = 1$ and

eigenvector $\frac{1}{\sqrt{2}}\begin{pmatrix} 1 \\ -1 \end{pmatrix}$ =|−1> with *eigenvalue* $\lambda = -1$.

- Let us consider a given state

$$|\psi> = \frac{1}{\sqrt{5}}\begin{pmatrix} 2 \\ 1 \end{pmatrix}, \quad <\psi|\psi> =1 \text{ (normalized)}$$

This can be expanded in the basis of these orthonormalized eigenkets|1 > , | −1>
as follows

$$|\psi> = c_1|1> + c_2|-1>$$

$$\frac{1}{\sqrt{5}}\begin{pmatrix} 2 \\ 1 \end{pmatrix} = c_1\frac{1}{\sqrt{2}}\begin{pmatrix} 1 \\ 1 \end{pmatrix} + c_2\frac{1}{\sqrt{2}}\begin{pmatrix} 1 \\ -1 \end{pmatrix} \Rightarrow \begin{pmatrix} \frac{2}{\sqrt{5}} \\ \frac{1}{\sqrt{5}} \end{pmatrix} = \begin{pmatrix} \frac{c_1}{\sqrt{2}} + \frac{c_2}{\sqrt{2}} \\ \frac{c_1}{\sqrt{2}} - \frac{c_2}{\sqrt{2}} \end{pmatrix}$$

Solving we get : $c_1 = \frac{3}{\sqrt{10}}, c_2 = \frac{1}{\sqrt{10}}$.

So if we make a measurement and get result $\lambda = 1$ (the probability of which is
$|c_1|^2 = (\frac{3}{\sqrt{10}})^2 = \frac{9}{10}$) then the *post-measurement state* will be |1 > $=\frac{1}{\sqrt{2}}\begin{pmatrix} 1 \\ 1 \end{pmatrix}$.

✓Let us recast the problem using projection operators. Projection operators are

$$\hat{P}_1 = |1 > <1| = \frac{1}{\sqrt{2}}\begin{pmatrix} 1 \\ 1 \end{pmatrix}\frac{1}{\sqrt{2}}(1 \ \ 1) = \frac{1}{2}\begin{pmatrix} 1 & 1 \\ 1 & 1 \end{pmatrix}$$

$$\hat{P}_{-1} = |-1 > <-1| = \frac{1}{\sqrt{2}}\begin{pmatrix} 1 \\ -1 \end{pmatrix}\frac{1}{\sqrt{2}}(1 \ \ -1) = \frac{1}{2}\begin{pmatrix} 1 & -1 \\ -1 & 1 \end{pmatrix}$$

$$\hat{P}_1 + \hat{P}_{-1} = \frac{1}{2}\begin{pmatrix} 1 & 1 \\ 1 & 1 \end{pmatrix} + \frac{1}{2}\begin{pmatrix} 1 & -1 \\ -1 & 1 \end{pmatrix} = \frac{1}{2}\begin{pmatrix} 2 & 0 \\ 0 & 2 \end{pmatrix} = \begin{pmatrix} 1 & 0 \\ 0 & 1 \end{pmatrix} = \hat{I}$$

$$\hat{P}_1|\psi> = \frac{1}{2}\begin{pmatrix} 1 & 1 \\ 1 & 1 \end{pmatrix}\frac{1}{\sqrt{5}}\begin{pmatrix} 2 \\ 1 \end{pmatrix} = \frac{1}{2\sqrt{5}}\begin{pmatrix} 3 \\ 3 \end{pmatrix} = \frac{3}{2\sqrt{5}}\begin{pmatrix} 1 \\ 1 \end{pmatrix} = \frac{3}{\sqrt{10}}\frac{1}{\sqrt{2}}\begin{pmatrix} 1 \\ 1 \end{pmatrix} = c_1|1>$$

$$\hat{P}_{-1}|\psi> = \frac{1}{2}\begin{pmatrix} 1 & -1 \\ -1 & 1 \end{pmatrix}\frac{1}{\sqrt{5}}\begin{pmatrix} 2 \\ 1 \end{pmatrix} = \frac{1}{2\sqrt{5}}\begin{pmatrix} 1 \\ -1 \end{pmatrix} = \frac{1}{\sqrt{10}}\frac{1}{\sqrt{2}}\begin{pmatrix} 1 \\ -1 \end{pmatrix} = c_2|-1>$$

Hence $\hat{P}_1|\psi > +\hat{P}_{-1}|\psi > =c_1|1 > +c_2|-1 > =|\psi>$

After getting a result $\lambda = 1$ (the probability of which is $|c_1|^2 = (\frac{3}{\sqrt{10}})^2 = \frac{9}{10}$) the
post-measurement state will be

$$\frac{\hat{P}_1|\psi>}{\sqrt{<\psi|\hat{P}_1|\psi>}} = \frac{\frac{3}{\sqrt{10}}\frac{1}{\sqrt{2}}\binom{1}{1}}{\sqrt{\frac{1}{\sqrt{5}}(2\ \ 1)\frac{3}{\sqrt{10}}\frac{1}{\sqrt{2}}\binom{1}{1}}} = \frac{\frac{3}{\sqrt{10}}\frac{1}{\sqrt{2}}}{\sqrt{\frac{1}{\sqrt{5}}\frac{3}{\sqrt{10}}\frac{1}{\sqrt{2}}\cdot 3}}\binom{1}{1} = \frac{1}{\sqrt{2}}\binom{1}{1} = |1>$$

Clearly the *post-measurement state* can also be written as

$$\frac{\hat{P}_1|\psi>}{\sqrt{<\psi|\hat{P}_1|\psi>}} = \frac{c_1|1>}{\sqrt{<\psi|c_1|1>}} = \frac{c_1|1>}{\sqrt{c_1<\psi|1>}}$$

$$= \frac{c_1|1>}{\sqrt{c_1}\sqrt{(c_1^*<1|+c_2^*<-1|)|1>}} = \frac{c_1|1>}{\sqrt{c_1}\sqrt{c_1^*}} = \frac{c_1|1>}{\sqrt{|c_1|^2}} = |1>$$

- In general

$$\frac{\hat{P}_i|\psi>}{\sqrt{<\psi|\hat{P}_i|\psi>}} = \frac{c_i|n_i>}{\sqrt{<\psi|c_i|n_i>}} = \frac{c_i|n_i>}{\sqrt{c_i}\sqrt{<\psi|n_i>}} = \frac{c_i|n_i>}{\sqrt{c_i}\sqrt{c_i^*}} = |n_i>$$

To extract information out of a quantum system we have to perform a physical measurement. Upon measuring a quantum system the very state we are trying to measure is automatically changed. We obtain, in general, a random result, which may be different from the original state.

Expectation value

When a large number of measurements of an observable A, the corresponding operator being $\hat{A}$, having *eigenvalue equation* $\hat{A}|n_i> =n_i|n_i>$, are performed upon a system in normalized state $|\psi>$ we get results n_i (which are real numbers) $(i = 1, 2, 3, ...)$ and are identified with eigenvalues n_i of the eigenstates $|n_i>$ $(i = 1, 2, 3, ...)$ of the operator $\hat{A}$. The average of these results is called expectation value. It is a *real number* and is defined as

$$<\hat{A}> = <\psi|\hat{A}|\psi>$$

With $|\psi> = \sum_i c_i|n_i>$, $<\psi| = \sum_j c_j^* <n_j|$

we have

$$<\hat{A}> = \sum_i c_i^* <n_i|\hat{A} \sum_j c_j|n_j> = \sum_i \sum_j c_i^* c_j <n_i|\hat{A}|n_j>$$

$$= \sum_i \sum_j c_i^* c_j <n_i|n_j|n_j> = \sum_i \sum_j c_i^* c_j n_j <n_i|n_j>$$

$$= \sum_i \sum_j c_i^* c_j n_j \delta_{ij}$$

$$<\hat{A}> = <\psi|\hat{A}|\psi> = \sum_i n_i \ |c_i|^2 = \sum(\text{result})(\text{probability})$$

2.4 Fourth postulate: Time evolution of a state

All quantum processes occur or develop *unitarily* and this is ensured by using unitary operators. Unitary operators decide time evolution of a state when measurement is not done. Unitary operator *preserves norm of the state vector*. The *process of measurement* makes it a *discontinuous* process as the state *collapses* to an eigenstate of the system when we make a measurement.

Explanation

To find the time evolution of the state $|\psi>$ of a quantum system we have to construct an *equation* that is satisfied by the wave function $\psi(x, t)$ (called the Schrödinger equation) and solve it to find how a state changes or evolves with time.

Consider a particle of mass m, momentum p, energy E and Hamiltonian H moving *non-relativistically* in a *conservative* force field that can be derived from a potential V through the equation

$$\vec{F} = -\vec{\nabla}V = -\frac{dV}{dx}.$$

The potential V is *real* and *time independent*

$$V = V(x), \quad V \neq V(t).$$

Spin of particle is *not considered* while solving the equation. After solving we can attach the spin part to the wave function that we get after the solution.

In a *conservative* force field :

Hamiltonian of system = Energy of system

$$\Rightarrow H = E = K.\,E + P.\,E = \frac{p^2}{2m} + V(x)$$

($K.\,E$ = kinetic energy, $P.\,E$ = potential energy)

This is the *classical equation* describing a particle. Replacing observables by *linear Hermitian operators* we get

$$H \rightarrow \hat{H}, \ x \rightarrow \hat{x}, \ p \rightarrow \hat{p} = -i\hbar\frac{\partial}{\partial x}, \ E \rightarrow \hat{E} = i\hbar\frac{\partial}{\partial t}, \ V(x) \rightarrow \hat{V}(\hat{x}) = V(x)$$

Let $\psi(x, t)$ = wave function associated with the particle. The *Schrödinger equation* is

$$\hat{H}\psi(x, t) = \hat{E}\psi(x, t)$$

$$\left(\frac{\hat{p}^2}{2m} + \hat{V}(\hat{x})\right)\psi(x, t) = i\hbar\frac{\partial}{\partial t}\psi(x, t)$$

$$\left[\frac{1}{2m}\left(-i\hbar\frac{\partial}{\partial x}\right)\left(-i\hbar\frac{\partial}{\partial x}\right) + V(x)\right]\psi(x, t) = i\hbar\frac{\partial}{\partial t}\psi(x, t)$$

$$\left[-\frac{\hbar^2}{2m}\frac{\partial^2}{\partial x^2} + V(x) \right] \psi(x,\, t) = i\hbar\frac{\partial}{\partial t}\psi(x,\, t) \qquad (2.2a)$$

This is the *time dependent* Schrödinger equation.

For a 1D *free* particle $V = 0$ and we have

$$-\frac{\hbar^2}{2m}\frac{\partial^2}{\partial x^2}\psi(x,\, t) = i\hbar\frac{\partial}{\partial t}\psi(x,\, t)$$

Time dependent Schrödinger equation for 3D particle moving in a *conservative* force field $(V = 0)$ is given by

$$\left[-\frac{\hbar^2}{2m}\nabla^2 + V(\vec{r}) \right] \psi(\vec{r},\, t) = i\hbar\frac{\partial}{\partial t}\psi(\vec{r},\, t)$$

$$\Rightarrow -\frac{\hbar^2}{2m}\nabla^2\psi(\vec{r},\, t) = i\hbar\frac{\partial}{\partial t}\psi(\vec{r},\, t)$$

2.5 Solution of the Schrödinger equation

We can solve the Schrödinger equation by applying the *method of separation of variables*, assuming that

$$\psi(x,\, t) = \psi(x)T(t)$$

With this, the *time dependent* 1D Schrödinger equation with potential energy $V(x)$ becomes

$$\left[-\frac{\hbar^2}{2m}\frac{\partial^2}{\partial x^2} + V(x) \right]\psi(x)T(t) = i\hbar\frac{\partial}{\partial t}\psi(x)T(t)$$

$$-T(t)\frac{\hbar^2}{2m}\frac{d^2}{dx^2}\psi(x) + V(x)\,\psi(x)T(t) = i\hbar\psi(x)\frac{d}{dt}T(t)$$

Divide by $\psi(x,\, t) = \psi(x)T(t)$

$$-\frac{1}{\psi(x)}\frac{\hbar^2}{2m}\frac{d^2}{dx^2}\psi(x) + V(x) = i\hbar\frac{1}{T(t)}\frac{d}{dt}T(t) = E(\text{say}) = \text{separation constant}$$

This leads to two equations: time dependent part and time independent part as discussed now.

Time dependent part of Schrödinger equation

$$i\hbar\frac{1}{T(t)}\frac{d}{dt}T(t) = E$$

$$\Rightarrow \frac{d}{dt}T(t) = \frac{E}{i\hbar}T(t) = -\frac{iE}{\hbar}T(t) \quad \Rightarrow \frac{dT}{T} = -\frac{iE}{\hbar}dt$$

$$\Rightarrow \int d\ln T = -\frac{iE}{\hbar}\int dt + \ln A \quad (\ln A = \text{constant})$$

$$\ln T = -\frac{iE}{\hbar}t + \ln A \quad \Rightarrow \ln\frac{T}{A} = -\frac{iE}{\hbar}t$$

$$T = Ae^{-iEt/\hbar}$$

We note that the quantity $\dfrac{Et}{\hbar}$ should be dimensionless.

So E should have dimension of $[\dfrac{\hbar}{t}] \rightarrow \dfrac{J.\,s}{s} = J$. We thus identify E as total energy.

Space part or time independent part of the Schrödinger equation

$$-\frac{1}{\psi(x)}\frac{\hbar^2}{2m}\frac{d^2}{dx^2}\psi(x) + V(x) = E$$

Multiply by $\psi(x)$

$$-\frac{\hbar^2}{2m}\frac{d^2}{dx^2}\psi(x) + V(x)\psi(x) = E\psi(x)$$

$$\left[-\frac{\hbar^2}{2m}\frac{d^2}{dx^2} + V(x)\right]\psi(x) = E\psi(x) \tag{2.2b}$$

$$\hat{H}\psi(x) = E\psi(x) \quad (\text{since } \hat{H} = -\frac{\hbar^2}{2m}\frac{d^2}{dx^2} + V(x))$$

This is called the *time independent* Schrodinger equation.

This is an *eigenvalue equation* for the Hamiltonian operator with eigenvalue E and eigenfunction $\psi(x)$.

If the potential is known $V = V(x)$ we can solve equation (2.2) to find $\psi(x)$ and E.

An alternative popular form of the *time independent* Schrödinger equation viz. $\hat{H}\psi(x) = E\psi(x)$ can be arrived at by multiplying equation (2.2b) with $-\dfrac{2m}{\hbar^2}$. This gives

$$\frac{d^2}{dx^2}\psi(x) + \frac{2m}{\hbar^2}[E - V(x)]\psi(x) = 0$$

Solution of the *time dependent* Schrödinger equation (2.2a) is thus obtained to be

$$\psi(x, t) = \psi(x)T(t) = \psi(x)e^{-iEt/\hbar}$$

where $\psi(x)$ satisfies the eigenvalue equation of Hamiltonian $\hat{H}$(which is the time independent Schrödinger equation (2.2b)) viz.

$$\hat{H}\psi(x) = E\psi(x).$$

For the 3D time dependent Schrödinger equation

$$\left[-\frac{\hbar^2}{2m}\nabla^2 + V(\vec{r}) \right]\psi(\vec{r}, t) = i\hbar\frac{\partial}{\partial t}\psi(\vec{r}, t)$$

the solution is

$$\psi(\vec{r}, t) = \psi(\vec{r})e^{-iEt/\hbar}$$

where the space part $\psi(\vec{r})$ satisfies the time independent Schrödinger equation

$$\nabla^2\psi(\vec{r}) + \frac{2m}{\hbar^2}[\, E \,-\, V(\vec{r})\,]\,\psi(\vec{r}) = 0$$

where $V(\vec{r})$ = real time independent potential.

When we are not measuring or observing the system, the system develops in time following the time dependent Schrödinger equation

$$\hat{H}|\psi> = i\hbar\frac{\partial}{\partial t}|\psi>$$

and can be expressed as

$$|\psi(t)> = \hat{U}(t)|\psi(0)> = e^{-i\hat{H}t/\hbar}|\psi(0)>$$

where $\hat{U}(t)$ is the unitary operator that determines which way the state will develop or evolve, of course preserving normalization.

2.6 Unitary operator keeps the length of state vector constant

Suppose the system evolves from state $|\psi>$ to state $|\phi>$ and so we can write

$$\hat{U}|\psi> = |\phi> \quad \text{and so} \quad <\phi| = <\psi|\hat{U}^{\dagger}.$$

Hence

$$<\phi|\phi> = <\psi|\hat{U}^{\dagger}\hat{U}|\psi> .$$

As $\hat{U}^{\dagger}\hat{U} = I$ we have

$$<\phi|\phi> = <\psi|\psi> = 1 \Rightarrow \quad ||\phi|| \,=\, ||\psi|| = 1.$$

So length or norm is preserved by the unitary operator despite time evolution. Alternatively

$$||\hat{U}|\psi> ||^2 = \sum_j (\hat{U}\psi)^*_j(\hat{U}\psi)_j$$

$$= \sum_j \sum_k \sum_l (\hat{U}_{jk}\psi_k)^*(\hat{U}_{jl}\psi_l) = \sum_j \sum_k \sum_l \hat{U}_{kj}^* \psi_k^* \hat{U}_{jl}\psi_l$$

$$= \sum_k \sum_l \sum_j \hat{U}_{kj}^* \hat{U}_{jl}\psi_k^*\psi_l = \sum_k \sum_l \delta_{kl}\psi_k^*\psi_l$$

$$= \sum_k \psi_k^*\psi_k = \| \; |\psi> \; \|^2$$

2.7 Heisenberg's uncertainty principle or principle of indeterminism

Consider two *physical observables* A and B. Their commutator bracket is (section 1.25)

$$[\hat{A}, \hat{B}] = \hat{A}\hat{B} - \hat{B}\hat{A}$$

where $\hat{A}$ is the operator corresponding to the physical observable A and $\hat{B}$ is the operator corresponding to the physical observable B. This commutator bracket indicates whether the two observables *are simultaneously measurable* with *infinite accuracy* or not.

The *uncertainty* in A is defined as

$$\Delta\hat{A} = \sqrt{<(\hat{A} - <\hat{A}>)^2>} = \text{root mean square fluctuation/deviation(RMS)}$$

$$= \sqrt{<\hat{A}^2> - 2\hat{A}<\hat{A}> + <\hat{A}>^2} = \sqrt{<\hat{A}^2> - 2<\hat{A}><\hat{A}> + <\hat{A}>^2}$$

$$= \sqrt{<\hat{A}^2> - 2<\hat{A}>^2 + <\hat{A}>^2} = \sqrt{<\hat{A}^2> - <\hat{A}>^2}$$

and similarly the *uncertainty* in B is

$$\Delta\hat{B} = \sqrt{<(\hat{B} - <\hat{B}>)^2>} = \sqrt{<\hat{B}^2> - <\hat{B}>^2}$$

$$= \text{root mean square fluctuation/deviation(RMS)}$$

✓ If A and B are *commuting* observables (i.e. *not canonically conjugate*) then

$$[\hat{A}, \hat{B}] = \hat{A}\hat{B} - \hat{B}\hat{A} = 0$$

The observables are then *simultaneously measurable with infinite accuracy*. $\Delta\hat{A} = 0$, $\Delta\hat{B} = 0$ are simultaneously possible to achieve. Disturbance produced during measurement of one observable does not affect measurement of the other *commuting* or *canonically conjugate* observable.

- *Example* :

As $[\hat{x}, \hat{y}] = 0, \Delta x = 0, \Delta y = 0$; x and y are *simultaneously measurable* with *infinite accuracy*.

As $[\hat{p}_x, \hat{p}_y] = 0, \Delta p_x = 0, \Delta p_y = 0$; p_x and p_y are *simultaneously measurable* with *infinite accuracy*.

Other examples are $[\hat{x}, \hat{p}_y] = 0$, $[\hat{l}^2, \hat{l}_x] = 0$ meaning thereby that x, p_y are *simultaneously measurable* with *infinite accuracy*, l^2, l_x are also simultaneously measurable with infinite accuracy $(\hat{l}^2 = \hat{l}_x^2 + \hat{l}_y^2 + \hat{l}_z^2$ = square of orbital angular momentum).

✓ If A and B are *non-commuting* observables (i.e. *canonically conjugate*) then

$$[\hat{A}, \hat{B}] = \hat{A}\hat{B} - \hat{B}\hat{A} \neq 0$$

The observables are *not simultaneously measurable with infinite accuracy*. $\Delta\hat{A} = 0$, $\Delta\hat{B} = 0$ are simultaneously impossible to achieve. These uncertainties are related through *Heisenberg's uncertainty relation* as

$$\Delta\hat{A} \, \Delta\hat{B} \geqslant \frac{\hbar}{2}$$

where $\hbar = \dfrac{h}{2\pi}$, $h = 6.626 \times 10^{-34} J. \, s$ = Planck's constant.

- *Example* :

    $[\hat{x}, \hat{p}_x] = i\hbar\hat{I} \neq 0$ ($\hat{I}$ = unit operator) $\Delta x = 0$, $\Delta p_x = 0$ are not possible simultaneously. The position–momentum uncertainty relation is

$$\Delta x \, \Delta p_x \geqslant \frac{\hbar}{2}$$

where Δx is uncertainty in position x and Δp_x is uncertainty in momentum p_x. And x and p_x are *not simultaneously measurable* with *infinite accuracy*.

Other examples are $[\hat{y}, \hat{p}_y] = i\hbar\hat{I} \neq 0$, $[\hat{l}_x, \hat{l}_y] = i\hbar\hat{l}_z \neq 0$ meaning thereby that y, p_y are not *simultaneously measurable* with *infinite accuracy*. l_x, l_y are also not simultaneously measurable with infinite accuracy.

Quantum mechanics that is useful in quantum computing involves the use of a linear set of equations that can be solved. The main aspects are that we have a probabilistic theory and we have equation of motion (which is different from the classical case), like Schrödinger equation, Heisenberg equation (this form is used for one quantum system) or the *Liouville's equation* (this form is used for a set or ensemble of quantum systems). In the case of quantum computing, Liouville's equation becomes most important because we are actually trying to see how an entire set of problems can be solved simultaneously and the Liouville equation, where we have all the processes working together, gives an ensemble solution. This equation takes into account the effect of statistical mechanics.

Uncertainties creep in since we *cannot define* the path or trajectory of the observable. We cannot plot with definiteness the trajectories in phase space (x versus p plot) since x and p are not known with infinite precision.

We *do not* have a continuously evolving measurement process or path. Each time we measure we correlate with the classical analogue or concept.

Computing with a quantum system is possible because in a large number of measurements the quantum system will give the classical analogue and so *quantum system can be used as a computing device.* Based upon our knowledge of quantum mechanics we can move to a mathematical set which gives computational ability.

- *Example*

Let us try to connect the process of computation to the operation of converting the vector state $\begin{pmatrix} 0 \\ 1 \end{pmatrix}$ to the vector state $\begin{pmatrix} 1 \\ 0 \end{pmatrix}$.

This operation is actually an inversion operation also called a NOT operation, like switching over from $+x$ to $-x$. This switching of states or inversion of states occurs through the following operator equation

$$\begin{pmatrix} a & b \\ c & d \end{pmatrix} \begin{pmatrix} 0 \\ 1 \end{pmatrix} = \begin{pmatrix} 1 \\ 0 \end{pmatrix} \quad \text{and} \quad \begin{pmatrix} a & b \\ c & d \end{pmatrix} \begin{pmatrix} 1 \\ 0 \end{pmatrix} = \begin{pmatrix} 0 \\ 1 \end{pmatrix}$$

where the operator matrix is considered to be $\begin{pmatrix} a & b \\ c & d \end{pmatrix}$. It follows that

$$\begin{pmatrix} b \\ d \end{pmatrix} = \begin{pmatrix} 1 \\ 0 \end{pmatrix} \quad \text{and} \quad \begin{pmatrix} a \\ c \end{pmatrix} = \begin{pmatrix} 0 \\ 1 \end{pmatrix}$$

and so we identify $a = 0$, $b = 1$, $c = 1$, $d = 0$ to get the operator used in the computation of switching of states to have the form

$$\begin{pmatrix} a & b \\ c & d \end{pmatrix} = \begin{pmatrix} 0 & 1 \\ 1 & 0 \end{pmatrix} = \hat{X}$$

2.8 Further reading

[1] Guha J 2020 *Modern Physics* vols 1 and 2 (Kolkata: Techno Word)
[2] Guha J 2019 *Quantum Mechanics: Theory, Problems & Solutions* 3rd edn (Kolkata: Books and Allied (P) Ltd)
[3] Waghmare Y R 2014 *Fundamentals of Quantum Mechanics* (New Delhi: S. Chand)
[4] Zettili N 2009 *Quantum Mechanics: Concepts and Applications* (New York: Wiley)

2.9 Problems

1. *Why are Hermitian operators used in quantum mechanics?*

   Ans. According to the postulate of quantum mechanics, observables are replaced by corresponding operators which are identified with the act of measurement. The eigenvalues of the operator represent results of measurement. Since the result of measurement is real the eigenvalues of the operators have to be real. We thus have to choose Hermitian operators since they have real eigenvalues.

Hermitian operators are used to represent observables in quantum mechanics.

2. *What is the meaning of square integrable wave function. Give an example of a wave function that is square integrable and one that is not square integrable.*

Ans. Let ψ be a wave function. Square means we have to take $\psi^*\psi = |\psi|^2$ which represents position probability density of locating a particle at some x or $\vec{r}$ at time t.

Integrable means we should be able to integrate the quantity $\int_{-\infty}^{+\infty} \psi^*\psi dx$ (in the case of 1D) or $\int_{all\ space} \psi^*\psi\ d\tau$ (in the case of 3D) and get a finite value. This is a requirement for normalizing the wave function. In quantum mechanics we use only normalized wave functions and kets of unit length so that the total probability is set at unity.

- Example of square integrable wave function

  Consider wave function of linear harmonic oscillator in the ground state

$$\psi_{n=0} = \sqrt{\frac{\alpha}{\sqrt{\pi}}} e^{-z^2/2} = N e^{-z^2/2} \text{ where } z = \alpha x, \ \alpha = \sqrt{\frac{mw}{\hbar}}, \ N = \sqrt{\frac{\alpha}{\sqrt{\pi}}}$$

  $m \rightarrow$ oscillator mass, $w \rightarrow$ frequency of oscillation.

$$\text{Consider } \int_{-\infty}^{+\infty} \psi_0^*\psi_0 dx = \int_{-\infty}^{+\infty} [Ne^{-z^2/2}]^*[Ne^{-z^2/2}]dx = \frac{1}{\alpha} |N|^2 \int_{-\infty}^{+\infty} e^{-z^2} dz$$
$$(dx = \frac{dz}{\alpha})$$

$$= \frac{2}{\alpha} |N|^2 \int_0^{+\infty} e^{-z^2} dz = \frac{2}{\alpha} |N|^2 \frac{\sqrt{\pi}}{2} = \frac{2}{\alpha}\left[\sqrt{\frac{\alpha}{\sqrt{\pi}}}\right]^2 \frac{\sqrt{\pi}}{2} = 1$$

- Example of wave function that is not square integrable

  Consider plane wave $\psi = e^{ikx}$ or $e^{i\vec{k}.\vec{r}}$.

  Let us evaluate the integration of its square

$$\int_{-\infty}^{+\infty} \psi^*\psi dx = \int_{-\infty}^{+\infty} [e^{ikx}]^*[e^{ikx}]dx = \int_{-\infty}^{+\infty} e^{-ikx}e^{ikx}dx = \int_{-\infty}^{+\infty} dx \rightarrow \infty$$

  $\Leftarrow$ Not integrable

3. *Operators in quantum mechanics representing physical observables are*
   (a) commutative (b) Hermitian (c) associative (d) unitary.
   Ans. (b), (c)

4. *Operators in quantum mechanics (when measurement is not done) are*
   (a) commutative (b) Hermitian (c) associative (d) unitary.
   Ans. (c),(d) (Time evolution operator takes system from $|t = 0>$ to $|t>$).

5. *Prove that the expectation value of the square of an observable quantity is always positive.*

Ans. $<\hat{A}^2> \ = \ <\psi|\hat{A}^2|\psi> \ = \ <\psi|\hat{A}\hat{A}|\psi>$

Define $|\phi> \ = \hat{A}|\psi>$

$<\hat{A}^2> \ = <\psi|\hat{A}|\phi> \ = <\hat{A}^\dagger \psi|\phi> \ = <\hat{A}\psi|\phi>$ (as $\hat{A}^\dagger = \hat{A} = $ Hermitian)

$<\hat{A}^2> \ = \ <\phi|\phi>$ (since $|\phi> \ = \ \hat{A}|\psi> \ = |\hat{A}\psi>$, $<\phi| \ = \ <\hat{A}\psi|$ (as $\hat{A}$ is Hermitian)

$$<\hat{A}^2> \ = \int_{-\infty}^{+\infty} \phi^* \phi dx = \int_{-\infty}^{+\infty} |\phi|^2 \, dx = \int_{-\infty}^{+\infty} |\hat{A}\psi|^2 \, dx \ \geqslant 0$$

Clearly the average of the square of a Hermitian operator (representing an observable) is positive.

6. *If a system has two eigenstates $|\psi_1>$ and $|\psi_2>$ with eigenvalues E_1 and E_2, under what condition a linear combination $c_1|\psi_1> +c_2|\psi_2>$ is also an eigenstate.*

   Ans.

$$\hat{H}|\psi_1> \ = \ E_1|\psi_1> , \ \hat{H}|\psi_2> \ = E_2|\psi_2>$$
$$|\psi> \ = \ c_1|\psi_1> +c_2|\psi_2>$$
$$\hat{H}|\psi> \ = \ \hat{H}\left(c_1|\psi_1> +c_2|\psi_2>\right) = c_1\hat{H}|\psi_1> +c_2\hat{H}|\psi_2>$$
$$= \ c_1E_1|\psi_1> + c_2E_2|\psi_2> \ \neq \lambda|\psi>$$

$\Rightarrow |\psi>$ is not an eigenstate of $\hat{H}$.

If $E_1 = E_2 = E$(say) (doubly degenerate level)

$$\hat{H}|\psi> \ = c_1E_1|\psi_1> +c_2E_2|\psi_2> = E(\ c_1|\psi_1> +c_2|\psi_2> \ )=E|\psi>$$

$\Rightarrow |\psi> \ = c_1|\psi_1> +c_2|\psi_2>$ is an eigenstate of $\hat{H}$.

7. *Check Hermiticity of the operators $\hat{x}$, $\dfrac{d}{dx}$, $i\dfrac{d}{dx}$, $\hat{p}$*

Ans. Condition of Hermiticity of $\hat{A}$ is:

$$<\psi|\hat{A}\psi> \ =<\hat{A}\psi|\psi> \Rightarrow \hat{A} = \hat{A}^{\dagger,}$$

- Operator $\hat{x}$

  Consider $<\psi|\hat{x}\psi> \ = \int_{-\infty}^{+\infty} \psi^*(\hat{x}\psi)dx = \int_{-\infty}^{+\infty} \psi^* x\psi dx$ [as $\hat{x} = x$]

  $= \int_{-\infty}^{+\infty} \psi^* x^*\psi dx$ [as $x = x^* \to$ real]

  $= \int_{-\infty}^{+\infty} (x\psi)^*\psi dx = \ <\hat{x}\psi|\psi> \ = \ <\psi|\hat{x}^\dagger\psi>$

  $\Rightarrow \hat{x} = \hat{x}\dagger \Rightarrow \hat{x}$ (position operator) is Hermitian.

- Operator $\dfrac{d}{dx}$

  Consider $<\psi|\dfrac{d\psi}{dx}> = \int_{-\infty}^{+\infty} \psi^* \dfrac{d\psi}{dx} dx$

Integrate by parts to get

$$<\psi|\frac{d\psi}{dx}> = \psi^*\int\frac{d\psi}{dx}dx \mid_{-\infty}^{+\infty} - \int_{-\infty}^{+\infty}\frac{d\psi^*}{dx}\left(\int\frac{d\psi}{dx}dx\right) dx$$

$$<\psi|\frac{d\psi}{dx}> = \psi^*\psi|_{-\infty}^{+\infty} - \int_{-\infty}^{+\infty}\frac{d\psi^*}{dx}\psi \, dx$$

$$= 0 - \int_{-\infty}^{+\infty}\frac{d\psi^*}{dx}\psi \, dx \left[\text{as } \psi \to 0 \text{ at } x \to \pm\infty\right]$$

$$= -<\frac{d\psi}{dx}|\psi> = -<\psi|\frac{d}{dx}^{\dagger}\psi>$$

$$\Rightarrow\frac{d}{dx}^{\dagger} = -\frac{d}{dx} \Rightarrow \frac{d}{dx} \text{ is anti-Hermitian}$$

- Operator $i\frac{d}{dx}$

  Consider $\left(i\frac{d}{dx}\right)^{\dagger} = -i\frac{d}{dx}^{\dagger} = -i\left(-\frac{d}{dx}\right)=i\frac{d}{dx} \Rightarrow i\frac{d}{dx}$ is Hermitian

- Operator $\hat{p} = -i\hbar\frac{d}{dx}$

Consider $<\psi|\hat{p}\psi>=<\psi|-i\hbar\frac{d\psi}{dx}>=\int_{-\infty}^{+\infty}\psi^*\left(-i\hbar\frac{d\psi}{dx}\right)dx$

$$=-i\hbar\int_{-\infty}^{+\infty}\psi^*\frac{d\psi}{dx}dx$$

Integrate by parts to get

$$<\psi|\hat{p}\psi> = -i\hbar\left[\psi^*\int\frac{d\psi}{dx}dx \mid_{-\infty}^{+\infty} - \int_{-\infty}^{+\infty}\frac{d\psi^*}{dx}\left(\int\frac{d\psi}{dx}dx\right) dx \right]$$

$$<\psi|\hat{p}\psi> = -i\hbar\left[\psi^*\psi|_{-\infty}^{+\infty} - \int_{-\infty}^{+\infty}\frac{d\psi^*}{dx}\psi \, dx \right] = -i\hbar\left[0 - \int_{-\infty}^{+\infty}\frac{d\psi^*}{dx}\psi \, dx \right]$$

$$[\text{as}\psi \to 0 \text{ at } x \to \pm\infty]$$

$$<\psi|\hat{p}\psi> = i\hbar\int_{-\infty}^{+\infty}\frac{d\psi^*}{dx}\psi \, dx = \int_{-\infty}^{+\infty}\left(-i\hbar\frac{d\psi}{dx}\right)^*\psi \, dx = < -i\hbar\frac{d\psi}{dx}|\psi>$$

$$<\psi|\hat{p}\psi> = <\hat{p}\psi|\psi> = <\psi|\hat{p}^{\dagger}\psi>$$

$\Rightarrow\hat{p} = \hat{p}^{\dagger} \Rightarrow$ Momentum operator $\hat{p}$ is Hermitian.

8. *A three-level quantum system has energy eigenvalues 0,1,2 MeV. If the probabilities for the system, at time t, to be in the first two eigen states are 49% and 36%, respectively, write down the wave function for the system.*

Ans. $\hat{H}\to$ Hamiltonian with eigenstates $|\psi_1>, |\psi_2>, |\psi_3> \Leftarrow$ basis
With eigen values $E_1 = 0, E_2 = 1MeV, E_3 = 2MeV$
Eigen equations: $\hat{H}|\psi_1 > =E_1|\psi_1 > =0$

$$\hat{H}|\psi_2 > = E_2|\psi_2 > = |\psi_2>$$

$$\hat{H}|\psi_3 > = E_3|\psi_3 > = 2|\psi_3>$$

Construct linear combination of these stationary states with expansion coefficients c_1, c_2, c_3 as

$$|\psi(0) > = c_1|\psi_1 > + c_2|\psi_2 > + c_3|\psi_3>$$

And the time-evolved state is

$$|\psi(t) > = c_1|\psi_1 > e^{-iE_1t/\hbar} + c_2|\psi_2 > e^{-iE_2t/\hbar} + c_3|\psi_3 > e^{-iE_3t/\hbar}$$

$$|\psi(t) > = c_1|\psi_1 > + c_2|\psi_2 > e^{-it/\hbar} + c_3|\psi_3 > e^{-i2t/\hbar}$$

where $| c_1 |^2 =$ probability of first eigen state $= 0.49 = | 0.7 |^2$
$| c_2 |^2 =$ probability of second eigen state $= 0.36 = | 0.6 |^2$
$| c_3 |^2 =$ probability of third eigen state $1 - 0.49 - 0.36 = 0.15 = | \sqrt{0.15} |^2$

$$|\psi(t) > = 0.7|\psi_1 > + 0.6|\psi_2 > e^{-it/\hbar} + \sqrt{0.15}|\psi_3 > e^{-i2t/\hbar}$$

9. *Derive Heisenberg equation of motion (regarding time derivative of dynamical variable A in quantum mechanics) i.e. find $\frac{d}{dt}\langle \hat{A}\rangle$. Discuss cases where A is explicitly time independent, commutes with the Hamiltonian H of the system. What if A = H?*

Ans. In classical mechanics dynamical variables like coordinate x, momentum p are time varying. However, their operator representations in quantum mechanics do not change with time. The time variation of quantum mechanical states comes through the dependence of wave function $\psi(x, t)$ on time. In quantum mechanics it is the time derivatives of the expectation values of the dynamical variables that are studied.

Let A be a dynamical variable (observable)—the corresponding operator being $\hat{A}$. The expectation value of A in the normalized state $|\psi>$ is

$$<\hat{A} > = <\psi|\hat{A}|\psi > .$$

Time rate of change of $<\hat{A}>$ is

$$\frac{d}{dt} < \psi|\hat{A}|\psi > = <\frac{\partial}{\partial t}\psi|\hat{A}|\psi > + <\psi|\frac{\partial}{\partial t}\hat{A}|\psi > + <\psi|\hat{A}|\frac{\partial}{\partial t}\psi> \qquad (2.3)$$

The time dependent Schrödinger equation is (using $\hat{E} = i\hbar\frac{\partial}{\partial t}$, $\hat{H}$ = Hamiltonian operator)

$$\hat{H}|\psi> = \hat{E}|\psi>$$

$$\Rightarrow \hat{H}|\psi > = i\hbar\frac{\partial}{\partial t}|\psi> = i\hbar|\frac{\partial\psi}{\partial t}> \qquad (2.4)$$

$$\Rightarrow |\frac{\partial\psi}{\partial t} > = \frac{1}{i\hbar}\hat{H}|\psi> \qquad (2.5)$$

Take dagger (i.e. Hermitian conjugate) of equation (2.4)

$$<\psi|\hat{H}^{\dagger} = -i\hbar < \frac{\partial \psi}{\partial t}| \Rightarrow <\psi|\hat{H} = -i\hbar < \frac{\partial \psi}{\partial t}|\hat{H}(\text{as } \hat{H} \text{ is Hermitian } \hat{H}=\hat{H}^{\dagger})$$

(2.6)

$$<\frac{\partial \psi}{\partial t}| = -\frac{1}{i\hbar} < \psi|\hat{H}$$

Hence we have from equations (2.3), (2.4), (2.6)

$$\frac{d}{dt} < \psi|\hat{A}|\psi > = -\frac{1}{i\hbar} < \psi|\hat{H}\hat{A}|\psi > + <\psi|\frac{\partial}{\partial t}\hat{A}|\psi > + \frac{1}{i\hbar} < \psi|\hat{A}\hat{H}|\psi >$$

$$= \frac{1}{i\hbar} < \psi|\hat{A}\hat{H} - \hat{H}\hat{A}|\psi > + <\psi|\frac{\partial}{\partial t}\hat{A}|\psi >$$

$$= \frac{<\psi|[\hat{A}, \hat{H}]|\psi >}{i\hbar} + <\psi|\frac{\partial}{\partial t}\hat{A}|\psi >, \quad [\ ] = \text{commutator bracket}$$

$$\frac{d}{dt} < \hat{A}> = \frac{<[\hat{A}, \hat{H}]>}{i\hbar} + <\frac{\partial \hat{A}}{\partial t}> \Leftarrow \text{Heisenberg's equation of motion}$$

This is the time evolution of expectation value of an observable A.

- If $\hat{A}$ is explicitly time independent i.e. $\frac{\partial \hat{A}}{\partial t} = 0$, and if $\hat{A}$ commutes with the Hamiltonian $\hat{H}$ i.e. $[\hat{A}, \hat{H}] = 0$ then

$$\frac{d}{dt} < \hat{A} > = 0 \text{ i.e. } A \text{ is a constant of motion (conserved)}.$$

- If $\hat{A}$ is explicitly time independent i.e. $\frac{\partial \hat{A}}{\partial t} = 0$, then

$$\frac{d}{dt} < \hat{A} > = \frac{<[\hat{A}, \hat{H}]>}{i\hbar}$$

In a c.s.c.o (= complete set of commuting observables) generally Hamiltonian $\hat{H}$ is taken as a member. The other members commute with $\hat{H}$ and if they are explicitly time independent then they are constants of motion.

- If $\hat{A} = \hat{H}$ (= time independent Hamiltonian) then Heisenberg equation of motion becomes

$$\frac{d}{dt} < \hat{H}> = \frac{<[\hat{H}, \hat{H}]>}{i\hbar} + <\frac{\partial \hat{H}}{\partial t}> = 0 \Rightarrow <\hat{H}> = \text{conserved}.$$

This means, for a conservative system (for which Hamiltonian is time independent) energy is a constant of motion.

10. *Prove that if $\hat{A}$ and $\hat{B}$ are constants of motion and $\hat{H}$ is the Hamiltonian then $[\hat{A}, \hat{B}]$ is a constant of motion.*

Ans. Assume $\hat{A}, \hat{B}, \hat{H}$ to be explicitly time independent.

Consider $[\, [\, \hat{A}, \, \hat{B}], \, \hat{H}] = [(\hat{A}\hat{B} - \hat{B}\hat{A}), \, \hat{H}] = [\hat{A}\hat{B}, \, \hat{H}] - [\hat{B}\hat{A}, \, \hat{H}]$

$$= \hat{A}[\hat{B}, \, \hat{H}] + [\hat{A}, \, \hat{H}]\hat{B} - \hat{B}[\hat{A}, \, \hat{H}] - [\hat{B}, \, \hat{H}]\hat{A}$$

Since $[\hat{A}, \, \hat{H}] = 0, [\hat{B}, \, \hat{H}] = 0$ (as $\hat{A}, \, \hat{B}$ are constants of motion).

$$[[\hat{A}, \, \hat{B}], \, \hat{H}] = 0$$

So $[\, \hat{A}, \, \hat{B}]$ is a constant of motion.

11. Prove that trace of commutator bracket is zero.

Ans. $Tr[\hat{A}, \quad \hat{B} \,] = Tr \, [\hat{A}\hat{B} - \hat{B}\hat{A}] = Tr(\hat{A}\hat{B}) - Tr(\hat{B}\hat{A})$

$$= Tr(\hat{A}\hat{B}) - Tr(\hat{A}\hat{B}) = 0$$

IOP Publishing

Quantum Optics and Quantum Computation
An introduction
Dipankar Bhattacharyya and Jyotirmoy Guha

Chapter 3

Introduction to quantum computing

3.1 Introduction

Feynman showed that there are fundamental limitations in trying to *simulate* a complex quantum system on a conventional classical computer. (Simulation means imitation of a real-world process or system over time through a model based on mathematical formulae and equations.) However, the limitations could be overcome if a quantum computer is built based upon quantum mechanical postulates.

Quantum computing is a natural consequence or limit at which one is sure to arrive, as one proceeds with miniaturization indefinitely to avoid wastage of time, space and energy. This leads us to the fundamental quantum levels.

Though the laws of quantum world are *different* (compared to the laws of the classical world) we can exploit the quantum mechanical nature of a system or matter to achieve a process of computing—and this is given the name quantum computing. This is the reason why ideas of quantum mechanics come into the computing process.

3.2 Some basic ideas about classical and quantum computing

Unlike classical computing systems, quantum systems are unfortunately *fragile* and the important aspects of quantum systems like superposition and entanglement (sections 3.3.8 and 5.12) are *difficult to sustain*. So implementation of quantum computing is very tough.

Unlike classical computing, quantum computing is *reversible* with minimal energy loss (sections 3.3.5 and 8.2).

In classical computing we rely upon two components—the hardware (which is the circuitry) and the software (which contains algorithms and programs, based upon which system is made to function logically).

Classical computing is based on classical logic (i.e. on classical physics, Boolean algebra and mathematical logic). In modern classical computers integrated circuits

doi:10.1088/978-0-7503-2715-2ch3

are used which involve transistors and other electrical components. They work on quantum mechanical principles but the logic they follow is classical.

There are two important quantum mechanical postulates that are followed when quantum computing is done.

(1) The system evolves or *develops unitarily* (i.e. follows a unitary operation) as it passes through transistors and gates and we are not interested in getting intermediate results.

(2) When we are not making any measurement on a system the process of evolution of the *state vector is continuous with time*. The process of measurement, i.e. taking a reading *makes the process discontinuous* and the state collapses to one of the eigenstates of the operator corresponding to the observable we are interested in measuring but we cannot tell for sure which state the system would collapse to. We can, however, give a probabilistic prediction of the collapse for each of the possible states.

A traditional computer software is designed for serial computation. When an algorithm is written the logic-flow takes place from one point to another point over a time. In other words a particular process must be completed before another process is taken up in serial computation.

In traditional classical computation the concept of *parallel computation* means that a problem is broken up into independently computable parts or logics. Each part or logic can be processed by different *serially computing processors*. These serially computing processors run simultaneously in parallel so that the results of each part or logic can be combined.

For instance, when we take product of two 2×2 matrices

$$\begin{pmatrix} a & b \\ c & d \end{pmatrix}\begin{pmatrix} p & q \\ r & s \end{pmatrix} = \begin{pmatrix} ap + br & aq + bs \\ cp + dr & cq + ds \end{pmatrix}$$

the products of the elements of two rows of first matrix viz. $(a \ b)$ & $(c \ d)$ with the two columns of the second matrix $\begin{pmatrix} p \\ r \end{pmatrix}$ and $\begin{pmatrix} q \\ s \end{pmatrix}$, respectively, can be computed simultaneously using separate processors and each of these separate processors does serial computation. In other words, each of the computations, like multiplying the row $(a \ b)$ of first matrix with the column $\begin{pmatrix} p \\ r \end{pmatrix}$ of the second matrix etc, are done serially. But they are done at the same time in parallel mode so as to generate the elements $ap + br, aq + bs, cp + dr, cq + ds$ which are combined or assembled to get the product 2×2 matrix.

So a traditional parallel computer must have N number of processors, say, each being a serial processor. All the N processors take up (serial) processing jobs at the same time to deliver N outputs in parallel, that are ultimately combined to get a final result.

A classical register can store a particular state at a given time. In contrast a quantum register can store *a linear combination of states*. In other words a quantum computer can handle a multitude of states (or multiple inputs) and not just a single state. As a result, the quantum computer will be able to compute the output simultaneously for all the states, i.e. for the multiple inputs. The parallelism of a quantum computer is thus an inherent feature. In other words, a quantum computer exhibits an *inherent parallelism* (sections 3.3.4 and 8.1) and the same processor can perform operations on multiple inputs simultaneously.

Quantum computers, which exploit ideas of superposition, quantum entanglement (sections 3.3.8, 4.7, and 5.12), no-cloning theorem (sections 3.3.7 and 6.1), interference etc, will be *incredibly powerful* and would perform tasks that are considered virtually impossible for a classical computer (or even for a super computer).

A quantum computer, once operational, will be used to encrypt data (i.e. to conceal data by converting to a code, i.e. codifying data) with unconditional security, build precise climate models, help invent new drugs, and build complex financial models and risk-management models.

3.3 Definition of certain terms relating to quantum computing

We here define very briefly certain terms that are widely used in the study of quantum computing. Detailed explanation is given in later chapters.

3.3.1 Information

Information (section 10.1) means *meaningful facts* or details, i.e. coherent knowledge about a subject or something that we wish to know and communicate to someone else. And this meaningful and coherent information or facts or details, i.e. knowledge, is carried by a message that can be in a specified format or code or any language. Again information may be precise or information may involve uncertainty.

For instance the message that states 'today's temperature is 17 °C' is a precise statement or a piece of complete and unambiguous information, while the message that states 'today's temperature is around 17 °C' is an uncertain and imprecise statement or a piece of incomplete information.

3.3.2 Computability

Any statement cannot be considered as information. A statement can be considered as representing information if it is meaningful and coherent, useful and can be *processed*. Computation is based on valid classical information.

All information is not computable. A statement representing a universal fact or event carries information that is meaningful, but cannot be altered or changed since we cannot make the fact or event occur in a different way. In other words this information cannot be processed—computational activities cannot be carried out. Such an unchangeable universal fact or event or statement is not treated as useful computable information. For instance the message 'the Sun rises in the east and sets in the west' is meaningful information representing an unalterable truth and further we cannot interfere with it—cannot make the process faster or slower or make the sunrise or sunset happen in a different part of the horizon as we desire. So this is not a piece of computable information.

Statements or facts that can be subjected to an operation, as a result of which they change are treated as useful information. For instance if we are informed about a car moving from city A to city B over time we can construct the classical equation of motion, solve it with the given boundary conditions and compute the various parameters of the motion at various instants. So the information we get can be processed to get certain knowledge or data leading to many other possibilities emerging out of it. Another example is making weather predictions. Starting from a weather information chart or weather database, regarding condition of weather in a region, we can process information, analyze data and predict if the climate will be hot or cold or wet or humid. So based upon information we can set up or formulate equations characterizing the situation or problem and then find solution of equations and make meaningful predictions.

3.3.3 Algorithm

Algorithm (chapter 8) refers to a set or sequence of *well-defined* rules or instructions (a step-by-step guide) that are to be followed or implemented by a computer while performing a calculation, i.e. computational data-processing pertaining to a problem. We discuss various algorithms in chapter 8.

3.3.4 Quantum parallelism

A quantum computer can deal not only with individual qubit states (sections 4.5 and 4.1) but also with a superposition of these qubit states. So an operation on the superposition of states means each of the component states is being simultaneously operated upon. This means the component qubit states are processed in parallel. As a result, the speed of quantum computation becomes enormous.

3.3.5 Quantum reversible computing

Irreversible means a process that cannot be undone. In irreversible computing we get an output starting from an input but we do not get the input back starting from the output. In other words, once processed, we fail to regenerate the input.

In reversible computing we get back the input from the output, i.e. an operation is capable of being undone. Even after processing we can regenerate the input.

In classical computing *only* the NOT gate is reversible. For input A the output of NOT gate is $Y = \bar{A}$ (complement of A). We can start the other way round since for output $Y = \bar{A}$ the input can be calculated to be $\bar{\bar{A}} = A$ (this is the complement or invert of $\bar{A}$).

Classical gates other than the NOT gate are *not* reversible. For instance the input $A = 0$, $B = 1$ to an AND gate yields at the output $Y = 0$, but the output $Y = 0$ does not mean the input has necessarily to be $A = 0$, $B = 1$, since $A = 0$, $B = 0$ also gives $Y = 0$ and so does the combination $A = 1$, $B = 0$. This means we *cannot* go back or revert from output to input (as we *could* with NOT gate) (figure 8.1).

Normal classical computing is thus generally not reversible. We can, however, make it reversible by storing a lot of garbage information (garbage means bits generated that are not useful) but this will be energy inefficient. So we have to settle with irreversible classical computing except that of NOT computation (section 8.2).

A quantum computer has to be reversible. The evolution of quantum state occurs unitarily so that the length of the state vector is always unity and the total probability of all possible outcomes is normalized to unity. Therefore, any quantum gate must be implemented as a unitary operator. In other words, the unitarity requirement necessitates reversibility of quantum computing. Thus quantum computing must be reversible.

3.3.6 Ancilas (Ancila bits)

For performing quantum computation, which is a reversible computation, often it is necessary to use some extra qubits. We then have to employ *additional registers* called ancilas. Register refers to quickly accessible location available to the processor of a computer.

3.3.7 No-cloning theorem

No-cloning theorem (section 6.1) states that it is *impossible* to create an identical copy, i.e. to make a clone of an arbitrary unknown quantum state. The impossibility of cloning, i.e. impossibility of producing replicas is used in quantum cryptography (sections 3.3.10 and 12.1). No-cloning theorem was stated by Wootters, Zurek and Dieks in 1982. The no-cloning theorem is explained in detail in the chapter on teleportation, chapter 6.

3.3.8 Quantum entanglement

Entanglement means a mixed-up or an *entwined state* involving, say two members. Entanglement occurs when a pair of states is inextricably linked.

When a pair of quantum particles is intricately linked such that whatever happens to one immediately affects the other, regardless of their separation we say the particles are entangled quantum mechanically. Einstein described quantum entanglement as a '*spooky action at a distance*'. Quantum state of each particle in the pair cannot be described independently of the state of the other. In other words the key to entanglement is that the entangled state *cannot be decomposed into component states*. Particles which are far apart can still be in quantum entanglement. If one is disturbed through measurement the other is also disturbed. This is discussed elaborately in chapter 11.

3.3.9 Quantum teleportation

Teleport means to disappear and then to reappear at a distance. Teleportation means *transfer of information* from one point to another without transfer of particles through the physical space between them.

Quantum entanglement gives the idea of quantum teleportation. Quantum teleportation refers to transference of quantum information (quantum state) between quantum mechanically entangled sender and receiver (particles) through classical communication. Quantum teleportation is possible because particles which are far apart can still be in *quantum entanglement*.

Two photons or ions can be entangled in such a way that when quantum state of one is altered the state of the other also suffers alteration—as they are still connected. So if one of the quantum mechanically entangled photons or ions is far apart, quantum information can be teleported since one photon or ion suffers the *same alteration of state* as its entangled partner.

We discuss teleportation in chapter 6.

3.3.10 Quantum cryptography

Sometimes it is necessary to keep information secret. Any and all information is not for public view such as someone's ATM PIN (personal identification number, which may be a four digit code). Cryptography is the art of writing or solving codes. It is a science of *protecting information.*

A code actually carries some information. Since it is in coded form the information is not explicit or obvious to all and sundry—in other words, to protect information from being public it is codified or encrypted. This coded or encrypted information may be transmitted and/or stored and we expect the coded information to maintain its integrity (i.e. to maintain its secret nature throughout).

Secrecy is preserved only if the code possesses a strong and *secure format* or structure and can be decoded or decrypted using a unique key, i.e. by a method which cannot be easily guessed or worked out by a stranger or unauthorized person (eavesdropper). One would even wish that the coded information would not get revealed even if it passes through an insecure network (internet) or is stored in an unfaithful site or location.

Encryption means converting information into a code, i.e. *codification*, while decryption means conversion of encrypted data back to its original form, i.e. *decodification*. To encrypt and decrypt, sophisticated mathematical equations and algorithms are used. A simple example may be to encrypt a message by replacing the letters with symbols or numbers etc. Cryptographic operations refer to encryption and decryption— the latter requiring a secret password or secret key.

Quantum cryptography uses properties of quantum mechanics, no-cloning theorem, Heisenberg's uncertainty principle to perform cryptographic operations (i.e. encryption and decryption) which is impossible (or intractable) by conventional cryptography. For instance secrecy would be preserved even if any effort to copy or read encoded data is made. Through such acts, copies (i.e. replicas or clones) are not generated but the quantum state gets changed since clones cannot be built through the act of copying or reading. The information shared between two parties would be unconditionally secure if the decoding key (also called QKD = Quantum Key Distribution) is shared only between them without involvement of a third party.

In other words, secure and faithful communication is possible between two parties if only they share a secret key.

We discuss quantum cryptography in chapter 12.

3.3.11 Protocols of quantum computing

Protocol (section 12.10) refers to a *set of rules or guidelines* that are to be strictly followed during exchange or transmission of data between devices—like two or more computers. If the rules are not followed, the network will fail to transmit data.

- Example

  Hypertext Transfer Protocol (HTTP) is the protocol that was developed by Tim Berners-Lee and is followed for online transfer of data (web pages of different websites and small files) from one machine to another (say between server and client, i.e. between a centralized resource computer and a desktop computer). HTTP is a client server protocol that allows clients to request *web pages from web servers*.

  In 1984 Bennett and Brassard introduced the first protocol for quantum cryptography, the quantum key distribution scheme.

3.3.12 Quantum de-coherence

In the micro-world, particles behave like waves (their deBroglie wave length cannot be ignored as it is comparable to or may be even greater than particle size). Since waves can superpose or overlap upon each other to form a new single wave, quantum particles can exist in multiple overlapping states (waveforms) at the same time.

As long as there exists a definite phase relation between different states, the system is said to be coherent. For a perfectly isolated quantum system, coherence is maintained indefinitely.

But during a measurement, wave function collapses to a single eigenstate, the perfect *isolation is lost* as coherence is shared with the environment (or heat bath). This means the quantum superposition is lost with time. Coherence of a system is thus destroyed and we say de-coherence occurs. De-coherence is thus the loss of quantum coherence—a concept introduced by Dieter in 1970 and can be looked upon as *seepage or leak of information* from system to environment.

3.4 Journey towards quantum computing

While the concept of computing dates back to 3000 BC (invention of the Sumerian abacus), the slide rule was invented in 1622 (by William Oughtred) and in 1623 Wilhelm Schickard invented the calculating machine. In 1703, Leibniz invented the binary system which is the base of modern day computer systems or networks.

The idea of computing through an automated machine was first introduced by Turing and was later clarified by Church based upon what is called Church–Turing principle. It dictates all the classical computing principles which involve well defined problems that can be settled on a computational basis.

In 1936 Turing proved that a device with a limitless memory and a scanner to scan the memory backward and forward, capable of reading it symbol by symbol and writing additional symbols, can execute any computation (i.e. can process the information). This principle holds true for classical computation, and all present day classical computers (including the parallel computers) are equivalent to the universal Turing machine.

There are problems that cannot be efficiently handled by the universal Turing machine. For instance it cannot predict whether a program will terminate (called classical halting problem). Program compilers cannot anticipate all the possibilities of program run crashes. Also, they cannot address the complexity of some problems. There are problems that are solvable in a reasonable time but the inverse problem may be hard to solve in a reasonable time. For instance it is easy to multiply numbers so as to generate a parent number (which is obviously the product of the component numbers or factors). But starting from the parent number (i.e. the product) it is tough to obtain the component factors or numbers. In other words *multiplication is easy but factorization is hard.*

3.4.1 Moore's law (1965)

With miniaturization, the speed of computers increased. After invention of silicon transistors the power of the computer increased enormously while its size shrank equally remarkably (the dimension of VLSI is about 0.08 μm). This is embodied in Moore's law (figure 3.1) stated as follows.

The power of a computer system depends upon the number density of transistors that one can pack into a chip which doubles every two years—achieving more and more miniaturization. The number of transistors per chip increases almost linearly with each passing year.

Space between transistors reduces significantly and might reach atomic dimension. The size of the tiniest transistor is becoming the size of a single atom or molecule. Shrinking the size down further is impossible due the fact that quantum laws take over and operate in the micro world.

The deterministic approach of the classical concept fails. There is a physical limit to the miniaturization process. As we move into the micro world to avoid waste of

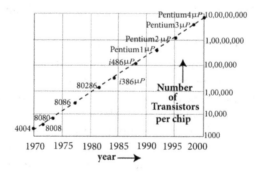

Figure 3.1. Moore's law.

time and energy and to increase speed and power we inevitably sail into the quantum world.

Moore's law is not a law in the strict sense like a law of physics but it is a rule of thumb to set a goal or target. And as per Moore's law, we hit the quantum level in around 2010–2020. Due to the atomic limit regarding density of packing, the time has come for quantum computers to take over and rule the computation world.

3.4.2 Feynman idea of quantum computer (1982)

The laws of quantum mechanics are more general and reduce to classical laws under the limit of $\hbar \to 0$, meaning thereby that any theory based upon quantum mechanical concepts cannot be mimicked in a classical computer and can *only* be implemented in a quantum computer—the conceptual model of which was pre-scribed by Feynman in 1982.

3.4.3 Deutsch algorithm

In 1985 Deutsch designed a simple algorithm according to which quantum computers could *solve certain computations* faster than classical computers. We discuss the Deutsch algorithm in chapter 8.

3.4.4 Shor's algorithm

In 1994 Shor developed a quantum algorithm, based on quantum principles for *factorization* of a large composite number (for which efficient classical algorithms did not exist). This showed that quantum computers would be much more powerful than classical computers. It is possible to break a number of cryptosystems used today with Shor's algorithm. We discuss Shor's algorithm in chapter 8.

3.4.5 Concept of qubit

Schumacher in 1995 defined quantum bit or qubit which is the analogue of classical bit—the *unit of information storage* or communication. We discuss quantum bits in chapter 4.

3.4.6 Grover's algorithm

Grover in 1996 developed a quantum algorithm for the search of a quantity or datum from an unsorted or unstructured database and proved that *quantum algorithms are faster* than their classical counterparts. The Grover algorithm is discussed in chapter 8.

3.4.7 Experimental demonstration of quantum teleportation

The first ever experimental demonstration of quantum teleportation was done using polarization of a photon as a qubit in 1997 by Bouwmeester and others. We discuss teleportation in chapter 6.

3.4.8 First quantum computer

In 1998, the first ever two-qubit (four-state) quantum computer was demonstrated by Chuang *et al* using Nuclear Magnetic Resonance (NMR) to manipulate the atomic nuclei of a chloroform molecule implementing Grover's algorithm.

3.4.9 Other milestones

- **Di Vincenzo criteria:**

  Di Vincenzo prepared the list of requirements or conditions that must be satisfied by any technology for construction of a quantum computer (like building a scalable system of well-defined qubits, reliable method of initialization of a quantum system, long coherence time, existence of universal gates, efficient measurement scheme etc). This is discussed in chapter 13.
- In 2003 Yuan, Gobby and Shields demonstrated first commercial use of quantum cryptography over fibres.
- In 2005 Haffner *et al* created the first qubyte using ion traps. (Quantum byte means a series of eight qubits.)
- In 2010 free space quantum teleportation was achieved by Xian-Min Jin *et al.*
- The Nobel Prize in Physics for 2012 was awarded to Haroche and Wineland for their ground-breaking experimental methods to measure and manipulate individual quantum systems that can help create a practical quantum computer. Wineland used a LASER to put an ion in its lowest energy state, thus enabling the study of quantum phenomena with the trapped ion. Haroche trapped a single photon in a small vacuous cavity at almost absolute zero over a duration ~0.1 s during which quantum manipulations can be performed. Their work established that direct observation and control (say counting) of quantum particles without destroying them (i.e. preserving their quantum nature) is possible.

While Wineland trapped electrically charged atoms or ions and controlled and measured them with light (photons), Haroche controlled and measured trapped photons by sending atoms through a trap.

Future quantum computers will revolutionize in both pace and precision while handling an extraordinarily huge amount of data.

Computing is based on classical concepts involving solving linear equations. Classical computing employs classical devices that use bits, for instance utilizes gates to perform Boolean operations. Classically all information is stored in the bits. Borrowing the letter 'q' from quantum mechanics the term qubit is coined (chapter 4).

The present day computer follows Newtonian mechanics which makes precise predictions, i.e. gives correct or deterministic and certain answers to a problem. We are able to predict the trajectory of two moving particles with certainty. So if we have two points A and B then we can follow their motion and sketch their pathway as well as telling how they would reach each other. This is the classical mechanics principle which is unambiguous/deterministic by nature and followed by modern day computers.

Starting from classical determinism, when we try to miniaturize the classical ideas we confront length scales (less than 1 nm) where classical mechanics does not work. We therefore have to move away from the idea of deterministic approach of a classical system to probabilistic approach of a quantum system.

3.5 Need for quantum computers

Quantum computer is needed to solve very hard problems for which we may require a result very fast, after analysis of numerous raw data that is virtually impossible even by a classical super computer.

With the achievement of miniaturization, the speed of classical computers has increased significantly. But when miniaturization reaches the limit of atomic dimension (i.e. when the distance between two transistors in a chip is of the order of atomic spacing) the results become no longer reliable due to Heisenberg's uncertainty principle, which plays a major role in the micro world—things are then not deterministic and we can then make only *probabilistic predictions*.

Further, the heat generation in a computing device depends on the volume occupied by the number of gates. Continuous heat removal is necessary from the surface. In the limit of miniaturization, for too-closely spaced components, heat produced by one component would affect the performance of a neighboring component. Also, a classical computing device, in the limit of miniaturization, cannot remove heat efficiently. So the performance of the classical computer would become unreliable due to this *heat problem*.

Most of the processes in classical computing are done irreversibly (except for NOT gate, all other classical gates are irreversible). By Landauer's principle (section 3.6) such irreversible computation will lead to increase of thermodynamic entropy which results in the loss of energy and the process becomes gradually inefficient with increase in number of components. Quantum computational tasks are thus needed to be carried out *reversibly* through use of *unitary operators*. Every logical step can be reversed resulting in negligible loss of energy. (Classical computing if done through reversible computation suffers from the problem of garbage bit production, garbage bit storage and garbage disposal.)

3.6 Landauer's principle

Any logically irreversible computation of information (e.g. erasing a bit) is accompanied with *irreversible release* of heat energy and entropy rise.

A logically *reversible* computation (in which there is no loss of information) involves *no release* of heat.

Landauer's limit

The *minimum* possible amount of energy required or generated to erase one bit of information is called Landauer's limit and is given by $kT \ln 2$, where $k = 1.38 \times 10^{-23} \text{JK}^{-1}$, T is temperature in Kelvin scale.

The thermodynamics of computing says that in order to get the best out of a computer we would like to have the computation to be run in the most energy efficient way, which is possible if the system works in a reversible manner. That is the reason why reversible computing is needed. A Turing machine is not reversible (once a bit is read it moves on and there is expenditure of energy involved in movement of the tape etc) and so it gets heated with time. This dissipation of heat means the device (Turing machine, classical computer) works irreversibly. But quantum systems are reversible since the time dependent quantum mechanical equation of motion is reversible.

We cannot however devise a completely new way of computation just because we are using quantum principles which vastly differ from classical principles. Our definition of computation will rely on the principles laid down by the Church and Turing principles. But the machine operations will not be exactly the same.

3.7 Quantum computing

Quantum computing refers to the manipulation of quantum mechanical systems for processing of information.

Quantum Computer

A quantum computer is a device that processes information in a quantum mechanically coherent fashion and is capable of executing certain calculations with speed and efficiency far superior to classical computers.

There exists a vast array of minute objects of micro dimension: electron, proton, neutron, photon etc, and quantum mechanics describes their behavior. They behave in classically unusual ways and their state is generally unknown at any given time and changes if we try to observe them. Properties of these systems can be manipulated and measured.

3.8 Bits 0 and 1

We work in binary situations involving bits 0 and 1. The bits 0 and 1 stand, respectively, for *low voltage* and *high voltage* in electronics. We can also use them in the binary sense to represent information. In the quantum formalism we can use them to represent *low energy state* and *high energy state* or spins in two different senses, namely *clockwise* spin and *anticlockwise* spin, i.e. up spin and down spin. A two-state quantum system will in general be in an arbitrary superposition of these states, i.e. not purely in 0 state or not purely in 1 state but in a mix of 0 state and 1 state. So a quantum system cannot be explained by looking only at one of the two states. Since arbitrary superposition of all states is involved, in general, all such states can be processed at once in quantum operations.

In the classical coin toss problem we only have head and tail. But in quantum mechanics we have superposition of states.

For instance, a mixture of spin up and spin down states is a quantum system. This means that not only are the up state or the down state taken care of, but in addition other possibilities (the precessional states or the mixed states) are also included. As long as we do not measure, all possible states are probable states and find a place in the superposed set of states. However, a measurement at any point of time will yield a particular state since measurement perturbs the system to collapse to that particular state.

Figure 3.2 highlights the difference between classical states |0> and |1> and a quantum state $\cos \frac{\theta}{2}|0> +e^{i\gamma} \sin \frac{\theta}{2}|1>$ that is a mixture of the states |0> and |1>.

3.9 A bit of Boolean algebra

In the binary number system the base (or radix) is 2, the first digit is 0 and last digit is Base $-1 = 2 - 1 = 1$. So the building blocks of the binary number system are the two bits 0,1. Computers work with this number system. An 8 bit string is called a byte which is the unit of information storage in a classical computer.

Digital devices based on electronic circuits function in a binary manner. Operating devices or circuits can exist in two distinct or discrete (quantized) states or levels, when the device is on (figure 3.3).

If output voltage is high (4 V ± 1 V) we call it ON state and denote by symbol 1.

If output voltage is low (0.2 V ± 0.2 V) we call it OFF state and denote by symbol 0.

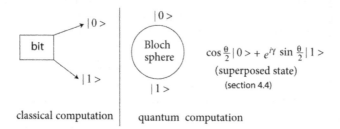

Figure 3.2. Superposed states are important in quantum computation. In classical computation states are say |0> and |1> only. But in quantum computation mixtures of states are important. (Bloch sphere is discussed in chapter 4)

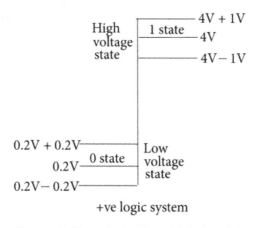

Figure 3.3. Positive logic system: defining 0 and 1.

No other value of output voltage is considered significant. Exact value of output voltage is not required—we just need to know whether it is above or below a reference value (level), i.e. whether in the range (4 V ± 1 V), denoted by 1 or in the range (0.2 V ± 0.2 V), denoted by 0.

This technique of mentioning two states only (1 or 0) is referred to as a logical statement.

For digital operation the system (say a transistor) is biased to stay either in high voltage state or in low voltage state. In other words, by proper biasing, the transistor is tied either in saturation region (ON state) or in cut off region (OFF state).

The rules of binary addition have been mentioned in table 3.1. This table also depicts the number of pieces of information (i.e. the information size) that the bits (1 bit, 2 bits, 3 bits ....) represent.

3.10 Gate

Gate is a logic circuit (an electronic circuit) that makes *logical decisions*. They have one or multiple inputs but single output. Gates are analysed with Boolean algebra.

Truth table

Tabular systematic representation of all the possibilities of output signal corresponding to various input signal-combinations is called a Truth table (which is an indicator table). In other words the truth table is indicative of the *performance* (or lack of performance) of a gate.

Figure 3.4 shows a diagrammatic representation of NOT gate, OR gate and the AND gate along with the respective truth table.

Basic or fundamental logic gates are NOT gate, OR gate and the AND gate. They are called fundamental because any digital system can be built with them. So they are looked upon as logically complete gates. The NOT operation is denoted by putting an overbar, OR operation is denoted by putting a plus '+' sign and AND

Table 3.1. Rules of binary addition.

Rules of Binary addition	Number of information/set of codes	
$0 + 0 = 0$ $0 + 1 = 1$ $1 + 0 = 1$ $1 + 1 = 10$ $1 + 1 + 1 = 1 + 10 = 11$	$\begin{pmatrix} 0 \\ 1 \end{pmatrix} = 2 = 2^1 \text{codes}$	$\begin{pmatrix} 0\,0\,0 \\ 0\,0\,1 \\ 0\,1\,0 \\ 0\,1\,1 \\ 1\,0\,0 \\ 1\,0\,1 \\ 1\,1\,0 \\ 1\,1\,1 \end{pmatrix} = 8 = 2^3 \text{codes}$
	$\begin{pmatrix} 0\,0 \\ 0\,1 \\ 1\,0 \\ 1\,1 \end{pmatrix} = 4 = 2^2 \text{codes}$	

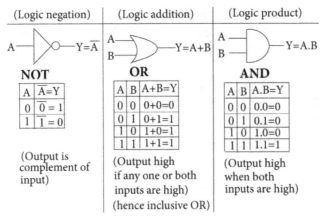

Figure 3.4. Logically complete gates: NOT gate, OR gate, AND gate.

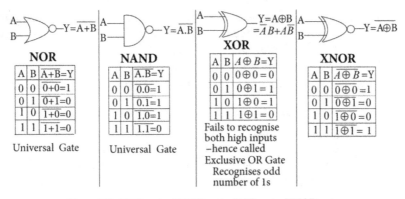

Figure 3.5. NOR gate, NAND gate, XOR gate, XNOR gate.

operation is denoted by putting a dot '. ' sign. OR gate is also called inclusive OR gate since it responds to $0 + 1 = 1$, $1 + 0 = 1$ (case of one high input) as well as to $1 + 1 = 1$ (case of both high inputs), i.e. it recognizes any one high and also includes both high cases.

There are many other gates that are commonly used such as the NOR gate, NAND gate, XOR gate, XNOR gate. These gates are called mixed gates (figure 3.5). The bubble sign in the block diagram of NOT, NOR, NAND, XNOR represent the operation of negation.

NOR gate is OR gate followed by NOT, NAND gate is AND followed by NOT. The XOR gate is the operation $\bar{A}B + A\bar{B}$ (where A, B are the inputs). XOR gate is also called exclusive OR gate since it responds to $0 \text{ XOR } 1 = \bar{0}. 1 + 0. \bar{1} = 1$, $1 \text{ XOR } 0 = \bar{1}. 0 + 1. \bar{0} = 1$ i.e. recognizes any one high but since $1 \text{ XOR } 1 = \bar{1}. 1 + 1. \bar{1} = 0$ it excludes, i.e. fails to recognize both high inputs. XNOR gate is XOR gate followed by NOT gate.

With NOR gate any basic logic gate can be built and hence the NOR gate is called the universal gate. Similarly, with the NAND gate any basic logic gate can also be built and hence the NAND gate is also called universal gate.

We mention some Boolean expressions for ready reference.

DeMorgan's theorem		
$\overline{A+B} = \overline{A}.\overline{B}$	$A + \overline{A} = 1$	$A.\overline{A} = 0$
	$A + A = A$	$A.A = A$
$\overline{A.B} = \overline{A} + \overline{B}$	$A + 0 = A$	$A.0 = 0$
	$A + 1 = 1$	$A.1 = A$

3.11 Computational complexity

Polynomial time versus exponential time

In computer science the complexity of a problem is determined by the *number of operations or steps* needed to execute or compute, i.e. perform a desired task (computation) using the input data. This is referred to as time of execution (computation). Obviously the complexity measured by the time of execution depends on the length of the input data string, i.e. number of bits involved (which let us denote by n).

In general, if we can express the complexity of a problem as a polynomial function of the length of string (say n^2) then this problem is said to have a complexity of polynomial time and is considered to be an *easy problem*.

If we cannot express the complexity of a problem as a polynomial function (but express as 2^n) then it is called exponential time problem and is considered to be a *hard problem*.

Factorization is a hard problem in classical computing algorithms since by classical computing the problem cannot be solved in polynomial time—it is solvable in exponential time. But if we employ Shor's algorithm (1994) it is solvable through quantum computer in polynomial time and is not a hard problem. This opens up amazing and novel possibilities in the world of computing. We discuss this in chapter 8.

Let us construct a plot of input size in bits (say n bit input) along abscissa and the number of operations or steps or state transitions which we call the time to execute or compute a particular task (computation). The shape of the growth curve in this diagram is a meaningful way to classify the complexity or hardness of the problem or task (figure 3.6).

If the plot is a straight line we say that the algorithm solves the task or problem in linear time (easy problem).

If the plot is a curve described by a polynomial say n^2 (or n^k) we say that the algorithm solves the task or problem in polynomial time (easy problem).

If the plot is a curve described by an exponential say 2^n (or k^n) we say that the algorithm solves the task or problem in exponential time (hard problem).

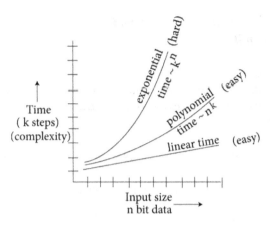

Figure 3.6. Polynomial time versus exponential time.

3.12 Further reading

[1] Ghosh D 2016 Online course on quantum information and computing (Bombay: Indian Institute of Technology)
[2] Nielsen M A and Chuang I L 2019 *Quantum Computation and Quantum Information* (Cambridge: Cambridge University Press)
[3] Nakahara N and Ohmi T 2017 *Quantum Computing* (Boca Raton, FL: CRC Press)
[4] Pathak A 2016 *Elements of Quantum Computing and Quantum Communication* (Boca Raton, FL: CRC Press)

3.13 Problems

1. *Which of the following gates can perform reversible operation?*
   (a) AND, (b) NOT, (c) NAND, (d) All.
   Ans. *(b).*

2. *Reversible quantum computing is needed because of the requirement of*
   (a) entanglement, (b) unitarity, (c) parallelism, (d) Landauer's principle
   Ans. *(b).*

3. *Identify which is applicable to quantum computing.*
   (a) Parallel computation, (b) Faster computation than classical algorithms,
   (c) Reversible computation (d) Easy reading of output.
   Ans. *(a), (b), (c).*

4. *Which statement is wrong?*
   (a) Erasing a bit leads to irreversible release of heat energy in logically irreversible computation,
   (b) Erasing a bit leads to no release of heat in logically reversible computation,
   (c) Erasing a bit leads to a minimum energy release of amount k ln 2,

(d) *Erasing a bit requires a minimum energy release which is called the Landauer's limit.*

Ans. *(c).*

5. *When does a quantum wave function collapse?*

   (a) During quantum entanglement, (b) During measurement,

   (c) During unitary development, (d) During creation in Hilbert space.

   Ans. *(b).*

6. *Factorization is*

   a) a hard problem in classical algorithm, (b) an easy problem in classical algorithm,

   (c) a hard problem in Shor's algorithm, (d) an easy problem in Shor's algorithm.

   Ans. *(a), (d).*

7. *Which of the following is not true for a valid quantum state of an electronic system?*

   (a) Spin up state, (b) Spin down state,

   (c) Superposition of spin up and spin down states, (d) Any of (a) or (b) or both.

   Ans. *(d).*

8. *Quantum teleportation is based on which of the following quantum phenomena?*

   (a) Parallel computing, (b) Reversible computing,

   (c) Principle of linear superposition, (d) Entangled pair of states.

   Ans. *(d).*

9. *HTTP protocol stands for*

   (a) hypertexting transfer protocol, (b) hypertextual transfer protocol,

   (c) hypertext transferring protocol, (d) hypertext transfer protocol.

   Ans. *(d).*

10. *Quantum cryptography uses which of the following?*

    (a) Quantum entanglement, (b) No-cloning theorem,

    (c) Quantum teleportation, (d) Quantum coherence.

    Ans. *(b).*

IOP Publishing

Quantum Optics and Quantum Computation
An introduction
Dipankar Bhattacharyya and Jyotirmoy Guha

Chapter 4

Quantum bits

In this chapter we shall discuss the essential features of quantum computing by focusing on the basics.

4.1 Qubits and comparison with classical bits

✓ The smallest unit of classical information in a classical computer is bit (or we may call it classical bit or cbit) which is the fundamental unit in classical computation.

✓ Bits represent actual physical (logic) systems.

✓ Bit, also called binary digit, takes values either 0 or 1. No linear combination or mixture of states 0 and 1 can be assumed.

✓ Bit value 0 or 1 represents two states. A high voltage state (4 V ± 1 V) is represented by bit 1 and low voltage state (0.2 V ± 0.2 V) is represented by bit 0. It does not make any sense to speak of linear combination of two such voltages (i.e. linear combination of high voltage and low voltage) (figure 3.3).

✓ Bit is like a coin having two states: H = Head = 0 state, T = Tail = 1 state. A mixture of H and T is an absurd idea.

✓ Classical computation is done using the cbits 0 and 1 or a collection of the cbits 0s and 1s.

• The fundamental unit of quantum information in a quantum computer is a quantum bit or qubit, which is the fundamental unit in quantum computation. In other words, qubits are the building blocks of a quantum computer.

• Qubits are abstract mathematical objects that have real and experimentally verifiable consequences.

• We use Dirac's Bra ket state symbol to represent qubit states.

• $|0>$ and $|1>$ are called computational base states which are orthonormalized (i.e. $<0|0> = <1|1> = 1$, $<0|1> = <1|0> = 0$ i.e. $<m|n> = \delta_{mn}$) and form a complete set (i.e. $|0><0| + |1><1| = \hat{I}$). They form a 2D complex vector space (called the Hilbert space, 2D in this case, we may denote by $H^2 or C^2$) where we can define a general state $|\phi>$ which is a general linear combination, i.e. superposition of the computational base states $|0> = \begin{pmatrix} 1 \\ 0 \end{pmatrix}$ and $|1> = \begin{pmatrix} 0 \\ 1 \end{pmatrix}$, as per the expansion postulate, and is called a general qubit given by

$$|\phi> = \alpha_0|0> + \alpha_1|1> = \alpha_0\begin{pmatrix} 1 \\ 0 \end{pmatrix} + \alpha_1\begin{pmatrix} 0 \\ 1 \end{pmatrix} = \begin{pmatrix} \alpha_0 \\ \alpha_1 \end{pmatrix}$$

where α_0, α_1 are complex numbers in general. Let us normalize it by introducing the normalization constant N. Let us define

$$|\psi> = N|\phi> = N\begin{pmatrix} \alpha_0 \\ \alpha_1 \end{pmatrix}$$

Demanding that $<\psi|\psi> = 1$ we get

$$<N\phi|N\phi> = 1 \qquad \Rightarrow N^*N < \phi|\phi> = 1$$

$$\Rightarrow N^*N(\alpha_0^* \;\; \alpha_1^*)\begin{pmatrix} \alpha_0 \\ \alpha_1 \end{pmatrix} = 1 \qquad \Rightarrow |N|^2(\alpha_0^*\alpha_0 + \alpha_1^*\alpha_1) = 1$$

$$\Rightarrow |N|^2(|\alpha_0|^2 + |\alpha_1|^2) = 1 \qquad \Rightarrow N = \frac{1}{\sqrt{|\alpha_0|^2 + |\alpha_1|^2}} = 1$$

$$|\psi> = N|\phi> = N \; [\alpha_0|0> + \alpha_1|1>]$$

$$|\psi> = \frac{\alpha_0}{\sqrt{|\alpha_0|^2 + |\alpha_1|^2}}|0> + \frac{\alpha_1}{\sqrt{|\alpha_0|^2 + |\alpha_1|^2}}|1>$$

If we measure the qubit in state $|\psi>$ we get (as per Copenhagen interpretation) either of the following:

(i) the result 0 with probability $= |\frac{\alpha_0}{\sqrt{|\alpha_0|^2 + |\alpha_1|^2}}|^2 = \frac{|\alpha_0|^2}{|\alpha_0|^2 + |\alpha_1|^2}$ and then post measurement state is $|\psi'> = |0>$

(ii) the result 1 with probability $= |\frac{\alpha_1}{\sqrt{|\alpha_0|^2 + |\alpha_1|^2}}|^2 = \frac{|\alpha_1|^2}{|\alpha_0|^2 + |\alpha_1|^2}$ and then post measurement state is $|\psi'> = |1>$.

Clearly though the qubit $|\psi>$ is an unobservable state, the states $|0>$, $|1>$ are observable states. Also all $|0>$, $|1>$, $|\psi>$ are of unit length.

• Such linear combination $|\psi> = \alpha_0|0> + \alpha_1|1>$ provides the quantum computer with enormous computing capability. For a quantum state to exist in the linear combination of $|\psi>$ simultaneously in the basis states $|0>$ and $|1>$ means that in the linear combination the amount of information that the state has is much more than if it were in the state either $|0>$ or in the state $|1>$ (though we are unable to read the information in state $|\psi>$).

Illustration:

(1) Consider a qubit state in the computational basis |0> and |1> given as

$$|+> = \frac{1}{\sqrt{2}}|0> + \frac{1}{\sqrt{2}}|1>.$$

The qubit state can exist in a continuum of states between |0> and |1> until measurements are made. When the qubit is measured it collapses to the state |0>, $|\frac{1}{\sqrt{2}}|^2 = 0.5 \rightarrow 50\%$ of the time and to the state |1>, $|\frac{1}{\sqrt{2}}|^2 = 0.5 \rightarrow 50\%$ of the time.

(2) Consider electronic states in an atom.

|0> = ground state and |1> = excited state, as shown in figure 4.1.

Transitions like |0>→|1> or |1 > →|0> are possible. Also, an electron in |0> (say) can be moved to |ψ>, which is halfway between |0> and |1>, i.e. in the qubit state given by

$$|\psi> = \alpha_0|0> + \alpha_1|1> \qquad \text{(taking } |\alpha_0|^2 + |\alpha_1|^2 = 1).$$

4.2 Qubit model applied to the Stern–Gerlach experiment

Let us try to describe the spin state by the qubit model using the Stern–Gerlach set up. We use a bistable quantum system (which can exist in one of the two possible states at a time), e.g. spin of electron, proton, neutron, silver atom, hydrogen atom etc.

If a spin magnetic dipole moment $\vec{\mu}$ is placed in a magnetic induction $\vec{B}$ it gains potential energy

$$U = -\vec{\mu}.\vec{B} = -\mu B \cos \theta \ [\theta = \text{angle between } \vec{\mu} \text{ and } \vec{B}]$$

and the force is

$$\vec{F} = -\vec{\nabla}U = -\vec{\nabla}(-\vec{\mu}.\vec{B}) = \vec{\nabla}(\vec{\mu}.\vec{B}) = \vec{\nabla}(\mu B \cos \theta)$$

Taking inhomogeneous magnetic induction $\vec{B} = B(z)\hat{z} = (Bz)\hat{z}$, so that $\frac{\partial B}{\partial z} \neq 0$

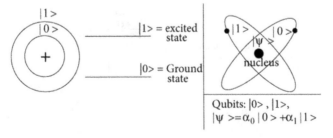

Figure 4.1. Electronic states in an atom.

$$\vec{F} = \hat{z}\frac{d}{dz}[\mu(Bz)\cos\theta] = \hat{z}\mu B \cos\theta$$

If $\vec{\mu}$ is along $\vec{B}$ (spin up state) then

$\theta = 0$, $\cos 0° = 1$ and $U = -\mu B$, $\vec{F} = \hat{z}\mu B \Rightarrow$ denote by $|+z>$

If $\vec{\mu}$ is against $\vec{B}$ (spin down state) then

$\theta = \pi$, $\cos \pi = -1$ and $U = \mu B$, $\vec{F} = -\hat{z}\mu B \Rightarrow$ denote by $|-z>$

Clearly we have shown that if the quantum state is put in a magnetic induction the output will show two peaks |up spin> and |down spin> denoted by $|+z>$, $|-z>$, respectively, when inhomogeneous magnetic induction $B = B(z)$ is along $\hat{z}$.

Consider a beam of atomic dipoles passing through the inhomogeneous magnetic induction $\vec{B} = B(z)\hat{z}$ of Stern–Gerlach apparatus (figure 4.2). The beam splits into two daughter beams viz. $|+z>$ (along the field gradient) and $|-z>$ (against the field gradient).

In the thought experiment let us block $|-z>$ beam but pass $|+z>$ beam again through another Stern–Gerlach apparatus having inhomogeneous magnetic induction $\vec{B} = B(z)\hat{z}$. We get one beam out viz. $|+z>$ (i.e. one central peak).

Now we pass $|+z>$ through another Stern–Gerlach apparatus having inhomogeneous magnetic induction $\vec{B} = B(x)\hat{x}$. We now get two peaks of equal intensity, $|+x>$ and $|-x>$. We can thus say that $|+z>$ state consists of equal portions of $|+x>$ and $|-x>$.

Blocking $|-x>$ we pass $|+x>$ through another Stern–Gerlach apparatus having inhomogeneous magnetic induction $\vec{B} = B(z)\hat{z}$. We now get two peaks of equal intensity $|+z>$ and $|-z>$. We can thus say that $|+x>$ state consists of equal portions of $|+z>$ and $|-z>$.

Let us use the qubit model to explain the result of passing a beam of atomic dipoles through the inhomogeneous magnetic induction in Stern–Gerlach apparatus.

(1) We explain the emergence of $|+z>$ and $|-z>$ from the fourth Stern–Gerlach set of apparatus having inhomogeneous magnetic induction $\vec{B} = B(z)\hat{z}$ from the input beam $|+x>$ in the qubit model.

We identify $|+z>=|0>$ and $|-z>=|1>$ as the computational basis states. Express $|+x> = |\psi>$ as a linear combination of computational basis states as

$$|\psi> = \alpha_0|0> + \alpha_1|1> \qquad (\text{i.e. } |+x> = \alpha_0|+z> + \alpha_1|-z>)$$

The output beams ($|+z>$ and $|-z>$) are of equal intensity $\alpha_0 = \frac{1}{\sqrt{2}}$, $\alpha_1 = \frac{1}{\sqrt{2}}$ which will lead to probability of spin state $|+z>$ to be $|\alpha_0|^2 = \frac{1}{2}$

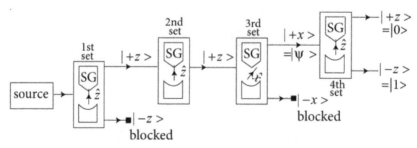

Figure 4.2. Cascaded Stern–Gerlach (SG) experiment.

and probability of spin state $|-z>$ to be $|\alpha_1|^2 = \frac{1}{2}$ and also ensuring that $|\alpha_0|^2 + |\alpha_1|^2 = \frac{1}{2} + \frac{1}{2} = 1$ (normalized to unity). Clearly the expansion

$$|+x> = \frac{1}{\sqrt{2}}|+z> + \frac{1}{\sqrt{2}}|-z>$$

in the qubit model, using computational basis, helps explain the observed facts.

(2) We explain the emergence of $|+x>$ and $|-x>$ from the third Stern–Gerlach set of apparatus having inhomogeneous magnetic induction $\vec{B} = B(x)\hat{x}$ from the input beam $|+z>$ in the qubit model.

Let us work in the basis $|+x>$ and $|-x>$ given by

$$|+x> = \frac{1}{\sqrt{2}}(|+z> + |-z>) = \frac{1}{\sqrt{2}}(|0> + |1>) \quad (as |+z> = |0>, |-z> = |1>$$

$$|-x> = \frac{1}{\sqrt{2}}(|+z> - |-z>) = \frac{1}{\sqrt{2}}(|0> - |1>)$$

Express $|+z>$ as a linear combination of basis states $|+x>$ and $|-x>$ as

$$|+z> = A|+x> + B|-x>$$

Let us find the values of A, B. Rewriting in terms of the computational base states $|0>, |1>$ we have

$$|0> = A\left[\frac{1}{\sqrt{2}}(|0> + |1>)\right] + B\left[\frac{1}{\sqrt{2}}(|0> - |1>)\right] = \frac{A+B}{\sqrt{2}}|0> + \frac{A-B}{\sqrt{2}}|1>$$

$$\Rightarrow 1|0> + 0|1> = \frac{A+B}{\sqrt{2}}|0> + \frac{A-B}{\sqrt{2}}|1> \quad \Rightarrow A - B = 0 \Rightarrow A = B$$

Also, $\frac{A+B}{\sqrt{2}} = 1 \Rightarrow \frac{2A}{\sqrt{2}} = 1 \Rightarrow A = \frac{1}{\sqrt{2}}, B = \frac{1}{\sqrt{2}}$.

Hence the expression of $|+z>$ as a linear combination of basis states $|+x>$ and $|-x>$ becomes

$$|+z> = \frac{1}{\sqrt{2}}|+x> + \frac{1}{\sqrt{2}}|-x>$$

This is consistent with the experimental observation of equally intense output beams $|+x>$ and $|-x>$, since probability of $|+x>$ is $|\frac{1}{\sqrt{2}}|^2 = \frac{1}{2}$ and probability of $|-x>$ is $|\frac{1}{\sqrt{2}}|^2 = \frac{1}{2}$ and also normalization to unity condition is ensured since $\frac{1}{2} + \frac{1}{2} = 1$. Clearly the expansion $|+z> = \frac{1}{\sqrt{2}}|+x> + \frac{1}{\sqrt{2}}|-x>$ in the qubit model, using a different basis (other than computational basis) helps explain the observed facts.

The qubit model properly predicts results of a cascaded Stern–Gerlach experiment.

4.3 Qubit model applied to polarized photon (computational and Hadamard basis introduced)

Consider a photon travelling along $\hat{z}$. Such a photon has its direction of polarization (i.e. the electric field direction) either along $\hat{x}$ or along $\hat{y}$.

We take as computational base states the x-polarized state $|x> = |0>$ and the y-polarized state $|y> = |1>$ and

$$|0> = 1|0> +0|1> = \begin{pmatrix} 1 \\ 0 \end{pmatrix}; \quad |1> = 0|0> +1|1> = \begin{pmatrix} 0 \\ 1 \end{pmatrix}$$

which is an orthonormalized basis $<m|n> = \delta_{mn}$.

We can also describe things in terms of a different basis which correspond to polarization directions 45°, 135°. We denote them as (figure 4.3)

$$|+>=|45°>, \quad |->=|135°>.$$

We can express this $|+>, |->$ basis in terms of the computational basis as follows:

The state $|+>=|45°>$ has equal probability of being in the x-direction as well in the y-direction. So we can write

$$|+>=A|x> +B|y>$$

As the 45° line is symmetrically placed, i.e. equally inclined w.r.t $|x>$ and $|y>$ we take $A = B$ and hence

$$|+>=A|0> +A|1> = A\begin{pmatrix} 1 \\ 0 \end{pmatrix} + A\begin{pmatrix} 0 \\ 1 \end{pmatrix} = A\begin{pmatrix} 1 \\ 1 \end{pmatrix}$$

Normalization gives

$$< +|+>=1 \quad \Rightarrow A^*(1 \ 1)A\begin{pmatrix} 1 \\ 1 \end{pmatrix} = 2|A|^2 = 1 \quad \Rightarrow A = \frac{1}{\sqrt{2}}$$

$$|+>=\frac{1}{\sqrt{2}}|x> +\frac{1}{\sqrt{2}}|y> =\frac{1}{\sqrt{2}}|0> +\frac{1}{\sqrt{2}}|1> =\frac{1}{\sqrt{2}}\begin{pmatrix} 1 \\ 1 \end{pmatrix}$$

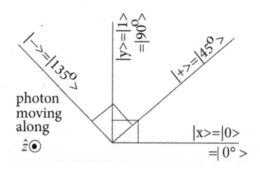

Figure 4.3. Photon polarization states.

As the 135° line is also symmetrically placed, i.e. equally inclined w.r.t $|x>$ and $|y>$ we can write for $|->=|135°>$

$$|->=C|x> -D|y> \text{ with } C = D.$$

$$| - >=C|0 > -C|1 > =C\begin{pmatrix}1\\0\end{pmatrix} - C\begin{pmatrix}0\\1\end{pmatrix} = C\begin{pmatrix}1\\-1\end{pmatrix}$$

Normalization gives

$$< - |->=1 \Rightarrow C^*(1 \ -1) \ C\begin{pmatrix}1\\-1\end{pmatrix} = 2|C|^2 = 1 \Rightarrow C = \tfrac{1}{\sqrt{2}}$$

$$| - >=\tfrac{1}{\sqrt{2}}|x > -\tfrac{1}{\sqrt{2}}|y > =\tfrac{1}{\sqrt{2}}|0 > -\tfrac{1}{\sqrt{2}}|1 > =\tfrac{1}{\sqrt{2}}\begin{pmatrix}1\\-1\end{pmatrix}$$

Consider light polarized in state $|i>$ passing through an f polarizer—that polarizes light along f direction. Transmitted light will be reduced in intensity by an amount $\cos^2 \theta$(which is Malus law) where θ is the angle between i and f directions. This result is related to the probability of i polarized photon passing through f polarizer, which is $|<f|i > |^2$. For instance, the probability of

(i) x polarized photon passing through x polarizer is

$$|<x|x > |^2 = | < 0|0 > |^2 = \left| (1 \ 0)\begin{pmatrix}1\\0\end{pmatrix}\right|^2 = 1 \Rightarrow 100\% \Rightarrow \text{Full Transmission}$$

(ii) x polarized photon passing through y polarizer is

$$|<y|x > |^2 = | < 1|0 > |^2 = \left| (0 \ 1)\begin{pmatrix}1\\0\end{pmatrix}\right|^2 = 0 \Rightarrow 0\% \Rightarrow \text{No Transmission}$$

(iii) x polarized photon passing through 45° polarizer is

$$|<45°|x > |^2 = | < +|0 > |^2 = \left| \tfrac{1}{\sqrt{2}}(1 \ 1)\begin{pmatrix}1\\0\end{pmatrix}\right|^2 = \tfrac{1}{2} \Rightarrow 50\% \Rightarrow 50\% \text{ Transmission}$$

We can also find the suitable Hermitian operators representing the corresponding experiment as follows.

Let the operator in the computational basis be $\hat{Z}=\begin{pmatrix}a & b\\c & d\end{pmatrix}$ having $|0>$ and $|1>$ as its eigenbasis with eigenvalues $+1$ and -1, respectively, i.e.

$$\hat{Z}|0> =|0 > , \ \hat{Z}|1 > =-|1 > .$$

$$\begin{pmatrix} a & b \\ c & d \end{pmatrix}\begin{pmatrix} 1 \\ 0 \end{pmatrix} = \begin{pmatrix} 1 \\ 0 \end{pmatrix} \ \Rightarrow \ \begin{pmatrix} a \\ c \end{pmatrix} = \begin{pmatrix} 1 \\ 0 \end{pmatrix} \ \Rightarrow \ a = 1, c = 0$$

$$\begin{pmatrix} a & b \\ c & d \end{pmatrix}\begin{pmatrix} 0 \\ 1 \end{pmatrix} = -\begin{pmatrix} 0 \\ 1 \end{pmatrix} \ \Rightarrow \ \begin{pmatrix} b \\ d \end{pmatrix} = -\begin{pmatrix} 0 \\ 1 \end{pmatrix} \ \Rightarrow \ b = 0, d = -1$$

$$\hat{Z} = \begin{pmatrix} 1 & 0 \\ 0 & -1 \end{pmatrix}$$

It is diagonal and hence it is in its eigen representation, i.e. in its eigenbasis $|0 >$, $|1>$.

Let the operator in the $|+>$, $| - >$ basis be $\hat{X} = \begin{pmatrix} p & q \\ r & s \end{pmatrix}$ having $|+>$ and $|->$ as its eigenbasis with eigenvalues $+1$ and -1, respectively, i.e.

$$\hat{X}|+>=| + >, \ \hat{X}|->=-| - >.$$

$$\begin{pmatrix} p & q \\ r & s \end{pmatrix}\frac{1}{\sqrt{2}}\begin{pmatrix} 1 \\ 1 \end{pmatrix} = \frac{1}{\sqrt{2}}\begin{pmatrix} 1 \\ 1 \end{pmatrix} \ \Rightarrow \ \begin{pmatrix} p + q \\ r + s \end{pmatrix} = \begin{pmatrix} 1 \\ 1 \end{pmatrix} \ \Rightarrow \ p + q = 1, \ r + s = 1$$

$$\begin{pmatrix} p & q \\ r & s \end{pmatrix}\frac{1}{\sqrt{2}}\begin{pmatrix} 1 \\ -1 \end{pmatrix} = -\frac{1}{\sqrt{2}}\begin{pmatrix} 1 \\ -1 \end{pmatrix} \ \Rightarrow \ \begin{pmatrix} p - q \\ r - s \end{pmatrix} = \begin{pmatrix} -1 \\ 1 \end{pmatrix} \ \Rightarrow \ p - q = -1, \ r - s = 1$$

Solving, we get $p = 0, q = 1, r = 1, s = 0$.

$$\hat{X} = \begin{pmatrix} 0 & 1 \\ 1 & 0 \end{pmatrix}$$

It is non-diagonal and hence it is not in its eigen representation.
- Note that: $\hat{X}|0 > =|1>$, $\hat{X}|1 > =|0>$.
- We mention that the basis $|0>$, $|1>$ is called computational basis while the basis $|+>$, $| - >$ is called the Hadamard basis.

Qutrit, Qudit
- For quantum information processing, we defined qubit which was a two-state quantum system. In principle for quantum information processing it is possible to think of three computational base states say triplet state of a system (atom). We then have qutrit (= three-state system).
- Similarly, for a D dimensional system the unit of information will be a qudit.

4.4 Bloch sphere representation of a qubit

The Bloch sphere is a three-dimensional unit sphere that geometrically represents qubit states as points on its surface, as shown in figure 4.4.

Qubits are represented as state vectors. Consider the qubit state

$$|\psi > =\alpha_0|0 > +\alpha_1|1>$$

where $|0>$ and $|1>$ are the orthonormalized computational basis and they form a complete set, i.e.

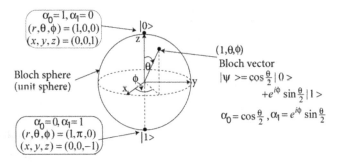

Figure 4.4. Bloch sphere representation of qubit.

$$|0><0| + |1><1| = I.$$

As $|\psi>$ is normalized we have

$$<\psi|\psi> = 1 \quad \Rightarrow \quad <\psi|\hat{I}|\psi> = 1$$
$$\Rightarrow \ <\psi| \ [|0><0| + |1><1|] \ |\psi> = 1$$
$$<\psi|0><0|\psi> +<\psi|1><1|\psi> = 1$$
$$\Rightarrow |<0|\psi>|^2 + |<1|\psi>|^2 = 1$$
$$\Rightarrow |\alpha_0|^2 + |\alpha_1|^2 = 1 \text{(called normalization)} \quad \text{(using } \alpha_0 = <0|\psi>, \ \alpha_1 = <1|\psi>\text{)}$$

Let us choose $\alpha_0 = \cos\frac{\theta}{2}$, $\alpha_1 = e^{i\phi}\sin\frac{\theta}{2}$ so that

$$\alpha_0^2 + \alpha_1^2 = \left| \cos\frac{\theta}{2} \right|^2 + \left| e^{i\phi}\sin\frac{\theta}{2} \right|^2 = \cos^2\frac{\theta}{2} + e^{-i\phi}e^{i\phi}\sin^2\frac{\theta}{2} = 1$$

and let $(r = 1, \theta, \phi)$ represent a point on 3D unit sphere which stands for the state vector of the qubit or the Bloch vector (figure 4.4).

$$|\psi> = \alpha_0|0> +\alpha_1|1> = \cos\frac{\theta}{2}|0> +e^{i\phi}\sin\frac{\theta}{2}|1>$$

$$= \cos\frac{\theta}{2}\begin{pmatrix}1\\0\end{pmatrix} + e^{i\phi}\sin\frac{\theta}{2}\begin{pmatrix}0\\1\end{pmatrix} = \begin{pmatrix}\cos\frac{\theta}{2}\\e^{i\phi}\sin\frac{\theta}{2}\end{pmatrix} = \begin{pmatrix}\alpha_0\\\alpha_1\end{pmatrix}$$

The angles θ, ϕ define the direction $\hat{r}$. The corresponding point on this unit sphere ($r = 1$) has Cartesian coordinates ($x = \sin\theta\cos\phi$, $y = \sin\theta\sin\phi$, $z = \cos\theta$) and represents the vector called Bloch vector (figure 4.4).

Bloch vectors corresponding to qubit $|0>$ are obtained as follows:

$$|\psi> = |0> \Rightarrow \alpha_0 = 1, \alpha_1 = 0. \quad \text{Hence} \quad \alpha_0 = \cos\frac{\theta}{2} = 1, \quad \alpha_1 = e^{i\phi}\sin\frac{\theta}{2} = 0. \quad \text{So}$$
$\theta = 0, \phi = 0$. Also, $x = \sin 0\cos 0 = 0$, $y = \sin 0\sin 0 = 0$, $z = \cos 0 = 1$. Hence,

$$|0> = |r = 1, \theta = 0, \phi = 0> = |x = 0, y = 0, z = 1>$$

Bloch vectors corresponding to qubit | 1> are obtained similarly

$|\psi> = |1> \Rightarrow \alpha_0 = 0, \alpha_1 = 1$.   Hence, $\alpha_0 = \cos\frac{\theta}{2} = 0$, $\alpha_1 = e^{i\phi}\sin\frac{\theta}{2} = 1$.   So

$\theta = \pi, \phi = 0$.   Also, $x = \sin\pi\cos 0 = 0$, $y = \sin\pi\sin 0 = 0, z = \cos\pi = -1$.
Hence,

$$|1> = |r = 1, \theta = \pi, \phi = 0> = |x = 0, y = 0, z = -1>$$

There are infinite points in the sphere and so we can think of an infinite number of qubits (corresponding to continuous range of θ s). But if we measure the qubit the state collapses to either |0> with probability $|\alpha_0|^2$ or to |1> with probability $|\alpha_1|^2$. This is as per the fundamental postulate of quantum mechanics (chapter 2).

Comment:
- Though on measurement we get only two pieces of information (|0> or |1>) corresponding to $\theta = 0$ and $\theta = \pi$ the qubit, corresponding to $\theta = 0$ to π can contain a wide range of information.
- All single qubits have a representation on Bloch sphere.
- Normalization of states is automatically ensured as Bloch sphere radius is unity and the state vectors lie on the surface of the unit sphere.
- If a single qubit state (i.e. a point on Bloch unit sphere) evolves with time unitarily it goes to another point but stays on the Bloch unit sphere. In other words, the unitary operator takes one single qubit state on a Bloch sphere to another point on the same Bloch sphere.
- Geometrically such transformation is a sequence of operations which are either rotation or reflection or both.
- There is no generalization of the Bloch sphere picture to multiple qubits.
- Opposite points in the Bloch sphere are orthogonal as we show in the following.

The state vector of a qubit corresponding to a point on Bloch sphere (θ, ϕ) is

$$|\psi> = \cos\frac{\theta}{2}|0> + e^{i\phi}\sin\frac{\theta}{2}|1>) = \begin{pmatrix} \cos\frac{\theta}{2} \\ e^{i\phi}\sin\frac{\theta}{2} \end{pmatrix}$$

The opposite point is obtained by reflection $(\theta, \phi) \rightarrow (\pi - \theta, \pi + \phi)$. So the Bloch vector corresponding to this opposite point is

$$|\psi'> = \cos\frac{\pi-\theta}{2}|0> + e^{i(\pi+\phi)}\sin\frac{\pi-\theta}{2}|1>)$$

$$= \cos\frac{\pi-\theta}{2}|0> - e^{i\phi}\sin\frac{\pi-\theta}{2}|1> = \begin{pmatrix} \cos\frac{\pi-\theta}{2} \\ -e^{i\phi}\sin\frac{\pi-\theta}{2} \end{pmatrix} = \begin{pmatrix} \sin\frac{\theta}{2} \\ -e^{i\phi}\cos\frac{\theta}{2} \end{pmatrix}$$

Consider the scalar product of $|\psi'>$ and $|\psi>$

$$<\psi'|\psi> = \left(\cos\frac{\pi-\theta}{2} \quad -e^{-i\phi}\sin\frac{\pi-\theta}{2}\right)\begin{pmatrix}\cos\dfrac{\theta}{2} \\ e^{i\phi}\sin\dfrac{\theta}{2}\end{pmatrix}$$

$$= \cos\frac{\pi-\theta}{2}\cos\frac{\theta}{2} - \sin\frac{\pi-\theta}{2}\sin\frac{\theta}{2} = \cos\left(\frac{\pi-\theta}{2}+\frac{\theta}{2}\right) = \cos\frac{\pi}{2} = 0$$

This establishes that the qubit states corresponding to opposite points of Bloch sphere are orthogonal.

- Show that the Bloch vectors $|0>$ and $|1>$ of figure 4.4 are orthogonal.

$$|\psi> = \begin{pmatrix}\cos\dfrac{\theta}{2} \\ e^{i\phi}\sin\dfrac{\theta}{2}\end{pmatrix} \xrightarrow{\theta=0°,\phi=0°} \begin{pmatrix}1 \\ 0\end{pmatrix} = |0>$$

and $$|\psi> = \begin{pmatrix}\cos\dfrac{\theta}{2} \\ e^{i\phi}\sin\dfrac{\theta}{2}\end{pmatrix} \xrightarrow{\theta=\pi,\phi=0°} \begin{pmatrix}0 \\ 1\end{pmatrix} = |1>$$

The scalar product is $<0|1> = (1 \quad 0)\begin{pmatrix}0 \\ 1\end{pmatrix} = 0 \Rightarrow |0>, |1>$ are orthogonal.

4.5 Multiple qubits

In classical computing, for one classical bit (0 or 1) we have $2^1 = 2$ states or codes, namely 0, 1. For 2 classical bits there will be $2^2 = 4$ states or codes, namely 00,01,10,11. For 3 classical bits there are $2^3 = 8$ states or codes namely 000,001,010,011, 100,101,110,111. For n classical bits there are 2^n states or codes. Only one state (i.e. a particular combination of bits) is effective at a time and there is nothing called superposition of states.

In quantum computing we can generalize from one qubit ($|0>$, $|1>$) to multiple qubits. For one qubit we have two computational base states $|0>$, $|1>$. And we need $2^1 = 2$ complex coefficients (α_0, α_1) to describe the quantum state as

$$|\psi> = \alpha_0|0> + \alpha_1|1> = \begin{pmatrix}\alpha_0 \\ \alpha_1\end{pmatrix}. \quad (\alpha_0, \alpha_1 \text{ are the expansion coefficients}).$$

Unlike the classical computational case, there is superposition of states in quantum computation.

For a 2-qubit system (e.g. 2 electrons in 2 atoms, figure 4.5) we would have $2^2 = 4$ computational 2-qubit base states viz. $|00>$, $|01>$, $|10>$, $|11>$. And we require $2^2 = 4$ complex coefficients (α_{00}, α_{01}, α_{10}, α_{11}) to describe the quantum state obtained through superposition of $2^2 = 4$ individual 2 qubit states as

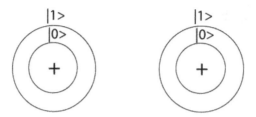

Figure   4.5. 2   electrons   in   2   atoms   can   lead   to   the   general   qubit
$|\psi> = \alpha_{00}|00> +\alpha_{01}|01> +\alpha_{10}|10> +\alpha_{11}|11>$.

$$|\psi> = \alpha_{00}|00> +\alpha_{01}|01> +\alpha_{10}|10> +\alpha_{11}|11> = \begin{pmatrix} \alpha_{00} \\ \alpha_{01} \\ \alpha_{10} \\ \alpha_{11} \end{pmatrix}$$

where α_{00}, α_{01}, α_{10}, α_{11} are the expansion coefficients. The probability of $|00>$ is $|\alpha_{00}|^2$, probability of $|01>$ is $|\alpha_{01}|^2$, probability of $|10>$ is $|\alpha_{10}|^2$ and probability of $|11>$ is $|\alpha_{11}|^2$ where

$$\sum_{ij}|\alpha_{ij}|^2 = 1$$

$\Rightarrow |\alpha_{00}|^2 + |\alpha_{01}|^2 + |\alpha_{10}|^2 + |\alpha_{11}|^2 = 1 =$ sum of probabilities (normalized to unity).

However, the measurement process will reveal 2 qubits of information (because any 2-qubit state is picked up, say either $|00>$ or $|01>$ or $|10>$ or $|11>$).

For 3 qubits we need $2^3 = 8$ basis states ($|000>$, $|001>$, $|010>$, $|011>$, $|100>$, $|101>$, $|110>$, $|111>$) and $2^3 = 8$ complex coefficients (α_{000}, α_{001}, α_{010}, α_{011}, α_{100}, α_{101}, α_{110}, α_{111}) to describe the superposed state through the expansion postulate. However, the measurement process will reveal 3 qubits of information (because any 3-qubit state is picked up, say either $|000>$ or $|001>$ or $|010>$ or $|011>$ or $|100>$ or $|101>$ or $|110>$ or $|111>$).

For n qubits we need 2^n computational basis states ($|00....0>$, $|00...1>$, ... $|11...1>$) and 2^n complex coefficients to describe the superposed quantum state through the expansion postulate. However, the measurement process will reveal n bits of information. (Note that in each ket of the computational basis there are n qubits.)

So most of the information remains hidden (in the superposed state which is a linear combination of 2^n qubit states) because when we measure, unitary development of quantum state is no longer there. Measurement is a discontinuous process as it disturbs the qubit and all the complex coefficients are lost. The system collapses to a particular n qubit state (revealing n qubit information).

Probability to measure the first qubit $|0>$ in a two-qubit state and identifying the post-measurement qubit state

Consider a two-qubit state

$$|\psi> = \alpha_{00}|00> +\alpha_{01}|01> +\alpha_{10}|10> +\alpha_{11}|11>$$

where α_{00}, α_{01}, α_{10}, α_{11} are the expansion coefficients with

$$\mid \alpha_{00} \mid^2 + \mid \alpha_{01} \mid^2 + \mid \alpha_{10} \mid^2 + \mid \alpha_{11} \mid^2 = 1.$$

If we make a measurement of a two-qubit state and get a first qubit value 0 then it means that the state of the qubit is either $|00>$ or $|01>$ (mutually exclusive states). So the probability of getting 0 is $\mid \alpha_{00} \mid^2 + \mid \alpha_{01} \mid^2$.

After this measurement the new state obtained is a mixture of $|00>$ and $|01>$ viz.
$|\psi'> = A[\alpha_{00}|00> + \alpha_{01}|01>]$

where A is normalization constant that can be obtained by demanding $<\psi'|\psi'> = 1$. This gives

$$A^*[\alpha_{00}^* < 00| + \alpha_{01}^* < 01|] \quad A[\alpha_{00}|00> + \alpha_{01}|01>] = 1$$

$$A^*A[|\alpha_{00}|^2 + |\alpha_{01}|^2] = 1 \quad \Rightarrow \quad |A|^2[|\alpha_{00}|^2 + |\alpha_{01}|^2] = 1$$

$$\Rightarrow A = \frac{1}{\sqrt{\mid \alpha_{00} \mid^2 + \mid \alpha_{01} \mid^2}}$$

So the new post-measurement state is

$$|\psi'> = \frac{\alpha_{00}|00> + \alpha_{01}|01>}{\sqrt{\mid \alpha_{00} \mid^2 + \mid \alpha_{01} \mid^2}}.$$

4.6 Explicit representation of the basis states

Computational basis (orthonormalized) for a single qubit state are

$$|0> = \begin{pmatrix} 1 \\ 0 \end{pmatrix}, \quad |1> = \begin{pmatrix} 0 \\ 1 \end{pmatrix}$$

Computational bases (orthonormalized) for a two-qubit state are
$|00>$, $|01>$, $|10>$, $|11>$, which are Kroneckar product (section 1.32) of the single qubit states and defined in state space $H \otimes H = H^{\otimes 2}$ (H is Hilbert space). Hence

$|00> = |0> \otimes |0>$

$$= \begin{pmatrix} 1 \\ 0 \end{pmatrix} \otimes \begin{pmatrix} 1 \\ 0 \end{pmatrix} = \begin{pmatrix} 1\begin{pmatrix} 1 \\ 0 \end{pmatrix} \\ 0\begin{pmatrix} 1 \\ 0 \end{pmatrix} \end{pmatrix} = \begin{pmatrix} 1 \\ 0 \\ 0 \\ 0 \end{pmatrix}$$

$|01> = |0> \otimes |1>$

$$= \begin{pmatrix} 1 \\ 0 \end{pmatrix} \otimes \begin{pmatrix} 0 \\ 1 \end{pmatrix} = \begin{pmatrix} 1\begin{pmatrix} 0 \\ 1 \end{pmatrix} \\ 0\begin{pmatrix} 0 \\ 1 \end{pmatrix} \end{pmatrix} = \begin{pmatrix} 0 \\ 1 \\ 0 \\ 0 \end{pmatrix}$$

$|10> = |1> \otimes |0>$

$$= \begin{pmatrix} 0 \\ 1 \end{pmatrix} \otimes \begin{pmatrix} 1 \\ 0 \end{pmatrix} = \begin{pmatrix} 0\begin{pmatrix} 1 \\ 0 \end{pmatrix} \\ 1\begin{pmatrix} 1 \\ 0 \end{pmatrix} \end{pmatrix} = \begin{pmatrix} 0 \\ 0 \\ 1 \\ 0 \end{pmatrix}$$

$|11> = |1> \otimes |1>$

$$= \begin{pmatrix} 0 \\ 1 \end{pmatrix} \otimes \begin{pmatrix} 0 \\ 1 \end{pmatrix} = \begin{pmatrix} 0\begin{pmatrix} 0 \\ 1 \end{pmatrix} \\ 1\begin{pmatrix} 0 \\ 1 \end{pmatrix} \end{pmatrix} = \begin{pmatrix} 0 \\ 0 \\ 0 \\ 1 \end{pmatrix}$$

So we have

$$|00> = \begin{pmatrix} 1 \\ 0 \\ 0 \\ 0 \end{pmatrix}, \; |01> = \begin{pmatrix} 0 \\ 1 \\ 0 \\ 0 \end{pmatrix}, \; |10> = \begin{pmatrix} 0 \\ 0 \\ 1 \\ 0 \end{pmatrix}, \; |11> = \begin{pmatrix} 0 \\ 0 \\ 0 \\ 1 \end{pmatrix}$$

The state space of a composite system (say a two qubit system $|\psi>$) is the tensor product of the state spaces of the component physical systems (i.e. single qubit states $|\psi_1>$, $|\psi_2>$, denoted by $H^{\otimes 2}$). Let us expand $|\psi_1>$, $|\psi_2>$ and write (α_1, β_1, α_2, β_2 are expansion coefficients)

$$|\psi_1> = \alpha_1 |0> + \beta_1 |1>$$
$$|\psi_2> = \alpha_2 |0> + \beta_2 |1>$$

and then expand $|\psi>$ as

$$|\psi> = |\psi_1> \otimes |\psi_2> = |\psi_1 \psi_2>$$
$$= \left(\alpha_1|0> + \beta_1|1> \right) \otimes \left(\alpha_2|0> + \beta_2|1> \right)$$
$$= \alpha_1\alpha_2|00> + \alpha_1\beta_2|01> + \beta_1\alpha_2|10> + \beta_1\beta_2|11>$$

In general, the two qubit state $|m\,n>$ is related to the component single qubit states $|m>$, $|n>$ through the direct product: $|m\,n> = |m> \otimes |n>$ (e.g. $|00> = |0> \otimes |0>$). But we mention that not all two-qubit states can be written as a product of two single-qubit states, e.g. the Bell state $|B_{00}> = \frac{1}{\sqrt{2}}(|00> + |11>)$ cannot be written as a product of two-qubit states due to entanglement property. We discuss Bell state in the following section.

4.7 Bell state or EPR pair (or state)

Bell states or EPR states (EPR standing for Einstein, Podolosky, Rosen) are quantum states of two qubits that demonstrate quantum entanglement (i.e. exhibit correlation at a distance). Bell states are defined as

$$|\text{Bell state}> = |B_{xy}> = \frac{1}{\sqrt{2}}(|0\,y> + (-1)^x|1\,\bar{y}>)$$

where $\bar{y}$ is the negation of y. Thus there are four Bell states corresponding to choices $(x, y) \to (0,0), (0,1), (1,0), (1,1)$ viz.

$$|B_{00}> = \frac{1}{\sqrt{2}}(|00> + |11>) \quad |B_{01}> = \frac{1}{\sqrt{2}}(|01> + |10>),$$
$$|B_{10}> = \frac{1}{\sqrt{2}}(|00> - |11>) \quad |B_{11}> = \frac{1}{\sqrt{2}}(|01> - |10>).$$

Explanation of entanglement property of Bell state

Consider any Bell state. Suppose we measure first qubit. We can get either 0 or 1 with equal probability. But when we measure this first qubit state, *the state of the second qubit is automatically determined.* This is entanglement. We illustrate this in the following.

(1) Consider the Bell state $|B_{00}> = \frac{1}{\sqrt{2}}(|00>+|11>)$. Probability of $|00>$ is $(\frac{1}{\sqrt{2}})^2 = \frac{1}{2}$, and probability of $|11>$ is also $(\frac{1}{\sqrt{2}})^2 = \frac{1}{2}$.

Measurement of the first qubit gives either 0 or 1.

Suppose we get 0 as the first qubit. Now 0 occurs only in $|00>$ (with probability $\frac{1}{2}$) leaving the post measurement state $|\psi' > = |00>$. The second qubit is clearly also 0 (same result as the first qubit).

Suppose we get 1 as the first qubit. Now 1 occurs only in $|11>$ (with probability $\frac{1}{2}$) leaving the post measurement state $|\psi' > = |11>$. The second qubit is clearly also 1 (same result as the first qubit).

Clearly as a result of measurement of first qubit, the second qubit always gives the same result as the first qubit. In other words when we measure the first qubit the second qubit is determined uniquely. So measurement outcomes are correlated or entangled in Bell state .

(2) Consider the Bell state $|B_{01}> = \frac{1}{\sqrt{2}}(|01 > +|10 >)$. Probability of $|01>$ is $(\frac{1}{\sqrt{2}})^2 = \frac{1}{2}$, and probability of $|10>$ is also $(\frac{1}{\sqrt{2}})^2 = \frac{1}{2}$.

Measurement of the first qubit gives either 0 or 1.

Suppose we get 0 as the first qubit. Now 0 in the first qubit occurs only in $|01>$ (with probability $\frac{1}{2}$) leaving the post measurement state $|\psi' > = |01>$. The second qubit is clearly 1 (the complement of the first qubit).

Suppose we get 1 as the first qubit. Now 1 in the first qubit occurs only in $|10>$ (with probability $\frac{1}{2}$) leaving the post measurement state $|\psi' > = |10>$. The second qubit is clearly 0 (the complement of the first qubit).

Clearly as a result of measurement of the first qubit, the second qubit always gives the complement of the first qubit. In other words when we measure the first qubit the second qubit is determined uniquely. So measurement outcomes are correlated or entangled in Bell state.

Comment:
- There exists a strong correlation between the two qubits. Such entanglement hints at the fact that quantum mechanics allows information processing beyond what is possible in the classical world.
- The state $|\psi > = \frac{1}{\sqrt{2}}(|00 > +|01 >)$ is not an entangled state.

Probability of $|00>$ is $(\frac{1}{\sqrt{2}})^2 = \frac{1}{2}$, and probability of $|01>$ is also $(\frac{1}{\sqrt{2}})^2 = \frac{1}{2}$.

Measurement of the first qubit gives 0. Now 0 in the first qubit occurs both in $|00>$ (with probability $\frac{1}{2}$) as well as in $|01>$ (with probability $\frac{1}{2}$). If the post measurement state is $|00>$ the second qubit is 0 (same as the first qubit) while if the post measurement state is $|01>$ the second qubit is 1 (complement of the first qubit). Clearly as a result of measurement of first qubit, the second qubit is not determined uniquely. So measurement outcomes are not correlated or entangled (as in Bell state).

- We can write the un-entangled state $|\psi> = \frac{1}{\sqrt{2}}(|00 > +|01>)$ as product of two component states as follows

$$|\psi> = \frac{1}{\sqrt{2}}(|00> +|01>) = |0> \otimes \frac{1}{\sqrt{2}}(|0 > +|1>)$$

explicitly showing the first qubit (which is 0) and the second qubits (which may be found to be either 0 or 1 on measurement).

Bell state cannot be written as a product state of two components states.

Let us try to write one Bell state, say $|B_{00}> = \frac{1}{\sqrt{2}}(|00 > +|11>)$ as a product state $|\psi_1\psi_2 > = |\psi_1 > \otimes |\psi_2>$. Expand

$$|\psi_1> = \alpha_1|0> +\beta_1|1>, \quad |\psi_2> = \alpha_2|0> +\beta_2|1>$$
$$|\psi_1\psi_2> = |\psi_1> \otimes |\psi_2> = (\alpha_1|0> +\beta_1|1>) \otimes (\alpha_2|0> +\beta_2|1>)$$
$$= \alpha_1\alpha_2|00> +\alpha_1\beta_2|01> +\beta_1\alpha_2|10> +\beta_1\beta_2|11>$$

For $|\psi_1\psi_2>$ to represent $|B_{00}>$ we have to choose

$$\alpha_1\alpha_2 = \frac{1}{\sqrt{2}}, \; \alpha_1\beta_2 = 0 \; \beta_1\alpha_2 = 0, \; \beta_1\beta_2 = \frac{1}{\sqrt{2}} \tag{4.1}$$

Now $\alpha_1\alpha_2 = \frac{1}{\sqrt{2}} \Rightarrow \alpha_1 \neq 0, \; \alpha_2 \neq 0$ and $\beta_1\beta_2 = \frac{1}{\sqrt{2}} \Rightarrow \beta_1 \neq 0, \; \beta_2 \neq 0$.

But $\alpha_1\beta_2 = 0, \; \beta_1\alpha_2 = 0 \Rightarrow$ at least some of $\alpha_1, \; \beta_1, \; \alpha_2, \; \beta_2$ are zero.

Clearly equation (4.1) does not lead to any meaningful solution for $\alpha_1, \; \beta_1, \; \alpha_2, \; \beta_2$. So

$$|B_{00} > \; \neq |\psi_1 > \otimes |\psi_2>$$

So we cannot represent the Bell state $|B_{00}>$ as a product state $|\psi_1 > \otimes |\psi_2>$. This is because Bell states are entangled. They are correlated in such a way that we cannot dissolve Bell state as product of two component states. They are always intricately coupled.

4.8 Global phase and relative phase

In the discussion we consider the single-qubit state. Consider an operator $\hat{A}$ and we wish to measure its expectation value in the states $|\psi>$ and $e^{i\theta}|\psi>$ (where θ is a real number, and $e^{i\theta}$ is a complex number of modulus unity). The expectation values are

$$<\psi|\hat{A}|\psi > \; and \; < \psi|e^{-i\theta}\hat{A}e^{i\theta}|\psi > \; = <\psi|\hat{A}|\psi > .$$

So from an observational point of view the two states $|\psi>$ and $e^{i\theta}|\psi>$ are identical. Also, since $|e^{i\theta}| = 1$ and $<\psi|\psi > \; = <\psi|e^{-i\theta}e^{i\theta}|\psi > \; = 1$ it follows that $|\psi>$ and $e^{i\theta}|\psi>$ are both normalized to unity (have the same length in Hilbert space). The two states $|\psi>$ and $e^{i\theta}|\psi>$ are said to differ by the phase factor $e^{i\theta}$ called global phase factor (and θ is called the global phase). This global phase factor is irrelevant and insignificant so far as the observed properties of a physical system are concerned. In other words states differing in global phase factor are physically equivalent.

Consider the single qubit state $|\psi>$ in the standard computational basis $|0>$, $|1>$ $|\psi> = \alpha_0|0> + \alpha_1|1>$ where α_0, α_1 are complex numbers.

Relative phase shift refers to the angle ψ in the complex plane between the two complex numbers α_0, α_1 as is clear from the following expression (θ being the global phase shift).

Writing the complex numbers as $\alpha_0 = |\alpha_0|e^{i\theta}$, $\alpha_1 = |\alpha_1|e^{i\theta_1}$ we get

$$|\psi> = \alpha_0|0> + \alpha_1|1> = |\alpha_0|e^{i\theta}|0> + |\alpha_1|e^{i\theta_1}|1>$$
$$= e^{i\theta}(|\alpha_0||0> + |\alpha_1|e^{i(\theta_1-\theta)}|1>) = e^{i\theta}(|\alpha_0||0> + |\alpha_1|e^{i\phi}|1>)$$

where $\phi = \theta_1 - \theta$

Now $e^{i\theta}$ is the insignificant overall phase factor called the global phase factor.

But $e^{i\phi}$ is the relative phase factor which is dependent on the choice of base. States which differ only by relative phases in some basis give rise to physically observable differences in measurement and so these states are not physically equivalent.

Accordingly the states

$$|\psi_1> = |\alpha_0||0> + |\alpha_1|e^{i\phi}|1> \text{ and } |\psi_2> = |\alpha_0||0> + |\alpha_1||1>$$

are not physically equivalent.

Another example:

The states $|\psi_1> = \frac{1}{\sqrt{2}}(|0>+|1>)$ and $|\psi_2> = \frac{1}{\sqrt{2}}(|0>+e^{i\phi}|1>)$ are not equivalent—they do not represent the same state.

Comment:
- Two states differing by a relative phase factor are different physical systems and evolve in different ways.
- Two states differing by a global phase factor are the same physical system.

4.9 Measurement depends on choice of basis

- Let us express the state $|\psi>$ in computational basis $|0>$, $|1>$

$$|\psi> = \frac{1}{\sqrt{2}}(|0> + |1>)$$

The probability of qubit $|0>$ is $|\frac{1}{\sqrt{2}}|^2 = \frac{1}{2}$ and the probability of qubit $|1>$ is $|\frac{1}{\sqrt{2}}|^2 = \frac{1}{2}$.

- Let us express the state $|\psi>$ in the basis

$$|+> = \frac{1}{\sqrt{2}}(|0>+|1>), \quad |->= \frac{1}{\sqrt{2}}(|0>-|1>)$$

$$|\psi> = \frac{1}{\sqrt{2}}(|0>+|1>) = |+> = 1|+> + 0|->.$$

The probability of the qubit $|+>$ is 1 and that of the qubit $|->$ is 0.

Clearly measurement depends on choice of basis.

Another example

Suppose in the two-qubit computational basis, the state $|\psi>$ is given by

$$|\psi> = \frac{1}{\sqrt{7}}|00> + \sqrt{\frac{2}{7}}|01> + \sqrt{\frac{3}{7}}|10> + \frac{1}{\sqrt{7}}|11> \qquad (4.2)$$

The probability of $|00>$ is $|\frac{1}{\sqrt{7}}|^2 = \frac{1}{7}$, probability of $|01>$ is $|\sqrt{\frac{2}{7}}|^2 = \frac{2}{7}$, probability of $|10>$ is $|\sqrt{\frac{3}{7}}|^2 = \frac{3}{7}$, probability of $|11>$ is $|\frac{1}{\sqrt{7}}|^2 = \frac{1}{7}$.

Probability of getting 0 in the first qubit is

$=$ Probability of 0 in $|00>+$ probability of 0 in $|01>=|\frac{1}{\sqrt{7}}|^2 + |\sqrt{\frac{2}{7}}|^2 = \frac{3}{7} = |\sqrt{\frac{3}{7}}|^2$

Probability of getting 1 in the first qubit is

$=$ Probability of 1 in $|10>+$ probability of 1 in $|11>= |\sqrt{\frac{3}{7}}|^2 + |\frac{1}{\sqrt{7}}|^2 = \frac{4}{7} = |\sqrt{\frac{4}{7}}|^2$

In view of this we can rewrite $|\psi>$ of equation (4.2) as

$$|\psi> = \left(\sqrt{\frac{3}{7} \cdot \frac{1}{3}}|00> + \sqrt{\frac{3}{7} \cdot \frac{2}{3}}|01> \right) + \left(\sqrt{\frac{4}{7} \cdot \frac{3}{4}}|10> + \sqrt{\frac{4}{7} \cdot \frac{1}{4}}|11> \right)$$

$$|\psi> = \sqrt{\frac{3}{7}}|0> \left(\sqrt{\frac{1}{3}}|0> + \sqrt{\frac{2}{3}}|1> \right) + \sqrt{\frac{4}{7}}|1> \left(\sqrt{\frac{3}{4}}|0> + \sqrt{\frac{1}{4}}|1> \right)$$

It is evident from the structure of $|\psi>$ that if measurement on the first qubit yields $|0>$ then the second qubit collapses to the state $\sqrt{\frac{1}{3}}|0> + \sqrt{\frac{2}{3}}|1>$. If measurement on the first qubit yields $|1>$ then the second qubit collapses to the state $\sqrt{\frac{3}{4}}|0> + \sqrt{\frac{1}{4}}|1>$.

4.10 Further reading

[1] Nielsen M A and Chuang I L 2019 *Quantum Computation and Quantum Information* (Cambridge: Cambridge University Press)
[2] Ghosh D 2016 Online Course on quantum information and computing (Bombay: Indian Institute of Technology)
[3] Nakahara M and Ohmi T 2016 *Quantum Computing* (Boca Raton, FL: CRC Press)
[4] Pathak A 2016 *Elements of Quantum Computing and Quantum Communication* (Boca Raton, FL: CRC Press)

4.11 Problems

1. *Identify Bell states*
    (a) $\frac{1}{\sqrt{2}}(|0> \pm |1>)$
    (b) $\frac{1}{\sqrt{2}}(|00> + |10>)$
    (c) $\frac{1}{\sqrt{2}}(|01> - |11>)$
    (d) $\frac{1}{\sqrt{2}}(|10> + |1>)$
    Ans. *(d)*

2. *Identify the qubits from the following states.*
    (i) $0.5|0> +0.5|1>$
    (ii) $\cos^2\theta|0> -\sin^2\theta|1>$
    (iii) $0.8|0> +0.6|1>$

Ans.

(i) Consider $|\psi> =0.5|0> +0.5|1>$.
   Since $|\,0.5\,|^2 + |\,0.5\,|^2 = 0.5 \neq 1$ it follows that this $|\psi>$ does not represent a qubit state. Normalization leads to the qubit state

$$|\psi> = \frac{0.5|0> +0.5|1>}{\sqrt{(0.5)^2 + (0.5)^2}} = \frac{1}{\sqrt{2}}(|0> +|1>)$$

(ii) Consider $|\psi> =\cos^2\theta|0> -\sin^2\theta|1>$.
   We demand that $|\cos^2\theta|^2 + |\sin^2\theta|^2 = 1 \Rightarrow \cos^4\theta + \sin^4\theta = 1$
   This holds for $\theta = 0, \pm\frac{\pi}{2}, \pm\pi$. So it follows that this $|\psi>$ represents a qubit state for these specified θ values.
   Normalization leads to the qubit state, valid for all θ as

$$\frac{\cos^2\theta|0> -\sin^2\theta|1>}{\sqrt{|\cos^2\theta|^2 + |\sin^2\theta|^2}} = \frac{\cos^2\theta|0> -\sin^2\theta|1>}{\sqrt{\cos^4\theta + \sin^4\theta}}$$

(iii) Consider $|\psi> =0.8|0> +0.6|1>$.
   Since $|\,0.8\,|^2 + |\,0.6\,|^2 = 1$, it follows that this $|\psi>$ represents a qubit state.

3. *Identify the local and global phase factors in the following expression*

$$|\psi> =e^{i\theta}(|\,\alpha_0\,|\,|0> +|\,\alpha_1\,|\,e^{i\phi}\,|1>)$$

    (a) θ, ϕ (b) ϕ, θ (c) $0, \phi$ (d) $\theta, 0$
    Ans. *(b)*

4. *If a measurement is made in the state $\frac{1}{\sqrt{13}}(3|0> -2i|1>)$ in the diagonal basis, with what probability we would get the state $|+>$?*
    Ans. $|\psi> =\frac{1}{\sqrt{13}}(3|0> -2i|1>)$, $|0> =\frac{|+>+|->}{\sqrt{2}}$, $|1> =\frac{|+>-|->}{\sqrt{2}}$

$$|\psi> =\frac{3}{\sqrt{13}}\frac{|+>+|->}{\sqrt{2}} - \frac{2i}{\sqrt{13}}\frac{|+>-|->}{\sqrt{2}} = \frac{1}{\sqrt{26}}[(3-2i)|+>+(3+2i)|->]$$

$$P(\,|+>\,) = \left|\frac{3-2i}{\sqrt{26}}\right|^2 = \frac{3-2i}{\sqrt{26}}\frac{3+2i}{\sqrt{26}} = \frac{3^2+2^2}{26} = \frac{13}{26} = \frac{1}{2}$$

5. *If a measurement is made in the state $\frac{1}{5}(4i|0> -3|1>)$ in the diagonal basis, with what probability would we get the state $|->$?*
    Ans: $|\psi> =\frac{1}{5}(4i|0> -3|1>)$, $|0> =\frac{|+>+|->}{\sqrt{2}}$, $|1> =\frac{|+>-|->}{\sqrt{2}}$

$$|\psi> = \frac{4i}{5}\frac{|+>+|->}{\sqrt{2}} - \frac{3}{5}\frac{|+>-|->}{\sqrt{2}} = \frac{1}{5\sqrt{2}}[(4i-3)|+>+(4i+3)|->]$$

$$P(|->) = \left| \frac{4i+3}{5\sqrt{2}} \right|^2 = \frac{4i+3}{5\sqrt{2}}\frac{4i-3}{5\sqrt{2}} = \frac{4^2+3^2}{26} = \frac{25}{50} = \frac{1}{2}$$

6. *Identify the state which cannot be written as a product of two component states.*

    (a) $\frac{1}{\sqrt{2}}(|00> + |01>)$ *(b)* $\frac{1}{\sqrt{2}}(|00> + |11>)$
    (c) $\frac{1}{\sqrt{2}}(|00> - |11>)$ *(d)* $\frac{1}{\sqrt{2}}(|01> + |10>)$.

    Ans. *(b),(c),(d) (Bell states).*

5. *In Bloch sphere |0> and |1> represent*

    (a) parallel vectors, (b) orthogonal vectors,
    (c) antiparallel vectors, (d) identical vectors.

    Ans. *(b).*

6. *In Bloch sphere |0> and |1> correspond to which coordinates in (r, θ, ϕ) system?*

    (a) $(0,0,1),\ \ (0,0,-1)$ *(b)* $(0,0,-1),\ \ (0,0,1)$
    (c) $(1,0,0),\ \ (1, \pi, 0)$ *(d)* $(1, \pi, 0),\ \ (1,0,0)$.

    Ans. *(c).*

7. *Which is not true?*

    (a) All two-qubit states can be written as a product of two single qubit states.

    (b) All two-qubit states cannot be written as a product of two single qubit states.

    (c) Only Bell states can be written as a product of two single-qubit states.

    (d) Bell states cannot be written as a product of two single-qubit states.

    Ans. *(a), (c).*

8. *Bell states cannot be written as product of two qubit states because of which of the following?*

    (a) Entanglement property, (b) Linear superposition principle,
    (c) Reversible computation, (d) Parallelism property.

    Ans. *(a).*

9. *What are used in quantum computing?*

    (a) cbit, (b) qubit, (c) both, (d) none.

    Ans. *(b).*

IOP Publishing

Quantum Optics and Quantum Computation
An introduction
Dipankar Bhattacharyya and Jyotirmoy Guha

Chapter 5

Quantum circuits

5.1 Quantum gate and quantum circuit

Classical computer is built from electrical circuits that contain wires which carry signal or information and classical logic gates (NOT, OR, AND, XOR, NOR, XNOR, NAND etc) that perform various operations and manipulations on the fed signal.

A quantum computer is built from a quantum circuit that contains quantum wires (to carry information) and quantum logic gates (to manipulate quantum information, i.e. to manipulate the quantum state of a qubit or a series of qubits).

A quantum gate is represented by a matrix that operates on a state or qubit. Quantum computation must be performed reversibly and the corresponding gates used must perform operation on a state unitarily.

In classical computation inputs are classical bits (say n classical bits) and outputs are also classical bits (say m classical bits).

In quantum computation we have to deal with quantum states (instead of classical bits), i.e. qubits (single or multiple). Logic flow involves arrival of an input qubit state (which is a quantum state) that is operated upon by a unitary operator (represented by a unitary quantum logic gate) and an output qubit state (i.e. output quantum state) is obtained, following the laws or postulates of quantum mechanics.

Constraint on matrix representation of quantum gate

We now go to the various quantum gates along with their matrix representation. We take the input and output quantum states to be both of the form $\begin{pmatrix} \alpha \\ \beta \end{pmatrix}$ and so we conclude that the operation of quantum gates on a single qubit is to be described by 2×2 matrices which are unitary matrices, the defining relation of a unitary matrix $\hat{U}$ being $\hat{U}^\dagger \hat{U} = \hat{U} \hat{U}^\dagger = \hat{I}$. This is for carrying out necessary unitary transformations that preserve length of quantum state ket to unity as set in the normalization process.

doi:10.1088/978-0-7503-2715-2ch5

So if input state is normalized then after operation by a unitary matrix the output state will also be normalized—a basic requirement of quantum postulate.

5.2 Single-qubit gates

A single-qubit gate is a unitary operator that transforms a single-qubit state $|\psi>$ into another single-qubit state $\hat{U}|\psi>$. The unitary operator $\hat{U}$ can be represented by a 2×2 matrix (figure 5.1).

- In classical theory of computation there are 2 single-bit logic gates viz. identity gate and the logical NOT gate.
- In quantum computational theory there are multiple single-qubit quantum gates.

5.3 Quantum NOT gate or Pauli $\hat{X}$ gate $(\hat{\sigma}_x)$

A quantum NOT gate, denoted by $\hat{X}$, flips (i.e. transforms) the base states (figure 5.2). It is analogous to the classical NOT gate that negates the input (i.e. flips the bits).

$$|0> = \binom{1}{0} \xrightarrow{\hat{X}} |1> = \binom{0}{1} \text{ and } |1> = \binom{1}{0} \xrightarrow{\hat{X}} |0> = \binom{0}{1}.$$

In other words,

$$\hat{X}|0> = |1>, \quad \hat{X}|1> = |0>$$

Thus the matrix elements of $\hat{X} = \hat{\sigma}_x$ are (in the computational basis):

$X_{00} = <0|\hat{X}|0> = <0|1> = 0$ $X_{01} = <0|\hat{X}|1> = <0|0> = 1$

$X_{10} = <1|\hat{X}|0> = <1|1> = 1$ $X_{11} = <1|\hat{X}|1> = <1|0> = 0$

The quantum NOT $\hat{X}$ gate is the following 2×2 matrix in the computational basis

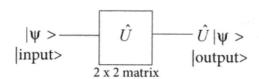

Figure 5.1. Arbitrary single-qubit gate $\hat{U}$.

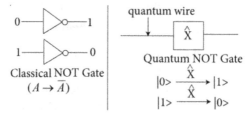

Figure 5.2. Classical NOT gate and quantum NOT gate.

$$\hat{X} = \begin{pmatrix} X_{00} & X_{01} \\ X_{10} & X_{11} \end{pmatrix} = \begin{pmatrix} 0 & 1 \\ 1 & 0 \end{pmatrix} = \hat{\sigma}_x$$

- Let us represent $\hat{X}$ in bra ket notation. Consider

$$\hat{X} = \begin{pmatrix} 0 & 1 \\ 1 & 0 \end{pmatrix} = \begin{pmatrix} 0 & 0 \\ 1 & 0 \end{pmatrix} + \begin{pmatrix} 0 & 1 \\ 0 & 0 \end{pmatrix} = \begin{pmatrix} 0 \\ 1 \end{pmatrix}(1 \ \ 0) + \begin{pmatrix} 1 \\ 0 \end{pmatrix}(0 \ \ 1)$$

$$\hat{X} = \ |1><0| + |0><1|$$

This is NOT gate in computational basis.

- The rule of writing follows from the fact that $|0> \xrightarrow{\hat{X}} |1>$ leads to the first term $|output><input|=|1><0|$ and similarly $|1> \xrightarrow{\hat{X}} |0>$ leads to the second term $|output><input|=|0><1|$

- With $\hat{X} = |1><0| + |0><1|$ let us have a check (using $<m|n> =\delta_{mn}$):

$$\hat{X}|0> =[|1><0|+|0><1| \ ] \ |0> =|1><0|0> +|0><1| \ 0> =|1>$$

$$\hat{X}|1> =[|1><0|+|0><1| \ ] \ |1> = \ |1><0|1> +|0><1| \ 1> = \ |0>$$

- The quantum NOT gate accepts and operates on the state $|\psi>$ given by

$|\psi> = \alpha_0|0> + \alpha_1|1> = \begin{pmatrix} \alpha_0 \\ \alpha_1 \end{pmatrix}$ as follows

$\hat{X}|\psi> = \alpha_0\hat{X}|0> +\alpha_1\hat{X}|1> =\alpha_0|1> + \alpha_1|0>$ or equivalently

$$\hat{X}|\psi> = \begin{pmatrix} 0 & 1 \\ 1 & 0 \end{pmatrix}\begin{pmatrix} \alpha_0 \\ \alpha_1 \end{pmatrix} = \begin{pmatrix} \alpha_1 \\ \alpha_0 \end{pmatrix}$$

Clearly the quantum NOT gate acts linearly on state $|\psi>$. In other words the linear combination of $|0>$ and $|1>$ is preserved though the roles of $|0>$ and $|1>$ have been flipped.

✓ A classical NOT gate, however, cannot accept such an input which is a linear superposition of 0 and 1.

✓ Quantum wire

Two quantum NOT gates in a row cause two flips to reproduce the original.

$$|0> \xrightarrow{\hat{X}} |1> \xrightarrow{\hat{X}} |0> , \ |1> \xrightarrow{\hat{X}} |0> \xrightarrow{\hat{X}} |1>$$

i.e. $|0> \xrightarrow{\hat{X}\hat{X}=\hat{X}^2} |0>, |1> \xrightarrow{\hat{X}\hat{X}=\hat{X}^2} |1>$

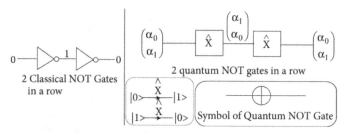

Figure 5.3. Two quantum NOT gates in a row is a quantum wire.

This corresponds to a quantum circuit which is equivalent to a quantum wire carrying a qubit (figure 5.3).

This is evident from the structure of the matrix $\hat{X}\hat{X} = \hat{X}^2$

$$\hat{X}\hat{X} = \begin{pmatrix} 0 & 1 \\ 1 & 0 \end{pmatrix}\begin{pmatrix} 0 & 1 \\ 1 & 0 \end{pmatrix} = \begin{pmatrix} 1 & 0 \\ 0 & 1 \end{pmatrix} = \hat{I}$$

where $\hat{I}$ = Identity matrix. Alternatively

$$\hat{X}^2|\psi> = \hat{X}\hat{X}|\psi> = \hat{X}\hat{X}(\alpha_0|0> +\alpha_1|1>) = \hat{X}(\alpha_0\hat{X}|0> +\alpha_1\hat{X}|1>)$$

$$= \hat{X}(\alpha_0|1> +\alpha_1|0>) = \alpha_0\hat{X}|1> +\alpha_1\hat{X}|0> =\alpha_0|0> +\alpha_1|1>$$

$$\Rightarrow \hat{X}^2|\psi> =|\psi> \text{ , i.e. original state } |\psi> \text{ is restored. Also}$$

$$\hat{X}^2|0> =\hat{X}|1> =|0> , \quad \hat{X}^2|1> =\hat{X}|0> =|1>$$

This is evident from matrix multiplication also since

$$\hat{X}^2\begin{pmatrix} \alpha_0 \\ \alpha_1 \end{pmatrix} = \begin{pmatrix} 0 & 1 \\ 1 & 0 \end{pmatrix}\begin{pmatrix} 0 & 1 \\ 1 & 0 \end{pmatrix}\begin{pmatrix} \alpha_0 \\ \alpha_1 \end{pmatrix} = \begin{pmatrix} 1 & 0 \\ 0 & 1 \end{pmatrix}\begin{pmatrix} \alpha_0 \\ \alpha_1 \end{pmatrix} = \begin{pmatrix} \alpha_0 \\ \alpha_1 \end{pmatrix}$$

Quantum NOT Gate $\hat{X}$ is unitary.

$$\hat{X} = \begin{pmatrix} 0 & 1 \\ 1 & 0 \end{pmatrix}, \quad \hat{X}{\dagger} = \begin{pmatrix} 0 & 1 \\ 1 & 0 \end{pmatrix} = \hat{X}$$

$$\Rightarrow \hat{X}{\dagger}\hat{X} = \hat{X}^2 = \begin{pmatrix} 0 & 1 \\ 1 & 0 \end{pmatrix}\begin{pmatrix} 0 & 1 \\ 1 & 0 \end{pmatrix} = \begin{pmatrix} 1 & 0 \\ 0 & 1 \end{pmatrix} = \hat{I}. \text{ So } \hat{X} \text{ is unitary.}$$

Comment

✓ Classical NOT Gate is the only classical single-bit gate.

✓ All other classical gates are multiple bit gates like AND, OR, XOR, NAND, NOR, XNOR.

✓ In quantum computing there are many single-qubit gates like $\hat{X}$ (Pauli $\hat{X}$ gate), $\hat{Y}$ (Pauli $\hat{Y}$ gate), $\hat{Z}$ (Pauli $\hat{Z}$ gate), $\hat{H}$ (Hadamard gate) as well as some multiple qubit quantum logic gates like Controlled-NOT gate or CNOT gate etc. We discuss these in this chapter.

✓ Reversible operation of $\hat{X}$ gate is clear since

$$|0> \xrightarrow{\hat{X}} |1> \quad |1> \xrightarrow{\hat{X}} |0> \quad \Rightarrow \quad |0> \xleftrightarrow{\hat{X}} |1>$$

NOT gate in the Hadamard basis

Let us denote the $\hat{X}$ gate (in the computational basis) as 'NOT' gate in the Hadamard basis $\{|+>, |->\}$. So NOT will transform $|+> \xleftrightarrow{\text{NOT}} |->$ and $|-> \xleftrightarrow{\text{NOT}} |+>$. We can thus write NOT gate in the bra ket notation as

$$\text{NOT} = |-><+| \quad + \quad |+><-|$$

Check:

$$\text{NOT}|+>=(|-><+| + |+><-|)|+> = |-><+|+> + |+><-|+> = |->$$

$$\text{NOT}|->=(|-><+| + |+><-|)|-> = |-><+|-> + |+><-|-> = |+>$$

In Hadamard basis we have $|+> = \frac{1}{\sqrt{2}}\binom{1}{1}, |->=\frac{1}{\sqrt{2}}\binom{1}{-1}$ (section 4.3). Hence the matrix representation of NOT is

$$\text{NOT} = |-><+| \quad + \quad |+><-| = \frac{1}{\sqrt{2}}\binom{1}{-1}\frac{1}{\sqrt{2}}(1 \ 1) + \frac{1}{\sqrt{2}}\binom{1}{1}\frac{1}{\sqrt{2}}(1 \ -1)$$

$$= \frac{1}{2}\begin{pmatrix} 1 & 1 \\ -1 & -1 \end{pmatrix} + \frac{1}{2}\begin{pmatrix} 1 & -1 \\ 1 & -1 \end{pmatrix} = \frac{1}{2}\begin{pmatrix} 2 & 0 \\ 0 & -2 \end{pmatrix} = \begin{pmatrix} 1 & 0 \\ 0 & -1 \end{pmatrix}$$

NOT gate is a diagonal matrix. So $\hat{X}$ is diagonal in the Hadamard basis which is its eigen basis. The diagonal elements are the eigenvalues of $\hat{X}$ and the NOT matrix is the eigen representation of $\hat{X}$ since

$$\hat{X}|+>=\begin{pmatrix} 0 & 1 \\ 1 & 0 \end{pmatrix}\frac{1}{\sqrt{2}}\binom{1}{1} = \frac{1}{\sqrt{2}}\binom{1}{1} = |+>$$

$$\hat{X}|->=\begin{pmatrix} 0 & 1 \\ 1 & 0 \end{pmatrix}\frac{1}{\sqrt{2}}\binom{1}{-1} = \frac{1}{\sqrt{2}}\binom{-1}{1} = -\frac{1}{\sqrt{2}}\binom{1}{-1} = -|->$$

So $\hat{X}$ has eigen kets $|+>$ with eigenvalue $+1$ and $|->$ with eigenvalue -1.

Hence $\hat{X}_{non-diagonal} = \begin{pmatrix} 0 & 1 \\ 1 & 0 \end{pmatrix} \xrightarrow{diagonalization} \text{NOT}_{diagonal} = \begin{pmatrix} 1 & 0 \\ 0 & -1 \end{pmatrix}$

• $\text{NOT}|+>=\begin{pmatrix} 1 & 0 \\ 0 & -1 \end{pmatrix}\frac{1}{\sqrt{2}}\binom{1}{1} = \frac{1}{\sqrt{2}}\binom{1}{-1} = |-> \Rightarrow$ flips $|+>$ to $|->$

$\text{NOT}|+>=\begin{pmatrix} 1 & 0 \\ 0 & -1 \end{pmatrix}\frac{1}{\sqrt{2}}\binom{1}{-1} = \frac{1}{\sqrt{2}}\binom{1}{1} = |+> \Rightarrow$ flips$|->$ to$|+>$

5.4 $\hat{Z}$ gate or Pauli $\hat{Z}$ gate ($\hat{\sigma}_z$)

A quantum $\hat{Z}$ gate or Pauli $\hat{Z}$ gate is defined through the following operation
$\hat{Z}|0> =|0>$, $\hat{Z}|1> =-|1>$ the base states being $|0> = \begin{pmatrix} 0 \\ 1 \end{pmatrix}$ and $|1> = \begin{pmatrix} 1 \\ 0 \end{pmatrix}$.

Clearly $\hat{Z}$ changes the sign of $|1>$ but leaves $|0>$ unchanged, i.e. selectively provides a phase to the qubit $|1>$ (figure 5.4).

Thus the matrix elements of $\hat{Z}$ gate or Pauli $\hat{Z}$ gate are

$$Z_{11} = <0|\hat{Z}|0> =<0|0> =1 \quad Z_{12} = <0|\hat{Z}|1> = -<0|1> =0$$

$$Z_{21} = <1|\hat{Z}|0> =<1|0> =0 \quad Z_{22} = <1|\hat{Z}|1> =-<1|1> =-1$$

The quantum $\hat{Z}$ gate is a 2×2 matrix (diagonal in the computational basis)

$$\hat{Z} = \begin{pmatrix} Z_{11} & Z_{12} \\ Z_{21} & Z_{22} \end{pmatrix} = \begin{pmatrix} 1 & 0 \\ 0 & -1 \end{pmatrix}$$

$\hat{Z}$ operates on an arbitrary quantum state $|\psi> =\alpha_0|0> +\alpha_1|1>$ as

$$\hat{Z}|\psi> =\hat{Z}(\alpha_0|0> +\alpha_1|1>) = \alpha_0\hat{Z}|0> +\alpha_1\hat{Z}|1>$$

$$=\alpha_0|0> -\alpha_1|1> =\alpha_0|0> +e^{i\pi}\alpha_1|1> \quad (\text{as } e^{i\pi} = -1)$$

The sole effect of $\hat{Z}$ is to flip the relative phase between the computational base qubits $|0>$ and $|1>$.

- In bra ket notation we can write (*each term is* $|output><input|$)

$$\hat{Z} = |0> <0| - |1> <1| = \begin{pmatrix} 1 \\ 0 \end{pmatrix}(1 \ 0) - \begin{pmatrix} 0 \\ 1 \end{pmatrix}(0 \ 1) = \begin{pmatrix} 1 & 0 \\ 0 & 0 \end{pmatrix} - \begin{pmatrix} 0 & 0 \\ 0 & 1 \end{pmatrix} = \begin{pmatrix} 1 & 0 \\ 0 & -1 \end{pmatrix}$$

5.5 Pauli $\hat{Y}$ gate or $\hat{\sigma}_y$

We define Pauli $\hat{Y}$ gate as

$$\hat{Y} = i\hat{X}\hat{Z} = i\begin{pmatrix} 0 & 1 \\ 1 & 0 \end{pmatrix}\begin{pmatrix} 1 & 0 \\ 0 & -1 \end{pmatrix} = i\begin{pmatrix} 0 & -1 \\ 1 & 0 \end{pmatrix} = \begin{pmatrix} 0 & -i \\ i & 0 \end{pmatrix}$$

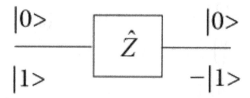

Figure 5.4. $\hat{Z}$ gate.

Let us see how $\hat{Y}$ gate operates on the computational base states

$$\hat{Y}|0> = \begin{pmatrix} 0 & -i \\ i & 0 \end{pmatrix}\begin{pmatrix} 1 \\ 0 \end{pmatrix} = \begin{pmatrix} 0 \\ i \end{pmatrix} = i\begin{pmatrix} 0 \\ 1 \end{pmatrix} = i|1> \Rightarrow |0> \xrightarrow{\hat{Y}} e^{i\pi/2}|1>$$

$$\hat{Y}|1> = \begin{pmatrix} 0 & -i \\ i & 0 \end{pmatrix}\begin{pmatrix} 0 \\ 1 \end{pmatrix} = \begin{pmatrix} -i \\ 0 \end{pmatrix} = -i\begin{pmatrix} 1 \\ 0 \end{pmatrix} = -i|0> \Rightarrow |1> \xrightarrow{\hat{Y}} - e^{i\pi/2}|1>$$

We note that $\hat{X}$ flips the qubit and $\hat{Z}$ flips the relative phase between the qubits. So $\hat{Y}$ is a combination of global phase factor of $i = e^{i\pi/2}$, bit flip (done by $\hat{X}$) and phase flip (done by $\hat{Z}$).

- In bra ket notation we can write (*each term is |output><input|*)

$$\hat{Y} = i|1> <0| - i|0> <1| = i\begin{pmatrix} 0 \\ 1 \end{pmatrix}(1 \quad 0) - i\begin{pmatrix} 1 \\ 0 \end{pmatrix}(0 \quad 1)$$

$$= i\begin{pmatrix} 0 & 0 \\ 1 & 0 \end{pmatrix} - i\begin{pmatrix} 0 & 1 \\ 0 & 0 \end{pmatrix} = i\begin{pmatrix} 0 & -1 \\ 1 & 0 \end{pmatrix} = \begin{pmatrix} 0 & -i \\ i & 0 \end{pmatrix}$$

- In general $\hat{X}$, $\hat{Y}$, $\hat{Z}$ are called Pauli gates.

5.6 Phase shift gates ($\hat{P}$ gate, $\hat{S}$ gate, $\hat{T}$ gate)

5.6.1 $\hat{P}$ gate

We introduce the phase shift gate called $\hat{P}$ gate that keeps $|0>$ unchanged but selectively imposes a phase shift of ϕ (i.e. introduces a phase factor $e^{i\phi}$) on $|1>$. $\hat{P}$ is also denoted as $\hat{P}(\phi)$.

$$\hat{P}|0> = |0> , \qquad \hat{P}|1> = e^{i\phi}|1> .$$

Thus the matrix elements are:

$$P_{11} = <0|\hat{P}|0> = <0|0> = 1 \quad P_{12} = <0|\hat{P}|1> = e^{i\phi} < 0|1> = 0,$$

$$P_{21} = <1|\hat{P}|0> = <1|0> = 0 \quad P_{22} = <1|\hat{P}|1> = e^{i\phi} < 1|1> = e^{i\phi}$$

$$\hat{P} = \hat{P}(\phi) = \begin{pmatrix} P_{11} & P_{12} \\ P_{21} & P_{22} \end{pmatrix} = \begin{pmatrix} 1 & 0 \\ 0 & e^{i\phi} \end{pmatrix}$$

- In bra ket notation we can write

$$\hat{P}(\phi) = |0> <0| + e^{i\phi}|1> <1| = \begin{pmatrix} 1 \\ 0 \end{pmatrix}(1 \quad 0) + e^{i\phi}\begin{pmatrix} 0 \\ 1 \end{pmatrix}(0 \quad 1) = \begin{pmatrix} 1 & 0 \\ 0 & e^{i\phi} \end{pmatrix}$$

As ϕ can assume any value, we can build any number of single-qubit gates (unlike the classical case where only 2 single-bit operations are needed viz. NOT gate and identity operation).

5.6.2 $\hat{S}$ gate

We can define a quantum gate for $\phi = \dfrac{\pi}{2}$ called $\hat{S}$ gate, i.e. $\hat{S} = \hat{P}(\dfrac{\pi}{2})$. Hence

$$\hat{S} = \hat{P}\Big(\frac{\pi}{2}\Big) = \begin{pmatrix} 1 & 0 \\ 0 & e^{i\phi} \end{pmatrix}_{\phi=\frac{\pi}{2}} = \begin{pmatrix} 1 & 0 \\ 0 & e^{i\frac{\pi}{2}} \end{pmatrix} = \begin{pmatrix} 1 & 0 \\ 0 & i \end{pmatrix}$$

$$\hat{S} = e^{i\frac{\pi}{4}}\begin{pmatrix} e^{-i\frac{\pi}{4}} & 0 \\ 0 & e^{i\frac{\pi}{4}} \end{pmatrix} = \hat{S} \text{ Phase gate.}$$

Let us see how $\hat{S}$ gate operates on the computational base states

$$\hat{S}|0> = \begin{pmatrix} 1 & 0 \\ 0 & i \end{pmatrix}\begin{pmatrix} 1 \\ 0 \end{pmatrix} = \begin{pmatrix} 1 \\ 0 \end{pmatrix} = |0>, \quad \hat{S}|1> = \begin{pmatrix} 1 & 0 \\ 0 & i \end{pmatrix}\begin{pmatrix} 0 \\ 1 \end{pmatrix} = \begin{pmatrix} 0 \\ i \end{pmatrix} = i\begin{pmatrix} 0 \\ 1 \end{pmatrix} = i|1>,$$

- In bra ket notation we can write

$$\hat{S} = |0><0| + i|1><1| = \begin{pmatrix} 1 \\ 0 \end{pmatrix}(1 \ \ 0) + i\begin{pmatrix} 0 \\ 1 \end{pmatrix}(0 \ \ 1) = \begin{pmatrix} 1 & 0 \\ 0 & 0 \end{pmatrix} + i\begin{pmatrix} 0 & 0 \\ 0 & 1 \end{pmatrix} = \begin{pmatrix} 1 & 0 \\ 0 & i \end{pmatrix}$$

5.6.3 $\hat{T}$ gate or $\dfrac{\pi}{8}$ gate

We can define a quantum gate for $\phi = \dfrac{\pi}{4}$ called $\hat{T}$ gate, i.e. $\hat{T} = \hat{P}(\dfrac{\pi}{4})$. Hence

$$\hat{T} = \hat{P}\Big(\frac{\pi}{4}\Big) = \begin{pmatrix} 1 & 0 \\ 0 & e^{i\phi} \end{pmatrix}_{\phi=\frac{\pi}{4}} = \begin{pmatrix} 1 & 0 \\ 0 & e^{i\frac{\pi}{4}} \end{pmatrix}$$

$$\hat{T} = e^{i\frac{\pi}{8}}\begin{pmatrix} e^{-i\frac{\pi}{8}} & 0 \\ 0 & e^{i\frac{\pi}{8}} \end{pmatrix} = \text{called } \hat{T} \text{ gate or } \frac{\pi}{8} \text{ gate.}$$

(Global phase factor is $e^{i\pi/8}$)

Clearly $\hat{T}|0> = |0>$, $\qquad \hat{T}|1> = e^{i\frac{\pi}{4}}|1>$

- In bra ket notation

$$\hat{T} = |0><0| + e^{i\pi/4}|1><1| = \begin{pmatrix} 1 & 0 \\ 0 & e^{i\frac{\pi}{4}} \end{pmatrix}$$

5.6.4 $\hat{Z}$ gate as a phase shift gate

Consider $\hat{P}(\pi) = \begin{pmatrix} 1 & 0 \\ 0 & e^{i\phi} \end{pmatrix}_{\phi=\pi} = \begin{pmatrix} 1 & 0 \\ 0 & e^{i\pi} \end{pmatrix} = \begin{pmatrix} 1 & 0 \\ 0 & -1 \end{pmatrix} = \hat{Z}$

5.6.5 Special $\hat{P}(\phi)$ gate: $\hat{R}_k = \hat{P}(\dfrac{2\pi}{2^k})$

$$\hat{R}_k = \hat{P}\Big(\frac{2\pi}{2^k}\Big) = \begin{pmatrix} 1 & 0 \\ 0 & e^{i\frac{2\pi}{2^k}} \end{pmatrix}$$

5.6.6 Inter-relations

Relation between $\hat{S}$ and $\hat{T}$ gates : $(\hat{S} = \hat{T}^2)$

Consider $\hat{T}^2 = \begin{pmatrix} 1 & 0 \\ 0 & e^{i\frac{\pi}{4}} \end{pmatrix} \begin{pmatrix} 1 & 0 \\ 0 & e^{i\frac{\pi}{4}} \end{pmatrix} = \begin{pmatrix} 1 & 0 \\ 0 & e^{i\frac{\pi}{2}} \end{pmatrix} = \begin{pmatrix} 1 & 0 \\ 0 & i \end{pmatrix} = \hat{S}$

$\hat{T}$ is not self-inverse but $\hat{Z}$ is self-inverse.

$\hat{T}^2 = \hat{S} = \begin{pmatrix} 1 & 0 \\ 0 & i \end{pmatrix} \neq \hat{I} \Rightarrow \hat{T}$ is not self-inverse.

$$\hat{Z} = \begin{pmatrix} 1 & 0 \\ 0 & -1 \end{pmatrix}, \ \hat{Z}^2 = \begin{pmatrix} 1 & 0 \\ 0 & -1 \end{pmatrix}\begin{pmatrix} 1 & 0 \\ 0 & -1 \end{pmatrix} = \begin{pmatrix} 1 & 0 \\ 0 & 1 \end{pmatrix} = \hat{I}$$

$\hat{Z}^2 = \hat{I} \Rightarrow \hat{Z}\hat{Z} = \hat{I} \Rightarrow \hat{Z}^{-1} = \hat{Z} \Rightarrow \hat{Z}$ is self-inverse (it is inverse of itself).

5.7 Hadamard gate $\hat{H}$, Hadamard basis $|+>, |->$

Pauli $\hat{X}$ gate is $\hat{X} = \begin{pmatrix} 0 & 1 \\ 1 & 0 \end{pmatrix}$ and Pauli $\hat{Z}$ gate is $\hat{Z} = \begin{pmatrix} 1 & 0 \\ 0 & -1 \end{pmatrix}$

Let us build the matrix operator

$$\hat{H} = \frac{1}{\sqrt{2}}(\hat{X} + \hat{Z}) = \frac{1}{\sqrt{2}}\left[\begin{pmatrix} 0 & 1 \\ 1 & 0 \end{pmatrix} + \begin{pmatrix} 1 & 0 \\ 0 & -1 \end{pmatrix} \right] = \frac{1}{\sqrt{2}}\begin{pmatrix} 1 & 1 \\ 1 & -1 \end{pmatrix}$$

This matrix operator represents a quantum gate called Hadamard gate. Let us evaluate the effect of the Hadamard gate (matrix operator) on the computational base states:

$$\hat{H}|0> = \frac{1}{\sqrt{2}}\begin{pmatrix} 1 & 1 \\ 1 & -1 \end{pmatrix}\begin{pmatrix} 1 \\ 0 \end{pmatrix} = \frac{1}{\sqrt{2}}\begin{pmatrix} 1 \\ 1 \end{pmatrix} = \frac{1}{\sqrt{2}}\left[\begin{pmatrix} 1 \\ 0 \end{pmatrix} + \begin{pmatrix} 0 \\ 1 \end{pmatrix}\right] = \frac{1}{\sqrt{2}}(|0> +|1>) = |+>$$

$$\hat{H}|1> = \frac{1}{\sqrt{2}}\begin{pmatrix} 1 & 1 \\ 1 & -1 \end{pmatrix}\begin{pmatrix} 0 \\ 1 \end{pmatrix} = \frac{1}{\sqrt{2}}\begin{pmatrix} 1 \\ -1 \end{pmatrix} = \frac{1}{\sqrt{2}}\left[\begin{pmatrix} 1 \\ 0 \end{pmatrix} - \begin{pmatrix} 0 \\ 1 \end{pmatrix}\right] = \frac{1}{\sqrt{2}}(|0> -|1>) = |->$$

Clearly Hadamard gate acts on one qubit and places it in a superposition of $|0>$ and $|1>$.

Starting from the computational basis $|0>$, $|1>$ we have generated a new basis $|+>, |->$ called Hadamard basis.

Let us evaluate the effect of the Hadamard gate on a superposition of states viz.

$$|\psi> = \alpha_0|0> +\alpha_1|1>$$

$$\hat{H}|\psi> = \hat{H}(\alpha_0|0> +\alpha_1|1>)$$

$$= \alpha_0\hat{H}|0> +\alpha_1\hat{H}|1>$$

$$= \alpha_0\frac{1}{\sqrt{2}}(|0> +|1>) + \alpha_1\frac{1}{\sqrt{2}}(|0> -|1>)$$

$$= \frac{1}{\sqrt{2}}(\alpha_0 + \alpha_1)|0> +\frac{1}{\sqrt{2}}(\alpha_0 - \alpha_1)|1>$$

Let us evaluate the effect of Hadamard gate $\hat{H}$ on the states $|\pm>=\frac{1}{\sqrt{2}}(|0> \pm |1>)$.

$$\hat{H}|+>=\hat{H}\frac{1}{\sqrt{2}}(|0>+|1>) = \frac{1}{\sqrt{2}}(\hat{H}|0>+\hat{H}|1>)$$

$$=\frac{1}{\sqrt{2}}\left[\left(\frac{1}{\sqrt{2}}(|0>+|1>)\right)+\left(\frac{1}{\sqrt{2}}(|0>-|1>)\right)\right] = |0>$$

$$\hat{H}|->=\hat{H}\frac{1}{\sqrt{2}}(|0>-|1>) = \frac{1}{\sqrt{2}}(\hat{H}|0>-\hat{H}|1>)$$

$$=\frac{1}{\sqrt{2}}\left[\left(\frac{1}{\sqrt{2}}(|0>+|1>)\right)-\left(\frac{1}{\sqrt{2}}(|0>-|1>)\right)\right] = |1>$$

So Hadamard gate produces the computational base states $|0>$, $|1>$ upon operating on the Hadamard basis $|\pm>=\frac{1}{\sqrt{2}}(|0> \pm |1>)$. This is summarized in the diagram of figure 5.5.

- GIST: $|0> \xrightarrow{\hat{H}} |+>$ $|1> \xrightarrow{\hat{H}} |->$

$$|+> \xrightarrow{\hat{H}} |0> |-> \xrightarrow{\hat{H}} |1>$$

Computational basis $\xleftrightarrow{\hat{H}}$ Hadamard basis

Double operation by Hadamard gate

$$\hat{H}^2 = \hat{H}\hat{H} = \frac{1}{\sqrt{2}}\begin{pmatrix} 1 & 1 \\ 1 & -1 \end{pmatrix}\frac{1}{\sqrt{2}}\begin{pmatrix} 1 & 1 \\ 1 & -1 \end{pmatrix}$$

$$=\frac{1}{2}\begin{pmatrix} 2 & 0 \\ 0 & 2 \end{pmatrix} = \begin{pmatrix} 1 & 0 \\ 0 & 1 \end{pmatrix} = \hat{I} = \text{identity matrix}$$

$$\hat{H}^2|0> =|0>, \hat{H}^2|1> =|1>$$

Clearly $\hat{H}^2$ reproduces the original. This quantum circuit is thus equivalent to a quantum wire carrying a qubit (figure 5.6).

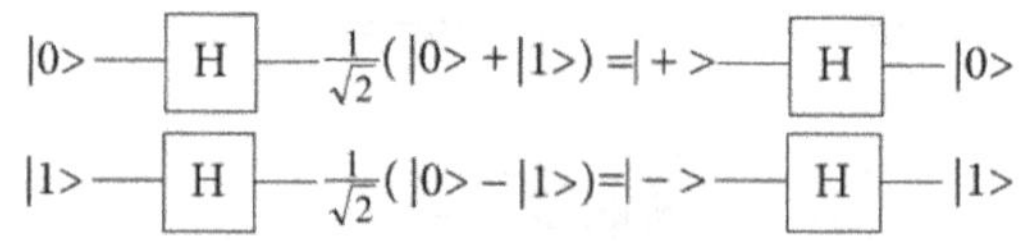

Figure 5.5. Operation by Hadamard gate.

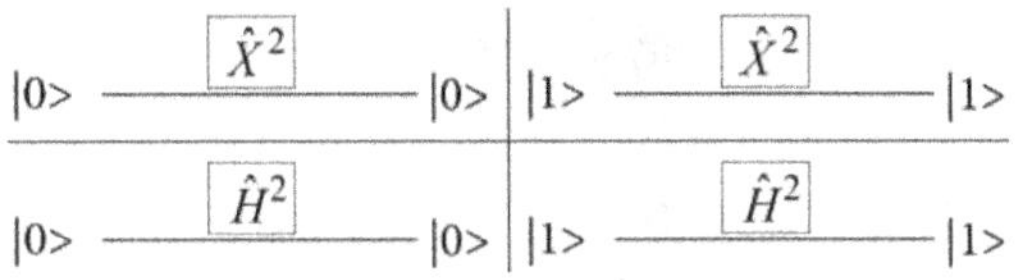

Figure 5.6. Quantum wire.

Comment

- A quantum wire carrying a qubit can be represented by two quantum NOT gates in a row (figure 5.3) or two Hadamard gates in a row (figure 5.6). They have been also shown in figure 5.6.
- Hadamard gate distinguishes the two states $|+>=\frac{1}{\sqrt{2}}(|0>+|1>)$ and $|->=\frac{1}{\sqrt{2}}(|0>-|1>)$ since the outputs are different viz.

$$\hat{H}|+>=\hat{H}\frac{1}{\sqrt{2}}(|0>+|1>)=|0>,\ \hat{H}|->=\hat{H}\frac{1}{\sqrt{2}}(|0>-|1>)=|1>$$

- **Hadamard operator in Bra ket notation**

The rule to build operator from qubit states is as follows. If an arbitrary single qubit gate $\hat{U}$ maps |0> to a single-qubit state $|\psi_0>$ and |1> to another single-qubit state $|\psi_1>$ then we can express $\hat{U}$ as (*each term is |output><input|*)

$$\hat{U}=|\psi_0><0|+|\psi_1><1|$$

Since $\hat{H}$ maps|0> to $\frac{1}{\sqrt{2}}(|0>+|1>)$ and |1> to $\frac{1}{\sqrt{2}}(|0>-|1>)$ we can express $\hat{H}$ as

$$\hat{H}=\frac{1}{\sqrt{2}}(|0>+|1>)<0|+\frac{1}{\sqrt{2}}(|0>-|1>)<1|$$

$$=\frac{1}{\sqrt{2}}(|0><0|+|1><0|+|0><1|-|1><1)|$$

$$=\frac{1}{\sqrt{2}}\left[\begin{pmatrix}1\\0\end{pmatrix}(1\ \ 0)+\begin{pmatrix}0\\1\end{pmatrix}(1\ \ 0)+\begin{pmatrix}1\\0\end{pmatrix}(0\ \ 1)-\begin{pmatrix}0\\1\end{pmatrix}(0\ \ 1)\right]$$

$$\hat{H}=\frac{1}{\sqrt{2}}\left[\begin{pmatrix}1&0\\0&0\end{pmatrix}+\begin{pmatrix}0&0\\1&0\end{pmatrix}+\begin{pmatrix}0&1\\0&0\end{pmatrix}-\begin{pmatrix}0&0\\0&1\end{pmatrix}\right]=\frac{1}{\sqrt{2}}\begin{pmatrix}1&1\\1&-1\end{pmatrix}$$

- **Hadamard operator can be recast as** $\hat{H}=\frac{1}{\sqrt{2}}\sum_{x=0}^{1}\sum_{y=0}^{1}(-1)^{xy}|x><y|$

  Consider

$$\hat{H}=\frac{1}{\sqrt{2}}\sum_{x=0}^{1}\sum_{y=0}^{1}(-1)^{xy}|x><y|=\frac{1}{\sqrt{2}}\sum_{x=0}^{1}[(-1)^{x.0}|x><0|+(-1)^{x.1}|x><1|]$$

$$= \frac{1}{\sqrt{2}}(-1)^{0.0}|0><0| + (-1)^{0.1}|0><1| + (-1)^{1.0}|1><0| + (-1)^{1.1}|1><1|$$

$$= \frac{1}{\sqrt{2}}(|0><0| + |0><1| + |1><0| - |1><1|)|$$

=Hadamard operator in Bra ket form

- **Compact notation for Hadamard operation**

$$\hat{H}|0> = \frac{1}{\sqrt{2}}(|0> +|1>) = \frac{1}{\sqrt{2}}[(-)^0|0> +|\bar{0}>]$$

$$\hat{H}|1> = \frac{1}{\sqrt{2}}(|0> -|1>) = \frac{1}{\sqrt{2}}(|\bar{1}> +(-)^1|1>) = \frac{1}{\sqrt{2}}[(-)^1|1> +|\bar{1}>]$$

Combining, we write

$$\hat{H}|x> = \frac{1}{\sqrt{2}}[(-)^x|x> +|\bar{x}>] \text{ where } x = 0,1$$

- **Can we choose $\hat{H}_1 = \frac{1}{\sqrt{2}}\begin{pmatrix} 1 & 1 \\ 1 & 1 \end{pmatrix}$ to represent quantum gate?**

$$\hat{H}_1|0> = \frac{1}{\sqrt{2}}\begin{pmatrix} 1 & 1 \\ 1 & 1 \end{pmatrix}\begin{pmatrix} 1 \\ 0 \end{pmatrix} = \frac{1}{\sqrt{2}}\begin{pmatrix} 1 \\ 1 \end{pmatrix} = \frac{1}{\sqrt{2}}\left[\begin{pmatrix} 1 \\ 0 \end{pmatrix} + \begin{pmatrix} 0 \\ 1 \end{pmatrix}\right] = \frac{1}{\sqrt{2}}(|0> +|1>)$$

$$\hat{H}_1|1> = \frac{1}{\sqrt{2}}\begin{pmatrix} 1 & 1 \\ 1 & 1 \end{pmatrix}\begin{pmatrix} 0 \\ 1 \end{pmatrix} = \frac{1}{\sqrt{2}}\begin{pmatrix} 1 \\ 1 \end{pmatrix} = \frac{1}{\sqrt{2}}\left[\begin{pmatrix} 1 \\ 0 \end{pmatrix} + \begin{pmatrix} 0 \\ 1 \end{pmatrix}\right] = \frac{1}{\sqrt{2}}(|0> +|1>)$$

Let us use these results to find

$$\hat{H}_1\frac{1}{\sqrt{2}}(|0> -|1>) = \frac{1}{\sqrt{2}}[H_1|0> -H_1|1>]$$

$$= \frac{1}{\sqrt{2}}\left[\frac{1}{\sqrt{2}}(|0> +|1>) - \frac{1}{\sqrt{2}}(|0> +|1>)\right] = 0$$

We get a null ket vector which is not of unit length, i.e. the requirement of normalization to unity is violated. So $\hat{H}_1$ will not be a legitimate choice as it leads to unnormalised vector.

- **Hadamard Gate is unitary**

$$\hat{H} = \frac{1}{\sqrt{2}}\begin{pmatrix} 1 & 1 \\ 1 & -1 \end{pmatrix}, \hat{H}^\dagger = \frac{1}{\sqrt{2}}\begin{pmatrix} 1 & 1 \\ 1 & -1 \end{pmatrix} = \hat{H}. \text{ So } \hat{H} \text{ is Hermitian.}$$

$$\Rightarrow \hat{H}^\dagger\hat{H} = \hat{H}^2 = \frac{1}{\sqrt{2}}\begin{pmatrix} 1 & 1 \\ 1 & -1 \end{pmatrix}\frac{1}{\sqrt{2}}\begin{pmatrix} 1 & 1 \\ 1 & -1 \end{pmatrix} = \begin{pmatrix} 1 & 0 \\ 0 & 1 \end{pmatrix} = \hat{I}. \text{ So } \hat{H} \text{ is unitary.}$$

- **Hadamard gate is self- inverse**

$$\hat{H}^2 = \hat{H}\hat{H} = \frac{1}{\sqrt{2}}\begin{pmatrix} 1 & 1 \\ 1 & -1 \end{pmatrix}\frac{1}{\sqrt{2}}\begin{pmatrix} 1 & 1 \\ 1 & -1 \end{pmatrix} = \begin{pmatrix} 1 & 0 \\ 0 & 1 \end{pmatrix} = \hat{I} \Rightarrow \hat{H}^{-1} = \hat{H}$$

- **Quantum gate** $\hat{V} = \sqrt{\hat{X}} \equiv \sqrt{NOT}$ **behaves as Hadamard Gate** $\hat{H}$

$$\hat{V} = \sqrt{\hat{X}} \equiv \sqrt{\text{NOT}} = \sqrt{\begin{pmatrix} 0 & 1 \\ 1 & 0 \end{pmatrix}}$$

Let us find the square root of $\hat{X}$.

$\hat{X} = \begin{pmatrix} 0 & 1 \\ 1 & 0 \end{pmatrix}$ has eigenvalues $+1$ with eigenvector $\begin{pmatrix} 1 \\ 1 \end{pmatrix}$ and -1 with eigenvector $\begin{pmatrix} 1 \\ -1 \end{pmatrix}$

(section 1.27). Construct the diagonal matrix $\hat{D} = \begin{pmatrix} 1 & 0 \\ 0 & -1 \end{pmatrix}$ and consider the matrix

$\hat{F} = \begin{pmatrix} 1 & 1 \\ 1 & -1 \end{pmatrix}$ built with the elements of the eigenvectors. Using the relation

$$\hat{X}^N = \hat{F}\hat{D}^N\hat{F}^{-1} \xrightarrow{N=1/2} \hat{X}^{1/2} = \hat{F}\hat{D}^{1/2}\hat{F}^{-1}$$

we proceed as follows

$$\hat{X}^{1/2} = \begin{pmatrix} 1 & 1 \\ 1 & -1 \end{pmatrix}\begin{pmatrix} 1 & 0 \\ 0 & -1 \end{pmatrix}^{1/2}\begin{pmatrix} 1 & 1 \\ 1 & -1 \end{pmatrix}^{-1} = \begin{pmatrix} 1 & 1 \\ 1 & -1 \end{pmatrix}\begin{pmatrix} \sqrt{1} & 0 \\ 0 & \sqrt{-1} \end{pmatrix}\left(-\frac{1}{2}\right)\begin{pmatrix} -1 & -1 \\ -1 & 1 \end{pmatrix}$$

$$\Rightarrow \sqrt{\hat{X}} = \frac{1}{2}\begin{pmatrix} 1 & 1 \\ 1 & -1 \end{pmatrix}\begin{pmatrix} 1 & 0 \\ 0 & i \end{pmatrix}\begin{pmatrix} 1 & 1 \\ 1 & -1 \end{pmatrix} = \frac{1}{2}\begin{pmatrix} 1 & 1 \\ 1 & -1 \end{pmatrix}\begin{pmatrix} 1 & 1 \\ i & -i \end{pmatrix}$$

$$\hat{V} = \sqrt{\hat{X}} = \frac{1}{2}\begin{pmatrix} 1+i & 1-i \\ 1-i & 1+i \end{pmatrix} = \sqrt{\text{NOT}}$$

Consider

$$\sqrt{\hat{X}}|0> = \frac{1}{2}\begin{pmatrix} 1+i & 1-i \\ 1-i & 1+i \end{pmatrix}\begin{pmatrix} 1 \\ 0 \end{pmatrix} = \frac{1}{2}\begin{pmatrix} 1+i \\ 1-i \end{pmatrix} = \frac{1}{2}\left[(1+i)\begin{pmatrix} 1 \\ 0 \end{pmatrix} + (1-i)\begin{pmatrix} 0 \\ 1 \end{pmatrix}\right]$$

$$= \frac{1}{2}[(1+i)|0> + (1-i)|1>] \Rightarrow \text{linear combination of } |0> \text{ and } |1>.$$

$$\sqrt{\hat{X}}|1> = \frac{1}{2}\begin{pmatrix} 1+i & 1-i \\ 1-i & 1+i \end{pmatrix}\begin{pmatrix} 0 \\ 1 \end{pmatrix} = \frac{1}{2}\begin{pmatrix} 1-i \\ 1+i \end{pmatrix} = \frac{1}{2}\left[(1-i)\begin{pmatrix} 1 \\ 0 \end{pmatrix} + (1+i)\begin{pmatrix} 0 \\ 1 \end{pmatrix}\right]$$

$$= \frac{1}{2}[(1-i)|0> + (1+i)|1>] \Rightarrow \text{linear combination of } |0> \text{ and } |1>.$$

Clearly $\sqrt{\hat{X}} = \sqrt{\text{NOT}}$ operates on $|0>$ or $|1>$ to produce superposition of states just like the Hadamard gate viz. $\hat{H}|0> = \frac{1}{\sqrt{2}}[|0> + |1>]$, $\hat{H}|1> = \frac{1}{\sqrt{2}}[0> - |1>]$.
The quantum gate $\hat{V} = \sqrt{\hat{X}} \equiv \sqrt{\text{NOT}}$ thus behaves as Hadamard Gate $\hat{H}$ but we note that

$$\hat{H}^2 = \left[\frac{1}{\sqrt{2}}\begin{pmatrix} 1 & 1 \\ 1 & -1 \end{pmatrix}\right]^2 = \hat{I} = \text{Identity matrix} \neq \hat{X} = \begin{pmatrix} 0 & 1 \\ 1 & 0 \end{pmatrix}$$

- **$\sqrt{\text{NOT}}$ is unitary**

$$\sqrt{\text{NOT}} \sqrt{\text{NOT}}^{\dagger} = \begin{pmatrix} \frac{1+i}{2} & \frac{1-i}{2} \\ \frac{1-i}{2} & \frac{1+i}{2} \end{pmatrix}\begin{pmatrix} \frac{1-i}{2} & \frac{1+i}{2} \\ \frac{1+i}{2} & \frac{1-i}{2} \end{pmatrix} = \begin{pmatrix} 1 & 0 \\ 0 & 1 \end{pmatrix} = \hat{I}$$

$$\sqrt{\text{NOT}}^{\dagger} \sqrt{\text{NOT}} = \begin{pmatrix} \frac{1-i}{2} & \frac{1+i}{2} \\ \frac{1+i}{2} & \frac{1-i}{2} \end{pmatrix}\begin{pmatrix} \frac{1+i}{2} & \frac{1-i}{2} \\ \frac{1-i}{2} & \frac{1+i}{2} \end{pmatrix} = \begin{pmatrix} 1 & 0 \\ 0 & 1 \end{pmatrix} = \hat{I}$$

So $\sqrt{\text{NOT}}^{\dagger}\sqrt{\text{NOT}} = \sqrt{\text{NOT}}\sqrt{\text{NOT}}^{\dagger} = \hat{I}$. So $\sqrt{\text{NOT}}$ is unitary.

- **$\sqrt{\text{NOT}}$ is not Hermitian**

$$\sqrt{\text{NOT}}^{\dagger} = \begin{pmatrix} \frac{1+i}{2} & \frac{1-i}{2} \\ \frac{1-i}{2} & \frac{1+i}{2} \end{pmatrix}^{\dagger} = \begin{pmatrix} \frac{1-i}{2} & \frac{1+i}{2} \\ \frac{1+i}{2} & \frac{1-i}{2} \end{pmatrix} \neq \begin{pmatrix} \frac{1+i}{2} & \frac{1-i}{2} \\ \frac{1-i}{2} & \frac{1+i}{2} \end{pmatrix} = \sqrt{\text{NOT}}$$

$$\Rightarrow \sqrt{\text{NOT}} \text{ is not Hermitian. } (\sqrt{\text{NOT}}^{\dagger} \neq \sqrt{\text{NOT}})$$

Comment

- Consider $\hat{V}^2 = \hat{V}\hat{V} = \frac{1}{2}\begin{pmatrix} 1+i & 1-i \\ 1-i & 1+i \end{pmatrix}\frac{1}{2}\begin{pmatrix} 1+i & 1-i \\ 1-i & 1+i \end{pmatrix}$

$$= \frac{1}{4}\begin{pmatrix} (1+i)^2 + (1-i)^2 & (1+i)(1-i) + (1-i)(1+i) \\ (1-i)(1+i) + (1+i)(1-i) & (1-i)^2 + (1+i)^2 \end{pmatrix}$$

$$= \frac{1}{4}\begin{pmatrix} 1 + i^2 + 2i + 1 + i^2 - 2i & 1 - i^2 + 1 - i^2 \\ 1 - i^2 + 1 - i^2 & 1 + i^2 - 2i + 1 + i^2 + 2i \end{pmatrix}$$

$$= \begin{pmatrix} 0 & 1 \\ 1 & 0 \end{pmatrix} = \hat{X} = \text{NOT}$$

For this reason $\hat{V}$ Gate is called square root of NOT gate and is denoted by $\sqrt{\text{NOT}}$.

- $\hat{V} = \sqrt{\hat{X}} = \frac{1}{2}\begin{pmatrix} 1+i & 1-i \\ 1-i & 1+i \end{pmatrix} = \sqrt{\text{NOT}}$, $\hat{V}^{\dagger} = \frac{1}{2}\begin{pmatrix} 1+i & 1-i \\ 1-i & 1+i \end{pmatrix}^{\dagger} = \frac{1}{2}\begin{pmatrix} 1-i & 1+i \\ 1+i & 1-i \end{pmatrix}$

- $\hat{H}\hat{X}\hat{H}$ and $\hat{H}\hat{Z}\hat{H}$

$$\hat{H}\hat{X}\hat{H} = \frac{1}{\sqrt{2}}\begin{pmatrix}1 & 1 \\ 1 & -1\end{pmatrix}\begin{pmatrix}0 & 1 \\ 1 & 0\end{pmatrix}\frac{1}{\sqrt{2}}\begin{pmatrix}1 & 1 \\ 1 & -1\end{pmatrix} = \frac{1}{2}\begin{pmatrix}1 & 1 \\ 1 & -1\end{pmatrix}\begin{pmatrix}1 & -1 \\ 1 & 1\end{pmatrix} = \begin{pmatrix}1 & 0 \\ 0 & -1\end{pmatrix} = \hat{Z}$$

$$\hat{H}\hat{Z}\hat{H} = \frac{1}{\sqrt{2}}\begin{pmatrix}1 & 1 \\ 1 & -1\end{pmatrix}\begin{pmatrix}1 & 0 \\ 0 & -1\end{pmatrix}\frac{1}{\sqrt{2}}\begin{pmatrix}1 & 1 \\ 1 & -1\end{pmatrix} = \frac{1}{2}\begin{pmatrix}1 & 1 \\ 1 & -1\end{pmatrix}\begin{pmatrix}1 & 1 \\ -1 & 1\end{pmatrix} = \begin{pmatrix}0 & 1 \\ 1 & 0\end{pmatrix} = \hat{X}$$

5.8 Unitary matrix as length preserving matrix

Consider the matrix $\hat{U} = \begin{pmatrix}e^{i\theta} & 0 \\ 0 & e^{i\phi}\end{pmatrix}$

$$\hat{U}|0> = \begin{pmatrix}e^{i\theta} & 0 \\ 0 & e^{i\phi}\end{pmatrix}\begin{pmatrix}1 \\ 0\end{pmatrix} = \begin{pmatrix}e^{i\theta} \\ 0\end{pmatrix} = e^{i\theta}\begin{pmatrix}1 \\ 0\end{pmatrix} = e^{i\theta}|0>$$

$$\hat{U}|1> = \begin{pmatrix}e^{i\theta} & 0 \\ 0 & e^{i\phi}\end{pmatrix}\begin{pmatrix}0 \\ 1\end{pmatrix} = \begin{pmatrix}0 \\ e^{i\phi}\end{pmatrix} = e^{i\phi}\begin{pmatrix}0 \\ 1\end{pmatrix} = e^{i\phi}|1>$$

$$\hat{U}(\alpha_0|0> +\alpha_1|1>) = \alpha_0\hat{U}|0> +\alpha_1\hat{U}|1> = \alpha_0 e^{i\theta}|0> +\alpha_1 e^{i\phi}|1>$$

The probability of getting the qubits $|0>$ and $|1>$ in the general state $|\psi> = \alpha_0|0> +\alpha_1|1>$ is, respectively, $|\alpha_0|^2$ and $|\alpha_1|^2$. Now consider the output state

$$|\psi'> = \hat{U}|\psi> = \hat{U}(\alpha_0|0> +\alpha_1|1>) = \alpha_0 e^{i\theta}|0> +\alpha_1 e^{i\phi}|1>$$

The probability of getting the qubits $|0>$ and $|1>$ in the output state $|\psi'> = \hat{U}|\psi>$ is $|\alpha_0 e^{i\theta}|^2 = |\alpha_0|^2$ and $|\alpha_1 e^{i\phi}|^2 = |\alpha_1|^2$, respectively.

As the probability of getting the qubits $|0>$ and $|1>$ in the general state $|\psi> = \alpha_0|0> +\alpha_1|1>$ and the output state $|\psi'> = \hat{U}|\psi>$ are the same, these two states viz. $\hat{U}|\psi>$ and $|\psi>$ appear to be indistinguishable, the unitary matrix $\hat{U}$ preserving the length of the state kets.

- **Effect of $\hat{U}$ on $|+> = \frac{1}{\sqrt{2}}(|0> +|1>)$ and $|-> = \frac{1}{\sqrt{2}}(|0> -|1>)$**

$$\hat{U}|+> = \hat{U}\frac{1}{\sqrt{2}}(|0> +|1>)$$

$$= \frac{1}{\sqrt{2}}(\hat{U}|0> +\hat{U}|1>)$$

$$= \frac{1}{\sqrt{2}}(e^{i\theta}|0> +e^{i\phi}|1>)\xrightarrow{\theta=0,\phi=\pi}\frac{1}{\sqrt{2}}(|0> -|1>) = |->$$

$$\hat{U}|-> = \hat{U}\frac{1}{\sqrt{2}}(|0> -|1>)$$

$$= \frac{1}{\sqrt{2}}(\hat{U}|0> - \hat{U}|1>)$$

$$= \frac{1}{\sqrt{2}}(e^{i\theta}|0> - e^{i\phi}|1>) \xrightarrow{\theta=0, \phi=\pi} \frac{1}{\sqrt{2}}(|0> + |1>) = |+>$$

Actually, the probability of getting $|0>$ and $|1>$ in both the states $|\pm> = \frac{1}{\sqrt{2}}(|0> \pm |1>)$ is $(\frac{1}{\sqrt{2}})^2 = \frac{1}{2}$.

- The bra ket notation of $\hat{U}$ is $(|0> \xrightarrow{\hat{U}} e^{i\theta}|0>, |1> \xrightarrow{\hat{U}} e^{i\phi}|1>)$

$$\hat{U} = e^{i\theta}|0> <0| + e^{i\phi}|1> <1| = e^{i\theta}\begin{pmatrix}1\\0\end{pmatrix}(1 \quad 0) + e^{i\phi}\begin{pmatrix}0\\1\end{pmatrix}(0 \quad 1)$$

$$= e^{i\theta}\begin{pmatrix}1 & 0\\0 & 0\end{pmatrix} + e^{i\phi}\begin{pmatrix}0 & 0\\0 & 1\end{pmatrix} = \begin{pmatrix}e^{i\theta} & 0\\0 & e^{i\phi}\end{pmatrix}$$

- $\hat{U}^\dagger \hat{U} = \begin{pmatrix}e^{-i\theta} & 0\\0 & e^{-i\phi}\end{pmatrix}\begin{pmatrix}e^{i\theta} & 0\\0 & e^{i\phi}\end{pmatrix} = \begin{pmatrix}1 & 0\\0 & 1\end{pmatrix} = \hat{I}$. Also, $\hat{U}\hat{U}^\dagger = 1$. $\hat{U}$ is unitary.

- **Phase gate $\hat{P}(\phi)$ from $\hat{U}$**

$$\hat{U} = \begin{pmatrix}e^{i\theta} & 0\\0 & e^{i\phi}\end{pmatrix} \xrightarrow{\theta=0} \begin{pmatrix}1 & 0\\0 & e^{i\phi}\end{pmatrix} = \hat{P}(\phi)$$

5.9 Rotation gates $\hat{R}_X(\theta)$, $\hat{R}_Y(\theta)$, $\hat{R}_Z(\theta)$

Consider the matrix $e^{i\hat{U}\theta}$, where $\hat{U}$ is a matrix such that $\hat{U}^2 = \hat{I}$. Now expanding we get

$$e^{i\hat{U}\theta} = \hat{I} + i\hat{U}\theta + \frac{1}{2!}(i\hat{U}\theta)^2 + \frac{1}{3!}(i\hat{U}\theta)^3 + \frac{1}{4!}(i\hat{U}\theta)^4 + \frac{1}{5!}(i\hat{U}\theta)^5 + \ldots\ldots$$

$$= \hat{I} + i\hat{U}\theta - \frac{1}{2!}\hat{U}\hat{U}\theta^2 - \frac{1}{3!}i\hat{U}\hat{U}\hat{U}\theta^3 + \frac{1}{4!}\hat{U}\hat{U}\hat{U}\hat{U}\theta^4 + \frac{1}{5!}i\hat{U}\hat{U}\hat{U}\hat{U}\hat{U}\theta^5 + \ldots\ldots$$

$$= \hat{I} + i\hat{U}\theta - \frac{1}{2!}\hat{I}\theta^2 - \frac{1}{3!}i\hat{U}\theta^3 + \frac{1}{4!}\hat{I}\theta^4 + \frac{1}{5!}i\hat{U}\theta^5 + \ldots\ldots$$

$$= \hat{I}\left(1 - \frac{\theta^2}{2!} + \frac{\theta^4}{4!} + \ldots\right) + i\hat{U}\left(\theta - \frac{\theta^3}{3!} + \frac{\theta^5}{5!} + \ldots\right)$$

$$e^{i\hat{U}\theta} = \hat{I}\cos\theta + i\hat{U}\sin\theta$$

since $\cos\theta = 1 - \frac{\theta^2}{2!} + \frac{\theta^4}{4!} + \ldots, \quad \sin\theta = \theta - \frac{\theta^3}{3!} + \frac{\theta^5}{5!} + \ldots$

Replace θ by $-\frac{\theta}{2}$

$$e^{i\hat{U}(-\frac{\theta}{2})} = \hat{I}\cos\left(-\frac{\theta}{2}\right) + i\hat{U}\sin\left(-\frac{\theta}{2}\right)$$

Using $\cos(-\frac{\theta}{2}) = \cos\frac{\theta}{2}$, $\sin(-\frac{\theta}{2}) = -\sin\frac{\theta}{2}$

$$e^{-i\hat{U}\theta/2} = \hat{I}\cos\frac{\theta}{2} - i\hat{U}\sin\frac{\theta}{2}$$

We define rotation gate $\hat{R}_X(\theta)$ by putting $\hat{U}=\hat{X}$ namely

$$\hat{R}_X(\theta) = e^{-i\hat{X}\frac{\theta}{2}} = \hat{I}\cos\frac{\theta}{2} - i\hat{X}\sin\frac{\theta}{2}$$

$$= \begin{pmatrix} 1 & 0 \\ 0 & 1 \end{pmatrix}\cos\frac{\theta}{2} - i\begin{pmatrix} 0 & 1 \\ 1 & 0 \end{pmatrix}\sin\frac{\theta}{2} = \begin{pmatrix} \cos\frac{\theta}{2} & -i\sin\frac{\theta}{2} \\ -i\sin\frac{\theta}{2} & \cos\frac{\theta}{2} \end{pmatrix}$$

We define rotation gate $\hat{R}_Y(\theta)$ by putting $\hat{U}=\hat{Y}$ namely

$$\hat{R}_Y(\theta) = e^{-i\hat{Y}\frac{\theta}{2}} = \hat{I}\cos\frac{\theta}{2} - i\hat{Y}\sin\frac{\theta}{2}$$

$$= \begin{pmatrix} 1 & 0 \\ 0 & 1 \end{pmatrix}\cos\frac{\theta}{2} - i\begin{pmatrix} 0 & -i \\ i & 0 \end{pmatrix}\sin\frac{\theta}{2} = \begin{pmatrix} \cos\frac{\theta}{2} & -\sin\frac{\theta}{2} \\ \sin\frac{\theta}{2} & \cos\frac{\theta}{2} \end{pmatrix}$$

We define rotation gate $\hat{R}_Z(\theta)$ by putting $\hat{U}=\hat{Z}$ namely

$$\hat{R}_Z(\theta) = e^{-i\hat{Z}\frac{\theta}{2}} = \hat{I}\cos\frac{\theta}{2} - i\hat{Z}\sin\frac{\theta}{2}$$

$$= \begin{pmatrix} 1 & 0 \\ 0 & 1 \end{pmatrix}\cos\frac{\theta}{2} - i\begin{pmatrix} 1 & 0 \\ 0 & -1 \end{pmatrix}\sin\frac{\theta}{2} = \begin{pmatrix} \cos\frac{\theta}{2} - i\sin\frac{\theta}{2} & 0 \\ 0 & \cos\frac{\theta}{2} + i\sin\frac{\theta}{2} \end{pmatrix}$$

$$= \begin{pmatrix} e^{-i\frac{\theta}{2}} & 0 \\ 0 & e^{i\frac{\theta}{2}} \end{pmatrix}$$

5.9.1 $\hat{R}_Z(\xi)$ represents rotation of the Bloch vector about Z-axis by ξ

An arbitrary qubit in the Bloch sphere is (section 4.4, figure 4.4)

$$|\psi> = \alpha_0|0> + \alpha_1|1> = \cos\frac{\theta}{2}|0> + e^{i\phi}\sin\frac{\theta}{2}|1> = \begin{pmatrix} \cos\frac{\theta}{2} \\ e^{i\phi}\sin\frac{\theta}{2} \end{pmatrix}$$

Let us consider

$$\hat{R}_Z(\xi)|\psi> = \begin{pmatrix} e^{-i\frac{\xi}{2}} & 0 \\ 0 & e^{i\frac{\xi}{2}} \end{pmatrix}\begin{pmatrix} \cos\frac{\theta}{2} \\ e^{i\phi}\sin\frac{\theta}{2} \end{pmatrix} = \begin{pmatrix} e^{-i\frac{\xi}{2}}\cos\frac{\theta}{2} \\ e^{i\frac{\xi}{2}}e^{i\phi}\sin\frac{\theta}{2} \end{pmatrix} = e^{-i\frac{\xi}{2}}\begin{pmatrix} \cos\frac{\theta}{2} \\ e^{i(\phi+\xi)}\sin\frac{\theta}{2} \end{pmatrix}$$

$$=e^{-i\frac{\xi}{2}}\left[\cos\frac{\theta}{2}\binom{1}{0} + e^{i(\phi+\xi)} \sin\frac{\theta}{2}\binom{0}{1} \right] = |\psi'>$$

$$=e^{-i\frac{\xi}{2}}\left[\cos\frac{\theta}{2}|0> + e^{i(\phi+\xi)} \sin\frac{\theta}{2}|1> \right] = |\psi'>$$

The factor $e^{-i\frac{\xi}{2}}$ represents an overall phase and is called global phase factor. Since $\left| e^{-i\frac{\xi}{2}} \right|^2 = 1$, this global phase does not contribute and can be ignored. Hence we can write

$$\hat{R}_Z(\xi)|\psi> = \cos\frac{\theta}{2}|0> + e^{i(\phi+\xi)} \sin\frac{\theta}{2}|1> = \begin{pmatrix} \cos\frac{\theta}{2} \\ e^{i(\phi+\xi)} \sin\frac{\theta}{2} \end{pmatrix}$$

Clearly thus the effect of $\hat{R}_Z(\xi)$ is to rotate Bloch vector from (r, θ, ϕ) :$(1, \theta, \phi) \rightarrow (1, \theta, \phi + \xi)$, i.e. by an angle ξ about the Z-axis. In other words

$$|\psi> = \begin{pmatrix} \cos\frac{\theta}{2} \\ e^{i\phi} \sin\frac{\theta}{2} \end{pmatrix} \xrightarrow{\text{rotate by } \xi \text{ about Z axis}} \hat{R}_Z(\xi)|\psi> = \begin{pmatrix} \cos\frac{\theta}{2} \\ e^{i(\phi+\xi)} \sin\frac{\theta}{2} \end{pmatrix}$$

5.9.2 Rotation about an arbitrary axis $(\hat{n})$

Consider the matrix $e^{i\hat{U}\theta}$, where $\hat{U}$ is a matrix such that $\hat{U}^2 = \hat{I}$. Expanding we get
$e^{i\hat{U}\theta} = \hat{I} \cos\theta + i\hat{U} \sin\theta$

Putting $\frac{\theta}{2}$ for θ and taking $\hat{U} = -\hat{\vec{\sigma}} \cdot \hat{n}$ we get

$$e^{i [-\hat{\vec{\sigma}}.\hat{n}]\frac{\theta}{2}} = \hat{I} \cos\frac{\theta}{2} + i [-\hat{\vec{\sigma}}. \hat{n}]\sin\frac{\theta}{2}$$

$$\Rightarrow \quad e^{-i\hat{\vec{\sigma}}.\hat{n}\frac{\theta}{2}} = \hat{I} \cos\frac{\theta}{2} - i\hat{\vec{\sigma}}. \hat{n} \sin\frac{\theta}{2}$$

where $\hat{\vec{\sigma}} = (\hat{\sigma}_x, \hat{\sigma}_y, \hat{\sigma}_z) = (\hat{X}, \hat{Y}, \hat{Z})$ = Pauli matrices; $\hat{n} = (n_x, n_y, n_z)$ = unit vector. (Cap or hat symbol in $\hat{\vec{\sigma}}$ represents operator while cap symbol in $\hat{n}$ represents unit vector.)

5.9.3 An arbitrary single qubit unitary operator can be converted into a Hadamard gate

Consider an arbitrary single qubit unitary operator

$$\hat{U} = e^{i\alpha}\hat{R}_n(\theta)$$

where $\hat{R}_n(\theta) = e^{-i\hat{\vec{\sigma}}.\hat{n}\frac{\theta}{2}} = \hat{I} \cos\frac{\theta}{2} - i\hat{\vec{\sigma}}. \hat{n} \sin\frac{\theta}{2}$. So we write

$$\hat{U} = e^{i\alpha}\left[\hat{I} \cos\frac{\theta}{2} - i\hat{\vec{\sigma}}. \hat{n} \sin\frac{\theta}{2}\right]$$

Using

$$\alpha = \frac{\pi}{2}, \ \theta = \pi, \ \vec{\hat{\sigma}} = (\hat{X}, \ \hat{Y}, \ \hat{Z}) = \text{Pauli matrices}, \ \hat{n} = (\frac{1}{\sqrt{2}}, \ 0, \ \frac{1}{\sqrt{2}}) = \text{unit vector}$$

$$\text{and } \vec{\sigma}. \ \hat{n} = \left[\hat{i}\begin{pmatrix} 0 & 1 \\ 1 & 0 \end{pmatrix} + \hat{j}\begin{pmatrix} 0 & -i \\ i & 0 \end{pmatrix} + \hat{k}\begin{pmatrix} 1 & 0 \\ 0 & -1 \end{pmatrix} \right] \cdot \left[\hat{i}\frac{1}{\sqrt{2}} + \hat{j}0 + \hat{k}\frac{1}{\sqrt{2}} \right]$$

$$= \left[\frac{1}{\sqrt{2}}\begin{pmatrix} 0 & 1 \\ 1 & 0 \end{pmatrix} + \frac{1}{\sqrt{2}}\begin{pmatrix} 1 & 0 \\ 0 & -1 \end{pmatrix} \right] = \frac{1}{\sqrt{2}}\begin{pmatrix} 1 & 1 \\ 1 & -1 \end{pmatrix}$$

$$\hat{U} = e^{i\frac{\pi}{2}}[\hat{I} \cos \frac{\pi}{2} - i(\vec{\sigma}. \ \hat{n})\sin \frac{\pi}{2}] = i[-i\vec{\sigma}. \ \hat{n}] = \vec{\sigma}. \ \hat{n}$$

$$= \frac{1}{\sqrt{2}}\begin{pmatrix} 1 & 1 \\ 1 & -1 \end{pmatrix} = \hat{H} = \text{Hadamard operator}$$

i.e. $\hat{U} = e^{i\alpha}\hat{R}_n(\theta) \xrightarrow{\alpha = \frac{\pi}{2}, \theta = \pi, \hat{n} = \frac{1}{\sqrt{2}}(\hat{i}+\hat{k})} \hat{H}$

5.9.4 $\hat{R}_X(\pi)\hat{R}_Y(\frac{\pi}{2})$ is a Hadamard operator

$$\hat{R}_X(\pi)\hat{R}_Y\left(\frac{\pi}{2}\right) = \begin{pmatrix} \cos \frac{\pi}{2} & -i \sin \frac{\pi}{2} \\ -i \sin \frac{\pi}{2} & \cos \frac{\pi}{2} \end{pmatrix}\begin{pmatrix} \cos \frac{\pi}{4} & -\sin \frac{\pi}{4} \\ \sin \frac{\pi}{4} & \cos \frac{\pi}{4} \end{pmatrix}$$

$$= \begin{pmatrix} 0 & -i \\ -i & 0 \end{pmatrix}\begin{pmatrix} \frac{1}{\sqrt{2}} & -\frac{1}{\sqrt{2}} \\ \frac{1}{\sqrt{2}} & \frac{1}{\sqrt{2}} \end{pmatrix} = \begin{pmatrix} -\frac{i}{\sqrt{2}} & -\frac{i}{\sqrt{2}} \\ -\frac{i}{\sqrt{2}} & \frac{i}{\sqrt{2}} \end{pmatrix} = -i\frac{1}{\sqrt{2}}\begin{pmatrix} 1 & 1 \\ 1 & -1 \end{pmatrix}$$

$$= e^{-i\frac{\pi}{2}}\frac{1}{\sqrt{2}}\begin{pmatrix} 1 & 1 \\ 1 & -1 \end{pmatrix} = e^{-i\frac{\pi}{2}}\hat{H} \xrightarrow{\text{overlook global phase factor}} \hat{H}$$

5.9.5 Evaluation of $\hat{X}\hat{Y}\hat{X}, \quad \hat{X}\hat{R}_Y(\theta)\hat{X}$

$$\hat{X}\hat{Y}\hat{X} = \begin{pmatrix} 0 & 1 \\ 1 & 0 \end{pmatrix}\begin{pmatrix} 0 & -i \\ i & 0 \end{pmatrix}\begin{pmatrix} 0 & 1 \\ 1 & 0 \end{pmatrix} = \begin{pmatrix} 0 & 1 \\ 1 & 0 \end{pmatrix}\begin{pmatrix} -i & 0 \\ 0 & i \end{pmatrix}$$

$$= \begin{pmatrix} 0 & i \\ -i & 0 \end{pmatrix} = -\begin{pmatrix} 0 & -i \\ i & 0 \end{pmatrix} = -\hat{Y}$$

$$\hat{X}\hat{R}_Y(\theta)\hat{X} = \hat{X}e^{-i\hat{Y}\frac{\theta}{2}}\hat{X}$$

Using $\hat{R}_Y(\theta) = e^{-i\hat{Y}\frac{\theta}{2}} = \hat{I} \cos \frac{\theta}{2} - i\hat{Y} \sin \frac{\theta}{2}$ we get

$$\hat{X}\hat{R}_Y(\theta)\hat{X} = \hat{X}\left(\hat{I} \cos \frac{\theta}{2} - i\hat{Y} \sin \frac{\theta}{2}\right)\hat{X} = \hat{X}\hat{I} \cos \frac{\theta}{2}\hat{X} - \hat{X}i\hat{Y} \sin \frac{\theta}{2}\hat{X}$$

$$= \hat{X}\hat{X} \cos \frac{\theta}{2} - i\hat{X}\hat{Y}\hat{X} \sin \frac{\theta}{2} = \hat{I} \cos \frac{\theta}{2} - i(-\hat{Y})\sin \frac{\theta}{2}$$

(as $\hat{X}^2 = \hat{I}$, $\hat{X}\hat{Y}\hat{X} = -\hat{X}\hat{X}\hat{Y} = -\hat{X}^2\hat{Y} = -\hat{Y}$ since Pauli matrices anti-commute.)

$$\hat{X}\hat{R}_Y(\theta)\hat{X} = \hat{I}\,\cos\left(-\frac{\theta}{2}\right) - i\hat{Y}\,\sin\left(-\frac{\theta}{2}\right) = \hat{R}_y(-\theta)$$

5.9.6 Hadamard operation is equivalent to rotation on Bloch sphere about Y-axis by 90° followed by rotation about X-axis by 180°

The state vector of the qubit on unit 3D Bloch sphere is described as

$$|\psi> = \alpha_0|0> + \alpha_1|1> = \cos\frac{\theta}{2}|0> + e^{i\phi}\sin\frac{\theta}{2}|1> = \begin{pmatrix} \cos\frac{\theta}{2} \\ e^{i\phi}\sin\frac{\theta}{2} \end{pmatrix} = |\psi(\theta,\phi)>$$

and represented by the point $(r = 1, \theta, \phi)$.
- Compare $|\psi(\theta, \phi)>$ with $|+>$ to get

$$\cos\frac{\theta}{2}|0> + e^{i\phi}\sin\frac{\theta}{2}|1> \; = \; |+> \; = \; \frac{1}{\sqrt{2}}(|0> +|1>)$$

$$\Rightarrow \cos\frac{\theta}{2} = \frac{1}{\sqrt{2}}, \quad e^{i\phi}\sin\frac{\theta}{2} = \frac{1}{\sqrt{2}} \quad \Rightarrow \phi = 0°, \frac{\theta}{2} = 45° \quad \Rightarrow \quad \theta = 90°$$

$$|+> = |\psi(90°, 0°)>$$

Bloch vector $= |r, \theta, \phi > = |1, 90°, 0°> =$ point a on the X-axis at unit distance from origin O as shown in figure 5.7. (Compare with the Bloch vector of $|->$ as shown in figure 5.32.)
- Compare $|\psi(\theta, \phi)>$ with $|1>$ to get

$$\cos\frac{\theta}{2}|0> + e^{i\phi}\sin\frac{\theta}{2}|1> \; = \; |1>$$

$$\Rightarrow \cos\frac{\theta}{2} = 0, \quad e^{i\phi}\sin\frac{\theta}{2} = 1 \quad \Rightarrow \phi = 0°, \frac{\theta}{2} = 90° \quad \Rightarrow \quad \theta = 180°$$

$$|1> = |\psi(180°, 0°)>$$

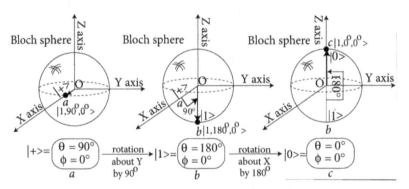

Figure 5.7. Hadamard gate can be considered as two rotations, one following the other $|$point $a> = |+> = |1, 90°, 0°> \xrightarrow{\hat{H}} |$point $c> = |0> = |1, 0°, 0°>$.

Bloch vector $=|r, \theta, \phi> =|1,180°, 0°>=$ point b on $-Z$ axis at unit distance from origin O as shown in figure 5.7.

- Compare $|\psi(\theta, \phi)>$ with $|0>$ to get

$$\cos\frac{\theta}{2}|0> +e^{i\phi}\sin\frac{\theta}{2}|1> \; = \; |0>$$

$$\Rightarrow \cos\frac{\theta}{2} = 1, \quad e^{i\phi}\sin\frac{\theta}{2} = 0 \quad \Rightarrow \phi = 0°, \quad \theta = 0°$$

$$|1> =|\psi(0°, 0°)>$$

Bloch vector $=|r, \theta, \phi> =|1,0°, 0°>=$ point c on Z axis at unit distance from origin O as shown in figure 5.7.

Now $\hat{H}|+>=\hat{H}\frac{1}{\sqrt{2}}(|0> +|1>) = |0>$ means that $\hat{H}$ takes the state $|+>=\frac{1}{\sqrt{2}}(|0> +|1>)$ (denoted by point a on X axis) to $|0>$ (denoted by point c on the Z-axis), i.e. from $|\psi> |_{\theta=90°, \phi=0°}= |+>$ to $|\psi> |_{\theta=0°, \phi=0°}=|0>$ as shown in the diagram in figure 5.7. This can be interpreted as anticlockwise rotation about Y-axis by 90° from point a to point $b(\; |+> \rightarrow |1>\;)$ followed by a rotation about the X-axis by 180° from point b to point c ($|1> \rightarrow|0>$).

5.10 Multi-qubit gates

In order to entangle two qubits we have to extend our discussion of quantum gates to two-qubit gates (like Controlled NOT gate, Controlled U gate) as well as to three qubit gates (like Toffli gate (CCNOT gate), Fredkin gate (CSWAP gate), Deutsch gate). All quantum gates may be made to be of arbitrary degree of precision by one- and two-qubit gates alone. The three-qubit gates used in quantum circuits are either controlled two-qubit gates or controlled single-qubit gates.

The state space of a composite physical system is the tensor product of the state spaces of the component physical systems. Tensor product embodies the essence of superposition. If system A is in state $|A>$ and system B is in state $|B>$ then their tensor product involves a little of A and a little of B.

If system i is in state $|\psi_i>$, $i = 1 .... n$ then the state of the composite system will be $|\psi> =|\psi_1> \otimes |\psi_2> .... \otimes |\psi_n>$.

- **Example**

Suppose $|\psi_1> =A|0> +B|1>$ and $|\psi_2> =C|0> +D|1>$ then

$$|\psi> =|\psi_1> \otimes |\psi_2> =|\psi_1\psi_2>$$

$$=A. C|0> \otimes |0> +A. D|0> \otimes |1> +B. C|1> \otimes |0> +B. D|1> \otimes |1>$$

$$=AC|00> +AD|01> +BC|10> +BD|11>$$

In a classical digital computer, information is physically stored as bits (i.e. in terms of 0, 1). The stability of the stored information and reliability of manipulations with data depend upon classical physics. One can make observations (examinations) without affecting the contents. But if information is stored at the microphysical level where quantum mechanics prevails, a storage qubit can represent both 0 and 1 simultaneously. Measurement and manipulations of the qubits can be modeled through matrix operations.

We show the generalization of a one-qubit system to a n qubit system in table 5.1.

[$x \in \{00,01,10,11\}$ means x takes on values $00,01,10,11$ while the same thing is denoted also by the expression $x \in \{0,1\}^2$ which refers to the set of qubit strings of length 2 (i.e. two qubit strings) with each qubit being 0 or 1.]

NOTE

If an arbitrary two-qubit gate $\hat{U}$ maps $|00 >$, $|01 >$, $|10 >$, $|11>$ to two qubit states $|\psi_{00}>$, $|\psi_{01}>$, $|\psi_{10}>$ and $|\psi_{11} >$, respectively, then the operator $\hat{U}$ can be expressed in bra ket notation as

$$\hat{U} = |\psi_{00} > <00| + |\psi_{01} > <01| + |\psi_{10} > <10| + |\psi_{11} > <11|$$

This rule can be extended to n qubit gates.

5.11 Controlled-NOT gate or CNOT gate

Controlled NOT gate or CNOT gate is a two-qubit gate and resembles the classical XOR gate.

It has 2 input qubits $|A>$, $|B>$ called control qubit and target qubit, respectively. The symbol of the quantum circuit of CNOT has been depicted in figure 5.8. The

Table 5.1. Generalisation of 1-qubit to n qubit system.

	$n = 1$ qubit system	$n = 2$ qubit system	n qubit system
Computational basis (2^n states)	$|0>,|1>$	$|0> \otimes |0>=|00>$ $|0> \otimes |1>=|01>$ $|1> \otimes |0>=|10>$ $|1> \otimes |1>=|11>$	$|00.........00>$ $|00.........01>$ $|00.........10>$ $|11.........11>$
State space	H	$H \otimes H = H^{\otimes 2}$	$H^{\otimes n}$
Arbitrary state	$|\Psi>=\alpha_0|0>+\alpha_1|1>$	$|\Psi> = \alpha_{00}|00> + \alpha_{01}|01>$ $+ \alpha_{10}|10> + \alpha_{11}|11>$ $= \sum_{x \in \{00,01,10,11\}} \alpha_x|x\rangle$ $= \sum_{x \in \{0,1\}^2} \alpha_x|x\rangle$	$|\psi\rangle = \sum_{x \in \{0,1\}^n} \alpha_x|x\rangle$
Closure relation	$|\alpha_0|^2+|\alpha_1|^2=1$	$|\alpha_{00}|^2+|\alpha_{01}|^2+|\alpha_{10}|^2+|\alpha_{11}|^2=1$ i.e. $\sum_{x \in \{00,01,10,11\}} |\alpha_x|^2 = 1$ i.e. $\sum_{x \in \{0,1\}^2} |\alpha_x|^2 = 1$	$\sum_{x \in \{0,1\}^n} |\alpha_x|^2 = 1$

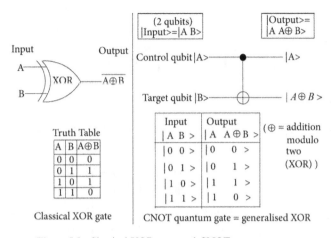

Figure 5.8. Classical XOR gate and CNOT quantum gate.

upper quantum wire with a dot • carries the control qubit while the lower quantum wire with an encircled cross ⊕ carries the target qubit.

If control qubit is $A = 0$ then target qubit is unchanged. So CNOT of $|A = 0\ B>$ leads to itself, i.e. $|0\ \ B> \xrightarrow{\text{CNOT}} |0\ \ B>$. Hence $|00> \rightarrow |00>$, $|01> \rightarrow |01>$.

And if control qubit is $A = 1$ then target qubit B is flipped. So CNOT of $|A = 1\ \ B>$ can be expressed as

$|1\ \ B> \xrightarrow{\text{CNOT}} |1\ \ \overline{B}>$ i.e. $|10> \rightarrow |11>$, $|11> \rightarrow |10>$.

Clearly the CNOT operation can be represented as $|A\ B> \xrightarrow{\text{CNOT}} |A\ A \oplus B>$. In other words, the target qubit B, after CNOT, becomes the XOR of the inputs A,B, i.e. $A \oplus B$ (⊕ represents addition modulo 2 operation or the XOR operation) (section 8.3). The quantum wires shown are directions of logic flow. The CNOT operation can be represented by a unitary matrix $\hat{U}_{\text{CNOT}}$. In Dirac's bra ket notation CNOT is given by

(considering that $|00> \xrightarrow{\text{CNOT}} |00>$, $|01> \xrightarrow{\text{CNOT}} |01>$, $|10> \xrightarrow{\text{CNOT}} |11>$, $|11> \xrightarrow{\text{CNOT}} |10>$ where the first qubit is the control qubit and the second qubit is the target qubit)

$$\hat{U}_{\text{CNOT}} = |00> <00| + |01> <01| + |11> <10| + |10> <11|$$

$$= \begin{pmatrix} 1 \\ 0 \\ 0 \\ 0 \end{pmatrix}(1\ 0\ 0\ 0) + \begin{pmatrix} 0 \\ 1 \\ 0 \\ 0 \end{pmatrix}(0\ 1\ 0\ 0) + \begin{pmatrix} 0 \\ 0 \\ 0 \\ 1 \end{pmatrix}(0\ 0\ 1\ 0) + \begin{pmatrix} 0 \\ 0 \\ 1 \\ 0 \end{pmatrix}(0\ 0\ 0\ 1)$$

$$= \begin{pmatrix} 1 & 0 & 0 & 0 \\ 0 & 0 & 0 & 0 \\ 0 & 0 & 0 & 0 \\ 0 & 0 & 0 & 0 \end{pmatrix} + \begin{pmatrix} 0 & 0 & 0 & 0 \\ 0 & 1 & 0 & 0 \\ 0 & 0 & 0 & 0 \\ 0 & 0 & 0 & 0 \end{pmatrix} + \begin{pmatrix} 0 & 0 & 0 & 0 \\ 0 & 0 & 0 & 0 \\ 0 & 0 & 0 & 0 \\ 0 & 0 & 1 & 0 \end{pmatrix} + \begin{pmatrix} 0 & 0 & 0 & 0 \\ 0 & 0 & 0 & 0 \\ 0 & 0 & 0 & 1 \\ 0 & 0 & 0 & 0 \end{pmatrix}$$

$$= \begin{pmatrix} 1 & 0 & 0 & 0 \\ 0 & 1 & 0 & 0 \\ 0 & 0 & 0 & 1 \\ 0 & 0 & 1 & 0 \end{pmatrix} = \begin{pmatrix} \hat{I} & 0 \\ 0 & \hat{X} \end{pmatrix}$$

We can write

$$\hat{U}_{CNOT}|A\ B> =|A\ A \oplus B> \text{ (i. e. } A \otimes B = |A\ B> \xrightarrow{\text{CNOT}} |A\ A \oplus B>)$$

$$\begin{pmatrix} 1 & 0 & 0 & 0 \\ 0 & 1 & 0 & 0 \\ 0 & 0 & 0 & 1 \\ 0 & 0 & 1 & 0 \end{pmatrix} \begin{pmatrix} |00> \\ |01> \\ |10> \\ |11> \end{pmatrix} = \begin{pmatrix} |00> \\ |01> \\ |11> \\ |10> \end{pmatrix},$$

- $\hat{U}_{CNOT}\hat{U}_{CNOT}^{-1} = \hat{I}$.

CNOT gate is used in many quantum circuits of significance.

With a suitable combination of all single-qubit gates and the CNOT gate we can construct all possible quantum circuits.

5.11.1 CNOT gate is Hermitian

$$\hat{U}_{CNOT} = |00><00| + |01><01| + |11><10| + |10><11|$$

$$\hat{U}_{CNOT}^{\dagger} = [|00><00| + |01><01| + |11><10| + |10><11|]^{\dagger}$$

$$= |00><00| + |01><01| + |10><11| + |11><10| = \hat{U}_{CNOT}$$

So CNOT gate is Hermitian.

5.12 Preparing Bell states

We shall explain preparation of Bell state $\frac{1}{\sqrt{2}}(|00> +|11>)$. The quantum circuit that prepares this Bell state has been shown in figure 5.9.

The input is $|0> \otimes|0>$. Let qubit $|0>$ be applied to a Hadamard gate $\hat{H}$ and what we get is $\hat{H}|0> =| + >=\frac{1}{\sqrt{2}}(|0> +|1>)$. This generates the control qubit.

The input to the CNOT gate is thus $\frac{1}{\sqrt{2}}(|0> +|1>) \otimes |0> =\frac{1}{\sqrt{2}}(|00> +|10>)$. The first qubit is the control qubit and the second qubit is the target qubit.

The control qubit 0 leaves the target qubit 0 unchanged, as $0 \oplus 0 = 0$ (so we get the state $|00>$), while the control qubit 1 changes the target qubit 0 to 1 since $0\oplus1=1$ (so that we get the state $|11>$). Hence

$$\frac{1}{\sqrt{2}}(|00> +|10>) \xrightarrow{\text{CNOT}} \frac{1}{\sqrt{2}}(|00> +|11>) = \text{Bell state.}$$

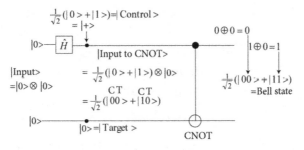

Figure 5.9. Preparing Bell state $\frac{1}{\sqrt{2}}(|00>+11>)$ (C = Control, T = Target).

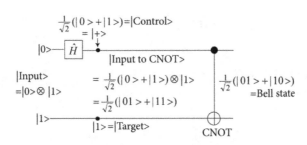

Figure 5.10. Preparing Bell state $\frac{1}{\sqrt{2}}(|01>+|10>)$.

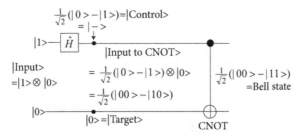

Figure 5.11. Preparing Bell state $\frac{1}{\sqrt{2}}(|00> - |11>)$.

Other Bell states can be similarly generated, as shown in the diagrams of figures 5.10, 5.11 and 5.12 as per the following schemes.

- Preparation of Bell state $=\frac{1}{\sqrt{2}}(|01> +|10>)$ (figure 5.10).

Input $|0 > \otimes|1>$.

$$\hat{H}|0> =| + >=\frac{1}{\sqrt{2}}(|0> +|1>).$$

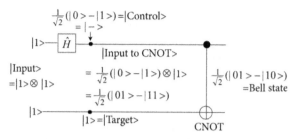

Figure 5.12. Preparing Bell state $\frac{1}{\sqrt{2}}(|01> - |10>)$.

Input to the CNOT gate $= \frac{1}{\sqrt{2}}(|0> + |1>) \otimes |1> = \frac{1}{\sqrt{2}}(|01> + |11>)$.
(The first qubit is the control qubit and the second qubit is target qubit)

$$\frac{1}{\sqrt{2}}(|01> + |11>) \xrightarrow{\text{CNOT}} \frac{1}{\sqrt{2}}(|01> + |10>)$$

- Preparation of Bell state $= \frac{1}{\sqrt{2}}(|00> - |11>)$.

Input $|1> \otimes |0>$.

$$\hat{H}|1> = |->= \frac{1}{\sqrt{2}}(|0> - |1>).$$

Input to the CNOT gate $= \frac{1}{\sqrt{2}}(|0> - |1>) \otimes |0> = \frac{1}{\sqrt{2}}(|00> - |10>)$.
(The first qubit is the control qubit and the second qubit is target qubit)

$$\frac{1}{\sqrt{2}}(|00> - |10>) \xrightarrow{\text{CNOT}} \frac{1}{\sqrt{2}}(|00> - |11>)$$

- Preparation of Bell state $= \frac{1}{\sqrt{2}}(|01> - |10>)$.

Input $|1> \otimes |1>$.

$$\hat{H}|1> = |+>= \frac{1}{\sqrt{2}}(|0> - |1>).$$

Input to the CNOT gate $= \frac{1}{\sqrt{2}}(|0> - |1>) \otimes |1> = \frac{1}{\sqrt{2}}(|01> - |11>)$.
(The first qubit is the control qubit and the second qubit is target qubit)

$$\frac{1}{\sqrt{2}}(|01> - |11>) \xrightarrow{\text{CNOT}} \frac{1}{\sqrt{2}}(|01> - |10>).$$

5.13 Swap gate

The swap gate shown in figure 5.13 is represented by the operator $\hat{U}_{\text{SWAP}}$. It swaps two qubits, i.e. exchanges two states without entanglement.

$$\hat{U}_{\text{SWAP}}(|\psi> \otimes |\phi>) = \hat{U}_{\text{SWAP}}|\psi \; \phi> = |\phi \; \psi>$$

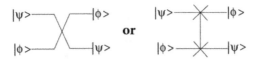

Figure 5.13. SWAP gate.

Clearly then $\hat{U}_{\text{SWAP}}$ maps $|00>$ to $|00>$, $|01>$ to $|10>$, $|10>$ to $|01>$, $|11>$ to $|11>$. Hence the operator representation of $\hat{U}_{\text{SWAP}}$ in bra ket notation will be

$$\hat{U}_{\text{SWAP}} = |00><00| + |10><01| + |01><10| + |11><11|$$

$$= \begin{pmatrix} 1 \\ 0 \\ 0 \\ 0 \end{pmatrix}(1\ 0\ 0\ 0) + \begin{pmatrix} 0 \\ 0 \\ 1 \\ 0 \end{pmatrix}(0\ 1\ 0\ 0) + \begin{pmatrix} 0 \\ 1 \\ 0 \\ 0 \end{pmatrix}(0\ 0\ 1\ 0) + \begin{pmatrix} 0 \\ 0 \\ 0 \\ 1 \end{pmatrix}(0\ 0\ 0\ 1)$$

$$= \begin{pmatrix} 1 & 0 & 0 & 0 \\ 0 & 0 & 0 & 0 \\ 0 & 0 & 0 & 0 \\ 0 & 0 & 0 & 0 \end{pmatrix} + \begin{pmatrix} 0 & 0 & 0 & 0 \\ 0 & 0 & 0 & 0 \\ 0 & 1 & 0 & 0 \\ 0 & 0 & 0 & 0 \end{pmatrix} + \begin{pmatrix} 0 & 0 & 0 & 0 \\ 0 & 0 & 1 & 0 \\ 0 & 0 & 0 & 0 \\ 0 & 0 & 0 & 0 \end{pmatrix} + \begin{pmatrix} 0 & 0 & 0 & 0 \\ 0 & 0 & 0 & 0 \\ 0 & 0 & 0 & 0 \\ 0 & 0 & 0 & 1 \end{pmatrix}$$

$$= \begin{pmatrix} 1 & 0 & 0 & 0 \\ 0 & 0 & 1 & 0 \\ 0 & 1 & 0 & 0 \\ 0 & 0 & 0 & 1 \end{pmatrix}$$

- $\hat{U}_{\text{SWAP}}^{\dagger}\hat{U}_{\text{SWAP}} = \hat{U}_{\text{SWAP}}\hat{U}_{\text{SWAP}}^{\dagger} = \hat{I}$

Alternatively we can express $\hat{U}_{\text{SWAP}}$ in terms of Pauli operators as follows
Let us consider the operator equation

$$\hat{I} \otimes \hat{I} + \hat{X} \otimes \hat{X} + \hat{Y} \otimes \hat{Y} + \hat{Z} \otimes \hat{Z}$$

$$= \begin{pmatrix} 1 & 0 \\ 0 & 1 \end{pmatrix} \otimes \begin{pmatrix} 1 & 0 \\ 0 & 1 \end{pmatrix} + \begin{pmatrix} 0 & 1 \\ 1 & 0 \end{pmatrix} \otimes \begin{pmatrix} 0 & 1 \\ 1 & 0 \end{pmatrix} + \begin{pmatrix} 0 & -i \\ i & 0 \end{pmatrix} \otimes \begin{pmatrix} 0 & -i \\ i & 0 \end{pmatrix} + \begin{pmatrix} 1 & 0 \\ 0 & -1 \end{pmatrix} \otimes \begin{pmatrix} 1 & 0 \\ 0 & -1 \end{pmatrix}$$

$$= \begin{pmatrix} 1 & 0 & 0 & 0 \\ 0 & 1 & 0 & 0 \\ 0 & 0 & 1 & 0 \\ 0 & 0 & 0 & 1 \end{pmatrix} + \begin{pmatrix} 0 & 0 & 0 & 1 \\ 0 & 0 & 1 & 0 \\ 0 & 1 & 0 & 0 \\ 1 & 0 & 0 & 0 \end{pmatrix} + \begin{pmatrix} 0 & 0 & 0 & -1 \\ 0 & 0 & 1 & 0 \\ 0 & 1 & 0 & 0 \\ -1 & 0 & 0 & 0 \end{pmatrix} + \begin{pmatrix} 1 & 0 & 0 & 0 \\ 0 & -1 & 0 & 0 \\ 0 & 0 & -1 & 0 \\ 0 & 0 & 0 & 1 \end{pmatrix}$$

$$= \begin{pmatrix} 2 & 0 & 0 & 0 \\ 0 & 0 & 2 & 0 \\ 0 & 2 & 0 & 0 \\ 0 & 0 & 0 & 2 \end{pmatrix} = 2 \begin{pmatrix} 1 & 0 & 0 & 0 \\ 0 & 0 & 1 & 0 \\ 0 & 1 & 0 & 0 \\ 0 & 0 & 0 & 1 \end{pmatrix}$$

$$=2\hat{U}_{\text{SWAP}}$$

Hence $\hat{U}_{\text{SWAP}} = \frac{1}{2}[\hat{I} \otimes \hat{I} + \hat{X} \otimes \hat{X} + \hat{Y} \otimes \hat{Y} + \hat{Z} \otimes \hat{Z}]$

5.13.1 Construction of SWAP gate using three CNOT gates

It is possible to build a SWAP gate employing three CNOT gates as hinted at in figure 5.14.

Figure 5.15 explains direct and reverse connections of CNOT gate.

Figure 5.16 explains the process of operation of SWAP gate that employs three CNOT gates (two connected directly and one connected in the reverse). It is found that $|00>$ is mapped to $|00>$, $|01>$ is mapped to $|10>$, $|10>$ is mapped to $|01>$ and $|11>$ is mapped to $|11>$.

5.14 Controlled U gates

Consider $\hat{U}$ to be an arbitrary single qubit (unitary) gate represented by the matrix

$$\hat{U} = \begin{pmatrix} U_{00} & U_{01} \\ U_{10} & U_{11} \end{pmatrix}$$

With it we can construct a two-qubit controlled gate denoted by say $\hat{U}_{\text{contrl}}$ as follows (figure 5.17)

$\hat{U}$ operates on target qubit only if control qubit is $|1>$.

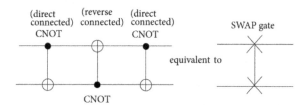

Figure 5.14. Construction of SWAP gate with 3 CNOT gates.

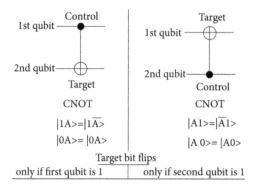

Figure 5.15. Explanation of direct and reverse connected CNOT.

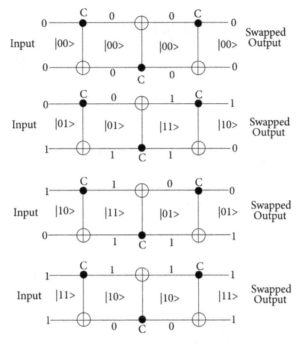

Figure 5.16. Explanation of working of SWAP using 3 CNOTs (C = Control).

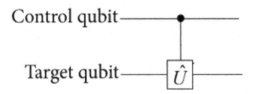

Figure 5.17. Controlled U gate.

If control qubit is 0 the $\hat{U}$ will not operate on target. Hence

$$\hat{U}_{\text{contrl}}|00> =|00> \qquad \hat{U}_{\text{contrl}}|01> =|01>$$

If control qubit is 1 then $\hat{U}$ will operate on target. Hence

- $\hat{U}_{\text{contrl}}|10> =|1> \otimes \hat{U}|0> =|1> \otimes \begin{pmatrix} U_{00} & U_{01} \\ U_{10} & U_{11} \end{pmatrix}\begin{pmatrix} 1 \\ 0 \end{pmatrix} = |1> \otimes \begin{pmatrix} U_{00} \\ U_{10} \end{pmatrix}$

$$=|1> \otimes \left[U_{00}\begin{pmatrix} 1 \\ 0 \end{pmatrix} + U_{10}\begin{pmatrix} 1 \\ 0 \end{pmatrix} \right] = |1> \otimes(U_{00}|0> +U_{10}|1>)$$

- $\hat{U}_{\text{contrl}}|11> =|1> \otimes \hat{U}|1> =|1> \otimes \begin{pmatrix} U_{00} & U_{01} \\ U_{10} & U_{11} \end{pmatrix}\begin{pmatrix} 0 \\ 1 \end{pmatrix} = |1> \otimes \begin{pmatrix} U_{01} \\ U_{11} \end{pmatrix}$

$$= |1> \otimes \left[U_{01}\begin{pmatrix}1\\0\end{pmatrix} + U_{11}\begin{pmatrix}0\\1\end{pmatrix} \right] = |1> \otimes(U_{01}|0> + U_{11}|1>)$$

In Dirac's bra ket notation the explicit form of $\hat{U}_{contrl}$ is (noting that

$$|00> \xrightarrow{\hat{U}_{contrl}} |00>, \quad |01> \xrightarrow{\hat{U}_{contrl}} |01>,$$

$$|10> \xrightarrow{\hat{U}_{contrl}} \hat{U}_{contrl}|10> = |1> \otimes \begin{pmatrix}U_{00}\\U_{10}\end{pmatrix},$$

$$|11> \xrightarrow{\hat{U}_{contrl}} \hat{U}_{contrl}|11> = 1> \otimes \begin{pmatrix}U_{01}\\U_{11}\end{pmatrix}$$

$$\hat{U}_{contrl} = |00><00| + |01><01| + \left(|1> \otimes \begin{pmatrix}U_{00}\\U_{10}\end{pmatrix}\right)<10| + \left(|1> \otimes \begin{pmatrix}U_{01}\\U_{11}\end{pmatrix}\right)<11|$$

$$= \begin{pmatrix}1\\0\\0\\0\end{pmatrix}(1\ 0\ 0\ 0) + \begin{pmatrix}0\\1\\0\\0\end{pmatrix}(0\ 1\ 0\ 0)$$

$$+ \left(\begin{pmatrix}0\\1\end{pmatrix} \otimes \begin{pmatrix}U_{00}\\U_{10}\end{pmatrix}\right)(0\ 0\ 1\ 0) + \left(\begin{pmatrix}0\\1\end{pmatrix} \otimes \begin{pmatrix}U_{01}\\U_{11}\end{pmatrix}\right)(0\ 0\ 0\ 1)$$

$$= \begin{pmatrix}1&0&0&0\\0&0&0&0\\0&0&0&0\\0&0&0&0\end{pmatrix} + \begin{pmatrix}0&0&0&0\\0&1&0&0\\0&0&0&0\\0&0&0&0\end{pmatrix} + \begin{pmatrix}0\\0\\U_{00}\\U_{10}\end{pmatrix}(0\ 0\ 1\ 0) + \begin{pmatrix}0\\0\\U_{01}\\U_{11}\end{pmatrix}(0\ 0\ 0\ 1)$$

$$\hat{U}_{contrl} = \begin{pmatrix}1&0&0&0\\0&0&0&0\\0&0&0&0\\0&0&0&0\end{pmatrix} + \begin{pmatrix}0&0&0&0\\0&1&0&0\\0&0&0&0\\0&0&0&0\end{pmatrix} + \begin{pmatrix}0&0&0&0\\0&0&0&0\\0&0&U_{00}&0\\0&0&U_{10}&0\end{pmatrix} + \begin{pmatrix}0&0&0&0\\0&0&0&0\\0&0&0&U_{01}\\0&0&0&U_{11}\end{pmatrix}$$

$$\hat{U}_{contrl} = \begin{pmatrix}1&0&0&0\\0&1&0&0\\0&0&U_{00}&U_{01}\\0&0&U_{10}&U_{11}\end{pmatrix} = \begin{pmatrix}\hat{I}&0\\0&\hat{U}\end{pmatrix} \quad \text{where } \hat{I} = \begin{pmatrix}1&0\\0&1\end{pmatrix}$$

5.14.1 Note on controlled $\hat{X}$, $\hat{Y}$, $\hat{Z}$ gates

For $\hat{U} = \hat{X}$, $\hat{U}_{CX} = \begin{pmatrix} \hat{I} & 0 \\ 0 & \hat{X} \end{pmatrix}$ and so controlled $\hat{U}$ gate reduces to controlled NOT gate or controlled $\hat{X}$ gate or CX gate.

For $\hat{U} = \hat{Y}$, $\hat{U}_{CY} = \begin{pmatrix} \hat{I} & 0 \\ 0 & \hat{Y} \end{pmatrix}$ and so controlled $\hat{U}$ gate reduces to controlled $\hat{Y}$ gate.

For $\hat{U} = \hat{Z}$, $\hat{U}_{CZ} = \begin{pmatrix} \hat{I} & 0 \\ 0 & \hat{Z} \end{pmatrix}$ and so controlled $\hat{U}$ gate reduces to controlled $\hat{Z}$ gate or CZ gate.

Figure 5.18 shows controlled $\hat{X}$, $\hat{Y}$, $\hat{Z}$ gates.

Similarly, we can construct controlled $\hat{T}$ gate $\hat{U}_{CT} = \begin{pmatrix} \hat{I} & 0 \\ 0 & \hat{T} \end{pmatrix}$, controlled $\hat{P}$ gate $\hat{U}_{CP} = \begin{pmatrix} \hat{I} & 0 \\ 0 & \hat{P} \end{pmatrix}$, controlled $\hat{R}_x$ gate $\hat{U}_{CR_x} = \begin{pmatrix} \hat{I} & 0 \\ 0 & \hat{R}_x \end{pmatrix}$, controlled $\hat{V}$ gate $\hat{U}_{CV} = \begin{pmatrix} \hat{I} & 0 \\ 0 & \hat{V} \end{pmatrix}$, controlled $\hat{V}^{\dagger}$ gate $\hat{U}_{CV^{\dagger}} = \begin{pmatrix} \hat{I} & 0 \\ 0 & \hat{V}^{\dagger} \end{pmatrix}$

NCV quantum gate library

NOT, CNOT, controlled $\hat{V}$, controlled $\hat{V}^{\dagger}$ form a universal quantum gate library for reversible circuits referred to as NCV quantum gate library.

5.15 Toffoli quantum gate or CCNOT gate (controlled controlled NOT gate)

Toffoli gate or CCNOT gate is a three-qubit gate. It has three input qubits A,B,C of which two (say A,B) are called control qubits and one (i.e. C) is the target qubit. The CCNOT quantum gate has been depicted in figure 5.19.

If $A = B = 1$ (i.e. if $AB = 1.1=1$) then C is flipped, i.e. $|1\ 1\ C> \rightarrow |1\ 1\ \overline{C}>$ e.g. $|110> \rightarrow |111>$, $|111> \rightarrow |110>$.

For other inputs of A, B (i.e. for $A \neq 1$, $B \neq 1$) the qubit C is unaffected, i.e. $|A \neq 1\ B \neq 1\ C> \rightarrow |A\ B\ C>$ e.g. $|100> \rightarrow |100>$, $|010> \rightarrow |010>$, $|000> \rightarrow |000>$.

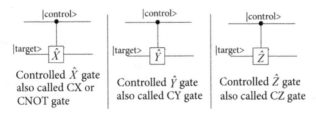

Controlled $\hat{X}$ gate also called CX or CNOT gate

Controlled $\hat{Y}$ gate also called CY gate

Controlled $\hat{Z}$ gate also called CZ gate

Figure 5.18. Controlled $\hat{X}$, $\hat{Y}$, $\hat{Z}$ gates.

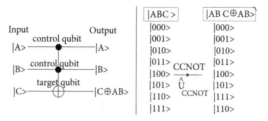

Figure 5.19. Toffoli gate or CCNOT quantum gate.

The target qubit C thus gives an output given by

$$C \oplus AB = \begin{cases} C \oplus 0 = C \text{ (if } AB = 0 \text{ (both } A, B \text{ not 1))} \\ C \oplus 0 = \bar{C} \text{ (if } AB = 1 \text{ (both } A, B, \text{ are 1))} \end{cases}$$

Toffoli gate or CCNOT gate operations can equivalently be written as

$$|A\ B\ C> \xrightarrow{\text{Toffli Gate}} |A\ B\ C \oplus AB>$$

The bra ket representation of Toffoli gate or CCNOT gate will thus be (figure 5.19)

$$\hat{U}_{\text{CCNOT}} = |000><000| + |001><001| + |010><010| + |011><011|$$

$$+|100><100| + |101><101| + |111><110| + |110><111|$$

The matrix representation of $\hat{U}_{\text{CCNOT}}$ is the following unitary matrix

$$|A\ B\ C \oplus AB> = \hat{U}_{\text{CCNOT}}|ABC>$$

$$\begin{pmatrix} |000> \\ |001> \\ |010> \\ |011> \\ |100> \\ |101> \\ |111> \\ |110> \end{pmatrix} = \begin{pmatrix} 1 & 0 & 0 & 0 & 0 & 0 & 0 & 0 \\ 0 & 1 & 0 & 0 & 0 & 0 & 0 & 0 \\ 0 & 0 & 1 & 0 & 0 & 0 & 0 & 0 \\ 0 & 0 & 0 & 1 & 0 & 0 & 0 & 0 \\ 0 & 0 & 0 & 0 & 1 & 0 & 0 & 0 \\ 0 & 0 & 0 & 0 & 0 & 1 & 0 & 0 \\ 0 & 0 & 0 & 0 & 0 & 0 & 0 & 1 \\ 0 & 0 & 0 & 0 & 0 & 0 & 1 & 0 \end{pmatrix} \begin{pmatrix} |000> \\ |001> \\ |010> \\ |011> \\ |100> \\ |101> \\ |110> \\ |111> \end{pmatrix}$$

5.16 Controlled SWAP gate or CS gate or Fredkin gate

Fredkin gate is a three-qubit gate, shown in figure 5.20, that performs a controlled swap. It has three inputs and three outputs. The first bit A is the control bit. The first bit A transmits unchanged, i.e. $|A'> = |A>$.

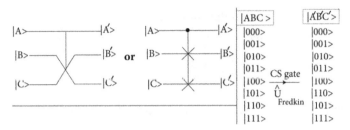

Figure 5.20. Controlled SWAP gate or Fredkin gate.

If A is 0, other bits B, C remain unchanged, i.e. If $A = 0$ then $|B' > =|B>$, $|C' > =|C>$ or $|0 \ B \ C > \rightarrow |0 \ B \ C>$.

If A is 1, other bits B, C get swapped, i.e. if $A = 1$ then $|B' > =|C>, |C' > =|B>$ or $|1 \ B \ C > \rightarrow |1 \ C \ B>$.

The bra ket representation of Fredkin gate will thus be (figure 5.20)

$$\hat{U}_{\text{Fredkin}} = |000 > <000| + |001 > <001| + |010 > <010| + |011 > <011|$$

$$+|100 > <100| + |110 > <101| + |101 > <110| + |111 > <111|$$

The matrix representation of $\hat{U}_{\text{Fredkin}}$ is the following unitary matrix

$$|A' \ B' \ C' > =\hat{U}_{\text{Fredkin}}|ABC>$$

$$
\begin{pmatrix} |000> \\ |001> \\ |010> \\ |011> \\ |100> \\ |111> \\ |101> \\ |111> \end{pmatrix} = \begin{pmatrix} 1 & 0 & 0 & 0 & 0 & 0 & 0 & 0 \\ 0 & 1 & 0 & 0 & 0 & 0 & 0 & 0 \\ 0 & 0 & 1 & 0 & 0 & 0 & 0 & 0 \\ 0 & 0 & 0 & 1 & 0 & 0 & 0 & 0 \\ 0 & 0 & 0 & 0 & 1 & 0 & 0 & 0 \\ 0 & 0 & 0 & 0 & 0 & 0 & 1 & 0 \\ 0 & 0 & 0 & 0 & 0 & 1 & 0 & 1 \\ 0 & 0 & 0 & 0 & 0 & 0 & 0 & 1 \end{pmatrix} \begin{pmatrix} |000> \\ |001> \\ |010> \\ |011> \\ |100> \\ |101> \\ |110> \\ |111> \end{pmatrix}
$$

5.17 Deutsch gate

The Deutsch gate (a three-qubit gate) was introduced by Deutsch as a universal quantum gate. The matrix representation of the Deutsch gate is given by

$$
\hat{D}(\theta) = \begin{pmatrix} 1 & 0 & 0 & 0 & 0 & 0 & 0 & 0 \\ 0 & 1 & 0 & 0 & 0 & 0 & 0 & 0 \\ 0 & 0 & 1 & 0 & 0 & 0 & 0 & 0 \\ 0 & 0 & 0 & 1 & 0 & 0 & 0 & 0 \\ 0 & 0 & 0 & 0 & 1 & 0 & 0 & 0 \\ 0 & 0 & 0 & 0 & 0 & 1 & 0 & 0 \\ 0 & 0 & 0 & 0 & 0 & 0 & i\cos\theta & \sin\theta \\ 0 & 0 & 0 & 0 & 0 & 0 & \sin\theta & i\cos\theta \end{pmatrix}
$$

where θ is a constant angle such that $\frac{2\theta}{\pi}$ is an irrational number. It forms a universal quantum gate library. However, this gate is not popular because the circuits built using it are not efficient.

- Note that $\hat{D}(\frac{\pi}{2}) = \hat{U}_{\text{CCNOT}} =$ Toffli gate (section 5.15).

5.18 Implementing classical computation by quantum gates

Implementation of some classical operations has been discussed using quantum gates.

5.18.1 Implementation of NOT operation by quantum gate

Classical NOT operation (which is reversible) can be performed

- using Pauli $\hat{X}$ gate $\hat{X}a = \bar{a}$, i.e. $a \xrightarrow{\hat{X}} \bar{a}$ (figure 5.21) or
- using CCNOT (i.e. Toffli) gate (figure 5.21) with control input qubits 1,1, i.e. $|1\ 1\ C> \rightarrow |1\ 1\ \bar{C}>$, i.e. $C \xrightarrow{\text{CCNOT}} \bar{C}$.

5.18.2 Implementation of XOR operation

XOR is addition modulo 2 (chapter 8). For a two-bit input, only if one of the bits is 0 and the other bit is 1 do we get 1, otherwise we get 0. XOR operation can be performed

- using CNOT gate (figure 5.22). So $|a\ b> \xrightarrow{\text{CNOT}} |a\ a \oplus b>$
- using CCNOT gate (figure 5.22). One of the controls is set to 1 while another control is set to x. Thus the target y is flipped only for $x = 1$. So $|1\ x\ y> \xrightarrow{\text{CCNOT}} |1\ x\ y \oplus x>$.

5.18.3 Implementation of AND operation

Implementation of AND operation is possible by a CCNOT gate by setting the target bit $c = 0$. The CCNOT operation is $|a\ b\ c> \xrightarrow{\text{CCNOT}} |a\ b\ c \oplus ab>$. For

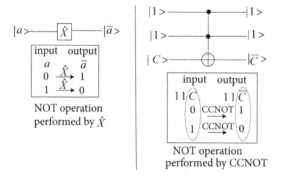

Figure 5.21. Implementation of NOT operation by quantum gate.

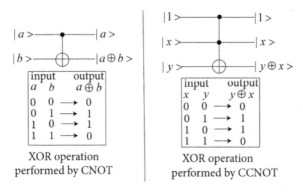

Figure 5.22. Implementation of XOR operation by quantum gate.

$c = 0$ we have $|a\ b\ 0 > \xrightarrow{\text{CCNOT}} |a\ b\ 0 \oplus ab >$. Let us study the value of $|0 \oplus ab>$ in details.

$$|0\oplus 0.0 > = |0\oplus 0 > = |0 > = \ |0.0>$$

$$|0\oplus 0.1 > = |0\oplus 0 > = |0 > = \ |0.1>$$

$$|0\oplus 1.0 > = |0\oplus 0 > = |0 > = \ |1.0>$$

$$|0\oplus 1.1 > \ = |0\oplus 1 > = |1 > \ = \ \ |1.1>$$

Clearly we can write $|0 \oplus ab > = |ab>$.

So $|a\ b\ 0 > \xrightarrow{\text{CCNOT}} |a\ b\ ab >$. This is AND operation, as depicted in figure 5.23.

5.19 Plan of a quantum circuit

A quantum circuit contains logic gates connected by straight lines which do not represent physical wires but indicate the direction of logic flow with time. 'Earlier time' is indicated to the left, i.e. time flows from left to right.

Input is shown in the left extreme part of a circuit and output is placed in the right. In a quantum circuit, inputs are qubits, generally in computational basis and the outputs are qubits also.

In a classical algorithm, looping is allowed, i.e. controlled repetition of a particular logic in a classical circuit can occur. This is not allowed in a quantum circuit since here there is flow of logic with time and looping will necessitate travelling back in time which is not possible. In other words, quantum circuits are acyclic, meaning that feedback from one part of the quantum circuit to another part is not allowed.

In classical circuits wires carrying bits are joined together. The resulting single wire contains OR of input bits, i.e. several inputs give rise to some output. This operation is called Fan-in (and by Fan-in we refer to the maximum number of digital inputs that a single logic gate can accept). But such an operation is not reversible, nor unitary. So Fan-in is not allowed in quantum circuits (figure 5.24).

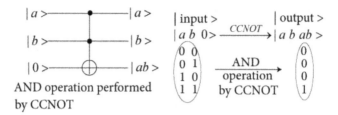

Figure 5.23. Implementation of AND operation by quantum gate.

Fan - in not allowed   Fan - out not allowed

Figure 5.24. Fan-in and fan-out are not allowed in a quantum circuit.

By the term Fan-out we refer to the maximum number of logic gates that can be driven by the output of a logic gate. Clearly Fan-out is an inverse operation to Fan-in, whereby several copies of the bit are fed to several logic gates. Just like Fan-in, its inverse Fan-out is also not permitted in a quantum circuit. Actually, quantum mechanics forbids copying of a qubit.

We now discuss some specific quantum circuits.

5.20 Quantum half adder circuit

A half adder (which is a two-bit binary adder) performs binary addition of two bits viz.

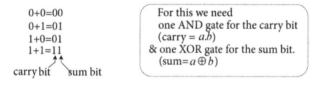

Quantum half adder has been depicted and explained in figure 5.25. (The plan of classical half adder has also been shown in the figure.)

Quantum half adder comprises

- CCNOT gate to perform the carry part : $| a \ b \ c \rangle \xrightarrow{\text{CCNOT}} | a \ b \ c \oplus ab \rangle$
  The target qubit of CCNOT is set to $c = 0$. So we have

$$| a \ b \ 0 \rangle \xrightarrow{\text{CCNOT}} | a \ b \ 0 \oplus ab \rangle = | a \ b \ ab \rangle \rightarrow | ab \rangle = | \text{carry} \rangle$$

- CNOT gate to perform the sum part : $| a \ b \rangle \xrightarrow{\text{CNOT}} | a \ a \oplus b \rangle \rightarrow | a \oplus b \rangle = | \text{sum} \rangle$

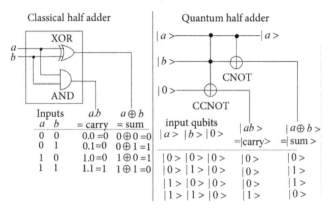

Figure 5.25. Plan of a quantum half adder.

In CCNOT gate, for any of the control qubits $|a>$, $|b>$ equal to $|0>$ the target qubit $|0>$ remains unchanged. So $|carry> =|0>$ in the first three rows. Again, if the control qubits both are $|a> =|b> =|1>$ then the target qubit $|0>$ flips. So $|carry> =|1>$ in the last row.

In CNOT gate for the control qubit $|a> =|0>$ the target qubit $|b>$ remains unchanged. So $|sum> =|0>$ for $|b> =|0>$ (first row) and $|sum> =|1>$ for $|b> =|1>$ (second row).

Again if control qubit $|a> =|1>$ then the target qubit flips. So $|sum> =|1>$ for $|b> =|0>$ (third row) and $|sum> =|0>$ for $|b> =|1>$ (fourth row).

5.21 Quantum full adder circuit

A classical full adder (which is a three-bit binary adder) performs binary addition of 3 bits viz.

		For this we need
(all 0)	0+0+0=00	three AND gates
(two 0)	0+0+1=01	& one OR gate for the carry bit
(one 0)	0+1+1=10	& one XOR gate for the sum bit.
(no 0)	1+1+1=11	

carry bit   sum bit

Quantum full adder has been depicted and explained in figure 5.26. (The plan of classical full adder has also been shown in the figure.)

Quantum full adder circuit uses

- CCNOT gate: $| a\ b\ c > \xrightarrow{\text{CCNOT}} |a\ b\ c \oplus ab > \rightarrow |c \oplus ab>$

  $\xrightarrow{c=0} |a\ b\ 0 \oplus ab > = |a\ b\ ab > \rightarrow |ab>$

- CNOT gate: $|a\ b > \xrightarrow{\text{CNOT}} |a\ a \oplus b > \rightarrow |a \oplus b>$

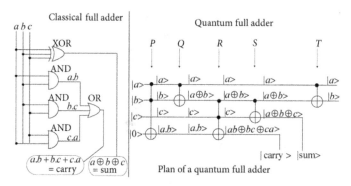

Figure 5.26. Full adder (classical and quantum).

Quantum full adder starts with four qubit registers labeled as $|a>$, $|b>$, $|c>$ and the fourth register is in state $|0>$. We move from left to right.

At point P

Apply CCNOT acting as AND gate: Controls are $|a>$ and $|b>$ and target is $|0>$. State obtained is $|a.b>$

At point Q

Apply CNOT acting as XOR: Control is $|a>$. Target is $|b>$.

State obtained is $|a \oplus b>$.

At point R

Apply CCNOT: Controls are $|a \oplus b>$ and $|c>$. Target is $|a.b>$.

State obtained is $|(a.b) \oplus [(a \oplus b).c]> = |(a.b) \oplus [a.c \oplus b.c]>$

$$=|a.b \oplus b.c \oplus c.a> =|carry>$$

At point S

Apply CNOT acting as XOR: Control is $|a \oplus b>$. Target is $|c>$.

State obtained is $|a \oplus b \oplus c> =|sum>$

At point T

Apply CNOT acting as XOR: Control is $|a>$. Target is $|a \oplus b>$.

State obtained is $|a \oplus (a \oplus b)> = |a \oplus a \oplus b> = |0 \oplus b> = |b>$ (we get back the inputs)

Clearly $|a.b \oplus b.c \oplus c.a> = |carry>$, $|a \oplus b \oplus c> = |sum>$.

5.22 Oracle (black box) in quantum computer

If we want to compute a function in classical computing we give a subroutine call. Subroutine refers to a set of instructions that are employed repeatedly in a program. One copy of this instruction is stored in the memory and, in general, it may be called many times, i.e. repeatedly called during execution of a program.

In quantum computing, estimation of computational cost of an algorithm is done in terms of number of calls to a computational mathematical device called the oracle (figure 5.27).

Oracle is basically a black box computation. It receives an input, does computation and gives an output. The term black box means that we cannot see inside of it

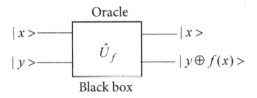

Figure 5.27. Oracle in quantum computing.

and we do not know what is going on inside of it. We just interrogate the oracle and discover its properties. So the oracle represents operations that we do not discuss in detail. An output is generated by the quantum oracle.

A typical model of quantum computation consists of input register, target register (where output will be stored), a black box or a quantum oracle which would do a computational task.

The output might be a linear combination of states corresponding to the linear combination of inputs given. If we measure the output there will not be any information about every output but one random result will be measured.

If $y = 0$ then output is $|0 \oplus f(x)> = |f(x)>$.

If $y = 1$ then output is $|1 \oplus f(x)> = |\overline{f}(x)>$.

These results will be easier to comprehend from the following XOR chart.

XOR operation

0 XOR 0 = 0	1 XOR 0 = 1=$\overline{0}$
0 XOR 1 = 1	1 XOR 1 = 0=$\overline{1}$
0 XOR f(x) = f(x)	1 XOR f(x) = $\overline{f(x)}$

5.23 Hadamard transformation on each of n qubits leads to a linear superposition of 2^n states

When one qubit $|0>$ is passed through one Hadamard gate we write as follows

$$\hat{H}^{\otimes 1}|0>^{\otimes 1} = \hat{H}|0> = \frac{1}{\sqrt{2}} (|0> +|1>)$$

What we get is a linear superposition of $2^1 = 2$ basis states of equal strength.

When two qubits $|0>^{\otimes 2} = |0> \otimes |0>$ are passed through two Hadamard gates (denoted by $\hat{H}^{\otimes 2}$) we have

$$\hat{H}^{\otimes 2}|0>^{\otimes 2} = \hat{H}|0> \otimes \hat{H}|0> = \frac{1}{\sqrt{2}} (|0> +|1>) \otimes \frac{1}{\sqrt{2}} (|0> +|1>)$$

$$= \frac{1}{\sqrt{2^2}}[|00> +|01> +|10> +|11>]$$

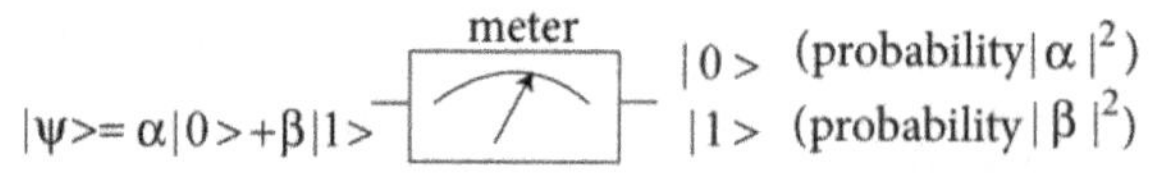

Figure 5.28. Circuit symbol for measurement of quantum state.

This is a linear superposition of $2^2 = 4$ basis states of equal strength.

When n qubits $|0>^{\otimes n} = |0> \otimes|0> \otimes.......$ are passed through n Hadamard gates (denoted by $\hat{H}^{\otimes n}$) we have

$$\hat{H}^{\otimes n}|0>^{\otimes n} = \frac{1}{\sqrt{2^n}}\sum_x |x>$$

where $|x>$ represents uniform superposition of n qubit basis states. So $\hat{H}^{\otimes n}|0>^{\otimes n}$ creates an equal superposition of 2^n possible logic states.

5.24 Process of measurement

Measurement process is generally in the computational basis and is represented by a meter-like symbol as shown in figure 5.28.

There is certain probability to measure various states at the output. But when we make measurement, out of the various possible output states only one output state is projected out.

A complete quantum circuit consists of input, the logic gate, oracle and the measurement meter.

5.25 Quantum coin flipping

Consider a game being played between two persons whom we, in quantum computing discussion, generally give the name Alice and Bob.

Suppose Bob has a quantum coin and he starts with a particular quantum state either say $|Head > = |H > = |0>$ or say $|Tail > = |T > = |1>$. The quantum coin now is passed over to Alice who has the liberty to flip the state or not flip the state. After her action on the quantum coin (flip or no flip) it is returned to Bob. If Bob measures and finds the state he started with he wins the game.

It can be argued (using principle of superposition) that irrespective of whether Alice has flipped the quantum state or not, Bob is sure to win (i.e. Bob has 100% chance of winning) this quantum coin flipping game (though classically Bob has only 50% chance of winning the game.).

Argument

Suppose Bob keeps the initial state of the quantum coin as $|Head> = |0>$. And he applies a Hadamard gate to this quantum state

$$\hat{H}|0> = \frac{1}{\sqrt{2}}(|0> +|1>) = |+>$$

This creates an equal superposition of the $|0>$ and $|1>$ states (as probability of measuring $|0>$ is $\left|\frac{1}{\sqrt{2}}\right|^2 = \frac{1}{2} = 50\%$ and that of $|1>$ is also $\left|\frac{1}{\sqrt{2}}\right|^2 = \frac{1}{2} = 50\%$).

- Bob passes the quantum coin to Alice who has the choice to flip the qubit by applying $\hat{X}$ gate and suppose she flips
  $\hat{X}|+>=\hat{X}\frac{1}{\sqrt{2}}(|0> +|1 >)=\frac{1}{\sqrt{2}}(|1 > +|0 >) = |+>$ ($\Rightarrow$ no change of quantum state occurs).
- On the other hand if Alice prefers not to flip, that means she does not apply $\hat{X}$ gate and the state is still $|+>$ ($\Rightarrow$ no change of quantum state occurs).

As the quantum state passes back to Bob he applies a Hadamard gate

$$\hat{H}|+>=|0>$$

which is the initial state he started with. So Bob wins the game. Clearly Bob has 100% chance to win the quantum coin flipping game and it does not depend upon flip or no flip by Alice.

Classically if Bob starts with Head H and Alice flips it to Tail T then Bob loses. Again, if Alice does not flip then the state remains as Head H and Bob wins. So Bob has only 50% chance of winning a classical coin flip game.

5.26 Further reading

[1] Pathak A 2016 *Elements of Quantum Computing and Quantum Communication* (Boca Raton, FL: CRC Press)
[2] Nielsen M A and Chuang I L 2019 *Quantum Computation and Quantum Information* (Cambridge: Cambridge University Press)
[3] Nakahara M and Ohmi T 2016 *Quantum Computing* (Boca Raton, FL: CRC Press)
[4] Ghosh D Online course on quantum information and computing (Bombay: Indian Institute of Technology)

5.27 Problems

1 *Which of the following operators represent a NOT gate?*
   (a)$|0 > <1|$, *(b)* $|0 > <0| + |1 > <1|$,
   (c) $|0 > <1| + |1 > <0|$, *(d)* $|1 > <0|$.
   Ans. (c).

2 *The effect of $\hat{X}$ gate on $|\pm>$ is correctly expressed in which of the following?*
   (a) $|+>\rightarrow| + >$, $|->\rightarrow-| - >$, *(b)* $|\pm>\rightarrow| \mp >$,
   (c) $|+>\rightarrow-| - >$, $|->\rightarrow| - >$, *(d)* $|+>\rightarrow-| - >$, $|->\rightarrow-| - >$.
   Ans. (*a*).

3 *Depict the state $\frac{\sqrt{3}}{2}|0 > +\frac{1}{2}|1>$ on the Bloch sphere before and after application of NOT gate.*

   Ans. $\frac{\sqrt{3}}{2}|0 > +\frac{1}{2}|1 > \xrightarrow{\text{NOT}} \frac{\sqrt{3}}{2}|1 > +\frac{1}{2}|0 > .$

   Comparing with $|\psi(\theta, \phi) > = \cos \frac{\theta}{2}|0 > +e^{i\phi} \sin \frac{\theta}{2}|1>$

   $|\text{initial} > \rightarrow \cos \frac{\theta}{2} = \frac{\sqrt{3}}{2}$, $e^{i\phi} \sin \frac{\theta}{2} = \frac{1}{2} \Rightarrow e^{i\phi} = 1$, $\cos \frac{\theta}{2} = \frac{\sqrt{3}}{2}$, $\sin \frac{\theta}{2} = \frac{1}{2}$

$$\Rightarrow \phi = 0°, \frac{\theta}{2} = 30°, \theta = 60°$$

$$|final> \rightarrow \cos\frac{\theta}{2} = \frac{1}{2}, e^{i\phi}\sin\frac{\theta}{2} = \frac{\sqrt{3}}{2} \Rightarrow e^{i\phi} = 1, \cos\frac{\theta}{2} = \frac{1}{2}, \sin\frac{\theta}{2} = \frac{\sqrt{3}}{2}$$

$$\Rightarrow \phi = 0°, \frac{\theta}{2} = 60°, \theta = 120°$$

The initial state and the final state have been depicted on Bloch sphere in figure 5.29.

$$(r, \theta, \phi) : (1,60°, 0°) \rightarrow (1,120°, 0°)$$

4 *What is the output of the circuit of figure 5.30?*

Ans. Input $=|1> \otimes |0> = |10>$

Hadamard to qubit 1 gives $|1> \xrightarrow{H} \frac{|0> - |1>}{\sqrt{2}}$

Input to CNOT is $\frac{|0> - |1>}{\sqrt{2}} \otimes |0> = \frac{|00> - |10>}{\sqrt{2}}$

CNOT is applied. Control = first qubit, Target = second qubit.

$$\frac{|00> - |10>}{\sqrt{2}} \xrightarrow{CNOT} \frac{|00> - |11>}{\sqrt{2}}$$

$\hat{X}$ gate on first qubit $(|0> \xrightarrow{\hat{X}} |1>, |1> \xrightarrow{\hat{X}} |0>)$ and $\hat{Z}$ gate on second

qubit $(|0> \xrightarrow{\hat{Z}} |0>, |1> \xrightarrow{\hat{Z}} -|1>)$ gives

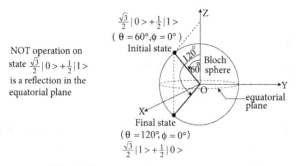

Figure 5.29. Depiction of state $\frac{\sqrt{3}}{\sqrt{2}}|0>+\frac{\sqrt{1}}{\sqrt{2}}|1>$ on the Bloch sphere before and after application of NOT gate.

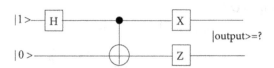

Figure 5.30. Circuit of problem 4.

$$\frac{|00> - |11>}{\sqrt{2}} \xrightarrow{\hat{X} \text{ on first qubit}} \frac{|10> - |01>}{\sqrt{2}} \xrightarrow{\hat{Z} \text{ on second qubit}} \frac{|10> + |01>}{\sqrt{2}} = |\text{output}>$$

5 *Obtain matrix representation of the circuit of figure 5.31.*

Ans. We study the output for the following inputs: $|00>$, $|01>$, $|10 >$, $|11>$

Circuit operation
represented by $\hat{U}$
(C=Control , T= Target)

$	00>$		1st	$	00>$	2nd		$	00>$
$	01>$		CNOT	$	01>$	CNOT		$	11>$
$	10>$		[C=1st qubit	$	11>$	[C=2nd qubit		$	01>$
$	11>$		T=2nd qubit]	$	10>$	T=1st qubit]		$	10>$

The matrix representation of the circuit can be obtained by writing $\hat{U}$ in the bra ket notation as

$$\hat{U} = |00 > <00| + |11 > <01| + |01 > <10| + |10 > <11|$$

$$= \begin{pmatrix} 1 \\ 0 \\ 0 \\ 0 \end{pmatrix} (1 \ 0 \ 0 \ 0) + \begin{pmatrix} 0 \\ 0 \\ 0 \\ 1 \end{pmatrix} (0 \ 1 \ 0 \ 0) + \begin{pmatrix} 0 \\ 1 \\ 0 \\ 0 \end{pmatrix} (0 \ 0 \ 1 \ 0) + \begin{pmatrix} 0 \\ 0 \\ 1 \\ 0 \end{pmatrix} (0 \ 0 \ 0 \ 1)$$

$$= \begin{pmatrix} 1 & 0 & 0 & 0 \\ 0 & 0 & 0 & 0 \\ 0 & 0 & 0 & 0 \\ 0 & 0 & 0 & 0 \end{pmatrix} + \begin{pmatrix} 0 & 0 & 0 & 0 \\ 0 & 0 & 0 & 0 \\ 0 & 0 & 0 & 0 \\ 0 & 1 & 0 & 0 \end{pmatrix} + \begin{pmatrix} 0 & 0 & 0 & 0 \\ 0 & 0 & 1 & 0 \\ 0 & 0 & 0 & 0 \\ 0 & 0 & 0 & 0 \end{pmatrix} + \begin{pmatrix} 0 & 0 & 0 & 0 \\ 0 & 0 & 0 & 0 \\ 0 & 0 & 0 & 1 \\ 0 & 0 & 0 & 0 \end{pmatrix}$$

$$= \begin{pmatrix} 1 & 0 & 0 & 0 \\ 0 & 0 & 1 & 0 \\ 0 & 0 & 0 & 1 \\ 0 & 1 & 0 & 0 \end{pmatrix}$$

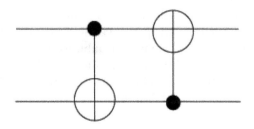

Figure 5.31. Circuit of problem 5.

NOTE:

$$
\begin{array}{cccc}
|00\rangle & |01\rangle & |10\rangle & |11\rangle
\end{array}
$$

$$
\begin{pmatrix}
1 & 0 & 0 & 0 \\
0 & 1 & 0 & 0 \\
0 & 0 & 1 & 0 \\
0 & 0 & 0 & 1
\end{pmatrix}
\xrightarrow[\text{of bases}]{\text{The transformation}}
\begin{array}{cccc}
|00\rangle & |11\rangle & |01\rangle & |10\rangle
\end{array}
\begin{pmatrix}
1 & 0 & 0 & 0 \\
0 & 0 & 1 & 0 \\
0 & 0 & 0 & 1 \\
0 & 1 & 0 & 0
\end{pmatrix}
$$

6 *How will the state* $\frac{1}{\sqrt{2}}(|00\rangle - |11\rangle)$ *change under the action of* $\hat{X} \otimes \hat{Z}$?

Ans. $\hat{X} \otimes \hat{Z} = \begin{pmatrix} 0 & 1 \\ 1 & 0 \end{pmatrix} \otimes \begin{pmatrix} 1 & 0 \\ 0 & -1 \end{pmatrix}$

$$
= \begin{pmatrix}
(0)\begin{pmatrix} 1 & 0 \\ 0 & -1 \end{pmatrix} & (1)\begin{pmatrix} 1 & 0 \\ 0 & -1 \end{pmatrix} \\
(1)\begin{pmatrix} 1 & 0 \\ 0 & -1 \end{pmatrix} & (0)\begin{pmatrix} 1 & 0 \\ 0 & -1 \end{pmatrix}
\end{pmatrix}
= \begin{pmatrix}
0 & 0 & 1 & 0 \\
0 & 0 & 0 & -1 \\
1 & 0 & 0 & 0 \\
0 & -1 & 0 & 0
\end{pmatrix}
$$

$$
\frac{1}{\sqrt{2}}(|00\rangle - |11\rangle) = \frac{1}{\sqrt{2}}\left[\begin{pmatrix} 1 \\ 0 \\ 0 \\ 0 \end{pmatrix} - \begin{pmatrix} 0 \\ 0 \\ 0 \\ 1 \end{pmatrix}\right] = \frac{1}{\sqrt{2}}\begin{pmatrix} 1 \\ 0 \\ 0 \\ -1 \end{pmatrix}
$$

$$
\hat{X} \otimes \hat{Z} \quad \frac{1}{\sqrt{2}}(|00\rangle - |11\rangle) = \begin{pmatrix}
0 & 0 & 1 & 0 \\
0 & 0 & 0 & -1 \\
1 & 0 & 0 & 0 \\
0 & -1 & 0 & 0
\end{pmatrix} \frac{1}{\sqrt{2}}\begin{pmatrix} 1 \\ 0 \\ 0 \\ -1 \end{pmatrix} = \frac{1}{\sqrt{2}}\begin{pmatrix} 0 \\ 1 \\ 1 \\ 0 \end{pmatrix}
$$

$$
= \frac{1}{\sqrt{2}}\left[\begin{pmatrix} 0 \\ 1 \\ 0 \\ 0 \end{pmatrix} + \begin{pmatrix} 0 \\ 0 \\ 1 \\ 0 \end{pmatrix}\right] = \frac{1}{\sqrt{2}}(|01\rangle + |10\rangle)
$$

7 *Consider the state* $\frac{1}{13}[12|001\rangle + 3|010\rangle + 4|100\rangle]$. *Find the probability of measuring the second qubit as* $|0\rangle$ *if a preceding measurement of the first qubit gave the state* $|0\rangle$.

   Ans. Measurement was done on the state $\frac{1}{13}[12|001\rangle + 3|010\rangle + 4|100\rangle]$. It gave the first qubit $|0\rangle$. So this state collapses to the state (consisting only of states having first qubit 0), namely

   $N[12|001\rangle + 3|010\rangle]$ where N is a normalizing constant.

   Probability of obtaining second qubit as 0 will be $|12N|^2$.

   Now normalization condition is $|12N|^2 + |3N|^2 = 1 \Rightarrow N = \dfrac{1}{\sqrt{12^2 + 3^2}}$

Hence probability $= |12N|^2 = \dfrac{12^2}{12^2 + 3^2} = \dfrac{144}{144+9} = \dfrac{16.9}{153} = \dfrac{16}{17}$.

8 If $|a> = \dfrac{1}{\sqrt{2}}\begin{pmatrix} -1 \\ 1 \end{pmatrix}$, $|a> \otimes |b> = \dfrac{1}{\sqrt{10}}\begin{pmatrix} -2 \\ 1 \\ 2 \\ -1 \end{pmatrix}$ find $|b>$.

Ans. Assume $|b> = \begin{pmatrix} x \\ y \end{pmatrix}$. Then

$$|a> \otimes |b> = \dfrac{1}{\sqrt{2}}\begin{pmatrix} -1 \\ 1 \end{pmatrix} \otimes \begin{pmatrix} x \\ y \end{pmatrix} = \dfrac{1}{\sqrt{2}}\begin{pmatrix} (-1)\begin{pmatrix} x \\ y \end{pmatrix} \\ (1)\begin{pmatrix} x \\ y \end{pmatrix} \end{pmatrix} = \dfrac{1}{\sqrt{2}}\begin{pmatrix} -x \\ -y \\ x \\ y \end{pmatrix} = \dfrac{1}{\sqrt{10}}\begin{pmatrix} -2 \\ 1 \\ 2 \\ -1 \end{pmatrix}$$

Comparison gives

$$-\dfrac{x}{\sqrt{2}} = -\dfrac{2}{\sqrt{10}} \Rightarrow x = \dfrac{2}{\sqrt{5}}, \quad -\dfrac{y}{\sqrt{2}} = \dfrac{1}{\sqrt{10}} \Rightarrow y = -\dfrac{1}{\sqrt{5}}$$

Hence $|b> = \begin{pmatrix} x \\ y \end{pmatrix} = \begin{pmatrix} \dfrac{2}{\sqrt{5}} \\ -\dfrac{1}{\sqrt{5}} \end{pmatrix} = \dfrac{1}{\sqrt{5}}\begin{pmatrix} 2 \\ -1 \end{pmatrix}$.

9 *Consider the state $\dfrac{1}{\sqrt{2}}(|0> -|1>)$. Represent it in Bloch sphere. Compare its position w.r.t $\dfrac{1}{\sqrt{2}}(|0> +|1>)$.*

Ans. Given state can be compared with the general state

$$|\psi(\theta, \phi)> = \cos\dfrac{\theta}{2}|0> + e^{i\phi}\sin\dfrac{\theta}{2}|1> = \dfrac{1}{\sqrt{2}}(|0> -|1>)$$

to get

$$\cos\dfrac{\theta}{2} = \dfrac{1}{\sqrt{2}}, \ e^{i\phi}\sin\dfrac{\theta}{2} = -\dfrac{1}{\sqrt{2}} \Rightarrow \quad \cos\dfrac{\theta}{2} = \dfrac{1}{\sqrt{2}}, \ \sin\dfrac{\theta}{2} = \dfrac{1}{\sqrt{2}}, \ e^{i\phi} = -1$$

$\dfrac{\theta}{2} = 45° \Rightarrow \theta = 90°$, $\phi = 180°$ (as $e^{i\pi} = -1$)

Clearly $|->= |\psi(\theta = 90°, \phi = 180°)>$ or equivalently

$$|->= |r = 1, \theta = 90°, \phi = 180°>.$$

$\theta = 90°$ means that $|->$ lies in the equatorial plane and $\phi = 180°$ means that $|->$ lies on $-X$ axis, as shown in figure 5.32.

The Bloch vector corresponding to $|+>= \dfrac{1}{\sqrt{2}}(|0> +|1>) = |1, 90°, 0°>$ has been discussed in section 5.9.6 and shown in figure 5.7 as well as in figure 5.32.

We note that $|+>$ and $|->$ are orthogonal since

$$<+|-> = \dfrac{1}{\sqrt{2}}(<0| + <1|)\dfrac{1}{\sqrt{2}}(|0> -|1>)$$

$$= \dfrac{1}{2}(<0|0> -<0|1> +<1|0> -<1|1>)$$

$$= \dfrac{1}{2}(1-0+0-1) = 0$$

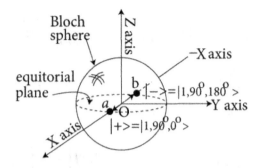

Figure 5.32. Showing |+> and |-> on Bloch sphere.
|point a> = |+> = $|1,90°,0°>$ and |point b> = |-> = $|1, 90°, 180°>$.

10 *Verify the following operator relations.*

(a) $\text{CNOT}^2 = I$

(b) $\text{CNOT}^2 = \text{CCNOT}$

(c) $\hat{X}^2 = I$

(d) $\hat{Z}^2 = \hat{Z}$

(e) $\hat{X}\hat{Z}\hat{X} = \hat{Z}$

(f) $\hat{H}\hat{Z}\hat{X} = \hat{X}$

(g) $\hat{Z}\hat{X}\hat{Z} = -\hat{X}$

(h) $\hat{H}\hat{X}\hat{H} = \hat{Z}$.

Ans.

(a) $\text{CNOT}^2 = \begin{pmatrix} 1 & 0 & 0 & 0 \\ 0 & 1 & 0 & 0 \\ 0 & 0 & 0 & 1 \\ 0 & 0 & 1 & 0 \end{pmatrix}\begin{pmatrix} 1 & 0 & 0 & 0 \\ 0 & 1 & 0 & 0 \\ 0 & 0 & 0 & 1 \\ 0 & 0 & 1 & 0 \end{pmatrix} = \begin{pmatrix} 1 & 0 & 0 & 0 \\ 0 & 1 & 0 & 0 \\ 0 & 0 & 1 & 0 \\ 0 & 0 & 0 & 1 \end{pmatrix} = I$, *(a)* true.

(b) $\text{CNOT}^2 \neq \text{CCNOT}$, (b) false.

(c) $\hat{X}^2 = \begin{pmatrix} 0 & 1 \\ 1 & 0 \end{pmatrix}\begin{pmatrix} 0 & 1 \\ 1 & 0 \end{pmatrix} = \begin{pmatrix} 1 & 0 \\ 0 & 1 \end{pmatrix} = \hat{I}$, (c) true.

(d) $\hat{Z}^2 = \begin{pmatrix} 1 & 0 \\ 0 & -1 \end{pmatrix}\begin{pmatrix} 1 & 0 \\ 0 & -1 \end{pmatrix} = \begin{pmatrix} 1 & 0 \\ 0 & 1 \end{pmatrix} = I \neq \hat{Z}$, (d) false.

(e) $\hat{X}\hat{Z}\hat{X} = \begin{pmatrix} 0 & 1 \\ 1 & 0 \end{pmatrix}\begin{pmatrix} 1 & 0 \\ 0 & -1 \end{pmatrix}\begin{pmatrix} 0 & 1 \\ 1 & 0 \end{pmatrix} = \begin{pmatrix} 0 & 1 \\ 1 & 0 \end{pmatrix}\begin{pmatrix} 0 & 1 \\ -1 & 0 \end{pmatrix} = \begin{pmatrix} -1 & 0 \\ 0 & 1 \end{pmatrix} = -\begin{pmatrix} 1 & 0 \\ 0 & -1 \end{pmatrix} = -\hat{Z}$, (e) false.

(f) $\hat{H}\hat{Z}\hat{X} = \frac{1}{\sqrt{2}}\begin{pmatrix} 1 & 1 \\ 1 & -1 \end{pmatrix}\begin{pmatrix} 1 & 0 \\ 0 & -1 \end{pmatrix}\begin{pmatrix} 0 & 1 \\ 1 & 0 \end{pmatrix} = \frac{1}{\sqrt{2}}\begin{pmatrix} 1 & 1 \\ 1 & -1 \end{pmatrix}\begin{pmatrix} 0 & 1 \\ -1 & 0 \end{pmatrix} = \begin{pmatrix} -1 & 1 \\ 1 & 1 \end{pmatrix} \neq \hat{X}$, (f) false.

(g) $\hat{Z}\hat{X}\hat{Z} = \begin{pmatrix} 1 & 0 \\ 0 & -1 \end{pmatrix}\begin{pmatrix} 0 & 1 \\ 1 & 0 \end{pmatrix}\begin{pmatrix} 1 & 0 \\ 0 & -1 \end{pmatrix} = \begin{pmatrix} 1 & 0 \\ 0 & -1 \end{pmatrix}\begin{pmatrix} 0 & -1 \\ 1 & 0 \end{pmatrix}$,

$$= \begin{pmatrix} 0 & -1 \\ -1 & 0 \end{pmatrix} = -\begin{pmatrix} 0 & 1 \\ 1 & 0 \end{pmatrix} = -\hat{X}$$

(g) true

(h) $\hat{H}\hat{X}\hat{H} = \frac{1}{\sqrt{2}}\begin{pmatrix} 1 & 1 \\ 1 & -1 \end{pmatrix}\begin{pmatrix} 0 & 1 \\ 1 & 0 \end{pmatrix}\frac{1}{\sqrt{2}}\begin{pmatrix} 1 & 1 \\ 1 & -1 \end{pmatrix} = \frac{1}{2}\begin{pmatrix} 1 & 1 \\ 1 & -1 \end{pmatrix}\begin{pmatrix} 1 & -1 \\ 1 & 1 \end{pmatrix}$

$$= \frac{1}{2}\begin{pmatrix} 2 & 0 \\ 0 & -2 \end{pmatrix} = \begin{pmatrix} 1 & 0 \\ 0 & -1 \end{pmatrix} = \hat{Z}$$

(h) is true.

11 *Verify the following for Hadamard operator.*
  (a) Hermitian,
  (b) unitary,
  (c) self-inverse,
  (d) preserves norm.

  Ans.

  (a) $\hat{H} = \frac{1}{\sqrt{2}}\begin{pmatrix} 1 & 1 \\ 1 & -1 \end{pmatrix}$, $\hat{H}^{\dagger} = \frac{1}{\sqrt{2}}\begin{pmatrix} 1 & 1 \\ 1 & -1 \end{pmatrix}^{*T} = \frac{1}{\sqrt{2}}\begin{pmatrix} 1 & 1 \\ 1 & -1 \end{pmatrix} = \hat{H} \Rightarrow$ Hermitian

  (b) $\hat{H}\hat{H}^{\dagger} = \frac{1}{\sqrt{2}}\begin{pmatrix} 1 & 1 \\ 1 & -1 \end{pmatrix}\frac{1}{\sqrt{2}}\begin{pmatrix} 1 & 1 \\ 1 & -1 \end{pmatrix} = \frac{1}{2}\begin{pmatrix} 2 & 0 \\ 0 & 2 \end{pmatrix} = \begin{pmatrix} 1 & 0 \\ 0 & 1 \end{pmatrix} = I$

  $\Rightarrow \hat{H}^{\dagger} = \hat{H}^{-1} \Rightarrow$ Unitary

  (c) As $\hat{H} = \hat{H}^{\dagger} = \hat{H}^{-1} \Rightarrow \hat{H}^2 = \hat{H}\hat{H} = \hat{H}\hat{H}^{-1} = I \Rightarrow$ Self inverse

  (d) Consider an arbitrary state $\begin{pmatrix} a \\ b \end{pmatrix}$, norm $= \sqrt{a^2 + b^2}$

  $$\hat{H}\begin{pmatrix} a \\ b \end{pmatrix} = \frac{1}{\sqrt{2}}\begin{pmatrix} 1 & 1 \\ 1 & -1 \end{pmatrix}\begin{pmatrix} a \\ b \end{pmatrix} = \frac{1}{\sqrt{2}}\begin{pmatrix} a + b \\ a - b \end{pmatrix}$$

  Norm $= \sqrt{(\frac{a+b}{\sqrt{2}})^2 + (\frac{a-b}{\sqrt{2}})^2} = \sqrt{\frac{a^2+b^2+2ab}{2} + \frac{a^2+b^2-2ab}{2}} = \sqrt{a^2 + b^2}$
  = same as that of the previous state (before application of Hadamard)
  $\Rightarrow$ Preserves norm.

22 *$|+>$ is flipped to $|->$ and $|->$ is flipped to $|+>$ by*
  (a) $\hat{X}$ in computational basis,
  (b) $\hat{X}$ in Hadamard basis,
  (c) $\hat{X}$ in its eigenbasis,
  (d) $\hat{X}$ in non-diagonal basis.

  Ans.
  $\hat{X}$ in computational basis $= \hat{X} = \begin{pmatrix} 0 & 1 \\ 1 & 0 \end{pmatrix}$ which is a non-diagonal basis,
  and $\hat{X}$ in Hadamard basis $=$ NOT $= \begin{pmatrix} 1 & 0 \\ 0 & -1 \end{pmatrix}$ which is diagonal and hence eigenbasis.
  Now $\hat{X}|+> = |+>$, $\hat{X}|-> = -|-> \Rightarrow$ No flipping by $\hat{X}$.
  But NOT$|+> = |->$, NOT$|-> = |+> \Rightarrow$ Flipping by NOT.
  Ans. *(b), (c).*

23 *Which gate is diagonal in computational basis, diagonal in Hadamard basis?*

(a) $\hat{X}$, $\hat{Z}$

(b) $\hat{Z}$, $\hat{Y}$

(c) $\hat{Z}$, $\hat{X}$

(d) $\hat{Y}$, $\hat{X}$

Ans. *(c)*.

IOP Publishing

Quantum Optics and Quantum Computation
An introduction
Dipankar Bhattacharyya and Jyotirmoy Guha

Chapter 6

Teleportation and super dense coding

6.1 Quantum no-cloning theorem

It is not possible to find a unitary transformation which will duplicate a given quantum state and write it onto a quantum blank state—i.e. xeroxing of a given quantum state is not possible.

Explanation

In the classical case of xeroxing we start with two things: one is the original page or material that needs to be copied and the other is the blank page on which the copying act is to be executed. And after execution of the act of copying the original page is returned back along with the copied page. Also the contents of the original page and the copied page are identical.

In the quantum case, corresponding to the blank page we have a standard quantum state, say single-qubit state $|0>$ and the original quantum state is $|\psi>$ that is to be copied.

So we start with a state $|0> \otimes |\psi>$. We can think of a unitary transformation $\hat{U}$ that operates on the input state $|0> \otimes |\psi>$ and gives rise to output $|\psi> \otimes |\psi>$ (one $|\psi>$ is the original state and another $|\psi>$ is the copied state). The quantum no-cloning theorem states that there is no unitary transformation that performs this task. This is proved as follows.

Suppose duplication or cloning is possible by the unitary transformation $\hat{U}$ on two arbitrary states $|\psi>$ and $|\phi>$. Hence

$$\hat{U}|0> \otimes |\psi> = |\psi> \otimes |\psi> \quad \Rightarrow \quad \hat{U}|0, \psi> = |\psi, \psi>$$

$$\hat{U}|0> \otimes |\phi> = |\phi> \otimes |\phi> \quad \Rightarrow \quad \hat{U}|0, \phi> = |\phi, \phi>$$

Consider

$$<\psi|\phi> = <\psi|\hat{I}|\phi> = <\psi| <0|0> |\phi> \quad (\text{as } <0|0> = \hat{I})$$

$$<\psi|\phi> \ =<\psi, 0|0, \phi> \quad \text{(defining } <\psi, 0|=<\psi| <0|, \ |0, \phi> \ =|0> |\phi> \text{)}$$

$$<\psi|\phi> \ =<\psi, 0|\hat{U}^\dagger \hat{U}|0, \phi> \quad \text{(as } \hat{U}^\dagger \hat{U} = \hat{I})$$

Since

$$\hat{U}|0, \phi> \ =|\phi, \phi>$$

$$\hat{U}|0, \psi> \ =|\psi, \psi> \ \Rightarrow \ <\psi, 0|\hat{U}^\dagger = <\psi, \psi|$$

We thus get

$$<\psi|\phi> \ =<\psi, \psi|\phi, \phi>$$

$$<\psi|\phi> \ =| <\psi|\phi> |^2$$

This can be possible if $|\psi>$ and $|\phi>$ are identical (when both sides are equal to unity as $<\psi|\phi> \ =<\psi|\psi> \ =1$) or if $|\psi>$ and $|\phi>$ are orthogonal (when both sides are zero as then $<\psi|\phi> \ =0$). This shows that $|\psi>$ and $|\phi>$ cannot be arbitrary states as we assumed in the beginning if cloning is possible. It follows therefore that duplication or cloning of the quantum state is not possible.

6.2 Teleportation

Teleportation is the technique of sending a quantum state from one point in space to another point in space. It is conventional, in quantum computing to give those points names like Alice and Bob and say that we are discussing the possibility of teleportation between Alice and Bob. Consider the teleportation circuit of figure 6.1.

Alice has a quantum state $|\psi> \ =\alpha|0> +\beta|1>$ which she wishes to send to Bob (with whom she pre-shares an entangled pair namely a $|\text{Bell state}>=\frac{1}{\sqrt{2}}(|00> +|11>$)). Alice does not know the details of the quantum state, i.e. she does not know about the complex coefficients α, β. Alice cannot make a measurement also since if she did she would get either $|0>$ (the probability of which

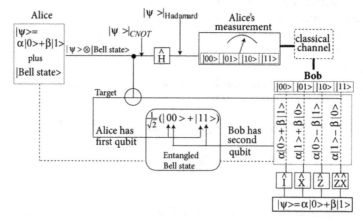

Figure 6.1. Teleportation circuit.

is $|\alpha|^2$) or |1>(the probability of which is $|\beta|^2$) and as she gets any one quantum state the original preparation will be destroyed, which she does not want.

Let Alice and Bob share one qubit each of the Bell state $\frac{1}{\sqrt{2}}(|00 > +|11 >)$ which is an entangled state.

The first qubit is of Alice and the second qubit is of Bob, i.e. we can rewrite the Bell state as

$$\frac{1}{\sqrt{2}}(|0_A 0_B > +|1_A 1_B >) \ (0_A = \text{Alice's qubit}, \ 0_B = \text{Bob's qubit})$$

Alice's encoding and measurement

Alice makes $|\psi > = \alpha|0 > +\beta|1>$ to interact with her first bit in the Bell state. This yields the quantum state

$$|\psi > \otimes |\text{Bell state}>$$

$$=(\alpha|0 > +\beta|1 >) \otimes \frac{1}{\sqrt{2}}(|00 > +|11 >)$$

$$=\alpha|0 > \otimes \frac{1}{\sqrt{2}}(|00 > +|11 >) + \beta|1 > \otimes \frac{1}{\sqrt{2}}(|00 > +|11 >)$$

We rewrite identifying original qubits of $|\psi>$ as 0_o, 1_o and the Bell state qubits as 0_A, 1_A of Alice and 0_B, 1_B of Bob. Hence

$$|\psi > \otimes |\text{Bell state}>$$

$$=(\alpha|0_o > +\beta|1_o >) \otimes \frac{1}{\sqrt{2}}(|0_A 0_B > +|1_A 1_B >)$$

$$=\alpha|0_o > \otimes \frac{1}{\sqrt{2}}(|0_A 0_B > +|1_A 1_B >) + \beta|1_o > \otimes \frac{1}{\sqrt{2}}(|0_A 0_B > +|1_A 1_B >)$$

$$=\frac{1}{\sqrt{2}}[\alpha|0_o 0_A > |0_B > +\alpha|0_o 1_A > |1_B > +\beta|1_o 0_A > |0_B > +\beta|1_o 1_A > |1_B >]$$

We now rewrite the above doing away with the subscripts

$$|\psi > \otimes |\text{Bell state}>$$

$$=\frac{1}{\sqrt{2}}[\alpha|00 > |0 > +\alpha|01 > |1 > +\beta|10 > |0 > +\beta|11 > |1 >]$$

$$\rightarrow |\text{original qubit Alice's qubit} > |\text{Bob's qubit}>$$

(this is the representation used)

Alice applies CNOT to $|\psi > \otimes |\text{Bell state}>$ treating the original qubit as the control qubit and the first Bell qubit of Alice as the target qubit, i.e.

$|\psi> \otimes|$Bell state $> \rightarrow \;|$original qubit   Alice's qubit $> |$Bob's qubit$>$

$\rightarrow \;|$Control qubit   Target qubit $> |$Bob's qubit$>$

If control qubit is 1 target qubit flips, otherwise no change occurs to target qubit (i.e. $|0\,0> \rightarrow|00>, |0\,1> \rightarrow|01>, |1\,0> \rightarrow|11>, |1\,1> \rightarrow|10>$). Hence the quantum state after CNOT gate is (we denote by $|\psi>|_{\text{CNOT}}$)

$$|\psi>|_{\text{CNOT}} \;=\; \frac{1}{\sqrt{2}}[\alpha|00>|0>+\alpha|01>|1>+\beta|11>|0>+\beta|10>|1>]$$

Alice applies Hadamard gate to the first qubit, as a result the first qubit changes as

$$|0> \rightarrow\frac{1}{\sqrt{2}}(|0>+|1>) \qquad |1> \rightarrow\frac{1}{\sqrt{2}}(|0>-|1>)$$

The quantum state thus obtained is (we denote by $|\psi>|_{\text{Hadamard}}$)

$$|\psi>|_{\text{Hadamard}} = \frac{1}{\sqrt{2}}[\alpha(\tfrac{|0>+|1>}{\sqrt{2}})|0>|0>+\alpha(\tfrac{|0>+|1>}{\sqrt{2}})|1>|1>$$

$$+\beta(\tfrac{|0>-|1>}{\sqrt{2}})|1>|0>+\beta(\tfrac{|0>-|1>}{\sqrt{2}})|0>|1>]$$

$$=\tfrac{1}{2}[\alpha|00>|0>+\alpha|10>|0>+\alpha|01>|1>+\alpha|11>|1>$$

$$+\beta|01>|0>-\beta|11>|0>+\beta|00>|1>-\beta|10>|1>]$$

$$|\psi>|_{\text{Hadamard}} = \tfrac{1}{2}[|00>(\alpha|0>+\beta|1>)+|01>(\alpha|1>+\beta|0>)$$

$$+|10>(\alpha|0>-\beta|1>)+|11>(\alpha|1>-\beta|0>)$$

Clearly Alice gets any of the 4 quantum states viz. $|00>$, $|01>$, $|10>$, $|11>$ while Bob has any of the quantum states $\alpha|0>+\beta|1>$, $\alpha|1>+\beta|0>$, $\alpha|0>-\beta|1>$, $\alpha|1>-\beta|0>$.

Alice had started with the quantum state $|\psi> =\alpha|0>+\beta|1>$. But depending upon what she got after passing through CNOT gate and a Hadamard gate she will now pick up a classical communication channel (like telephone) and tell Bob about what she measured, i.e. she will send 2 bits of classical information. On receiving this information Bob can recreate the state that Alice had started with viz. $|\psi> =\alpha|0>+\beta|1>$ by passing the quantum states he has, through an appropriate Pauli gate say $\hat{X}$, $\hat{Z}$ or through both the Pauli gates $\hat{Z}\hat{X}$. This is explained as follows.

a) Suppose Alice tells Bob that she got $|00>$. Then Bob directly has the quantum state $\alpha|0>+\beta|1> =|\psi>$ ($\Leftarrow$ original state of Alice obtained after passing through identity operator $\hat{I}$ or a quantum wire.)

b) Suppose Alice tells Bob that she got $|01>$. Then Bob has to pass the quantum state he has viz. $\alpha|1>+\beta|0>$ through $\hat{X}$ to get
$\hat{X}(\alpha|1>+\beta|0>) = \alpha\,\hat{X}|1>+\beta\,\hat{X}|0> =\alpha|0>+\beta|1> =|\psi>$($\Leftarrow$ original state of Alice obtained)

c) Suppose Alice tells Bob that she got $|10>$. Then Bob has to pass the quantum state he has viz. $\alpha|0> -\beta|1>$ through $\hat{Z}$ to get

$\hat{Z}(\alpha|0> -\beta|1 >) = \alpha \hat{Z}|0> -\beta \hat{Z}|1 > =\alpha|0> +\beta|1 > =|\psi>(\Leftarrow$ original state of Alice obtained)

d) Suppose Alice tells Bob that she got $|11>$. Then Bob has to pass the quantum state he has viz. $\alpha|1> -\beta|0>$ through $\hat{X}$ and then through $\hat{Z}$ (i.e. through $\hat{Z}\hat{X}$) to get

$$\hat{Z}\hat{X}(\alpha|1> -\beta|0 >) = \hat{Z}(\alpha\hat{X}|1> -\beta\hat{X}|0 >) = \hat{Z}(\alpha|0> -\beta|1 >)$$

$$=\alpha\hat{Z}|0> -\beta\hat{Z}|1>$$

$$=|\alpha|0> +\beta|1 > =|\psi > (\Leftarrow \text{original state of Alice obtained})$$

Obviously Bob, who shares a Bell state with Alice, is able to reconstruct the original state of Alice by choosing appropriate Pauli gates. This is the process of teleportation.

We note the following.

- As classical information is communicated, Bob and Alice cannot be separated by space-like distance. Information sharing has to be according to the special theory of relativity, so they are separated by time-like distance which is traversed by classical signal with speed not exceeding the speed of light in free space.
- Bob recreated the state which Alice started with, but this is not cloning. This is because of the following. Alice and Bob share a copy of the same state (Bell state) as they are an entangled pair. At Alice's end, after measurement, original state ($|\psi> =\alpha|0> +\beta|1 >$) has to collapse to either $|0>$ or $|1>$ (so the original state is destroyed). At Bob's end as he reconstructed the state of Alice ($|\psi> =\alpha|0> +\beta|1 >$) he also loses the state he started with. So the basic requirement of cloning (retaining an identical copy of quantum state) is not respected and so it is not cloning.
- In fact because of the restriction imposed by the no cloning theorem, a simple transportation of a quantum state from one place to another place is not possible. This can, however, be achieved in a different way through the process of quantum teleportation.
- Quantum teleportation has an important role in quantum cryptography (chapter 12).

6.3 Super dense coding (or dense coding) (of Bennett and Wiesner)

In quantum teleportation Alice sent 2 bits of *classical* information (00,01,10 or 11) and Bob constructed or extracted a *quantum* state $|\psi> =\alpha|0> +\beta|1>$.

Super dense coding is the reverse process of teleportation.

In super dense coding Alice sends one qubit of *quantum* information and Bob constructs or extracts two bits of *classical* information. Hence the name dense coding.

Alice and Bob pre-share the Bell entangled state to achieve this. Figure 6.2 shows a super dense coding circuit. We explain the working of the circuit as follows.

Alice starts with 2 qubits $|0> |0>$. One $|0>$ is passed through Hadamard gate to get $|0> \rightarrow \frac{1}{\sqrt{2}}(|0> + |1>)$. So the input to the CNOT is

$$|\psi>|_{\text{input}} = \frac{1}{\sqrt{2}}(|0> + |1>) \otimes |0> = \frac{1}{\sqrt{2}}(|00> + |10>)$$

Taking the first qubit as control qubit and the second qubit as target qubit $|\psi>|_{\text{input}} = \frac{1}{\sqrt{2}}(|00> + |10>)$ is passed through CNOT gate. Target qubit changes only if control qubit is 1 (i.e. $|00> \rightarrow |00>$, $|10> \rightarrow |11>$). Clearly thus a Bell state is generated as

$$|\psi>|_{\text{input}} = \frac{1}{\sqrt{2}}(|00> + |10>) \xrightarrow{\text{CNOT}} \frac{1}{\sqrt{2}}(|00> + |11>) = |\text{Bell state} >.$$

The Bell state generated is shared by Alice and Bob and so they are entangled. The first qubit of the entangled state is with Alice and the second qubit is with Bob. Alice works on her qubit, i.e. on the first qubit in the Bell state and can do nothing on the second qubit in the Bell state since that corresponds to Bob.

Encoding by Alice

Alice encodes on her qubit (i.e. on the first qubit of the Bell state) 2 bits of classical information (any of 00, 01, 10, 11) and we call the resulting quantum state with encoded classical information $|\psi>|_{\text{quantum}}$. For encoding the 2 bits of classical information Alice uses quantum unitary gates, denoted by $\hat{U}$ (00 is encoded using unit matrix $\hat{I}$, 01 is encoded using $\hat{X}$, 10 is encoded using $i\hat{Y}$, 11 is encoded using $\hat{Z}$). This is explained in the following.

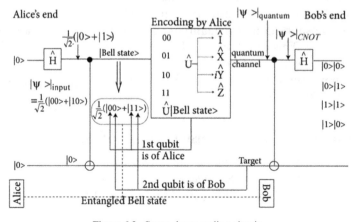

Figure 6.2. Super dense coding circuit.

00 is encoded by applying $\hat{I}$ (which is the identity operator that represents doing nothing): $\hat{I}\frac{1}{\sqrt{2}}(|00> +|11 >) = \hat{I}|$1st qubit of Bell state $> =\frac{1}{\sqrt{2}}(|00> +|11 >)$

$$=|\psi > |_{\text{quantum}}$$

01 is encoded by applying $\hat{X}$ (since $\hat{X}|0> =|1 >$, $\hat{X}|1 > =|0>$ Alice's qubit suffers flip): $\hat{X}\frac{1}{\sqrt{2}}(|00> +|11 >) = \hat{X}|$1st qubit of Bell state $> =\frac{1}{\sqrt{2}}(|10> +|01 >)$

$$=|\psi > |_{\text{quantum}}$$

10 is encoded by applying $i\hat{Y}$ (using $\hat{Y}|0> =i|1 >$, $\hat{Y}|1 > =-i|0>$) and so encoding occurs as : $i\hat{Y}\frac{1}{\sqrt{2}}(|00> +|11 >) = i\hat{Y}|$1st qubit of Bell state$>$

$$=\frac{i}{\sqrt{2}}(\hat{Y}|00> + \hat{Y}|11 >) = \frac{i}{\sqrt{2}}(i|10> -i|01 >)$$

$$=\frac{1}{\sqrt{2}}(-|10> +|01 >) = |\psi > |_{\text{quantum}}$$

11 is encoded by applying $\hat{Z}$ (using $\hat{Z}|0> =|0 >$, $\hat{Z}|1 > =-|1>$) and so encoding occurs as: $\hat{Z}\frac{1}{\sqrt{2}}(|00> +|11 >) = \hat{Z}|$1st qubit of Bell state $> =\frac{1}{\sqrt{2}}(|00> -|11 >)$

$$=|\psi > |_{\text{quantum}}$$

All these quantum states $|\psi > |_{\text{quantum}}$ carrying classical information are actually Bell states and are sent over a quantum channel to Bob.

Bob thus receives four different Bell states all denoted by $|\psi > |_{\text{quantum}}$ and generated by Alice (carrying the classical information). Bob uses a CNOT gate treating Alice's qubit (i.e. the first qubit) as control qubit and his qubit (i.e. the second qubit) as target qubit. The effect of Bob's CNOT is shown in table 6.1:

$$|\psi > |_{\text{quantum}} \xrightarrow{\text{CNOT}} |\psi > |_{\text{CNOT}}$$

Table 6.1. Effect of CNOT on $|\psi > |_{\text{quantum}}$ received by Bob over the quantum channel.

Encoding by Alice	Encoded qubit received by Bob over quantum channel and CNOT applied $|\psi >|_{\text{quantum}} \xrightarrow{CNOT} |\psi >|_{CNOT} =|$1st qubit 2nd qubit $>$
00	$\frac{1}{\sqrt{2}}(\|00> +\|11>) \xrightarrow{CNOT} \frac{1}{\sqrt{2}}(\|00>+\|10>) = \frac{1}{\sqrt{2}}(\|0>+\|1>)\|0>$
01	$\frac{1}{\sqrt{2}}(\|10> +\|01>) \xrightarrow{CNOT} \frac{1}{\sqrt{2}}(\|11>+\|01>) = \frac{1}{\sqrt{2}}(\|1>+\|0>)\|1>$
10	$\frac{1}{\sqrt{2}}(-\|10>+\|01>) \xrightarrow{CNOT} \frac{1}{\sqrt{2}}(-\|11>+\|01>)$ $= \frac{1}{\sqrt{2}}(\|01>-\|11>)=\frac{1}{\sqrt{2}}(\|0>-\|1>)\|1>$
11	$\frac{1}{\sqrt{2}}(\|00> -\|11>) \xrightarrow{CNOT} \frac{1}{\sqrt{2}}(\|00>-\|10>) = \frac{1}{\sqrt{2}}(\|0>-\|1>)\|0>$

Table 6.2. Effect of Hadamard on first qubit of $|\psi> |_{\text{CNOT}}$ by Bob.

Encoding by Alice	Application of Hadamard on first qubit of $|\psi>|_{CNOT} = |$ 1st qubit 2nd qubit $>$ by Bob
✓ 00	$[\ \hat{H}\frac{1}{\sqrt{2}}(|0>+|1>)\ ]|0>=|0>|0>$ ✓
✓ 01	$[\ \hat{H}\frac{1}{\sqrt{2}}(|1>+|0>)\ ]|1>=|0>|1>$ ✓
✓✓ 10	$[\ \hat{H}\frac{1}{\sqrt{2}}(|0>-|1>)\ ]|1>=|1>|1>$ ✓✓
✓✓ 11	$[\ \hat{H}\frac{1}{\sqrt{2}}(|0>-|1>)\ ]|0>=|1>|0>$ ✓✓

Table 6.3. Picking up the result by Bob.

2nd qubit	1st qubit	The logic [U means intersection deriving the common data]			Result is the common set of bits which is the classical information sent by Alice to Bob.
		Data for 2nd qubit	U	Data for 1st qubit	
0	0	{ 00 , 11 } U		{ 00 , 01 }	00
0	1	{ 00 , 11 } U		{ 10 , 11 }	11
1	0	{ 01 , 10 } U		{ 00 , 01 }	01
1	1	{ 01 , 10 } U		{ 10 , 11 }	10

The quantum state at Bob's end after the passage of encoded qubit through CNOT is written in the representation |Alice>|Bob> =|1st qubit>|2nd qubit>.

Bob measures his qubit (the second qubit).

If it is |0> then Alice sent him either 00 or 11. ✓

If it is |1> then Alice sent him either 01 or 10. ✓✓

Measurement of the 2nd qubit at Bob's end means the 2nd qubit has collapsed (to either |0> or |1>), but the 1st qubit is still in a linear combination of states.

Bob now applies Hadamard gate to the first qubit. The result is compiled in table 6.2.

Bob now measures the first qubit.

If it is |0> then Alice sent him either 00 or 01. ✓

If it is |1> then Alice sent him either 10 or 11. ✓✓

Now a combination or intersection of information gathered from measurement of the first qubit and second qubit will enable Bob to decide which state was sent to him by Alice. The common combination of bits is to be picked up as the result. This is shown in table 6.3.

The essence of super coding is that, by sending a smaller number of qubits through a quantum channel, Alice is able to communicate to Bob a large number of classical bits of information, provided Alice and Bob share an entangled state.

6.4 Further reading

[1] Ghosh D 2016 Online course on quantum information and computing (Bombay: Indian Institute of Technology)

[2] Nielsen M A and Chuang I L 2019 *Quantum Computation and Quantum Information* (Cambridge: Cambridge University Press)

[3] Nakahara M and Ohmi T 2017 *Quantum Computing* (Boca Raton, FL: CRC Press)

[4] Pathak A 2016 *Elements of Quantum Computing and Quantum Communication* (Boca Raton, FL: CRC Press)

6.5 Problems

1. *Refer to the teleportation circuit of* figure 6.1. *Suppose Alice and Bob teleport sharing the Bell state* $\frac{1}{\sqrt{2}}(|01> -|10>)$. *If Alice's measurement gives* $|10>$ *what action should be taken by Bob to generate the state* $\alpha|0> +\beta|1>$?

   Ans. Refer to section 6.2 and figure 6.1. Alice has quantum state $|\psi> =\alpha|0> +\beta|1>$. Bell state used $\frac{1}{\sqrt{2}}(|01> -|10>)$. So input to CNOT is

   $$|\psi> \otimes|\text{Bell state}>$$

   $$=(\alpha|0> +\beta|1>) \otimes \frac{1}{\sqrt{2}}(|01> -|10>)$$

   $$= \frac{1}{\sqrt{2}}[\alpha(|001> -|010>) + \beta(|101> -|110>)]$$

   $$=\frac{1}{\sqrt{2}}[\alpha(|00> |1> -|01> |0>) + \beta(|10> |1> -|11> |0>)]$$

   Alice applies CNOT (Control = first qubit, target = second qubit) on

   $$|\psi> \otimes|\text{Bell state}> =|\text{Control bit Target bit}> |\text{Bob's bit}>$$

   to get $|\psi> |_{\text{CNOT}}$ given by

   $$|\psi> |_{\text{CNOT}} = \frac{1}{\sqrt{2}}[\alpha(|00> |1> -|01> |0>) + \beta(|11> |1> -|10> |0>)]$$

   $$=\frac{1}{\sqrt{2}}[\alpha(|0> |01> -|0> |10>) + \beta(|1> |11> -|1> |00>)]$$

   Apply H gate on first qubit $[\ |0> \xrightarrow{H} \frac{|0>+|1>}{\sqrt{2}}, |1> \xrightarrow{H} \frac{|0>-|1>}{\sqrt{2}}]$ to get $|\psi> |_{\text{Hadamard}}$ as

$$|\psi> |_{\text{Hadamard}} = \frac{1}{\sqrt{2}}\left[\alpha(\frac{|0>+|1>}{\sqrt{2}}|01> - \frac{|0>+|1>}{\sqrt{2}}|10>)\right.$$
$$\left. + \beta(\frac{|0>-|1>}{\sqrt{2}}|11> - \frac{|0>-|1>}{\sqrt{2}}|00>)\right]$$

$$=\frac{1}{2}[\alpha(|001>+|101> -|010> -|110>) + \beta(|011> -|111> -|000> +|100>)]$$

Rearranging

$$|\psi> |_{\text{Hadamard}} = \frac{1}{2}[(\alpha|001> -\beta|000>)+(-\alpha|010> +\beta|011>)]$$

$$+(\alpha|101> +\beta|100>) + (-\alpha|110> -\beta|111>)$$

$$|\psi> |_{\text{Hadamard}} = \frac{1}{2}[|00> \otimes(\alpha|1> -\beta|0>) + |01> \otimes(-\alpha|0> +\beta|1>)]$$

$$+|10> \otimes(\alpha|1> +\beta|0>) + |11> \otimes(-\alpha|0> -\beta|1>)$$

Alice's measurement gives $|10>$. So the state that Bob has is $\alpha|1> +\beta|0>$. But the state to be generated is $\alpha|0> +\beta|1>$. Comparing, we see that Bob has to apply $\hat{X}$ gate (i.e. NOT gate) for the purpose since

$$\alpha|1> +\beta|0> \xrightarrow{\hat{X}} \alpha|0> +\beta|1>.$$

2. *In super dense coding a single qubit is communicated in place of two classical bits during processing of a quantum state that Alice sends to Bob. She intends to communicate classical bits 11 and Bell state shared is $\frac{|00>+|11>}{\sqrt{2}}$. Verify that the following states are encountered by Bob during various stages of processing:*

   (a) $\frac{|00>-|10>}{\sqrt{2}}$, (b) $\frac{|0>-|1>}{\sqrt{2}} \otimes |0>$, (c) $|1> \otimes|0>$.

   Ans. Refer to section 6.3 and figure 6.2. As 11 is communicated encoding is done using $\hat{Z}$. 11 is encoded by applying $\hat{Z}$(using $\hat{Z}|0> =|0>$, $\hat{Z}|1> =-|1>$) and so encoding occurs as:

$$\hat{Z}\frac{1}{\sqrt{2}}(|00> +|11>) = \hat{Z}|1\text{st qubit of Bell state}> =\frac{1}{\sqrt{2}}(|00> -|11>) = |\psi> |_{\text{quantum}}$$

$$\Rightarrow \text{option}(a)$$

Now a CNOT is applied. (Control = 1st qubit, Target = 2nd qubit)

$$|\psi> |_{\text{quantum}} = \frac{1}{\sqrt{2}}(|00> -|11>) \xrightarrow{\text{CNOT}} \frac{1}{\sqrt{2}}(|00> -|10>)$$

$$= (\frac{|0> -|1>}{\sqrt{2}}) \otimes |0> \quad \Rightarrow \text{option}(b)$$

Apply Hadamard on first qubit $|0> \xrightarrow{\hat{H}} \frac{|0>+|1>}{\sqrt{2}}$, $|1> \xrightarrow{\hat{H}} \frac{|0>-|1>}{\sqrt{2}}$

$$(\frac{|0> -|1>}{\sqrt{2}}) \otimes |0> \xrightarrow{\hat{H} \text{ (on first qubit)}} \frac{1}{\sqrt{2}}[\frac{|0> +|1>}{\sqrt{2}} - \frac{|0> -|1>}{\sqrt{2}}] \otimes |0>$$

Final state is $\frac{1}{\sqrt{2}}[2. \frac{|1>}{\sqrt{2}}] \otimes |0> = |1> \otimes |0> = |10> \Rightarrow$ option *(c)*.

3. *Quantum no-cloning theorem says that*

 (a) Some states can be cloned;
 (b) No state can be cloned;
 (c) Only normalized states can be cloned;
 (d) Only orthonormalized states can be cloned.

Ans. *(b)*.

4. *Super dense coding is*

 (a) reverse process of entanglement;
 (b) reverse process of teleportation;
 (c) reverse process of low density coding;
 (d) coding of orthonormalized states.

Ans. *(b)*.

5. *Identify the correct statement from the following in which Alice sends a signal to Bob during quantum teleportation.*

 (a) Alice sends classical information and Bob extracts a classical state.
 (b) Alice sends quantum information and Bob extracts a classical state.
 (c) Alice sends classical information and Bob extracts a quantum state.
 (d) Alice sends quantum information and Bob extracts a quantum state.

Ans. *(c)*.

6. *Identify the correct statement from the following in which Alice sends a signal to Bob during super dense coding.*

 (a) Alice sends classical information and Bob extracts a classical state.
 (b) Alice sends quantum information and Bob extracts a classical state.
 (c) Alice sends classical information and Bob extracts a quantum state.
 (d) Alice sends quantum information and Bob extracts a quantum state.

Ans. *(b)*.

IOP Publishing

Quantum Optics and Quantum Computation
An introduction
Dipankar Bhattacharyya and Jyotirmoy Guha

Chapter 7

Pure and mixed state

The postulates of quantum mechanics (previously mentioned and explained in chapter 2) refer to an isolated, closed, self-contained pure system, i.e. a system which is there, by itself and does not interact with outside systems.

7.1 Pure state

A quantum system with a state vector $|\psi>$ with unit length, $<\psi|\psi> = 1$, in a complex Hilbert space is called a pure state (previously we called it just a state). Any other ket having the phase shifted form $e^{i\theta}|\psi>$ (with θ arbitrary) also represents the same state since $|e^{i\theta}|^2 = e^{-i\theta}e^{i\theta} = 1$. It is a quantum state that cannot be written as a mixture of other states. Pure states are completely defined states. (A pure state can be written as a superposition of other pure states (changing computational basis).)

Generally, we consider collection of a large number of systems (that are macroscopically identical but microscopically different) called an ensemble of systems. We may consider a physical system which may not be an isolated system but may be part of a bigger system interacting with the environment. Now when we make a measurement on an ensemble of systems it is not proper to expect that we are picking up the same state all the time. For them some of the postulates may not hold good due to the non-closed nature.

7.2 Mixed state

For an ensemble of states we cannot say definitely which state the system is in. We can say that the system is in state $|\psi_i>$ with probability p_i. A system whose state cannot be defined uniquely by a single state vector is said to be in a mixed state.

Different probabilistic distributions of pure states or a mixture of quantum states is again a quantum state and is considered a mixed state.

A mixed state represents our ignorance about a physical state. So a mixed state cannot be described by a wave function. A mixed state cannot be written as a

doi:10.1088/978-0-7503-2715-2ch7

superposition of pure states. A mixed state does not have a state vector but is described by a density matrix.

7.3 Density operator (introduced by Von Neumann)

Density operator or density matrix denoted by $\hat{\rho}$ is used to describe the statistical state of a quantum system. It is a mathematical tool with which we can distinguish between pure state and mixed state. It is a Hermitian operator for it to be physically acceptable.

7.4 Density operator for a pure state

A pure state is described by a state vector, say $|\psi>$. The density matrix or density operator for the pure state is defined as

$$\hat{\rho} = |\psi> <\psi| = \hat{\rho}^\dagger$$

Clearly for a pure state $|\psi>$

$$\hat{\rho}|\psi> = |\psi> <\psi|\psi> = |\psi>$$
$$\hat{\rho}^2 = \hat{\rho}\hat{\rho} = |\psi> <\psi|\psi> <\psi| = |\psi> <\psi| = \hat{\rho} \qquad (\text{as } <\psi|\psi> = 1)$$

We can expand a pure state $|\psi>$ in Hilbert space as a linear combination of basis kets $|n>$ of eigen operator $\hat{A}$ (i.e. $\hat{A}|n> = n|n>$) as

$|\psi> = \sum_n c_n|n>$ where $c_n = <n|\psi>.=$ expansion coefficient.

The density operator becomes

$$\hat{\rho} = |\psi> <\psi| = \sum_n c_n|n> \sum_m c_m^* <m|$$
$$= \sum_n \sum_m c_m^* c_n|n> <m|$$

Using $c_n = <n|\psi>$, $\quad c_m^* = <m|\psi>^* = <\psi|m>$ and that scalars commute

$$c_m^* c_n = c_n c_m^* = <n|\psi> <\psi|m> = <n|(|\psi> <\psi|)|m>$$
$$= <n|\hat{\rho}|m> = \rho_{nm}$$

Hence we can write

$$\hat{\rho} = \sum_n \sum_m c_m^* c_n|n> <m| = \sum_n \sum_m \rho_{nm}|n> <m|$$

where $\rho_{nm} = nm$ th matrix element of the density operator $\hat{\rho}$ for pure state.

It is a matrix with diagonal elements given by

$\rho_{nn} = <n|\hat{\rho}|n> = c_n^* c_n = |c_n|^2 =$ probability (in the Born's interpretation) of the pure state $|\psi>$ collapsing to the base state $|n>$ in a measurement.

As the total probability is normalized to unity it follows that

$$\sum_n |c_n|^2 = 1.$$

This means the sum of diagonal elements of the $\hat{\rho}$ matrix is unity, i.e. $Tr\hat{\rho} = \sum_n \rho_{nn} = \sum_n |c_n|^2 = 1$. Also, $\sum_n <n|\hat{\rho}|n> =1$

As $\hat{\rho} = \hat{\rho}^2 \Rightarrow Tr\hat{\rho}^2 = Tr\hat{\rho} = 1$.

The diagonal elements give population of states.

Also, the diagonal elements $\rho_{nn} = c_n^* c_n = |c_n|^2$ of density matrix $\hat{\rho}$ are necessarily non-negative. It follows that density matrix operator is a positive operator. So the eigenvalues λ_i are non-negative and sum to unity ($\sum_i \lambda_i = 1$).

7.4.1 Coherence

If we write the amplitude of states in polar form

$$c_n = |c_n|e^{i\phi_n}$$

then the off-diagonal elements of the density matrix are

$$\rho_{nm} = c_m^* c_n = |c_m|e^{-i\phi_m}|c_n|e^{i\phi_n} = |c_m||c_n|e^{i(\phi_n - \phi_m)}$$

Clearly the off diagonal terms depend upon the relative phase difference between the states m and n and this results in interference. As the off-diagonal terms represent correlation between the states they are called coherence.

A pure state is the quantum state where we have exact information about the quantum system and are represented by a point on the Bloch sphere.

The relations $\hat{\rho}^2 = \hat{\rho}$ and $Tr\hat{\rho}^2 = 1$ are the testing relations or confirmatory expressions to check whether the state is pure state or not.

For pure state: $\hat{\rho}^2 = \hat{\rho}$ and $Tr\hat{\rho}^2 = 1$

For mixed state $\hat{\rho}^2 \neq \hat{\rho}$ and $Tr\hat{\rho}^2 < 1$.

$Tr\hat{\rho}^2$ is called purity of a state. A state is pure when purity $Tr\hat{\rho}^2 = 1$, otherwise it is mixed.

7.4.2 Decoherence

Decoherence refers to the loss of coherence, i.e. the change in relative phase $\phi_n - \phi_m$ between the amplitude of states. Obviously lack of coherence leads to incompatibility and loss of quantum information in a system.

7.5 Average

Average in the state $|\psi>$ is given by

$$<\hat{A}> = <\psi|\hat{A}|\psi>$$

Using $|\psi> = \sum_n c_n|n>, \quad <\psi| = \sum_m c_m^* < m|$

$$<\hat{A}> = \sum_m c_m^* < m|\hat{A} \sum_n c_n|n> = \sum_m \sum_n c_m^* c_n < m|\hat{A}|n>$$

Using $c_m = <m|\psi>, c_m^* = <\psi|m>, c_n = <n|\psi>$

$$<\hat{A}> = \sum_m \sum_n <\psi|m> <n|\psi><m|\hat{A}|n>$$

Interchanging the positions of the scalars $<\psi|m>$, $<n|\psi>$ (as they commute) we get

$$<\hat{A}> = \sum_m \sum_n <n|\psi> <\psi|m><m|\hat{A}|n>$$

Using $\hat{\rho} = |\psi> <\psi|$ we have

$$<\hat{A}> = \sum_m \sum_n <n|\hat{\rho}|m><m|\hat{A}|n> = \sum_n <n|\hat{\rho} \sum_m |m><m|\hat{A}|n>$$

Using $\sum_m |m> <m| = \hat{I}$ (completeness condition)

$$<\hat{A}> = \sum_n <n|\hat{\rho} \ \hat{I} \ \hat{A}|n> = \sum_n <n|\hat{\rho} \ \hat{A}|n> = \text{sum of diagonal elements}$$

$<\hat{A}> = Tr(\hat{\rho}\hat{A})$ (derived for pure state and holds for general case also).
Taking $\hat{A} = 1$ we get

$$<1> = Tr\hat{\rho}$$

$\Rightarrow Tr\hat{\rho} = 1$ (derived for pure state, holds for general case also).

7.6 Density operator of a mixed state (or an ensemble)

Consider a quantum state that is not in a pure state (i.e. not of the type $|\psi> = \sum_n c_n|n>$). We consider a state that is a mixture of states $|\psi_i>$. Not all of the N systems comprising the ensemble are in the same state and if N_i systems in the ensemble are in the same state $|\psi_i>$ then the probability to find the ith system of ensemble in the state $|\psi_i>$ would be $p_i = \frac{N_i}{N}<1$ and obviously $\sum_i p_i = 1$ and $\sum_i N_i = N$ should hold. The suffix i runs over all the members of the ensemble. Also, each $|\psi_i>$ has a different expansion in the basis $|n>$.

The density operator of such mixed state is given by the weighted sum

$\hat{\rho} = \sum_i p_i |\psi_i> <\psi_i| = \hat{\rho}^\dagger$ (this is the general definition of density matrix)

✓ If all of the N systems comprising the ensemble are in the same state (say jth state) then $p_i = 1$ for $i = j$ and $p_i = 0$ for $i \neq j$. Then

$$\hat{\rho} = \sum_{i \neq j} p_i |\psi_i> <\psi_i| + p_j |\psi_j> <\psi_j| \text{ (2nd term is the } i = j \text{ term)}$$

$$= |\psi_j> <\psi_j| \text{ (1st term = 0 as } p_{i \neq j} = 0 \text{ and 2nd term contributes)}$$

We then have a pure state.

- Consider square of density matrix operator

$$\hat{\rho}^2 = \hat{\rho}\hat{\rho} = \sum_i p_i |\psi_i> <\psi_i| \sum_j p_j |\psi_j> <\psi_j|$$

$$= \sum_i \sum_j p_i p_j |\psi_i> <\psi_i|\psi_j> <\psi_j|$$

$$= \sum_i \sum_j p_i p_j |\psi_i> \delta_{ij} <\psi_j| \qquad (\text{using } <\psi_i|\psi_j> = \delta_{ij})$$

$$\hat{\rho}^2 = \sum_i p_i^2 |\psi_i> <\psi_i| \neq \hat{\rho} \qquad \text{since} \hat{\rho} = \sum_i p_i |\psi_i> <\psi_i|$$

So for mixed state $\hat{\rho}^2 \neq \hat{\rho}$

Trace of density operator in mixed state

- Consider in the mixed state the trace of $\hat{\rho}\hat{A}$

$$Tr(\hat{\rho}\hat{A}) = Tr\left(\sum_i p_i |\psi_i> <\psi_i|\hat{A}\right) = \sum_n <n| \sum_i p_i \psi_i> <\psi_i|\hat{A} |n>$$

$$= \sum_n \sum_i p_i <n|\psi_i> <\psi_i|\hat{A}|n> = \sum_i p_i \sum_n <\psi_i|\hat{A}|n> <n|\psi_i>$$

(as scalars commute we can write scalars in any order)

$$Tr(\hat{\rho}\hat{A}) = \sum_i p_i <\psi_i|\hat{A}| \sum_n |n> <n| \psi_i>$$

Using $\sum |n> <n| = 1$ (completeness condition) we have

$$Tr(\hat{\rho}\hat{A}) = \sum_i p_i <\psi_i|\hat{A}| \psi_i> = <\hat{A}> \text{ (expectation value of } \hat{A} \text{ in mixed state)}$$

- Consider in the mixed state the trace of $\hat{\rho}$

$$Tr\hat{\rho} = Tr\left(\sum_i p_i |\psi_i> <\psi_i|\right) = \sum_n <n| \sum_i p_i \psi_i> <\psi_i |n>$$

$$= \sum_n \sum_i p_i <n|\psi_i> <\psi_i|n> = \sum_i p_i \sum_n <\psi_i|n> <n|\psi_i>$$

(as scalars commute we can write scalars in any order)

$$Tr\hat{\rho} = \sum_i p_i <\psi_i| \sum_n |n> <n| \psi_i>$$

Using $\sum_n |n> <n| = 1$ (completeness condition) we have

$Tr\hat{\rho} = \sum_i^n p_i < \psi_i| \psi_i > = \sum_i p_i = 1$ (unit trace). (as $<\psi_i|\psi_i > = 1$)

We get to this result if we put $\hat{A} = \hat{I} =$ unit operator in the result $Tr(\hat{\rho}\hat{A}) = <\hat{A}>$.

NOTE:

In situations where normalization has not been done $(\sum_i p_i \neq 1)$ the system average, i.e. the expectation value in the mixed state is given by

$$<\hat{A}> = \frac{\sum_i p_i < \psi_i| \hat{A} | \psi_i >}{\sum_i p_i} = \frac{Tr(\hat{\rho}\hat{A})}{Tr\hat{\rho}}$$

- Consider in the mixed state the trace of $\hat{\rho}^2$

$Tr\hat{\rho}^2 = \sum_n < n| \hat{\rho}^2 |n > = \sum_n < n| \hat{\rho}\hat{\rho} |n>$

$= \sum_n < n| \sum_i p_i |\psi_i > <\psi_i| \sum_j p_j |\psi_j > <\psi_j| n >$ (since $\hat{\rho} = \sum_i p_i |\psi_i > <\psi_i|$)

$= \sum_i p_i \sum_j p_j \sum_n < n| \psi_i > <\psi_i|\psi_j > <\psi_j|n>$

$Tr\hat{\rho}^2 = \sum_i \sum_j p_i p_j \sum_n < \psi_i|\psi_j > <\psi_j|n > <n|\psi_i >$ (scalars commute)

$= \sum_i \sum_j p_i p_j < \psi_i|\psi_j > <\psi_j| \sum_n |n > <n|\psi_i>$

$= \sum_i \sum_j p_i p_j < \psi_i|\psi_j > <\psi_j|\psi_i > = \sum_i \sum_j p_i p_j | < \psi_i|\psi_j > |^2$

$= \sum_i \sum_j p_i p_j \delta_{ij} = \sum_i p_i^2$

Since $p_i < 1$, $p_i^2 < p_i$ and so $\sum_i p_i^2 < \sum_i p_i = 1$

$Tr\hat{\rho}^2 < 1$

The relation $Tr\hat{\rho}^2 < 1$ is a good measure for testing the mixedness of an ensemble.

The density operator is a positive operator in the mixed state

For an arbitrary state vector $|\phi>$ we have

$$< \phi|\hat{\rho}|\phi> = < \phi| \sum_i p_i |\psi_i > <\psi_i|\phi > = \sum_i p_i < \phi|\psi_i > <\psi_i|\phi>$$

$$= \sum_i p_i |<\phi|\psi_i>|^2 \geqslant 0.$$

7.7 Quantum mechanics of an ensemble

Consider a two-dimensional system with the base states $\{|0>, |1>\}$. Let the ensemble have two types of states in it given as

$|\psi> = \alpha|0> + \beta|1>$ with a probability p
$|\phi> = \gamma|0> + \delta|1>$ with a probability $1 - p$

Let us find the probability of picking up one state, say $|0>$. This state $|0>$ belongs to both $|\psi>$ (where it occurs with Born probability $|\alpha|^2$) and $|\phi>$ (where it occurs with Born probability $|\gamma|^2$). Again the classical probability of picking up the state $|\psi>$ is p and the $|\phi>$ is $1 - p$.

Hence the probability of picking up $|0>$ is $p|\alpha|^2 + (1 - p)|\gamma|^2$ and the probability of picking up $|1>$ is $p|\beta|^2 + (1 - p)|\delta|^2$ (similarly). (The product of classical probability and Born probability is considered since they are independent.)

Consider a composite system $A \otimes B$ and the corresponding Hilbert space is $H_A \otimes H_B$. Suppose we have $|\psi_{AB}>$ as a state in this Hilbert space.

Let us suppose that $|\psi_{AB}>$ is factorizable as $|\psi_{AB}> = |\psi_A> \otimes |\psi_B>$. This means that $A \otimes B$ are not entangled.

Let $\hat{O}_A$ be an operator which measures some property of system A and $\hat{I}_B$ is an identity operator on B. Then its expectation value of $\hat{O}_A$ in the state $|\psi_{AB}>$ is

$$< \psi_{AB}|\hat{O}_A|\psi_{AB}> = <\psi_{AB}|\hat{O}_A \otimes \hat{I}_B|\psi_{AB}> = <\psi_A|\hat{O}_A|\psi_A> \otimes <\psi_B|\hat{I}_B|\psi_B>$$
$$< \psi_{AB}|\hat{O}_A|\psi_{AB}> = <\psi_A|\hat{O}_A|\psi_A> \text{ (as } <\psi_B|\hat{I}_B|\psi_B> = <\psi_B|\psi_B> = 1).$$

This composite system behaves as a pure state. In this case the A portion of the combined system is only involved in extracting information about A as it is possible to factor out the B part.

If the composite system does not allow such factorization (i.e. $|\psi_{AB}> \neq |\psi_A> \otimes |\psi_B>$), i.e. $A \otimes B$ is entangled then the system is said to be in the mixed state. In this case the combined system is to be involved to extract information about A.

Let us consider a composite system which is entangled in the basis $|0>, |1>$

$$|\psi_{AB}> = a|0>_A \otimes |0>_B + b|1>_A \otimes |1>_B = a|0>_A |0>_B + b|1>_A |1>_B$$

Consider an operator in the state space $\hat{M}_A \otimes \hat{I}_B$ which is a general measurement operator on subsystem A. The expectation value in the state $|\psi_{AB}>$ is

$$< \psi_{AB}|\hat{M}_A \otimes \hat{I}_B|\psi_{AB}> .$$
$$= [a^* < 0_A|<0_B| + b^* < 1_A|<1_B|] \,|\, (\hat{M}_A \otimes \hat{I}_B) \,|\, [a|0_A> |0_B> + b|1_A> |1_B>]$$
$$= |a|^2 < 0_A|\hat{M}_A|0_A> <0_B|\hat{I}_B|0_B> + |b|^2 < 1_A|\hat{M}_A|1_A> <1_B|\hat{I}_B|1_B>$$
$$a^*b < 0_A|\hat{M}_A|1_A> <0_B|\hat{I}_B|I_B> + ab^* < 1_A|\hat{M}_A|0_A> <1_B|\hat{I}_B|0_B>$$

Using $<0_B|\hat{I}_B|0_B> = <0_B|0_B> = 1, <1_B|\hat{I}_B|1_B> = <1_B|1_B> = 1)$

$$<0_B|\hat{I}_B|1_B> = <0_B|1_B> = 0, <1_B|\hat{I}_B|0_B> = <1_B|0_B> = 0$$

we have

$$<\psi_{AB}|\hat{M}_A \otimes \hat{I}_B|\psi_{AB}> = |a|^2 < 0_A|\hat{M}_A|0_A> + |b|^2 < 1_A|\hat{M}_A|1_A>$$

(the diagonal matrix elements are there).
Define density operator for subsystem A as

$$\hat{\rho}_A = |a|^2|0_A> <0_A| + |b|^2|1_A> <1_A| = \begin{pmatrix} |a|^2 & 0 \\ 0 & |b|^2 \end{pmatrix}$$

$$\hat{M}_A = \begin{pmatrix} <0_A | \hat{M}_A | 0_A> & 0 \\ 0 & <1_A | \hat{M}_A | 1_A> \end{pmatrix}$$

$$\hat{\rho}_A\hat{M}_A = \begin{pmatrix} |a|^2 & 0 \\ 0 & |b|^2 \end{pmatrix}\begin{pmatrix} <0_A | \hat{M}_A | 0_A> & 0 \\ 0 & <1_A | \hat{M}_A | 1_A> \end{pmatrix}$$

$$= \begin{pmatrix} |a|^2 < 0_A | \hat{M}_A | 0_A> & 0 \\ 0 & |b|^2 < 1_A | \hat{M}_A | 1_A> \end{pmatrix}$$

Hence

$$<\psi_{AB}|\hat{M}_A \otimes \hat{I}_B|\psi_{AB}> = Tr(\hat{\rho}_A\hat{M}_A) = |a|^2 < 0_A|\hat{M}_A|0_A> + |b|^2 < 1_A|\hat{M}_A|1_A>$$

7.8 Density matrix for a two-level spin system (Stern–Gerlach experiment)

Consider the Stern–Gerlach experiment as depicted in figure 7.1.

Suppose the Stern–Gerlach apparatus is oriented along $\hat{z}$. Let $\hat{s}_z$ be the operator for the z-component of spin $\vec{s}$ of beam particles. Then upper beam particles have $s_z = +\frac{\hbar}{2}$ and lower beam particles have $s_z = -\frac{\hbar}{2}$.

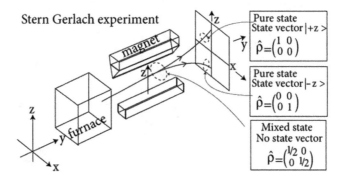

Figure 7.1. Stern Gerlach apparatus oriented along $\hat{z}$.

Particles in the upper beam are in the eigenstate $\lvert +z \rangle$ of operator $\hat{s}_z$. The density matrix is	Particles in the lower beam are in the eigenstate $\lvert -z \rangle$ of operator $\hat{s}_z$. The density matrix is
$$\hat{\rho} = \lvert +z \rangle \langle +z \rvert$$ $$= \begin{pmatrix} 1 \\ 0 \end{pmatrix}(1 \;\; 0) = \begin{pmatrix} 1 & 0 \\ 0 & 0 \end{pmatrix}$$	$$\hat{\rho} = \lvert -z \rangle \langle -z \rvert$$ $$= \begin{pmatrix} 0 \\ 1 \end{pmatrix}(0 \;\; 1) = \begin{pmatrix} 0 & 0 \\ 0 & 1 \end{pmatrix}$$
This is a pure state.	This is a pure state.
$$(Tr\hat{\rho}^2 = Tr\hat{\rho} = 1)$$	$$(Tr\hat{\rho}^2 = Tr\hat{\rho} = 1)$$

If we consider both streams taken together we have a mixed state comprising an equal mixture of particles in eigenstates $\lvert +z \rangle$ and $\lvert -z \rangle$. There is no state vector. The density matrix is ($\hat{\rho} = \sum_i p_i \lvert \psi_i \rangle \langle \psi_i \rvert$)

$$\hat{\rho} = \frac{1}{2}\lvert +z \rangle \langle +z \rvert + \frac{1}{2}\lvert -z \rangle \langle -z \rvert \text{ (probability of up and down states } \tfrac{1}{2})$$

$$= \frac{1}{2}\begin{pmatrix} 1 & 0 \\ 0 & 0 \end{pmatrix} + \frac{1}{2}\begin{pmatrix} 0 & 0 \\ 0 & 1 \end{pmatrix} = \begin{pmatrix} \frac{1}{2} & 0 \\ 0 & \frac{1}{2} \end{pmatrix}$$

$$Tr\hat{\rho}^2 = Tr\hat{\rho}\hat{\rho} = Tr\begin{pmatrix} \frac{1}{2} & 0 \\ 0 & \frac{1}{2} \end{pmatrix}\begin{pmatrix} \frac{1}{2} & 0 \\ 0 & \frac{1}{2} \end{pmatrix} = Tr\begin{pmatrix} \frac{1}{4} & 0 \\ 0 & \frac{1}{4} \end{pmatrix} = \frac{1}{4} + \frac{1}{4} = \frac{1}{2} < 1$$

Consider Stern–Gerlach experiment as depicted in figure 7.2.

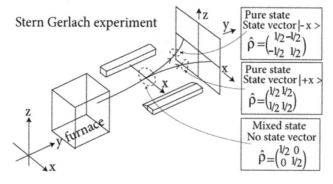

Figure 7.2. Stern–Gerlach apparatus oriented along $\hat{x}$.

Suppose Stern–Gerlach apparatus is oriented along $\hat{x}$. Let $\hat{s}_x$ be the operator for the x-component of spin $\vec{s}$ of beam particles. Then right beam particles have $s_x = +\frac{\hbar}{2}$ and left beam particles have $s_x = -\frac{\hbar}{2}$.

Particles in the +ve x beam are in the eigen state $|+x>$ of operator $\hat{s}_x$ and we can expand it and write

$$|+x> = \frac{1}{\sqrt{2}}(|+z> + |-z>)$$

$$= \frac{1}{\sqrt{2}}\left[\begin{pmatrix}1\\0\end{pmatrix} + \begin{pmatrix}0\\1\end{pmatrix}\right] = \begin{pmatrix}\frac{1}{\sqrt{2}}\\\frac{1}{\sqrt{2}}\end{pmatrix}$$

The density matrix is

$$\hat{\rho} = |+x><+x| = \begin{pmatrix}\frac{1}{\sqrt{2}}\\\frac{1}{\sqrt{2}}\end{pmatrix}\begin{pmatrix}\frac{1}{\sqrt{2}} & \frac{1}{\sqrt{2}}\end{pmatrix}$$

$$= \begin{pmatrix}\frac{1}{2} & \frac{1}{2}\\\frac{1}{2} & \frac{1}{2}\end{pmatrix}$$

This is a pure state.

$$(Tr\hat{\rho}^2 = Tr\hat{\rho} = 1)$$

Particles in the $-$ve x beam are in the eigen state $|-x>$ of operator $\hat{s}_x$ and we can expand it and write

$$|-x> = \frac{1}{\sqrt{2}}(|+z> - |-z>)$$

$$= \frac{1}{\sqrt{2}}\left[\begin{pmatrix}1\\0\end{pmatrix} - \begin{pmatrix}0\\1\end{pmatrix}\right] = \begin{pmatrix}\frac{1}{\sqrt{2}}\\-\frac{1}{\sqrt{2}}\end{pmatrix}.$$

The density matrix is

$$\hat{\rho} = |-x><-x| = \begin{pmatrix}\frac{1}{\sqrt{2}}\\-\frac{1}{\sqrt{2}}\end{pmatrix}\begin{pmatrix}\frac{1}{\sqrt{2}} & -\frac{1}{\sqrt{2}}\end{pmatrix}$$

$$= \begin{pmatrix}\frac{1}{2} & -\frac{1}{2}\\-\frac{1}{2} & \frac{1}{2}\end{pmatrix}$$

This is a pure state.

$$(Tr\hat{\rho}^2 = Tr\hat{\rho} = 1)$$

If we consider both streams taken together we have a mixed state comprising an equal mixture of particles in eigenstates $|+x>$ and $|-x>$. There is no state vector. The density matrix is ($\hat{\rho} = \sum_i p_i |\psi_i> <\psi_i|$)

$$\hat{\rho} = \frac{1}{2}|+x><+x| + \frac{1}{2}|-x><-x| \quad \left(\text{probability of } +x \text{ and } -x \text{ states is } \frac{1}{2}\right)$$

$$= \frac{1}{2}\left[\begin{pmatrix}\frac{1}{2} & \frac{1}{2}\\\frac{1}{2} & \frac{1}{2}\end{pmatrix} + \begin{pmatrix}\frac{1}{2} & -\frac{1}{2}\\-\frac{1}{2} & \frac{1}{2}\end{pmatrix}\right] = \begin{pmatrix}\frac{1}{2} & 0\\0 & \frac{1}{2}\end{pmatrix} \text{ and}$$

$$Tr\hat{\rho}^2 = Tr\begin{pmatrix}\frac{1}{4} & 0\\0 & \frac{1}{4}\end{pmatrix} = \frac{1}{2} < 1$$

7.9 Single-qubit density operator in terms of Pauli matrices

Representation of pure state and mixed state using Bloch sphere

The single-qubit state is

$$|\psi> = \cos\frac{\theta}{2}|0> + e^{i\phi}\sin\frac{\theta}{2}|1> = \begin{pmatrix} \cos\frac{\theta}{2} \\ e^{i\phi}\sin\frac{\theta}{2} \end{pmatrix} = \alpha_0|0> + \alpha_1|1> = \begin{pmatrix} \alpha_0 \\ \alpha_1 \end{pmatrix}$$

with $\alpha_0 = \cos\frac{\theta}{2}$, $\alpha_1 = e^{i\phi}\sin\frac{\theta}{2}$ the single-qubit density operator becomes

$$\hat{\rho} = |\psi> <\psi| = \begin{pmatrix} \alpha_0 \\ \alpha_1 \end{pmatrix}(\alpha_0^* \ \ \alpha_1^*)$$

$$\hat{\rho} = \begin{pmatrix} |\alpha_0|^2 & \alpha_0\alpha_1^* \\ \alpha_0^*\alpha_1 & |\alpha_1|^2 \end{pmatrix} = \begin{pmatrix} \cos^2\frac{\theta}{2} & \cos\frac{\theta}{2}e^{-i\phi}\sin\frac{\theta}{2} \\ \cos\frac{\theta}{2}e^{i\phi}\sin\frac{\theta}{2} & \sin^2\frac{\theta}{2} \end{pmatrix}$$

$$= \frac{1}{2}\begin{pmatrix} 2\cos^2\frac{\theta}{2} & 2\sin\frac{\theta}{2}\cos\frac{\theta}{2}e^{-i\phi} \\ 2\sin\frac{\theta}{2}\cos\frac{\theta}{2}e^{i\phi} & 2\sin^2\frac{\theta}{2} \end{pmatrix} = \frac{1}{2}\begin{pmatrix} 1+\cos\theta & \sin\theta\,e^{-i\phi} \\ \sin\theta\,e^{i\phi} & 1-\cos\theta \end{pmatrix}$$

$$\hat{\rho} = \frac{1}{2}\begin{pmatrix} 1+\cos\theta & \sin\theta\,(\cos\phi - i\sin\phi) \\ \sin\theta\,(\cos\phi + i\sin\phi) & 1-\cos\theta \end{pmatrix} \Rightarrow Tr\hat{\rho} = 1$$

$$\hat{\rho} = \frac{1}{2}\begin{pmatrix} 1+\cos\theta & \sin\theta\cos\phi - i\sin\theta\sin\phi \\ \sin\theta\cos\phi + i\sin\theta\sin\phi & 1-\cos\theta \end{pmatrix}$$

$$= \frac{1}{2}\begin{pmatrix} 1 & 0 \\ 0 & 1 \end{pmatrix} + \frac{\sin\theta\cos\phi}{2}\begin{pmatrix} 0 & 1 \\ 1 & 0 \end{pmatrix} + \frac{\sin\theta\sin\phi}{2}\begin{pmatrix} 0 & -i \\ i & 0 \end{pmatrix} + \frac{\cos\theta}{2}\begin{pmatrix} 1 & 0 \\ 0 & -1 \end{pmatrix}$$

$$\hat{\rho} = \frac{1}{2}(\hat{I} + \sin\theta\cos\phi\,\hat{\sigma}_1 + \sin\theta\sin\phi\,\hat{\sigma}_2 + \cos\theta\,\hat{\sigma}_3)$$

Clearly $\hat{\sigma}_x = \hat{\sigma}_1$, $\hat{\sigma}_y = \hat{\sigma}_2$, $\hat{\sigma}_z = \hat{\sigma}_3$ are the Pauli spin matrices or operators and are the components of the vector operator $\hat{\vec{\sigma}}$, i.e. $\hat{\vec{\sigma}} = (\hat{\sigma}_1, \hat{\sigma}_2, \hat{\sigma}_3)$.

Again we construct another vector

$$\vec{r} = (x, y, z) = (\sin\theta\cos\phi, \ \sin\theta\sin\phi, \ \cos\theta).$$

Clearly

$$x = \sin\theta\cos\phi, \ y = \sin\theta\sin\phi, \ z = \cos\theta \text{ and}$$

$$|\vec{r}| = r = \sqrt{x^2 + y^2 + z^2} = \sqrt{(\sin\theta\cos\phi)^2 + (\sin\theta\sin\phi)^2 + (\cos\theta)^2}$$

$$= \sqrt{\sin^2\theta(\cos^2\phi + \sin^2\phi)^2 + \cos^2\theta} = 1$$

As $\vec{r}$ has unit magnitude we can write it as a unit vector

$\vec{r} = \hat{n} = (\sin\theta\cos\phi, \ \sin\theta\sin\phi, \ \cos\theta) =$ unit vector along the particular direction (θ, ϕ). Hence the single qubit density operator in terms of Pauli spin operators is

$$\hat{\rho} = \tfrac{1}{2}(I + \hat{n}.\,\vec{\sigma}) = \tfrac{1}{2}(I + \vec{r}.\overset{\wedge}{\sigma})\left(\vec{r} = \hat{n} = \text{Bloch vector}\right)$$

$$\hat{\rho}^2 = \hat{\rho}\hat{\rho} = \tfrac{1}{2}(I + \vec{r}.\overset{\wedge}{\sigma})\tfrac{1}{2}(I + \vec{r}.\overset{\wedge}{\sigma}) = \tfrac{1}{4}[\hat{I} + 2\vec{r}.\overset{\wedge}{\sigma} + (\vec{r}.\overset{\wedge}{\sigma})^2]$$

$$= \tfrac{1}{4}[I + 2(x\hat{\sigma}_1 + y\hat{\sigma}_2 + z\hat{\sigma}_3) + (x\hat{\sigma}_1 + y\hat{\sigma}_2 + z\hat{\sigma}_3)^2]$$

$$= \tfrac{1}{4}[I + 2(x\hat{\sigma}_1 + y\hat{\sigma}_2 + z\hat{\sigma}_3) + (x^2\hat{\sigma}_1^2 + y^2\hat{\sigma}_2^2 + z^2\hat{\sigma}_3^2)$$

$$+ 2(xy\hat{\sigma}_1\hat{\sigma}_2 + yz\hat{\sigma}_2\hat{\sigma}_3 + xz\hat{\sigma}_3\hat{\sigma}_1)]$$

$$Tr\hat{\rho}^2 = \tfrac{1}{4}[Tr\hat{I} + 2(xTr\hat{\sigma}_1 + yTr\hat{\sigma}_2 + zTr\hat{\sigma}_3)$$

$$+ (x^2Tr\hat{\sigma}_1^2 + y^2Tr\hat{\sigma}_2^2 + z^2Tr\hat{\sigma}_3^2)$$

$$+ 2(xyTr\hat{\sigma}_1\hat{\sigma}_2 + yzTr\hat{\sigma}_2\hat{\sigma}_3 + xzTr\hat{\sigma}_3\hat{\sigma}_1)]$$

As $Tr\,\hat{I} = Tr\begin{pmatrix} 1 & 0 \\ 0 & 1 \end{pmatrix} = 2,\ Tr\hat{\sigma}_i = 0,\ Tr\hat{\sigma}_i^2 = Tr\,\hat{I} = 2$ (as $\hat{\sigma}_i^2 = \hat{I}$), $Tr(\hat{\sigma}_i\hat{\sigma}_j) = 0$

$$Tr\hat{\rho}^2 = \tfrac{1}{4}[2 + 2(0) + (2x^2 + 2y^2 + 2z^2) + 2(0)]$$

$$= \tfrac{1}{4}[2 + 2(x^2 + y^2 + z^2)] = \tfrac{1}{2}(1 + r^2)$$

✓ The Bloch sphere radius is unity. So $|\vec{r}| = 1$, i.e. $r^2 = 1$. So

$$Tr\hat{\rho}^2 = \tfrac{1}{2}(1 + r^2) = 1 (\Leftarrow \text{property of pure state})$$

Clearly the points on the surface of the Bloch sphere represent pure state. In other words $\vec{r} =$ Bloch vector with $|\vec{r}| = 1$ for pure state.

✓Again, for points inside the Bloch sphere $|\vec{r}| < 1$, $r^2 < 1$ and so

$$Tr\hat{\rho}^2 = \tfrac{1}{2}(1 + r^2) < 1 (\Leftarrow \text{property of pure state})$$

Clearly the points inside the surface of the Bloch sphere represent mixed state. The density matrix corresponding to mixed state would thus be

$$\hat{\rho} = \tfrac{1}{2}(\hat{I} + \vec{r}.\,\hat{\sigma}),\ \ |\vec{r}| < 1 \text{ (i.e. } \vec{r} = \text{Bloch vector with } |\vec{r}| < 1 \text{ for mixed state).}$$

- **GIST:**

   We have established that $Tr\hat{\rho}^2 = \tfrac{1}{2}(1 + |\vec{r}|^2)$ and related it to pure and mixed state:

   On surface of Bloch sphere $r = 1$, $Tr\hat{\rho}^2 = 1$ corresponding to pure state.

   Within Bloch sphere $r < 1$, $Tr\hat{\rho}^2 < 1$ corresponding to mixed state.

- **Example**

   Consider three points of the Bloch sphere: One point named a is within Bloch sphere at $r = z = \tfrac{1}{3}$, another point b is on the surface of Bloch sphere, i.e. at $r = z = 1$ along the direction $\hat{n} = \hat{z}(\theta = 0, \phi = 0)$(on which

$x = 0$, $y = 0$, $z = r$). And the point O is the center of Bloch sphere having $r = z = 0$. We find the density matrix and infer about the state it represents (figure 7.3).

$$\hat{\rho} = \tfrac{1}{2}(\hat{I} + \vec{r}.\,\hat{\vec{\sigma}}) = \tfrac{1}{2}\left(\hat{I} + x\hat{\sigma}_x + y\hat{\sigma}_y + z\hat{\sigma}_z\right) = \tfrac{1}{2}(\hat{I} + z\hat{\sigma}_z)$$

(as $x = 0$, $y = 0$ for all the 3 points)

✓ **For $r = z = \tfrac{1}{3}$ (point a within Bloch sphere,)**

$$\hat{\rho} = \tfrac{1}{2}(\hat{I} + z\hat{\sigma}_z) = \tfrac{1}{2}\left(\hat{I} + \tfrac{1}{3}\hat{\sigma}_z\right) = \tfrac{1}{2}\left[\begin{pmatrix} 1 & 0 \\ 0 & 1 \end{pmatrix} + \tfrac{1}{3}\begin{pmatrix} 1 & 0 \\ 0 & -1 \end{pmatrix}\right] = \begin{pmatrix} \tfrac{2}{3} & 0 \\ 0 & \tfrac{1}{3} \end{pmatrix}$$

$$= \tfrac{2}{3}\begin{pmatrix} 1 & 0 \\ 0 & 0 \end{pmatrix} + \tfrac{1}{3}\begin{pmatrix} 0 & 0 \\ 0 & 1 \end{pmatrix} = \tfrac{2}{3}\begin{pmatrix} 1 \\ 0 \end{pmatrix}(1 \quad 0) + \tfrac{1}{3}\begin{pmatrix} 0 \\ 1 \end{pmatrix}(0 \quad 1)$$

$$= \tfrac{2}{3}|0><0| + \tfrac{1}{3}|1><1|.$$

Here $Tr\hat{\rho} = Tr\begin{pmatrix} \tfrac{2}{3} & 0 \\ 0 & \tfrac{1}{3} \end{pmatrix} = 1$

$$\hat{\rho}^2 = \hat{\rho}\hat{\rho} = \begin{pmatrix} \tfrac{2}{3} & 0 \\ 0 & \tfrac{1}{3} \end{pmatrix}\begin{pmatrix} \tfrac{2}{3} & 0 \\ 0 & \tfrac{1}{3} \end{pmatrix} = \begin{pmatrix} \tfrac{4}{9} & 0 \\ 0 & \tfrac{1}{9} \end{pmatrix} \neq \hat{\rho} \text{ and } Tr\hat{\rho}^2 = Tr\begin{pmatrix} \tfrac{4}{9} & 0 \\ 0 & \tfrac{1}{9} \end{pmatrix} = \tfrac{5}{9} < 1 \text{ which}$$

means points within the Bloch sphere represent mixed state.

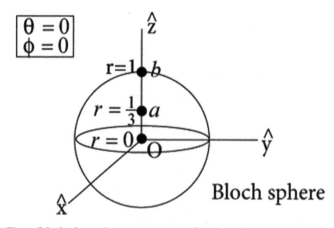

Figure 7.3. Surface points are pure states. Interior points are mixed states.

✓**For** $r = z = 0$ **(center O of Bloch sphere)** $(z = 0)$

$$\hat{\rho} = \tfrac{1}{2}(\hat{I} + z\hat{\sigma}_z) = \tfrac{1}{2}\hat{I} = \tfrac{1}{2}\begin{pmatrix} 1 & 0 \\ 0 & 1 \end{pmatrix} = \begin{pmatrix} \tfrac{1}{2} & 0 \\ 0 & \tfrac{1}{2} \end{pmatrix}$$

$$= \tfrac{1}{2}\begin{pmatrix} 1 & 0 \\ 0 & 0 \end{pmatrix} + \tfrac{1}{2}\begin{pmatrix} 0 & 0 \\ 0 & 1 \end{pmatrix} = \tfrac{1}{2}\begin{pmatrix} 1 \\ 0 \end{pmatrix}(1 \; 0) + \tfrac{1}{2}\begin{pmatrix} 0 \\ 1 \end{pmatrix}(0 \; 1)$$

$$= \tfrac{1}{2}|0><0| + \tfrac{1}{2}|1><1|.$$

Here $Tr\hat{\rho} = Tr\begin{pmatrix} \tfrac{1}{2} & 0 \\ 0 & \tfrac{1}{2} \end{pmatrix} = 1$

$$\hat{\rho}^2 = \hat{\rho}\hat{\rho} = \begin{pmatrix} \tfrac{1}{2} & 0 \\ 0 & \tfrac{1}{2} \end{pmatrix}\begin{pmatrix} \tfrac{1}{2} & 0 \\ 0 & \tfrac{1}{2} \end{pmatrix} = \begin{pmatrix} \tfrac{1}{4} & 0 \\ 0 & \tfrac{1}{4} \end{pmatrix} \neq \hat{\rho} \text{ and } Tr\hat{\rho}^2 = Tr\begin{pmatrix} \tfrac{1}{4} & 0 \\ 0 & \tfrac{1}{4} \end{pmatrix} = \tfrac{1}{2} < 1 \text{ which}$$

means the origin of Bloch sphere represents a mixed state.

✓**For** $r = z = 1$ **(point b on Bloch sphere surface)**

$$\hat{\rho} = \tfrac{1}{2}(\hat{I} + z\hat{\sigma}_z) = \tfrac{1}{2}(\hat{I} + \hat{\sigma}_z) = \tfrac{1}{2}\left[\begin{pmatrix} 1 & 0 \\ 0 & 1 \end{pmatrix} + \begin{pmatrix} 1 & 0 \\ 0 & -1 \end{pmatrix}\right] = \begin{pmatrix} 1 & 0 \\ 0 & 0 \end{pmatrix}$$

$$= \begin{pmatrix} 1 \\ 0 \end{pmatrix}(1 \; 0) = |0><0|.$$

Here $Tr\hat{\rho} = Tr\begin{pmatrix} 1 & 0 \\ 0 & 0 \end{pmatrix} = 1$

$$\hat{\rho}^2 = \hat{\rho}\hat{\rho} = \begin{pmatrix} 1 & 0 \\ 0 & 0 \end{pmatrix}\begin{pmatrix} 1 & 0 \\ 0 & 0 \end{pmatrix} = \begin{pmatrix} 1 & 0 \\ 0 & 0 \end{pmatrix} = \hat{\rho} \text{ and so } Tr\hat{\rho}^2 = 1 \text{ which means points on}$$

surface of the Bloch sphere represent pure state.

• Find $<\hat{\sigma}_x>$, $<\hat{\sigma}_y>$, $<\hat{\sigma}_z>$

Using the relation $<\hat{A}> = Tr(\hat{A}\hat{\rho})$ and $\hat{\rho} = \tfrac{1}{2}(\hat{I} + \vec{r}.\vec{\hat{\sigma}})$ we write

$$<\hat{\sigma}_x> = Tr(\hat{\sigma}_x\hat{\rho}) = Tr\left(\hat{\sigma}_x\tfrac{1}{2}(\hat{I} + \vec{r}.\vec{\hat{\sigma}})\right) = \tfrac{1}{2}Tr\left(\hat{\sigma}_x(\hat{I} + x\hat{\sigma}_x + y\hat{\sigma}_y + z\hat{\sigma}_z)\right)$$

$$= \tfrac{1}{2}Tr\left(\hat{\sigma}_x + x\hat{\sigma}_x^2 + y\hat{\sigma}_x\hat{\sigma}_y + z\hat{\sigma}_x\hat{\sigma}_z\right) = \tfrac{1}{2}Tr\left(\hat{\sigma}_x + x\hat{I} + yi\hat{\sigma}_z - zi\hat{\sigma}_y\right)$$

$$(\text{since } \hat{\sigma}_x^2 = \hat{I}, \; \hat{\sigma}_x\hat{\sigma}_y = i\hat{\sigma}_z, \; \hat{\sigma}_x\hat{\sigma}_z = -i\hat{\sigma}_y)$$

$$= \tfrac{1}{2}\left[Tr\hat{\sigma}_x + xTr\hat{I} + iyTr\hat{\sigma}_z - izTr\hat{\sigma}_y\right]$$

$<\hat{\sigma}_x> = \tfrac{1}{2}xTr\hat{I} = \tfrac{1}{2}x.2 = x$ (since $Tr\hat{\sigma}_i = 0$, $Tr\hat{I} = 2$)
Similarly $<\hat{\sigma}_y> = y$, $<\hat{\sigma}_z> = z$

7.10 Some illustration of density matrix for pure and mixed states

- Consider a state given by

$$|\psi> = \frac{1}{\sqrt{2}}|0> -\frac{i}{\sqrt{2}}|1> = \frac{1}{\sqrt{2}}\begin{pmatrix}1\\0\end{pmatrix} - \frac{i}{\sqrt{2}}\begin{pmatrix}0\\1\end{pmatrix} = \begin{pmatrix}\frac{1}{\sqrt{2}}\\-\frac{i}{\sqrt{2}}\end{pmatrix} = \frac{1}{\sqrt{2}}\begin{pmatrix}1\\-i\end{pmatrix}$$

As $<\psi|\psi> = \frac{1}{\sqrt{2}}(1 \ \ i)\frac{1}{\sqrt{2}}\begin{pmatrix}1\\-i\end{pmatrix} = \frac{1}{2}(1+1) = 1$ this means that $|\psi>$ is normalized.

Density operator is

$$\hat{\rho} = |\psi><\psi| = \frac{1}{\sqrt{2}}\begin{pmatrix}1\\-i\end{pmatrix}\frac{1}{\sqrt{2}}(1 \ \ i) = \frac{1}{2}\begin{pmatrix}1 & i\\-i & 1\end{pmatrix} = \begin{pmatrix}\frac{1}{2} & \frac{i}{2}\\-\frac{i}{2} & \frac{1}{2}\end{pmatrix}, \ Tr\hat{\rho} = \frac{1}{2} + \frac{1}{2} = 1$$

$$\hat{\rho}^2 = \frac{1}{2}\begin{pmatrix}1 & i\\-i & 1\end{pmatrix}\frac{1}{2}\begin{pmatrix}1 & i\\-i & 1\end{pmatrix} = \frac{1}{4}\begin{pmatrix}2 & 2i\\-2i & 2\end{pmatrix} = \frac{1}{2}\begin{pmatrix}1 & i\\-i & 1\end{pmatrix} = \hat{\rho}, \ Tr\hat{\rho}^2 = 1 \ \text{(pure state)}$$

- A coherent superposition of states is a pure state, e.g. in the basis $|0> = \begin{pmatrix}1\\0\end{pmatrix}$

and $|1> = \begin{pmatrix}0\\1\end{pmatrix}$ we express $|\psi> = \frac{1}{\sqrt{2}}(|0> +|1>) = \frac{1}{\sqrt{2}}(\begin{pmatrix}1\\0\end{pmatrix} + \begin{pmatrix}0\\1\end{pmatrix}) = \begin{pmatrix}\frac{1}{\sqrt{2}}\\\frac{1}{\sqrt{2}}\end{pmatrix}$

Let us find average or expectation of $\hat{\sigma}_x$ where

$$\hat{\sigma}_x|1> = |0> , \ \hat{\sigma}_x|0> = |1> .$$

✓ $<\hat{\sigma}_x> = <\psi|\hat{\sigma}_x|\psi> = \frac{1}{\sqrt{2}}(<0|+<1|)\hat{\sigma}_x\frac{1}{\sqrt{2}}(|0> +|1>)$

$$= \frac{1}{\sqrt{2}}(<0|+<1|)\frac{1}{\sqrt{2}}(\hat{\sigma}_x|0> +\hat{\sigma}_x|1>) = \frac{1}{\sqrt{2}}(<0|+<1|)\frac{1}{\sqrt{2}}(|1> +|0>)$$

$$= \frac{1}{2}(<0|1> +<0|0> +<1|1> +<1|0>) = 1$$

In this pure state case

$$\hat{\rho} = |\psi><\psi| = = \begin{pmatrix}\frac{1}{\sqrt{2}}\\\frac{1}{\sqrt{2}}\end{pmatrix}(\frac{1}{\sqrt{2}} \ \ \frac{1}{\sqrt{2}}) = \begin{pmatrix}\frac{1}{2} & \frac{1}{2}\\\frac{1}{2} & \frac{1}{2}\end{pmatrix}, \ Tr\hat{\rho} = \frac{1}{2} + \frac{1}{2} = 1$$

$$\hat{\rho}^2 = \begin{pmatrix}\frac{1}{2} & \frac{1}{2}\\\frac{1}{2} & \frac{1}{2}\end{pmatrix}\begin{pmatrix}\frac{1}{2} & \frac{1}{2}\\\frac{1}{2} & \frac{1}{2}\end{pmatrix} = \begin{pmatrix}\frac{1}{2} & \frac{1}{2}\\\frac{1}{2} & \frac{1}{2}\end{pmatrix} = \hat{\rho}, \ Tr\hat{\rho}^2 = \frac{1}{2} + \frac{1}{2} = 1$$

- Suppose there is an equal mixture of states $|0>$ and $|1>$. Let us calculate the expectation value of $\hat{\sigma}_x$.

✓ We can pick up either $|0>$ or $|1>$ with the same probability $\frac{1}{2}$.

$$<\hat{\sigma}_x> = (\text{Probability of picking up state} |0>) < 0|\hat{\sigma}_x|0>$$

$$+(\text{Probability of picking up state} |1>) < 1|\hat{\sigma}_x|1>$$

$$<\hat{\sigma}_x> = \frac{1}{2} < 0|\hat{\sigma}_x|0> + \frac{1}{2} < 1|\hat{\sigma}_x|1> = \frac{1}{2} < 0|1> + \frac{1}{2} < 1|0> = 0$$

In this mixed state case :

$$\hat{\rho} = (\text{Probability of picking} |0>)|0> <0|+(\text{Probability of picking} |1>)|1> <1|$$
$$= \frac{1}{2}|0> <0| + \frac{1}{2}|1> <1| = \frac{1}{2}[|0> <0| + |1> <1|]$$
$$= \frac{1}{2}(\text{completeness condition}) = \frac{1}{2}\hat{I}$$

Also in matrix form

$$\hat{\rho} = \frac{1}{2}|0> <0| + \frac{1}{2}|1> <1|$$

$$\hat{\rho} = \frac{1}{2}\begin{pmatrix}1\\0\end{pmatrix}(1\ \ 0) + \frac{1}{2}\begin{pmatrix}0\\1\end{pmatrix}(0\ \ 1) = \frac{1}{2}\begin{pmatrix}1 & 0\\0 & 1\end{pmatrix} = \begin{pmatrix}\frac{1}{2} & 0\\0 & \frac{1}{2}\end{pmatrix} = \frac{1}{2}\hat{I},\ Tr\hat{\rho} = \frac{1}{2} + \frac{1}{2} = 1$$

$$\hat{\rho}^2 = \hat{\rho}\hat{\rho} = \begin{pmatrix}\frac{1}{2} & 0\\0 & \frac{1}{2}\end{pmatrix}\begin{pmatrix}\frac{1}{2} & 0\\0 & \frac{1}{2}\end{pmatrix} = \begin{pmatrix}\frac{1}{4} & 0\\0 & \frac{1}{4}\end{pmatrix} \neq \hat{\rho},\ Tr\hat{\rho}^2 = \frac{1}{4} + \frac{1}{4} = \frac{1}{2} < 1$$

- Let us expand the pure state $|\psi>$ in terms of basis vectors $|0>$ and $|1>$ as

$$|\psi> = \alpha_0|0> + \alpha_1|1> = \alpha_0\begin{pmatrix}1\\0\end{pmatrix} + \alpha_1\begin{pmatrix}0\\1\end{pmatrix} = \begin{pmatrix}\alpha_0\\\alpha_1\end{pmatrix}$$

with

$$|\alpha_0|^2 + |\alpha_1|^2 = 1.$$

✓ The density matrix will be

$$\hat{\rho} = |\psi> <\psi| = \begin{pmatrix} \alpha_0 \\ \alpha_1 \end{pmatrix}(\alpha_0^* \ \ \alpha_1^*) = \begin{pmatrix} |\alpha_0|^2 & \alpha_0\alpha_1^* \\ \alpha_0^*\alpha_1 & |\alpha_1|^2 \end{pmatrix}. \ , \ Tr\hat{\rho} = |\alpha_0|^2 + |\alpha_1|^2 = 1$$

$$\hat{\rho}^2 = \hat{\rho}\hat{\rho} = \begin{pmatrix} |\alpha_0|^2 & \alpha_0\alpha_1^* \\ \alpha_0^*\alpha_1 & |\alpha_1|^2 \end{pmatrix}\begin{pmatrix} |\alpha_0|^2 & \alpha_0\alpha_1^* \\ \alpha_0^*\alpha_1 & |\alpha_1|^2 \end{pmatrix}$$

$$= \begin{pmatrix} |\alpha_0|^4 + \alpha_0\alpha_1^*\alpha_0^*\alpha_1 & |\alpha_0|^2 \alpha_0\alpha_1^* + \alpha_0\alpha_1^* |\alpha_1|^2 \\ \alpha_0^*\alpha_1 |\alpha_0|^2 + |\alpha_1|^2 \alpha_0^*\alpha_1 & \alpha_0^*\alpha_1\alpha_0\alpha_1^* + |\alpha_1|^4 \end{pmatrix}$$

$$= \begin{pmatrix} |\alpha_0|^2(|\alpha_0|^2 + |\alpha_1|^2) & \alpha_0\alpha_1^*(|\alpha_0|^2 + |\alpha_1|^2) \\ \alpha_0^*\alpha_1(|\alpha_0|^2 + |\alpha_1|^2) & |\alpha_1|^2(|\alpha_0|^2 + |\alpha_1|^2) \end{pmatrix} = \begin{pmatrix} |\alpha_0|^2 & \alpha_0\alpha_1^* \\ \alpha_0^*\alpha_1 & |\alpha_1|^2 \end{pmatrix} = \hat{\rho}$$

(using $|\alpha_0|^2 + |\alpha_1|^2 = 1$). Also $Tr\hat{\rho} = |\alpha_0|^2 + |\alpha_1|^2 = 1$

- Consider the state $|\psi> = \frac{1}{\sqrt{2}}(|\uparrow> +|\downarrow>) = \begin{pmatrix} \frac{1}{\sqrt{2}} \\ \frac{1}{\sqrt{2}} \end{pmatrix}$. Find the density matrix

and show that it is a pure state.

✓ $\hat{\rho} = |\psi> <\psi| = \begin{pmatrix} \frac{1}{\sqrt{2}} \\ \frac{1}{\sqrt{2}} \end{pmatrix}\begin{pmatrix} \frac{1}{\sqrt{2}} & \frac{1}{\sqrt{2}} \end{pmatrix} = \begin{pmatrix} \frac{1}{2} & \frac{1}{2} \\ \frac{1}{2} & \frac{1}{2} \end{pmatrix}, \ Tr\hat{\rho} = \frac{1}{2} + \frac{1}{2} = 1$

$$\hat{\rho}^2 = \begin{pmatrix} \frac{1}{2} & \frac{1}{2} \\ \frac{1}{2} & \frac{1}{2} \end{pmatrix}\begin{pmatrix} \frac{1}{2} & \frac{1}{2} \\ \frac{1}{2} & \frac{1}{2} \end{pmatrix} = \begin{pmatrix} \frac{1}{2} & \frac{1}{2} \\ \frac{1}{2} & \frac{1}{2} \end{pmatrix} = \hat{\rho}, \ Tr\hat{\rho}^2 = \frac{1}{2} + \frac{1}{2} = 1$$

So $|\psi> = \frac{1}{\sqrt{2}}(|\uparrow> +|\downarrow>)$ is a pure state.

- For the state $|\psi> = \sin\theta|0> + \cos\theta|1>$ find density matrix $\hat{\rho}$. Find $Tr\hat{\rho}$, $Tr\hat{\rho}^2$ and show that it is a pure state.

✓ $\hat{\rho} = |\psi> <\psi| = (\sin\theta|0> + \cos\theta|1>)(\sin\theta <0| + \cos\theta <1|)$

$$= \sin^2\theta|0><0| + \sin\theta\cos\theta|0><1| + \cos\theta\sin\theta|1><0| + \cos^2\theta|1><1|$$

Let us evaluate the matrix elements of the density matrix operator $\hat{\rho}$(using the orthonormalization relations $<0|0> =1$, $<1|1> =1$, $<0|1> =0$, $<1|0> =0$)

$$\rho_{11} = <0|\hat{\rho}|0> = \sin^2\theta \qquad \rho_{12} = <0|\hat{\rho}|1> = \sin\theta\cos\theta$$

$$\rho_{21} = <1|\hat{\rho}|0> = \cos\theta\sin\theta \qquad \rho_{22} = <1|\hat{\rho}|1> = \cos^2\theta$$

$$\hat{\rho} = \begin{pmatrix} \rho_{11} & \rho_{12} \\ \rho_{21} & \rho_{22} \end{pmatrix} = \begin{pmatrix} \sin^2\theta & \sin\theta\cos\theta \\ \cos\theta\sin\theta & \cos^2\theta \end{pmatrix}, \quad Tr\hat{\rho} = \sin^2\theta + \cos^2\theta = 1$$

$$\hat{\rho}^2 = \begin{pmatrix} \sin^2\theta & \sin\theta\cos\theta \\ \cos\theta\sin\theta & \cos^2\theta \end{pmatrix} \begin{pmatrix} \sin^2\theta & \sin\theta\cos\theta \\ \cos\theta\sin\theta & \cos^2\theta \end{pmatrix}$$

$$= \begin{pmatrix} \sin^4\theta + \sin^2\theta\cos^2\theta & \sin^3\theta\cos\theta + \sin\theta\cos^3\theta \\ \cos^3\theta\sin\theta + \cos\theta\sin^3\theta & \cos^2\theta\sin^2\theta + \cos^4\theta \end{pmatrix}$$

$$= \begin{pmatrix} \sin^2\theta(\sin^2\theta + \cos^2\theta) & \sin\theta\cos\theta(\sin^2\theta + \cos^2\theta) \\ \sin\theta\cos\theta(\cos^2\theta + \sin^2\theta) & \cos^2\theta(\sin^2\theta + \cos^2\theta) \end{pmatrix}$$

$$\hat{\rho}^2 = \begin{pmatrix} \sin^2\theta & \sin\theta\cos\theta \\ \sin\theta\cos\theta & \cos^2\theta \end{pmatrix} = \hat{\rho} \text{ and so } Tr\hat{\rho}^2 = 1.$$

So $|\psi> = \sin\theta|0> + \cos\theta|1>$ is a pure state.

- Consider a state with 50% $| \uparrow > = |up\ spin> = \begin{pmatrix} 1 \\ 0 \end{pmatrix}$ and 50% $| \downarrow > = | down\ spin>$
= $\begin{pmatrix} 0 \\ 1 \end{pmatrix}$. Find density matrix and show that it is a mixed state.

✓ $\hat{\rho} = \sum_i p_i |\psi_i> <\psi_i|$

$$\hat{\rho} = \tfrac{1}{2}| \uparrow > < \uparrow | + \tfrac{1}{2}| \downarrow > < \downarrow | \qquad \left[\text{coefficient}\tfrac{1}{2} = 50\%\text{probability}\right]$$

$$\rho = \tfrac{1}{2}\begin{pmatrix} 1 \\ 0 \end{pmatrix}(1\ \ 0) + \tfrac{1}{2}\begin{pmatrix} 0 \\ 1 \end{pmatrix}(0\ \ 1) \qquad \left(\text{since}| \uparrow > = \begin{pmatrix} 1 \\ 0 \end{pmatrix}, \ \ | \downarrow > = \begin{pmatrix} 0 \\ 1 \end{pmatrix}\right)$$

$$= \tfrac{1}{2}\begin{pmatrix} 1 & 0 \\ 0 & 0 \end{pmatrix} + \tfrac{1}{2}\begin{pmatrix} 0 & 0 \\ 0 & 1 \end{pmatrix} = \tfrac{1}{2}\begin{pmatrix} 1 & 0 \\ 0 & 1 \end{pmatrix} \quad \Rightarrow \quad Tr\hat{\rho} = 1$$

$$\hat{\rho}^2 = \tfrac{1}{2}\begin{pmatrix} 1 & 0 \\ 0 & 1 \end{pmatrix}\tfrac{1}{2}\begin{pmatrix} 1 & 0 \\ 0 & 1 \end{pmatrix} = \tfrac{1}{4}\begin{pmatrix} 1 & 0 \\ 0 & 1 \end{pmatrix} \neq \hat{\rho}. \text{ Also, } Tr\hat{\rho}^2 = \tfrac{1}{4} + \tfrac{1}{4} = \tfrac{1}{2} < 1.$$

So the state is a mixed state.

- Consider a state with 50% $|1> = \tfrac{1}{\sqrt{2}}(| \uparrow > +| \downarrow >) = \begin{pmatrix} \frac{1}{\sqrt{2}} \\ \frac{1}{\sqrt{2}} \end{pmatrix}$ and 50%

$|2> = \tfrac{1}{\sqrt{2}}(| \uparrow > -| \downarrow >) = \begin{pmatrix} \frac{1}{\sqrt{2}} \\ -\frac{1}{\sqrt{2}} \end{pmatrix}$. Find density matrix and show that it is

a mixed state.

✓ Since the base states are $| \uparrow > = \begin{pmatrix} 1 \\ 0 \end{pmatrix}$, $| \downarrow > = \begin{pmatrix} 0 \\ 1 \end{pmatrix}$ we have

$$|1 > = \frac{1}{\sqrt{2}}(| \uparrow > + | \downarrow >) = \frac{1}{\sqrt{2}}\begin{pmatrix} 1 \\ 0 \end{pmatrix} + \frac{1}{\sqrt{2}}\begin{pmatrix} 0 \\ 1 \end{pmatrix} = \begin{pmatrix} \frac{1}{\sqrt{2}} \\ \frac{1}{\sqrt{2}} \end{pmatrix}$$

$$|2 > = \frac{1}{\sqrt{2}}(| \uparrow > - | \downarrow >) = \frac{1}{\sqrt{2}}\begin{pmatrix} 1 \\ 0 \end{pmatrix} - \frac{1}{\sqrt{2}}\begin{pmatrix} 0 \\ 1 \end{pmatrix} = \begin{pmatrix} \frac{1}{\sqrt{2}} \\ -\frac{1}{\sqrt{2}} \end{pmatrix}$$

The density matrix is given by (taking the coefficient to represent $\frac{1}{2} = 50\%$ probability) $\hat{\rho} = \sum_i p_i |\psi_i> <\psi_i|$

$$\hat{\rho} = \frac{1}{2}|1> <1| + \frac{1}{2}|2> <2| = \frac{1}{2}\begin{pmatrix} \frac{1}{\sqrt{2}} \\ \frac{1}{\sqrt{2}} \end{pmatrix}\begin{pmatrix} \frac{1}{\sqrt{2}} & \frac{1}{\sqrt{2}} \end{pmatrix} + \frac{1}{2}\begin{pmatrix} \frac{1}{\sqrt{2}} \\ -\frac{1}{\sqrt{2}} \end{pmatrix}\begin{pmatrix} \frac{1}{\sqrt{2}} & -\frac{1}{\sqrt{2}} \end{pmatrix}$$

$$= \frac{1}{2}\begin{pmatrix} \frac{1}{2} & \frac{1}{2} \\ \frac{1}{2} & \frac{1}{2} \end{pmatrix} + \frac{1}{2}\begin{pmatrix} \frac{1}{2} & -\frac{1}{2} \\ -\frac{1}{2} & \frac{1}{2} \end{pmatrix} = \frac{1}{2}\begin{pmatrix} 1 & 0 \\ 0 & 1 \end{pmatrix}, \ Tr\hat{\rho} = \frac{1}{2} + \frac{1}{2} = 1,$$

$$\hat{\rho}^2 = \frac{1}{2}\begin{pmatrix} 1 & 0 \\ 0 & 1 \end{pmatrix}\frac{1}{2}\begin{pmatrix} 1 & 0 \\ 0 & 1 \end{pmatrix} = \frac{1}{4}\begin{pmatrix} 1 & 0 \\ 0 & 1 \end{pmatrix} \neq \hat{\rho}, \ Tr\hat{\rho}^2 = \frac{1}{4} + \frac{1}{4} = \frac{1}{2} < 1$$

So the state is a mixed state.
- Consider an ensemble of states with

(i) $|\psi_1> = |0>$, $|\psi_2> = |1>$ with probabilities 0.64, 0.36.
(ii) $|\psi_1> = 0.8|0> +0.6|1>$, $|\psi_2> = 0.8|0> -0.6|1>$ with probabilities 0.5, 0.5.

Obtain density operator for the two ensembles, $Tr\hat{\rho}^2$ and comment.
✓ Density operators

(i) $\hat{\rho} = \sum_i p_i |\psi_i> <\psi_i| = 0.64|0> <0| + 0.36|1> <1|$

$$= 0.64\begin{pmatrix} 1 \\ 0 \end{pmatrix}\begin{pmatrix} 1 & 0 \end{pmatrix} + 0.36\begin{pmatrix} 0 \\ 1 \end{pmatrix}\begin{pmatrix} 0 & 1 \end{pmatrix} = 0.64\begin{pmatrix} 1 & 0 \\ 0 & 0 \end{pmatrix} + 0.36\begin{pmatrix} 0 & 0 \\ 0 & 1 \end{pmatrix} = \begin{pmatrix} 0.64 & 0 \\ 0 & 0.36 \end{pmatrix}$$

(ii) $\hat{\rho} = \sum_i p_i |\psi_i> <\psi_i| = 0.5|\psi_1> <\psi_1| + 0.5|\psi_2> <\psi_2|$

$= 0.5[(0.8|0> + 0.6|1>)(0.8<0| + 0.6| < 1|)]$
$+ 0.5[(0.8|0> - 0.6|1>)(0.8<0| - 0.6| < 1|)]$

=2(0.5)[0.64|0 > <0| + 0.36|1 > <1|] (other terms are zero due to orthonormal-ization relation of the basis kets |0> and |1>)

$$\hat{\rho} = 0.64|0 > <0| + 0.36|1 > <1| = 0.64\begin{pmatrix} 1 & 0 \\ 0 & 0 \end{pmatrix} + 0.36\begin{pmatrix} 0 & 0 \\ 0 & 1 \end{pmatrix} = \begin{pmatrix} 0.64 & 0 \\ 0 & 0.36 \end{pmatrix}$$

$$\checkmark \hat{\rho}^2 = \hat{\rho}\hat{\rho} = \begin{pmatrix} 0.64 & 0 \\ 0 & 0.36 \end{pmatrix}\begin{pmatrix} 0.64 & 0 \\ 0 & 0.36 \end{pmatrix} = \begin{pmatrix} 0.4096 & 0 \\ 0 & 0.1296 \end{pmatrix} \neq \hat{\rho}$$

$$Tr\hat{\rho}^2 = 04096 + 0.1296 = 0.5392 < 1$$

✓Comment

It is a mixed state. An ensemble of states has a density operator. Two different ensembles may lead to the same density operator.

- Construct the density matrix for the following states

(a) having |0> with probability 1 and |1> with probability 0
(b) having −|0> with probability 1 and |1> with probability 0

$$\checkmark \hat{\rho} = \sum_i p_i |\psi_i > <\psi_i|$$

(a) $\hat{\rho} = (prob)|0 > <0| + (prob)|1 > <1|$

$$=1.|0 > <0| + 0.|1 > <1| = |0 > <0| = \begin{pmatrix} 1 \\ 0 \end{pmatrix}(1 \ \ 0) = \begin{pmatrix} 1 & 0 \\ 0 & 0 \end{pmatrix}$$

(b) $\hat{\rho} = (prob)(-|0 >)(-<0|) + (prob)|1 > <1|$

$$=1.|0 > <0| + 0.|1 > <1| = |0 > <0| = \begin{pmatrix} 1 \\ 0 \end{pmatrix}(1 \ \ 0) = \begin{pmatrix} 1 & 0 \\ 0 & 0 \end{pmatrix}$$

(a) and (b) are not distinguishable (i.e. are indistinguishable) as they are represented by the same density matrix. Also, $Tr\hat{\rho} = 1$, $\hat{\rho}^2 = \hat{\rho}$, $Tr\hat{\rho}^2 = 1$ for $\hat{\rho} = \begin{pmatrix} 1 & 0 \\ 0 & 0 \end{pmatrix}$

- Construct the density matrix for the following mixed states

(a) having |0> with probability $\frac{3}{4}$ and |1> with probability $\frac{1}{4}$
(b) having $|p > = \frac{\sqrt{3}}{2}|0 > +\frac{1}{2}|1>$ with probability $\frac{1}{2}$ and $|q > = \frac{\sqrt{3}}{2}|0 > -\frac{1}{2}|1>$ with probability $\frac{1}{2}$

$$\checkmark \hat{\rho} = \sum_i p_i |\psi_i > <\psi_i|$$

(a) $\hat{\rho} = (prob)|0 > <0| + (prob)|1 > <1| = \frac{3}{4}.|0 > <0| + \frac{1}{4}.|1 > <1| = \begin{pmatrix} \frac{3}{4} & 0 \\ 0 & \frac{1}{4} \end{pmatrix}$

(b) $\hat{\rho} = (\text{prob})|p > <p| + (\text{prob})|q > <q|$

$$=\frac{1}{2}\left[\frac{\sqrt{3}}{2}|0> +\frac{1}{2}|1>\right]\left[\frac{\sqrt{3}}{2}<0|+\frac{1}{2}<1|\right] + \frac{1}{2}\left[\frac{\sqrt{3}}{2}|0> -\frac{1}{2}|1>\right]\left[\frac{\sqrt{3}}{2}<0|-\frac{1}{2}<1|\right]$$

$$=2\frac{1}{2}\left[\frac{3}{4}|0> <0| + \frac{1}{4}|1> <1|\right] = \frac{3}{4}|0> <0| + \frac{1}{4}|1> <1| = \begin{pmatrix} \frac{3}{4} & 0 \\ 0 & \frac{1}{4} \end{pmatrix}$$

(a) and (b) are not distinguishable (i.e. are indistinguishable) as they are represented by the same density matrix. Also, $Tr\hat{\rho} = 1$, $\hat{\rho}^2 \neq \hat{\rho}$, $Tr\hat{\rho}^2 < 1$ for

$$\hat{\rho} = \begin{pmatrix} \frac{3}{4} & 0 \\ 0 & \frac{1}{4} \end{pmatrix}$$

7.11 Partially mixed, completely mixed, maximally mixed states

(a) Consider a mixed state having $|0>$ with probability 50% and $\frac{1}{\sqrt{2}}(|0> +|1>)$ with probability 50%. It is represented by a density matrix in the 2×2 basis $\{|0>,|1>\}$ as follows.

$$\hat{\rho} = \sum_i p_i |\psi_i> <\psi_i|$$

$$\hat{\rho} = (\text{prob})|0> <0| + (\text{prob})\frac{1}{\sqrt{2}}(|0> +|1>)$$

$$=\frac{1}{2}|0> <0| + \frac{1}{2}\frac{1}{\sqrt{2}}(|0> +|1>)\frac{1}{\sqrt{2}}(<0|+<1|)$$

$$=\frac{1}{2}|0> <0| + \frac{1}{4}[|0> <0| + |0> <1| + |1> <0| + |1> <1|]$$

$$\hat{\rho} = \frac{3}{4}|0> <0| + \frac{1}{4}|0> <1| + \frac{1}{4}|1> <0| + \frac{1}{4}|1> <1| = \begin{pmatrix} \frac{3}{4} & \frac{1}{4} \\ \frac{1}{4} & \frac{1}{4} \end{pmatrix}, \; Tr\hat{\rho} = \frac{3}{4} + \frac{1}{4} = 1$$

$$\hat{\rho}^2 = \hat{\rho}\hat{\rho} = \begin{pmatrix} \frac{3}{4} & \frac{1}{4} \\ \frac{1}{4} & \frac{1}{4} \end{pmatrix}\begin{pmatrix} \frac{3}{4} & \frac{1}{4} \\ \frac{1}{4} & \frac{1}{4} \end{pmatrix} = \frac{1}{16}\begin{pmatrix} 10 & 4 \\ 4 & 2 \end{pmatrix} \neq \hat{\rho}, \; Tr\hat{\rho}^2 = \frac{10}{16} + \frac{2}{16} = \frac{12}{16} < 1$$

This is a mixed state. There are off-diagonal matrix elements here and the mixed state is referred to as partially mixed state. This is expected since $|0>$ occurs in both states with some probability.

(b) Consider a mixed state having $|0>$ with probability 75% and $|1>$ with probability 25%. It is represented by a density matrix in the 2×2 basis $\{|0>,|1>\}$ as follows.

$$\hat{\rho} = \sum_i p_i |\psi_i> <\psi_i|$$

$$\hat{\rho} = (\text{prob})(|0>)(<0|) + (\text{prob})(|1> <1|) = \tfrac{3}{4}|0> <0| + \tfrac{1}{4}|1> <1| = \begin{pmatrix} \tfrac{3}{4} & 0 \\ 0 & \tfrac{1}{4} \end{pmatrix}$$

$$Tr\hat{\rho} = \tfrac{3}{4} + \tfrac{1}{4} = 1$$

$$\hat{\rho}^2 = \hat{\rho}\hat{\rho} = \begin{pmatrix} \tfrac{3}{4} & 0 \\ 0 & \tfrac{1}{4} \end{pmatrix}\begin{pmatrix} \tfrac{3}{4} & 0 \\ 0 & \tfrac{1}{4} \end{pmatrix} = \begin{pmatrix} \tfrac{9}{16} & 0 \\ 0 & \tfrac{1}{16} \end{pmatrix} \neq \hat{\rho}, \; Tr\hat{\rho}^2 = \tfrac{9}{16} + \tfrac{1}{16} = \tfrac{10}{16} < 1$$

This is a mixed state. Clearly $\hat{\rho}$ is a diagonal matrix and diagonal elements are unequal. The mixed state is referred to as completely mixed state (as off- diagonal elements are zero).

(c) Consider a mixed state having $|0>$ with probability 50% and $|1>$ with probability 50%. It is represented by a density matrix in the 2×2 basis $\{|0>,|1>\}$ as follows.

$$\hat{\rho} = \sum_i p_i |\psi_i> <\psi_i|$$

$$\hat{\rho} = (\text{prob})|0> <0| + (\text{prob})(|1> <1|) = \tfrac{1}{2}|0> <0| + \tfrac{1}{2}|1> <1| = \begin{pmatrix} \tfrac{1}{2} & 0 \\ 0 & \tfrac{1}{2} \end{pmatrix}$$

$$Tr\hat{\rho} = \tfrac{1}{2} + \tfrac{1}{2} = 1$$

$$\hat{\rho}^2 = \hat{\rho}\hat{\rho} = \begin{pmatrix} \tfrac{1}{2} & 0 \\ 0 & \tfrac{1}{2} \end{pmatrix}\begin{pmatrix} \tfrac{1}{2} & 0 \\ 0 & \tfrac{1}{2} \end{pmatrix} = \begin{pmatrix} \tfrac{1}{4} & 0 \\ 0 & \tfrac{1}{4} \end{pmatrix} \neq \hat{\rho}, \; Tr\hat{\rho}^2 = \tfrac{1}{4} + \tfrac{1}{4} = \tfrac{1}{2} < 1$$

This is a mixed state. Clearly $\hat{\rho}$ is a diagonal matrix and diagonal elements are equal. The mixed state is referred to as maximally mixed state (as off-diagonal elements are zero and diagonal elements are equal).

(d) Consider a mixed state having $|+>$ with probability 50% and $|->$ with probability 50%. It is represented by density matrix in the 2×2 basis $\{|0>,|1>\}$ as follows.

$$|+> = \tfrac{1}{\sqrt{2}}(|0> +|1>) = \tfrac{1}{\sqrt{2}}\begin{pmatrix} 1 \\ 1 \end{pmatrix}, \; |->=\tfrac{1}{\sqrt{2}}(|0> -|1>) = \tfrac{1}{\sqrt{2}}\begin{pmatrix} 1 \\ -1 \end{pmatrix}$$

$$\hat{\rho} = \sum_i p_i |\psi_i> <\psi_i|$$

$$\hat{\rho} = (\text{prob})|+><+| + (\text{prob})|-><-|$$

$$= \frac{1}{2}\frac{1}{\sqrt{2}}\begin{pmatrix} 1 \\ 1 \end{pmatrix}\frac{1}{\sqrt{2}}(1 \quad 1) + \frac{1}{2}\frac{1}{\sqrt{2}}\begin{pmatrix} 1 \\ -1 \end{pmatrix}\frac{1}{\sqrt{2}}(1 \quad -1) = \frac{1}{4}\begin{pmatrix} 1 & 1 \\ 1 & 1 \end{pmatrix} + \frac{1}{4}\begin{pmatrix} 1 & -1 \\ -1 & 1 \end{pmatrix} = \begin{pmatrix} \frac{1}{2} & 0 \\ 0 & \frac{1}{2} \end{pmatrix}$$

$$Tr\hat{\rho} = 1, \hat{\rho}^2 \neq \hat{\rho}, Tr\hat{\rho}^2 < 1$$

This is a mixed state. Clearly $\hat{\rho}$ is a diagonal matrix and diagonal elements are equal. The mixed state is referred to as maximally mixed state (as off-diagonal elements are zero and diagonal elements are equal.

✓ In (c) and (d) the density matrices are identical but the mixed states are different.

Clearly it is possible for two different mixed states to have the same density matrix. Two mixed states can be distinguished if and only if the density matrices are different. Also $\hat{\rho}_1 = \hat{\rho}_2$ does not imply the same mixed state.

Comment

The postulates of quantum mechanics were for state vector and can be replaced by a corresponding postulate for density matrix since the ensemble is described by a density matrix which lies in Hilbert space.

The Schrödinger equation satisfied by state vector is replaced by Liouville's equation for the density operator.

Table 7.1 collects the essential properties and differences between pure state and mixed state.

7.12 Time evolution of density matrix: Liouville–Von Neumann equation

The equation of motion of density matrix follows from the definition of $\hat{\rho}$ and the time dependent Schrödinger equation

$$i\hbar\frac{\partial}{\partial t}|\psi> = \hat{H}|\psi> \quad \Rightarrow \quad \frac{\partial}{\partial t}|\psi> = \frac{1}{i\hbar}\hat{H}|\psi>$$

and its Hermitian conjugate (dagger) ($\hat{H} = \hat{H}^{\dagger} \rightarrow$ Hamiltonian)

Table 7.1. Pure state versus mixed state.

Pure state	Mixed state
$\hat{\rho} = |\psi><\psi|$	$\hat{\rho} = \sum_i p_i|\psi_i><\psi_i|$
$Tr\hat{\rho} = 1$	$Tr\hat{\rho} = 1$
$\hat{\rho}^2 = \hat{\rho}$	$\hat{\rho}^2 \neq \hat{\rho}$
$Tr\hat{\rho}^2 = 1$	$Tr\hat{\rho}^2 < 1$
$\hat{\rho} \rightarrow$ positive operator	$\hat{\rho} \rightarrow$ positive operator
$<\hat{A}> = Tr(\hat{\rho}\hat{A})$	$<\hat{A}> = Tr(\hat{\rho}\hat{A})$

$$-i\hbar\frac{\partial}{\partial t} <\psi| = <\psi|\hat{H} \quad\Rightarrow\quad \frac{\partial}{\partial t} <\psi| = -\frac{1}{i\hbar} <\psi|\hat{H}$$

Consider

$$\frac{\partial}{\partial t}\hat{\rho} = \frac{\partial}{\partial t}|\psi> <\psi| = \left(\frac{\partial}{\partial t}|\psi>\right) <\psi| + |\psi> \left(\frac{\partial}{\partial t} <\psi|\right)$$

$$= \left(\frac{1}{i\hbar}\hat{H}|\psi>\right) <\psi| + |\psi> \left(-\frac{1}{i\hbar} <\psi|\hat{H}\right)$$

$$= \frac{1}{i\hbar}[\ \hat{H}|\psi> <\psi| - |\psi> <\psi|\hat{H}]$$

Use $\hat{\rho} = |\psi> <\psi|$ to get

$$\frac{\partial}{\partial t}\hat{\rho} = \frac{1}{i\hbar}[\ \hat{H}\hat{\rho} - \hat{\rho}\hat{H}] = \frac{1}{i\hbar}[\ \hat{H},\ \hat{\rho}]$$

$$\Rightarrow\quad i\hbar\frac{\partial}{\partial t}\hat{\rho} = [\ \hat{H},\ \hat{\rho}]$$

This is the Liouville–Von Neumann equation.
- The Schrödinger equation describes how pure states evolve with time while the Liouville–Von Neumann equation describes how the density operator evolves with time.
- $\hat{\rho}$ does not satisfy the Heisenberg equation of motion $i\hbar\frac{d}{dt}\hat{A} = [\hat{A}, \hat{H}]$ because, though it has mathematical structure of an operator, it does not represent a physical observable.

7.13 Partial trace and the reduced density matrix

We deal with systems which are not closed systems. They interact with surroundings. For instance our system may be a one-qubit system interacting with another qubit which is not of our interest and can be treated as the environment. We thus have to deal with a composite system that comprises the system coupled with its environment though our interest lies with the system only. In other words, we have to extract information regarding the system only, though we have to deal with the composite system. This is done by averaging out the environment by taking partial trace and what we get after taking partial trace is called the reduced density matrix or reduced density operator.

Suppose the system of interest is A and that it is a part of a bigger system B which is the environment. (Composite system = A + B). Properties of A may be obtained by taking partial trace over the environment B.

Let $\{|a>\}$ be a basis in Hilbert space H_A and $\{|b>\}$ is a basis in Hilbert space H_B. The density matrix $\hat{\rho}_{AB}$ on a composite space $H_A \otimes H_B$ is

$$\hat{\rho}_{AB} = |a_1> <a_2| \otimes |b_1> <b_2|$$

Let us take partial trace Tr_B which is mapping from the density matrix $\hat{\rho}_{AB}$ (on $H_A \otimes H_B$) onto density matrix $\hat{\rho}_A$ (on H_A). We define reduced density operator as

$$\hat{\rho}_A = Tr_B \ \hat{\rho}_{AB} = Tr_B[|a_1> <a_2| \otimes |b_1> <b_2|]$$
$$= |a_1> <a_2|Tr_B[|b_1> <b_2|]$$

Now $Tr_B|b_1> <b_2| = \sum_i \text{diagonal elements} = \sum_i <i|b_1> <b_2|i>$ where $|i>$ is some

basis in B. As $<i|b_1>$, $<b_2|i>$ are scalars they commute

$$Tr_B|b_1> <b_2| = \sum_i <b_2|i> <i|b_1> = <b_2|\sum_i (|i> <i|)|b_1>$$

$$= <b_2|\hat{I}|b_1> = <b_2|b_1> \ (\text{since} \sum_i |i> <i| = 1 \text{ for completeness})$$

- From the relation $Tr_B|b_1> <b_2| = <b_2|b_1>$ it follows that the Trace of 'ket bra' is the scalar product of 'bra ket').
  We thus end up with the result

$$\hat{\rho}_A = Tr_B \ \hat{\rho}_{AB} = |a_1> <a_2| < b_2|b_1> = <b_2|b_1> |a_1> <a_2|$$

Partial trace essentially averages out the effect of environment and extracts the properties of the system of interest.

- **Example**

✓ Consider the two-qubit state $|\psi> = \frac{1}{\sqrt{2}}(|01> -|10>)$. The corresponding density matrix is

$$|\psi> <\psi| = \frac{|01> -|10>}{\sqrt{2}} \frac{<01|-<10|}{\sqrt{2}}$$

$$= \frac{1}{2}(|01> <01| - |01> <10| - |10> <01| + |10> <10|)$$

Taking partial trace over B (second qubits) we get

$$Tr_B|\psi> <\psi|$$

$$= \frac{1}{2}[(|0> <0|)Tr(|1> <1|) - |0> <1|Tr(|1> <0|)$$

$$- |1> <0|Tr(|0> <1|) + |1> <1|Tr(|0> <0|)]$$

$$= \frac{1}{2}(|0> <0| <1|1> -|0> <1|<0|1> -|1> <0| <1|0> +|1> <1|<0|0>)$$

The first qubits form the operator while the second qubits form the scalar product. Using $<0|0> =1$, $<1|1> =1$, $<0|1> =0$, $<1|0> =0$ we have

$$Tr_B|\psi> <\psi| = \frac{1}{2}(|0> <0| + |1> <1|)$$

$$= \frac{1}{2}\left[\begin{pmatrix}1\\0\end{pmatrix}(1 \ \ 0) + \begin{pmatrix}0\\1\end{pmatrix}(0 \ \ 1) \right] = \frac{1}{2}\begin{pmatrix}1 & 0\\0 & 1\end{pmatrix} = \frac{1}{2}\hat{I} = \hat{\rho}_A$$

- An entangled pure state, on being traced over one of the components may give a mixed state.

✓ Consider the entangled Bell state $|\psi> = \frac{1}{\sqrt{2}}(|00> +|11>)$ (which is a pure state). The density matrix is

$$\hat{\rho} = |\psi> <\psi| = \frac{1}{\sqrt{2}}(|00> +|11>)\frac{1}{\sqrt{2}}(<00|+<11|)$$

$$\hat{\rho} = \frac{1}{2}[|00> <00| + |00> <11| + |11> <00| + |11> <11|]$$

Suppose we are interested in the first qubit. Taking trace over the second qubit we get

$$\hat{\rho}_1 = Tr_2\, \hat{\rho}$$

$$\hat{\rho}_1 = \frac{1}{2}[|0> <0|Tr(|0> <0|) + |0> <1|Tr(|0> <1|)$$

$$+ |1> <0|Tr(|1> <0|) + |1> <1|Tr(|1> <1|)]$$

Since trace of 'ket bra' is the scalar product of 'bra ket', i.e.

$$Tr\ |a> <b|\ = <b|a>$$

we have

$$\hat{\rho}_1 = \frac{1}{2}[|0> <0| < 0|0> +|0> <1|<1|0> +|1> <0| < 0|1> +|1> <1|<1|1>$$

$$= \frac{1}{2}|0> <0| + |1> <1| = \frac{1}{2}\begin{pmatrix} 1 & 0 \\ 0 & 1 \end{pmatrix} = \begin{pmatrix} \frac{1}{2} & 0 \\ 0 & \frac{1}{2} \end{pmatrix}$$

= mixed state (as $\hat{\rho}_1^2 \neq \hat{\rho}_1$, $Tr\hat{\rho}_1^2 < 1$)

It is clear that we started with a pure state and after taking average over the second qubit we end up with a mixed state.

Purification

The reduced density matrix of a two-qubit state gives a mixed state. The converse of this holds and is referred to as purification. In other words, from a reduced density matrix for a single qubit, one can construct a two-qubit pure state whose partial trace would produce the given density matrix. This is known as purification.

7.14 Measurement theory of mixed states

For a mixed state the density matrix is

$$\hat{\rho} = \sum_i p_i |\psi_i> <\psi_i|$$

If the states are expressed in a basis $\{|j>\}$ we have $|\psi_i> = \sum_j c_{ij}|j>$ with

$$<j|\psi> = <j|\sum_{j'} c_{ij'}|j'> = \sum_{j'} c_{ij'} <j|j'> = \sum_{j'} c_{ij'}\delta_{jj'} = c_{ij}$$

The Born probability to observe $|\psi_i>$ in the basis state $|j>$ is $|<j|\psi_i>|^2$, and $p_i =$ classical probability to pick up state $|\psi_i>$.

Probability to observe $|\psi>$ in the basis state $|j>$ is

$$\sum (\text{classical probability})(\text{Born probability})$$

$$= \sum_i p_i |<j|\psi_i>|^2 = \sum_i p_i < \psi_i|j> <j|\psi_i> = \sum_i p_i <j|\psi_i> <\psi_i|j>$$

$$= <j| \left(\sum_i p_i |\psi_i> <\psi_i| \right) |j> = <j|\hat{\rho}|j>$$

= matrix element of the density matrix in the basis $|j>$.

A generalized quantum measurement is described by a set of measurement operators $\{\hat{M}_m\}$ where $\hat{M}_m$ is a measurement operator and m the measurement outcome that might occur in a measurement on a system described by density matrix $\hat{\rho}$. The probability that result m occurs is given by $Tr(\hat{M}_m^\dagger \hat{M}_m \hat{\rho})$ and the density matrix after the measurement is $\dfrac{\hat{M}_m \hat{\rho} \hat{M}_m^\dagger}{Tr(\hat{M}_m^\dagger \hat{M}_m \hat{\rho})}$ (as per reformulated quantum postulate in terms of density operator instead of state vector).

The measurement operators satisfy the completeness relation $\sum_m \hat{M}_m^\dagger \hat{M}_m = \hat{I}$

Also

$$Tr\left(\hat{M}_m^\dagger \hat{M}_m \hat{\rho}\right) = Tr\left[\hat{M}_m^\dagger \hat{M}_m \sum_i p_i |\psi_i> <\psi_i| \right]$$

$$= \sum_i p_i \, Tr\left(\hat{M}_m^\dagger \hat{M}_m |\psi_i> <\psi_i|\right) = \sum_i p_i \, Tr\left(|\phi_i> <\psi_i|\right)$$

where $|\phi_i> = \hat{M}_m^\dagger \hat{M}_m |\psi_i>$

Using $Tr \ |a> <b| = <b|a>$ we have

$$Tr\left(\hat{M}_m^\dagger \hat{M}_m \hat{\rho}\right) = \sum_i p_i <\psi_i|\phi_i> = \sum_i p_i <\psi_i|\hat{M}_m^\dagger \hat{M}_m|\psi_i>$$

- **Example**

    Take one qubit state $|\psi> = \alpha|0> + \beta|1>$

    Suppose we measure z-component of spin

$$\hat{\sigma}_z = \begin{pmatrix} 1 & 0 \\ 0 & -1 \end{pmatrix} = |0> <0| - |1> <1| \ (\hat{\sigma}_z \text{ has eigenvalues } m_0 \rightarrow \pm 1).$$

Thus $\hat{M}_0 = |0> <0|$ (for eigenvalue $m_0 = 1$)

$$\hat{\rho} = |\psi> <\psi| = (\alpha|0> + \beta|1>)(\alpha^* <0| + \beta^* <1|)$$

Let us evaluate

$$Tr\left(\hat{M}_0^\dagger \hat{M}_0 \hat{\rho}\right) = Tr[|0> <0|0> <0|(\alpha|0> + \beta|1>)(\alpha^* <0| + \beta^* <1|)]$$

$= Tr[|0> <0|(\alpha|0> + \beta|1>)(\alpha^* <0| + \beta^* <1|)]$ using $<0|0> = 1$

$= Tr[|0> (\alpha <0|0> + \beta <0|1>)(\alpha^* <0| + \beta^* <1|)]$

$= Tr[|0> \alpha(\alpha^* <0| + \beta^* <1|)]$ using $<0|0> = 1, \ <0|1> = 0$

$= Tr[\alpha\alpha^*|0> <0| + \alpha\beta^*|0> <1|]$

$= \alpha\alpha^* Tr|0> <0| + \alpha\beta^* Tr|0> <1|$

$= \alpha\alpha^* <0|0> + \alpha\beta^* <1|0>$ using $Tr \ |a> <b| \ = \ <b|a>$

$$Tr\left(\hat{M}_0^\dagger \hat{M}_0 \hat{\rho}\right) = |\alpha|^2, \quad \text{using } <0|0> = 1, \ <1|0> = 0$$

Let us evaluate

$$\hat{M}_0 \hat{\rho} \hat{M}_0^\dagger = |0> <0|[(\alpha|0> + \beta|1>)(\alpha^* <0| + \beta^*| <1|)]|0> <0|$$

$$= |0> [(\alpha <0|0> + \beta <0|1>)(\alpha^* <0|0> + \beta^*|<1|0>)] <0|$$

$$= |0> \alpha\alpha^* <0| = |0> |\alpha|^2 <0| = |\alpha|^2|0> <0|$$

The post measurement density matrix will be

$$\frac{\hat{M}_m \hat{\rho} \hat{M}_m^\dagger}{Tr(\hat{M}_m^\dagger \hat{M}_m \hat{\rho})} = \frac{\hat{M}_0 \hat{\rho} \hat{M}_0^\dagger}{Tr(\hat{M}_0^\dagger \hat{M}_0 \hat{\rho})} = \frac{|\alpha|^2|0> <0|}{|\alpha|^2} = |0> <0| = \text{ corresponding density matrix}$$

Repeating a measurement

When a measurement is made, the state collapses to a particular eigenstate (say $|0>$). If we repeat the measurement, the state will not remain in the same state. Let us discuss the effect of changing basis after measurement, say computational basis $\{|0>, |1>\}$ and diagonal or Hadamard basis $\{|+>, |->\}$.

- Let $|\psi> = \alpha|0> + \beta|1>$.

If we measure in the $\{|0>, |1>\}$ basis we get $|0>$ with probability $|\alpha|^2$ and the state collapses to $|0>$. If we measure it now in $\{|+>, |->\}$ basis in which $|0> = \frac{1}{\sqrt{2}}(|0> + |1>)$, we get $|+>$ with probability $|\frac{1}{\sqrt{2}}|^2 = \frac{1}{2}$. So the probability of getting 0 and then + is

$P(0, +) = (|\alpha|^2)(\frac{1}{2}) = \frac{|\alpha|^2}{2}$. (probabilities are multiplied as events are independent).

Let us measure first in $\{|+>, |->\}$ basis and then in $\{|0>, |1>\}$ basis.

If we measure in the $\{|+>, |->\}$ basis in which

$$|\psi> = \alpha|0> + \beta|1> = \alpha\left(\frac{|+> + |->}{\sqrt{2}}\right) + \beta\left(\frac{|+> - |->}{\sqrt{2}}\right)$$

$$= \frac{\alpha + \beta}{\sqrt{2}}|+> + \frac{\alpha - \beta}{\sqrt{2}}|->$$

we get $|+>$ with probability $|\frac{\alpha+\beta}{\sqrt{2}}|^2 = \frac{|\alpha+\beta|^2}{2}$ and the state collapses to $|+>$. If we measure it now in $\{|0>, |1>\}$ basis in which $|+> = \frac{1}{\sqrt{2}}(|0> + |1>)$, we get $|0>$ with probability $|\frac{1}{\sqrt{2}}|^2 = \frac{1}{2}$. So the probability of getting + and then 0 is

$$P(+, 0) = (\frac{|\alpha+\beta|^2}{2})(\frac{1}{2}) = \frac{|\alpha+\beta|^2}{4} \text{(probabilities are multiplied as events are}$$
independent).

Clearly thus

$$P(0, +) \neq P(+, 0)$$

7.15 Positive Operator Valued Measure (POVM)

POVM refers to a non-projective measurement technique to distinguish non-orthogonal states.

POVM involves defining a complete set of positive operators $\hat{E}_m = \hat{M}_m^\dagger \hat{M}_m$ that satisfy the completeness relation $\sum_m \hat{E}_m = \hat{I}$, may or may not commute and their number may exceed the dimension of Hilbert space and they need not be orthogonal.

The probability of outcome of event m is $p(m) = <\psi|\hat{E}_m|\psi> = Tr(\hat{\rho}\hat{E}_m)$.

The post measurement state is described by $\hat{\rho} \rightarrow \hat{\rho}' = \frac{\hat{M}_m \hat{\rho} \hat{M}_m^\dagger}{Tr(\hat{\rho}\hat{E}_m)}$ if the state is read enforcing the collapse and it is a linear combination $\sum_m \hat{M}_m \hat{\rho} \hat{M}_m^\dagger$ if state is not read.

We explain the technique of POVM through an example.

✓ Consider the non-orthogonal states

$$|\psi_1> = |0>, \ |\psi_2> = \frac{1}{\sqrt{2}}(|0> + |1>)$$

with $<\psi_1|\psi_2> = <0|\frac{1}{\sqrt{2}}(|0> + |1>) = \frac{1}{\sqrt{2}} \neq 0$.

We cannot distinguish these states $|\psi_1>, |\psi_2>$ by orthogonal projective measurement since the state $|\psi_2> = \frac{|0> + |1>}{\sqrt{2}}$ has a projection both along $|0>$ and along $|1>$.

However, these states can be distinguished by the POVM $\{E_1, E_2, E_3\}$.

Consider a POVM containing three elements

$$\hat{E}_1 = \frac{\sqrt{2}}{1+\sqrt{2}}|1> <1|, \ \hat{E}_2 = \frac{\sqrt{2}}{1+\sqrt{2}}\frac{(|0> - |1>)(<0| - <1|)}{2}, \ \hat{E}_3 = \hat{I} - \hat{E}_1 - \hat{E}_2$$

We check if they satisfy the requirements or conditions:

$$\sum_m \hat{E}_m = \hat{E}_1 + \hat{E}_2 + \hat{E}_3 = \hat{I} \text{ (they are complete)}$$

They are positive operators. Also,

$$\hat{E}_1\hat{E}_2 = \left(\frac{\sqrt{2}}{1+\sqrt{2}}|1><1|\right)\left(\frac{\sqrt{2}}{1+\sqrt{2}}\frac{(|0>-|1>)(<0|-<1|)}{2}\right)$$

$$= \left(\frac{\sqrt{2}}{1+\sqrt{2}}\right)^2|1>\frac{(<1|0>-<1|1>)(<0|-<1|)}{2} = -\frac{1}{(1+\sqrt{2})^2}(|1><0|-|1><1|)$$

$$\hat{E}_2\hat{E}_1 = \left(\frac{\sqrt{2}}{1+\sqrt{2}}\frac{(|0>-|1>)(<0|-<1|)}{2}\right)\left(\frac{\sqrt{2}}{1+\sqrt{2}}|1><1|\right)$$

$$= \left(\frac{\sqrt{2}}{1+\sqrt{2}}\right)^2\frac{(|0>-|1>)(<0|1>-<1|1>)}{2}<1| = -\frac{1}{(1+\sqrt{2})^2}(|0><1|-|1><1|)$$

Clearly $\hat{E}_1\hat{E}_2 \neq \hat{E}_2\hat{E}_1$, i.e. $[\hat{E}_1, \hat{E}_2] \neq 0$. So they do not commute.

This POVM describes single-qubit measurements and works on two-dimensional Hilbert space, but it contains three elements. Clearly thus they are non-orthogonal projectors.

Suppose Bob gets two states from Alice and he does not know whether it is $|\psi_1> = |0>$ or $|\psi_2> = \frac{1}{\sqrt{2}}(|0>+|1>)$.

Now if Bob gets the result E_1 then the state received cannot be $|\psi_1>$ since

$$<\psi_1|\hat{E}_1|\psi_1> = <\psi_1|\frac{\sqrt{2}}{1+\sqrt{2}}|1><1|0> = 0$$

As $|\psi_1>$ cannot generate E_1 the state Bob received has to be $|\psi_2>$ since

$$<\psi_2|\hat{E}_1|\psi_2> = \frac{<0|+<1|}{\sqrt{2}}|\frac{\sqrt{2}}{1+\sqrt{2}}|1><1|\frac{|0>+|1>}{\sqrt{2}}$$

$$= \frac{\sqrt{2}}{1+\sqrt{2}}\frac{<0|1>+<1|1>}{\sqrt{2}}\frac{<1|0>+<1|1>}{\sqrt{2}} = \frac{\sqrt{2}}{1+\sqrt{2}}\frac{1}{2} \neq 0$$

Again if Bob gets the result E_2 then the state received cannot be $|\psi_2>$ since

$$<\psi_2|\hat{E}_2|\psi_2> = <\psi_2|\frac{\sqrt{2}}{1+\sqrt{2}}\frac{(|0>-|1>)(<0|-<1|)}{2}|\frac{1}{\sqrt{2}}(|0>+|1>)$$

$$= <\psi_2|\frac{1}{1+\sqrt{2}}\frac{(|0>-|1>)}{2}|(<0|0>+<0|1>-<1|0>-<1|1>) = 0$$

As $|\psi_2>$ cannot generate E_2 the state Bob received has to be $|\psi_1>$ since

$$<\psi_1|\hat{E}_2|\psi_1> = <0|\frac{\sqrt{2}}{1+\sqrt{2}}\frac{(|0>-|1>)(<0|-<1|)}{2}|0>$$

$$= \frac{\sqrt{2}}{1+\sqrt{2}}\frac{1}{2}(<0|0>-<0|1>)(<0|0>-<1|0>) = \frac{\sqrt{2}}{1+\sqrt{2}}\frac{1}{2} \neq 0$$

Again, if Bob gets the result E_3 then the state received cannot be identified.

Comment:

It follows thus that if measurement outcome is E_1 then the state is $|\psi_2>$ and if measurement outcome is E_2 then the state is $|\psi_1>$ and if measurement outcome is E_3 then the state is unidentified. In a nutshell, through this POVM method we never make a mistake in identifying the state (though sometimes there is no information about identity of the state).

7.16 Further reading

[1] Ghosh D 2016 Online course on quantum information and computing (Bombay: Indian Institute of Technology)

[2] Nielsen M A and Chuang I L 2019 *Quantum Computation and Quantum Information* (Cambridge: Cambridge University Press)

[3] Nakahara M and Ohmi T 2017 *Quantum Computing* (Boca Raton, FL: CRC Press)

[4] Pathak A 2016 *Elements of Quantum Computing and Quantum Communication* (Boca Raton, FL: CRC Press)

7.17 Problems

1. *Density matrix satisfies which of the following properties?*
   (a) Hermitian, (b) Unit trace, (c) Unitary, (d) Positive matrix.
   Ans. *(a),(b),(d).*

2. *Justify if $\hat{p} = \dfrac{1}{4}\begin{pmatrix} 2 & 1 & 1 \\ 1 & 0 & 0 \\ 1 & 0 & 1 \end{pmatrix}$ is a density matrix representing an ensemble.*

   Ans. $Tr\hat{p} = \dfrac{2}{4} + 0 + \dfrac{1}{4} = \dfrac{3}{4} \neq 1$. So $\hat{p}$ is not a valid density matrix.

3. *Which of the following is true regarding distance from centre of Bloch sphere in the case of mixed states?*
   (a) 0, (b) equal to one, (c) more than one, (d) less than one.
   Ans. *(d).*

4. *Find the density matrix for the quantum state given by $|\psi> = \dfrac{3}{5}|0> + \dfrac{4i}{5}|1>$.*

   Ans. $|\psi> = \dfrac{3}{5}|0> + \dfrac{4i}{5}|1> = \dfrac{3}{5}\begin{pmatrix}1\\0\end{pmatrix} + \dfrac{4i}{5}\begin{pmatrix}0\\1\end{pmatrix} = \dfrac{1}{5}\begin{pmatrix}3\\4i\end{pmatrix}$

   $\hat{\rho} = |\psi> <\psi| = \dfrac{1}{5}\begin{pmatrix}3\\4i\end{pmatrix}\dfrac{1}{5}(3 - 4i) = \dfrac{1}{25}\begin{pmatrix}9 & -12i\\12i & 16\end{pmatrix}$, $Tr\hat{\rho} = \dfrac{9}{25} + \dfrac{16}{25} = 1.$

5. *Consider the density matrix corresponding to the state $\dfrac{|00>+|01>+|11>}{\sqrt{3}}$ of which the first qubit is with Alice and second with Bob. Obtain the reduced density matrix for Alice.*

   Ans. $|\psi> = \dfrac{|00>+|01>+|11>}{\sqrt{3}} = \dfrac{|0>_A|0>_B+|0>_A|1>_B+|1>_A|1>_B}{\sqrt{3}}$ ($A \rightarrow$ Alice, $B \rightarrow$ Bob)

   $\hat{\rho} = |\psi> <\psi| = \dfrac{|0>_A|0>_B+|0>_A|1>_B+|1>_A|1>_B}{\sqrt{3}}\dfrac{_A<0|_B<0|+_A<0|_B<1|+_A<1|_B<1|}{\sqrt{3}}.$

Reduced density matrix for Alice for first qubit is the trace of the density matrix w.r.t the second qubit. So we evaluate

$$_B<0\,|\hat{\rho}|0>_B +_B<1|\hat{\rho}|1>_B$$

$$=_B<0\,|\frac{|0>_A|0>_B+|0>_A|1>_B+|1>_A|1>_B}{\sqrt{3}}\frac{_A<0\,|_B<0\,|+_A<0\,|_B<1\,|+_A<1\,|_B<1\,|}{\sqrt{3}}|0>_B$$

$$+_B<1\,|\frac{|0>_A|0>_B+|0>_A|1>_B+|1>_A|1>_B}{\sqrt{3}}\frac{_A<0\,|_B<0\,|+_A<0\,|_B<1\,|+_A<1\,|_B<1\,|}{\sqrt{3}}|1>_B$$

$$=\frac{|0>_A\,_B<0\,|0>_B+|0>_A\,_B<0\,|1>_B+|1>_A\,_B<0\,|1>_B}{\sqrt{3}}$$

$$\frac{_A<0\,|_B<0\,|0>_B+_A<0\,|_B<1\,|0>_B+_A<1\,|_B<1\,|0>_B}{\sqrt{3}}$$

$$+\frac{|0>_A\,_B<1\,|0>_B+|0>_A\,_B<1\,|1>_B+|1>_A\,_B<1\,|1>_B}{\sqrt{3}}$$

$$\frac{_A<0\,|_B<0\,|1>_B+_A<0\,|_B<1\,|1>_B+_A<1\,|_B<1\,|1>_B}{\sqrt{3}}$$

Since $_B<0|0>_B = 1$, $_B<0|1>_B = 0, _B<1|1>_B = 1$, $_B<1|0>_B = 0$ we have

$$_B<0\,|\hat{\rho}|0>_B +_B<1|\hat{\rho}|1>_B$$

$$=\frac{|0>_A}{\sqrt{3}}\frac{_A<0\,|}{\sqrt{3}}+\frac{|0>_A+|1>_A}{\sqrt{3}}\frac{_A<0\,|+_A<1\,|}{\sqrt{3}}=\frac{|0>_A\,_A<0\,|}{3}+\left[\frac{|0>_A+|1>_A}{\sqrt{3}}\frac{_A<0\,|+_A<1\,|}{\sqrt{3}}\right]$$

$$=\frac{|0>_A\,_A<0\,|}{3}+\frac{|0>_A\,_A<0\,|+|0>_A\,_A<1\,|+|1>_A\,_A<0\,|+|1>_A\,_A<1\,|}{3}$$

$$=\frac{2}{3}|0>_A\,_A<0\,|+\frac{1}{3}|0>_A\,_A<1\,|+\frac{1}{3}|1>_A\,_A<0\,|+\frac{1}{3}|1>_A\,_A<1\,|=\begin{pmatrix}\frac{2}{3}&\frac{1}{3}\\\frac{1}{3}&\frac{1}{3}\end{pmatrix}.$$

6. *Consider a mixture of states in the diagonal basis which is 25% in $|->$ and 75% in $|+>$. Find the probability that a measurement in computational basis will find it in $|0>$. Write down the density matrix of the system after measurement is done.*

Ans. Density matrix is $\hat{\rho} = \sum_i p_i|\psi_i><\psi_i|$

$$\hat{\rho}=\frac{3}{4}|+><+|+\frac{1}{4}|-><-|\quad\left(75\%=\frac{3}{4}, 25\%=\frac{1}{4}\right)$$

Using $|+>=\frac{1}{\sqrt{2}}\begin{pmatrix}1\\1\end{pmatrix}$, $|->=\frac{1}{\sqrt{2}}\begin{pmatrix}1\\-1\end{pmatrix}$

$$\hat{\rho}=\frac{3}{4}\frac{1}{\sqrt{2}}\begin{pmatrix}1\\1\end{pmatrix}\frac{1}{\sqrt{2}}(1\ \ 1)+\frac{1}{4}\frac{1}{\sqrt{2}}\begin{pmatrix}1\\-1\end{pmatrix}\frac{1}{\sqrt{2}}(1\ -1)$$

$$=\frac{3}{8}\begin{pmatrix}1&1\\1&1\end{pmatrix}+\frac{1}{8}\begin{pmatrix}1&-1\\-1&1\end{pmatrix}=\begin{pmatrix}\frac{1}{2}&\frac{1}{4}\\\frac{1}{4}&\frac{1}{2}\end{pmatrix}\text{ (in computational basis)}$$

Probability that measurement in computational basis gives $|0>$ is $<0|\hat{\rho}|0> =\frac{1}{2}$ (1–1 element of $\hat{\rho}$).

After measurement is done, the system is in state $|\psi> =|0>$ (pure state).

Corresponding density matrix is $\hat{\rho}' = |\psi><\psi| = |0><0|=\begin{pmatrix}1&0\\0&0\end{pmatrix}.$

7. *Consider the qutrit that can be in three states* $|+>$, $|0>$, $|->$ *which are the eigenstates of angular momentum operator for* $J = 1$ *with respective eigenvalues* $J_z = +1, 0, -1$. *Find the average value and standard deviation if the density matrix is given by* $\frac{1}{4}\begin{pmatrix} 2 & 1 & 1 \\ 1 & 1 & 0 \\ 1 & 0 & 1 \end{pmatrix}$.

Ans. $\langle \hat{J}_z \rangle = Tr(\hat{J}_z \hat{\rho})$ where in its eigenbasis $\hat{J}_z = \begin{pmatrix} 1 & 0 & 0 \\ 0 & 0 & 0 \\ 0 & 0 & -1 \end{pmatrix}$, the diagonal populated by its eigenvalues $+1, 0, -1$.

$$\langle \hat{J}_z \rangle = Tr(\hat{J}_z \hat{\rho}) = Tr\left[\begin{pmatrix} 1 & 0 & 0 \\ 0 & 0 & 0 \\ 0 & 0 & -1 \end{pmatrix} \frac{1}{4} \begin{pmatrix} 2 & 1 & 1 \\ 1 & 1 & 0 \\ 1 & 0 & 1 \end{pmatrix} \right] = Tr\left[\frac{1}{4} \begin{pmatrix} 2 & 1 & 1 \\ 0 & 0 & 0 \\ -1 & 0 & -1 \end{pmatrix} \right] = \frac{1}{4}$$

Standard deviation is $\sqrt{\langle \hat{J}_z^2 \rangle - \langle \hat{J}_z \rangle^2}$

$$\langle \hat{J}_z^2 \rangle = Tr\left(\hat{J}_z^2 \hat{\rho} \right) = Tr\left[\begin{pmatrix} 1 & 0 & 0 \\ 0 & 0 & 0 \\ 0 & 0 & -1 \end{pmatrix} \begin{pmatrix} 1 & 0 & 0 \\ 0 & 0 & 0 \\ 0 & 0 & -1 \end{pmatrix} \frac{1}{4} \begin{pmatrix} 2 & 1 & 1 \\ 1 & 1 & 0 \\ 1 & 0 & 1 \end{pmatrix} \right]$$

$$= Tr\left[\frac{1}{4} \begin{pmatrix} 1 & 0 & 0 \\ 0 & 0 & 0 \\ 0 & 0 & -1 \end{pmatrix} \begin{pmatrix} 2 & 1 & 1 \\ 0 & 0 & 0 \\ -1 & 0 & -1 \end{pmatrix} \right] = Tr\left[\frac{1}{4} \begin{pmatrix} 2 & 1 & 1 \\ 0 & 0 & 0 \\ 1 & 0 & 1 \end{pmatrix} \right] = \frac{3}{4}$$

$$\sqrt{\langle \hat{J}_z^2 \rangle - \langle \hat{J}_z \rangle^2} = \sqrt{\frac{3}{4} - \left(\frac{1}{4}\right)^2} = \frac{\sqrt{11}}{4}$$

8. *Density matrix for a mixed state satisfies which of the following properties?*
   (a) Unit trace, (b) Trace of square less than unity,
   (c) Non-negative eigenvalues, (d) For any operator $\langle \hat{A} \rangle = Tr(\hat{A}\hat{\rho})$.
   Ans. *(a),(b),(c),(d).*

9. *Prove that* $\rho_{mn}\rho_{nm} = \rho_{mm}\rho_{nn}$ *for all m, n for a pure state.*
   Ans. $\rho_{mn}\rho_{nm} = <m|\hat{\rho}|n> <n|\hat{\rho}|m>$

$$\hat{\rho} = |\psi> <\psi| \text{ (pure state)}$$

$$\rho_{mn}\rho_{nm} = <m|\psi> <\psi|n> <n|\psi> <\psi|m>$$

Rearranging $\rho_{mn}\rho_{nm} = <m|\psi> <\psi|m> <n|\psi> <\psi|n>$
(since scalars can be placed in any order)

$$\rho_{mn}\rho_{nm} = <m|\hat{\rho}|m> <n|\hat{\rho}|n> = \rho_{mm}\rho_{nn}.$$

10. *Consider the density matrix of a system to be known as* $\frac{1}{5}\begin{pmatrix} 1 & 2 \\ 2 & 4 \end{pmatrix}$.

    Show that it is a pure state. Find the probability that a measurement of the system will find it in state $|0>$. *Find expectation value of* $\hat{\sigma}_x$ *in this state.*

Ans. $\hat{\rho} = \frac{1}{5}\begin{pmatrix} 1 & 2 \\ 2 & 4 \end{pmatrix}$, $Tr\hat{\rho} = Tr\frac{1}{5}\begin{pmatrix} 1 & 2 \\ 2 & 4 \end{pmatrix} = 1$

$\hat{\rho}^2 = \frac{1}{5}\begin{pmatrix} 1 & 2 \\ 2 & 4 \end{pmatrix}\frac{1}{5}\begin{pmatrix} 1 & 2 \\ 2 & 4 \end{pmatrix} = \frac{1}{25}\begin{pmatrix} 5 & 10 \\ 10 & 20 \end{pmatrix} \neq \hat{\rho}$, $Tr\hat{\rho}^2 = \frac{5}{25} + \frac{20}{25} = 1 \Rightarrow$ pure state.

Probability of finding system in state $|0>$ is given by the expectation value of $\hat{\rho}$ w.r.t $|0>$, i.e. $<0|\hat{\rho}|0>$ which is the 1–1 element of the density matrix $\hat{\rho}$ viz. $\frac{5}{25} = \frac{1}{5}$.

$$\langle \hat{\sigma}_x \rangle = Tr(\hat{\sigma}_x\hat{\rho}) = Tr\begin{pmatrix} 0 & 1 \\ 1 & 0 \end{pmatrix}\frac{1}{5}\begin{pmatrix} 1 & 2 \\ 2 & 4 \end{pmatrix} = Tr\frac{1}{5}\begin{pmatrix} 2 & 4 \\ 1 & 2 \end{pmatrix} = \frac{4}{5}.$$

11. *Density matrix for a pure state satisfies which of the following properties?*
    (a) Unitary, (b) $\rho_{mn}\rho_{nm} = \rho_{mm}\rho_{nn}$ for all m, n,
    (c) Exactly one non-zero eigenvalue, (d) $\hat{\rho} = \hat{\rho}^2$.
    Ans. *(a),(b),(c),(d).*

12. *Define $\hat{\rho}_0 = |0><0|$, $\hat{\rho}_1 = |1><1|$, $\hat{\sigma} = \frac{\hat{\rho}_0 + \hat{\rho}_1}{2}$. Which of the following is/ are true?*
    (a) $Tr\hat{\sigma} = 1$, (b) $Tr\hat{\sigma}^2 = 1$, (c) $\hat{\sigma} = $ pure state, (d) $\hat{\sigma} = $ mixed state.

    Ans.
    $\hat{\sigma} = \frac{\hat{\rho}_0 + \hat{\rho}_1}{2} = \frac{|0><0| + |1><1|}{2} = \frac{1}{2}\left[\begin{pmatrix} 1 \\ 0 \end{pmatrix}(1 \ \ 0) + \begin{pmatrix} 0 \\ 1 \end{pmatrix}(0 \ \ 1)\right] = \frac{1}{2}\begin{pmatrix} 1 & 0 \\ 0 & 1 \end{pmatrix}$

    $Tr\hat{\sigma} = Tr\frac{1}{2}\begin{pmatrix} 1 & 0 \\ 0 & 1 \end{pmatrix} = 1$, *(a) is true*

    $Tr\hat{\sigma}^2 = Tr[\ \frac{1}{2}\begin{pmatrix} 1 & 0 \\ 0 & 1 \end{pmatrix}\frac{1}{2}\begin{pmatrix} 1 & 0 \\ 0 & 1 \end{pmatrix}\ ] = Tr\frac{1}{4}\begin{pmatrix} 1 & 0 \\ 0 & 1 \end{pmatrix} = \frac{1}{4}$, (b) false

    $Tr\hat{\sigma}^2 < 1 \Rightarrow$ mixed state. *(c)* false, *(d)* true.

13. *Show that the following state has $\hat{\rho} = \frac{1}{4}\begin{pmatrix} 1 & 0 \\ 0 & 3 \end{pmatrix}$ as the reduced density matrix.*

    $$|\psi> = \frac{1}{2}\left[\ \begin{pmatrix} 1 \\ 0 \end{pmatrix} \otimes \begin{pmatrix} 1 \\ 0 \end{pmatrix} + \sqrt{3}\begin{pmatrix} 0 \\ 1 \end{pmatrix} \otimes \begin{pmatrix} 0 \\ 1 \end{pmatrix}\right],$$

    Ans.

$$|\psi> = \frac{1}{2}\left[\ \begin{pmatrix} 1 \\ 0 \end{pmatrix} \otimes \begin{pmatrix} 1 \\ 0 \end{pmatrix} + \sqrt{3}\begin{pmatrix} 0 \\ 1 \end{pmatrix} \otimes \begin{pmatrix} 0 \\ 1 \end{pmatrix}\right] = \frac{1}{2}[\ |0>_A |0>_B + \sqrt{3}\ |1>_A |1>_B\ ]$$

$$\hat{\rho}_{AB} = |\psi><\psi| = \frac{1}{2}[\ |0>_A |0>_B + \sqrt{3}\ |1>_A |1>_B\ ]\frac{1}{2}[\ _A<0|_B<0| + \sqrt{3}\ _A<1|_B<1|]$$

$$= \frac{1}{4}[|0>_A |0>_B + \sqrt{3}\ |1>_A |1>_B\ ][\ _A<0|_B<0| + \sqrt{3}\ _A<1|_B<1|]$$

Take reduced density matrix, i.e. take trace of $\hat{\rho}_{AB}$ w.r.t B

$$Tr_B(\hat{\rho}_{AB}) = {}_B < 0|\hat{\rho}|0 >_B + {}_B < 1|\hat{\rho}|1>_B$$

$$= {}_B < 0| \tfrac{1}{4}[|0 >_A |0 >_B + \sqrt{3} |1 >_A |1 >_B][{}_A < 0|_B < 0| + \sqrt{3} {}_A < 1|_B < 1|] |0>_B$$

$$+ {}_B < 1| \tfrac{1}{4}[|0 >_A |0 >_B + \sqrt{3} |1 >_A |1 >_B][{}_A < 0|_B < 0| + \sqrt{3} {}_A < 1|_B < 1|] |1>_B$$

$$= \tfrac{1}{4}[|0 >_A {}_B < 0|0 >_B + \sqrt{3} |1 >_A {}_B < 0|1 >_B]$$

$$[{}_A < 0|_B < 0|0 >_B + \sqrt{3} {}_A < 1|_B < 1|0 >_B]$$

$$+ \tfrac{1}{4}[|0 >_A {}_B < 1|0 >_B + \sqrt{3} |1 >_A {}_B < 1|1 >_B]$$

$$[{}_A < 0|_B < 0|1 >_B + \sqrt{3} {}_A < 1|_B < 1|1 >_B]$$

$$= \tfrac{1}{4}[|0 >_A][{}_A < 0|] + \tfrac{1}{4}[\sqrt{3} |1 >_A][\sqrt{3} {}_A < 1|]$$

$$= \tfrac{1}{4}|0 >_A {}_A < 0| + \tfrac{3}{4}|1 >_A {}_A < 1|$$

(using

$$_B < 0|0 >_B = 1, \quad _B < 1|1 >_B = 1, \quad _B < 0|1 >_B = 0_B < 1|0 >_B = 0)$$

$$Tr_B(\hat{\rho}_{AB}) = \begin{pmatrix} \tfrac{1}{4} & 0 \\ 0 & \tfrac{3}{4} \end{pmatrix} = \tfrac{1}{4}\begin{pmatrix} 1 & 0 \\ 0 & 3 \end{pmatrix} = \hat{\rho}$$

14. *Verify that the given POVM*

$$\hat{E}_1 = \frac{\sqrt{2}}{\sqrt{2}+1}|0><0|, \quad \hat{E}_2 = \frac{(|0>-|1>)(<0|-<1|)}{2+\sqrt{2}}, \quad \hat{E}_3 = I - \hat{E}_1 - \hat{E}_2$$

allows identification of the states $|1>$ *and* $\frac{1}{\sqrt{2}}(|0>+|1>)$.

Ans. $|\psi_1 > = |1 > , \ |\psi_2 > = \frac{1}{\sqrt{2}}(|0> +|1 >)$

$$<\psi_1|\hat{E}_1|\psi_1 > = <1|\frac{\sqrt{2}}{\sqrt{2}+1}|0><0|1 > = \frac{\sqrt{2}}{\sqrt{2}+1} < 1|0><0|1 > = 0$$

$$<\psi_2|\hat{E}_1|\psi_2 > = \frac{1}{\sqrt{2}}(<0|+<1|)\frac{\sqrt{2}}{\sqrt{2}+1}|0><0|\frac{1}{\sqrt{2}}(|0> +|1 >)$$

$$= \frac{\sqrt{2}}{\sqrt{2}+1}\frac{1}{\sqrt{2}}\frac{1}{\sqrt{2}}(<0|0> +<1|0 >)(<0|0> +<0|1 >) = \frac{1}{\sqrt{2}+1}\frac{1}{\sqrt{2}} \neq 0$$

So $\hat{E}_1$ is able to distinguish between $|\psi_1>$ and $|\psi_2>$. If measurement outcome is E_1 then state is $|\psi_2>$.

Also,

$$<\psi_1|\hat{E}_2|\psi_1> \; = \; <1|\frac{(|0>-|1>)(<0|-<1|)}{2+\sqrt{2}}|1>$$

$$=\frac{(<1|0>-<1|1>)(<0|1>-<1|1>)}{2+\sqrt{2}} = \frac{1}{2+\sqrt{2}} \neq 0$$

$$<\psi_2|\hat{E}_2|\psi_2> \; = \; \frac{(<0|+<1|)}{\sqrt{2}}|\frac{(|0>-|1>)(<0|-<1|)}{2+\sqrt{2}}|\frac{(|0>+|1>)}{\sqrt{2}}$$

$$=\frac{1}{2+\sqrt{2}}\frac{1}{\sqrt{2}}\frac{1}{\sqrt{2}}(<0|0>-<0|1>+<1|0>-<1|1>)$$

$$(<0|0>+<0|1>-<1|0>-<1|1>)$$

$$=\frac{1}{2+\sqrt{2}}\frac{1}{\sqrt{2}}\frac{1}{\sqrt{2}}(1-1)(1-1)=0 \text{ (in fact } \frac{1}{\sqrt{2}}(|0>\pm|1>) \text{) are orthogonal states)}$$

So $\hat{E}_2$ is able to distinguish between $|\psi_1>$ and $|\psi_2>$. If measurement outcome is E_2 then state is $|\psi_1>$.

IOP Publishing

Quantum Optics and Quantum Computation
An introduction
Dipankar Bhattacharyya and Jyotirmoy Guha

Chapter 8

Quantum algorithms

A step-by-step procedure or systematic sequence of instructions needed to perform a computational task is called an algorithm. Each step is to be performed on a computer.

The algorithm is called a quantum algorithm if it is performed on a quantum computer and if it performs computational task with the help of quantum effects.

All classical algorithms run on a quantum computer.

Quantum computers are designed to outperform standard classical computers by running quantum algorithms. Quantum algorithms execute a job faster than their classical counterpart. A quantum algorithm finds applications in various fields such as cryptography, simulation of quantum systems, solving large systems of linear equations etc.

Examples of quantum algorithms that we shall discuss are Deutsch algorithm (section 8.5), Deutsch–Jozsa (DJ) algorithm (section 8.6), Bernstein–Vazirani algorithm (section 8.7), Simon's algorithm (section 8.8), Grover's search algorithm (which is popular and used for unsorted data base search, section 8.9), Shor's algorithm (used for factorization, section 8.16) etc.

8.1 Quantum parallelism

Classical computers use registers. Similarly, a quantum computer has quantum memory registers corresponding to input and output. A classical register at a given time can only store a particular state. A quantum register can store a linear combination or superposition of base states. This is called quantum parallelism. This is the advantage of quantum computing over classical deterministic computing.

According to quantum parallelism the computation of $|f(x)\rangle$ corresponding to x is not restricted to one particular value of x, i.e. not locked to one particular state. Instead of a single state we can have a linear combination of states. Then $|f(x)\rangle$ will be computed for every component of that linear superposition.

- **Example:**

A three-bit classical register can store at a time a three-bit string like 001 or 110 etc. But a quantum register of 3 qubits can store at any time a quantum state defined as

$$|\psi> = a|000> + b|001> + c|010> + d|011> + e|100> + f|101> + g|110> + h|111>$$

where a, b, c, d, e, f, g, h are complex numbers. So the quantum state can simultaneously exist in all the states $|000>$, $|001>$, $|010>$, $|011>$, $|100>$, $|101>$, $|110>$, $|111>$ unlike the classical register. This is quantum parallelism.

A unitary operator can be applied to such quantum superposition of states at the input to transform the entire set of states in parallel. Such quantum parallelism permits quantum algorithms to achieve computational complexities superior to classical alternatives.

8.2 Reversibility

Consider a function $z = f(x, y)$. This means that if x, y are given we can find z uniquely. Reversibility requires that if z is given we should be able to find x, y uniquely.

Classical gates (except the NOT gate) are not reversible—as explained in figure 8.1. Quantum gates and circuits are reversible gates (where feedback, fan-in, fan-out are not allowed), information is lossless.

8.3 XOR is addition modulo 2

XOR on binary numbers is the same as addition modulo 2. This is proved in the table 8.1.

XOR logic operation (denoted by $\oplus$) for the binary inputs A and B is defined as $A \oplus B = A\bar{B} + \bar{A}B = 1$ for an odd number of 1s, i.e. for inputs 01, 10.

Addition modulo 2 means adding and then dividing by 2 and taking the remainder.

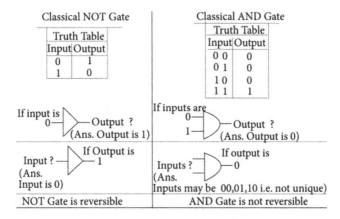

Figure 8.1. Except NOT Gate classical gates are not reversible.

Table 8.1. Comparing XOR and addition modulo 2.

Input A B	XOR operation A ⊕ B	Addition then divide by 2	Quotient	Addition modulo 2 operation (Remainder)
0 0	$0.\overline{0}+\overline{0}.0 = 0+0 = 0$	$\frac{0+0}{2}$	0	**0**
0 1	$0.\overline{1}+\overline{0}.1 = 0+1 = 1$	$\frac{0+1}{2}$	0	1
1 0	$1.\overline{0}+\overline{1}.0 = 1+0 = 1$	$\frac{1+0}{2}$	0	1
1 1	$1.\overline{1}+\overline{1}.1 = 0+0 = 0$	$\frac{1+1}{2}$	1	**0**

Table 8.2. Some examples of modulo operation (x = product sign).

Example: $a \bmod n = b$ $(nq + b = a)$	Example: $7 \bmod 3 = 1$ $(3 \times 2 + 1 = 7)$	Example: $19 \bmod 11 = 8$ $(11 \times 1 + 8 = 19)$	Example: $a \bmod 1 = 0$ $(1 \times a + 0 = a)$
n) a (q c b ← Remainder	3) 7 (2 6 1 ← Remainder	11) 19 (1 11 8 ← Remainder	1) a (a a 0 ← Remainder
(Dividng a by n we get quotient q and remainder b)	(Dividng 7 by 3 we get quotient 2 and remainder 1)	(Dividng 19 by 11 we get quotient 1 and remainder 8)	(Dividng a by 1 we get quotient a and remainder 0)

8.4 Quantum arithmetic and function evaluations

We describe how quantum computers actually compute (i.e. add and multiply numbers, evaluate Boolean functions through unitary operations). For the purpose we will use modular arithmetic discussed in the following. It is an arithmetic for integers developed by Gauss.

- **Modulo or mod operation**

If a and b are 2 positive numbers then
a modulo b = Remainder of the division of a by b [a = dividend, b = divisor]
We furnish examples in table 8.2.
- **NOTE**
  a mod 0 = undefined
- **Definition of congruence in modular arithmetic**

Consider the arithmetic relation $a = nq + b$ (a, b, n, q are all integers).

✓ Here we have divided a by n and the remainder is b which is called the residue of a mod n. We write this as $a = b$ mod n or $b = a$ mod n and say that a and b are congruent.

✓ If $a = nq + b$, i.e. a mod $n = b$ and $a' = nq' + b$, i.e. a' mod $n = b$ then we say that a' and a are congruent to each other w.r.t mod n or a mod $n = a'$ mod $n = b$. In other words, $a = a'$ (mod n) $= b$. We furnish some examples in table 8.3.

✓ Simplification by modular arithmetic

Modular exponentiation / mod of an exponential number

The modular exponentiation or finding the mod of an exponential number has been illustrated through examples given in table 8.4.

- **Need for modular arithmetic**

  Modular arithmetic is ideal for computers as it efficiently restricts the range of all intermediate results. For l-bit mod n, the intermediate results of any addition, subtraction or multiplication will not be more than $2l$ bits long.

  In quantum registers of size n, addition modulo 2^n is the most common operation for all $x \in \{0,1\}^n$ and for any $a \in \{0,1\}^n$, $|x > \; \to \; |(x + a)$ mod $2^n>$ is a unitary transformation.

- **Conversion of a positive number to modular arithmetic**

Table 8.5 shows some more examples of conversion of a positive number to modular arithmetic.

- **Conversion of a negative number to modular arithmetic**

Table 8.6 shows some examples of conversion of a negative number to modular arithmetic.

- **Multiplying in modular arithmetic**

We illustrate through the following example (dot means product)

✓ Prove that $8.11 = 2.2 = 1$(mod 3)

Table 8.3. Some examples of congruence in modulo
1 mod 7 = 8 mod 7 = 15 mod 7 = 50 mod 7 = 1 (same remainder 1)
1 = 8 = 15 = 50 (mod 7) i.e. 1, 8, 15, 50 are congruent in modulo 7.

Example:	Example:	Example:	Example:
1 mod 7 =1	8 mod 7 =1	15 mod 7 =1	50 mod 7 =1
(7×0 + 1 = 1)	(7×1 +1 =8)	(7×2 +1 =15)	(7×7 + 1 =50)
7) 1 (0 0 ——— 1 ← Remainder	7) 8 (1 7 ——— 1 ← Remainder	7) 15 (2 14 ——— 1 ← Remainder	7) 50 (7 49 ——— 1 ← Remainder

Table 8.4. Illustration of modular exponentiation.

Find 5^{40} mod 7

$\quad$ 7) $\underline{25}$ (3
$\qquad$ $\underline{21}$
$\qquad\quad$ 4

$5^1 = 5$ mod 7

$5^2 = 25$ mod 7 $= 4$ mod 7

$\quad$ 7) $\underline{16}$ (2
$\qquad$ $\underline{14}$
$\qquad\quad$ 2

$5^4 = (5^2)^2$ mod 7 $= 4^2$ mod 7 $= 16$ mod 7 $= 2$ mod 7

$\quad$ 7) $\underline{4}$ (0
$\qquad$ $\underline{0}$
$\qquad$ 4

$5^8 = (5^4)^2$ mod 7 $= 2^2$ mod 7 $= 4$ mod 7 $= 4$

$5^{16} = (5^8)^2$ mod 7 $= 4^2$ mod 7 $= 16$ mod 7 $= 2$ mod 7

$5^{32} = (5^{16})^2$ mod 7 $= 2^2$ mod 7 $= 4$ mod 7

$5^{40} = 5^{32+8}$ mod 7 $= 5^{32} 5^8$ mod 7

$\qquad\qquad = 4.4$ mod 7 $= 16$ mod 7 $= \boxed{2 \text{ mod } 7}$

Find 3^{200} mod 50

$3^1 = 3$ mod 50

$3^2 = 9$ mod 50

$3^4 = (3^2)^2$ mod 50 $= 9^2$ mod 50 $= 81$ mod 50 $= 31$ mod 50

$3^8 = (3^4)^2$ mod 50 $= 31^2$ mod 50 $= 961$ mod 50 $= 11$ mod 50

$3^{16} = (3^8)^2$ mod 50 $= 11^2$ mod 50 $= 121$ mod 50 $= 21$ mod 50

$3^{32} = (3^{16})^2$ mod 50 $= 21^2$ mod 50 $= 441$ mod 50 $= 41$ mod 50

$3^{64} = (3^{32})^2$ mod 50 $= 41^2$ mod 50 $= 1681$ mod 50 $= 31$ mod 50

$3^{128} = (3^{64})^2$ mod 50 $= 31^2$ mod 50 $= 961$ mod 50 $= 11$ mod 50

$3^{200} = 3^{128+64+8}$ mod 50 $= 3^{128} 3^{64} 3^8$ mod 50 $\quad$ (expressing the powers as sum of powers of 2)

$\qquad = 11.31.11$ mod 50 $= 3751$ mod 50

$\qquad = \boxed{1 \text{ mod } 50}$

Table 8.5. Conversion of positive number to modular arithmetic.

Example: $(m = qn + R)$	Example: $(51 = 5 \times 10 + 1)$	Example: $(27 = 4 \times 6 + 3)$	Example: $(3 = 0 \times 4 + 3)$
n) m (q $\quad$ s $\quad$ $\underline{\quad}$ $\quad$ R $\leftarrow$ $\quad$ Remainder	10) 51 (5 $\quad$ $\underline{50}$ $\quad$ 1 $\leftarrow$ $\quad$ Remainder	6) 27 (4 $\quad$ $\underline{24}$ $\quad$ 3 $\leftarrow$ $\quad$ Remainder	4) 3 (0 $\quad$ $\underline{0}$ $\quad$ 3 $\leftarrow$ $\quad$ Remainder
m = R (mod n)	51 = 1 (mod 10)	27 = 3 (mod 6)	3 = 3 (mod 4)

Table 8.6. Conversion of negative number to modular arithmetic

Example: $-51 = (-6)10 + 9$	Example: $-37 = (-8)5 + 3$	Example: $-m = -qn + R$
$\begin{array}{r} 10)-51(-6 \\ -60 \\ \hline 9 \leftarrow \\ \text{Remainder} \end{array}$	$\begin{array}{r} 5)-37(-8 \\ -40 \\ \hline 3 \leftarrow \\ \text{Remainder} \end{array}$	$\begin{array}{r} n)-m(-q \\ -n \\ \hline R \leftarrow \\ \text{Remainder} \end{array}$
$-51 = 9 \ (\text{mod } 10)$	$-37 = 3 \ (\text{mod } 5)$	$-m = R \ (\text{mod } n)$

$$8.11 = 88 = 29.3 + 1 = 1 (\text{mod } 3)$$

We also note that $a. \ b \ (\text{mod } n) = a(\text{mod } n) \ \ b(\text{mod } n)$ and hence

$$8.11 \ (\text{mod } 3) = 8 \ (\text{mod } 3) \ 11 \ (\text{mod } 3)$$

Now $8 = 2.3 + 2 = 2 \ (\text{mod } 3)$

$$11 = 3.3 + 2 = 2 \ (\text{mod } 3)$$

Hence

$$8.11 \ (\text{mod } 3) \ = 2 \ (\text{mod } 3) \ 2 \ (\text{mod } 3) \ = 2.2 \ (\text{mod } 3)$$

Again

$$2.2 = 4 = 1.3 + 1 = 1 \ (\text{mod } 3)$$

Hence it is established that $8.11 = 2.2 = 1 \ (\text{mod } 3)$.

- **Significance of congruence in modular arithmetic**

  ✓ The expression $a = R \ (\text{mod } n)$ means that a is congruent to R in mod n where R is positive whole number greater than 1. This means the following:
  ✓ a, R have the same remainder when divided by n.

Example: $10 \equiv 14 \ (\text{mod } 4)$

This follows since: $\begin{aligned} 10 &= 2.4 + 2 \\ 14 &= 3.4 + 2 \end{aligned}$

$$\begin{array}{r} 4)10(2 \\ 8 \\ \hline 2 \leftarrow \text{Remainder} \end{array} \qquad \begin{array}{r} 4)14(3 \\ 12 \\ \hline \rightarrow 2 \end{array}$$

✓ In $a \equiv R$ (mod n) we can write $a = qn + R$.

In $10 \equiv 14$ (mod 4) we can write $10 = (-1).4 + 14$

✓ Since for $a \equiv R$ (mod n) we can write $a = qn + R$. It follows that

$$a - R = qn \Rightarrow \frac{a - R}{n} = q$$

So $a - R$ is a multiple of n, the quotient being q.

In $10 \equiv 14$ (mod 4) we have $\frac{a-R}{n} = \frac{10-14}{4} = \frac{-4}{4} = -1$. So $10-14$ is a multiple of 4.

With this background let us spell out the significance of congruence in modular arithmetic through the following two illustrations.

(i) Consider the number line shown in figure 8.2 corresponding to mod 2. It is clear that the numbers ...$-5, -3, -1, 1, 3, 5$ ... are equivalent in mod 2 since they all have value 1, while ...$-4, -2, 0\ 2, 4$ ... are equivalent in mod 2 since they all have value 0. In mod 2 there are 2 values: $\{0,1\}$. We note that congruence or equivalence is achieved for a shift of 2. So any number x and $x \pm 2k$ are congruent (k = integer).

(ii) Consider the number line shown in figure 8.3 corresponding to mod 3. It is clear that the numbers ...$-6, -3, -0, 3, 6$ ... are equivalent in mod 3 since they all have value 0, while ...$-5, -2, 1, 4$...are equivalent in mod 3 since they all have value 1, while ...$-4, -1, 2, 5$... are equivalent in mod 3 since they all have value 2. In mod 3 there are 3 values: $\{0, 1, 2\}$. So any number x and $x \pm 3k$ are congruent (k = integer).

(iii) Mod n means there are n values starting from 0 to $n - 1$.

8.5 Deutsch algorithm

Let us introduce simple and intuitive algorithms that demonstrate that quantum information is superior to classical bit information.

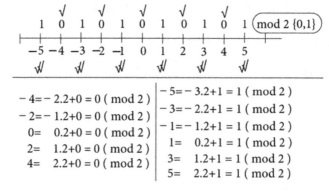

Figure 8.2. Equivalent numbers in the number line for mod 2.

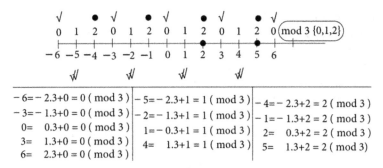

$$-6 = -2.3+0 = 0 \ (\bmod \ 3) \quad -5 = -2.3+1 = 1 \ (\bmod \ 3) \quad -4 = -2.3+2 = 2 \ (\bmod \ 3)$$
$$-3 = -1.3+0 = 0 \ (\bmod \ 3) \quad -2 = -1.3+1 = 1 \ (\bmod \ 3) \quad -1 = -1.3+2 = 2 \ (\bmod \ 3)$$
$$0 = \ \ \ 0.3+0 = 0 \ (\bmod \ 3) \quad 1 = -0.3+1 = 1 \ (\bmod \ 3) \quad 2 = \ \ \ 0.3+2 = 2 \ (\bmod \ 3)$$
$$3 = \ \ \ 1.3+0 = 0 \ (\bmod \ 3) \quad 4 = \ \ \ 1.3+1 = 1 \ (\bmod \ 3) \quad 5 = \ \ \ 1.3+2 = 2 \ (\bmod \ 3)$$
$$6 = \ \ \ 2.3+0 = 0 \ (\bmod \ 3) \qquad\qquad\qquad\qquad\qquad\qquad$$

Figure 8.3. Equivalent numbers in the number line for mod 3.

Let us consider a univariate function $f(x)$ where x and $f(x)$ can take values 0 and 1. In other words input x is one qubit (0 or 1) and output $f(x)$ is one qubit (0 or 1). We discuss the possibilities:

(i) Output $f(x)$ is independent of the input x. Output $f(x)$ is such that $f(0) = f(1)$. Such function $f(x)$ is called constant function. This corresponds to two possibilities:

(a) $f(0) = f(1) = 0$ or (b) $f(0) = f(1) = 1$.

(irrespective of the input, output is fixed, hence constant function 0 or 1)

(ii) Output $f(x)$ is dependent on the input x. Output $f(x)$ is such that $f(0) \neq f(1)$. Such function $f(x)$ is called balanced function. This corresponds to two possibilities:

(a) $f(0) = 0, f(1) = 1$ (i.e. inputs pass through f unchanged, since a $x = 0$ input gives $f(x = 0) = 0$ i.e. same output or a $x = 1$ input gives $f(x = 1) = 1$ i.e. same output) or

(b) $f(0) = 1, f(1) = 0$ (i.e. inputs get exchanged on passing through f, since $x = 0$ input gives $f(x = 0) = 1$ i.e. complemented output or a $x = 1$ input gives $f(x = 1) = 0$, i.e. complemented output).

We can express these possibilities alternatively as: $f(0) = \bar{f}(1)$, $f(1) = \bar{f}(0)$. In other words, half the inputs give 0 and the other half of inputs give 1, so it is a balanced function.

Illustration:

	$f(x)$: Output (one qubit)			
x: Input (one qubit)	Constant function (possible values)		Balanced function (possible values)	
0	$f(0) = 0$	$f(0) = 1$	$f(0) = 0$	$f(0) = 1$
1	$f(1) = 0$	$f(1) = 1$	$f(1) = 1$	$f(1) = 0$

The problem is to determine whether $f(x)$ is constant or balanced. Suppose the output is $f(0) = 0$. Classical computation would now need the second query

$f(1) = ?$ to determine whether $f(x)$ is constant or balanced. If $f(1) = 0$ it is a constant function and if $f(1) = 1$ it is a balanced function. In classical computation we thus need two queries to solve this problem viz. (i) To find $f(0)$ and (ii) to find $f(1)$. Now we have to check whether $f(0) = f(1)$ or $f(0) \neq f(1)$ and conclude whether it is a constant or balanced function.

In quantum computation we need a single query to solve this problem through the Deutsch algorithm.

Assume that a black box or an oracle is given to compute the value of the output function $f(x)$ for a given input x, i.e. to determine whether the output function $f(x)$ is constant or balanced. This problem can be symbolically represented as:

$$f(x): \{\text{inputs}\} \rightarrow \{\text{outputs}\} \text{ i.e. } f(x): \{0,1\} \rightarrow \{0,1\}.$$

And the constraint on $f(x)$ (also called promise) is that: $f(x)$ is either constant or balanced.

And the task is to determine whether $f(x)$ is constant or balanced.

Figure 8.4 shows the circuit for implementation of the Deutsch algorithm.

NOTE

In this chapter we represent the Hadamard operator, as well as other operators without the overhead cap sign.

Start with two qubits $|0>$ and $|1>$ at the input side.

$|0>$ is passed through a Hadamard Gate which gets transformed as $|0> \xrightarrow{H} \frac{1}{\sqrt{2}}(|0> +|1>) = |x>$. The input register is $|x>$.

The target qubit is set to $|1>$ that is passed through Hadamard Gate which is transformed as

$|1> \xrightarrow{H} \frac{1}{\sqrt{2}}(|0> - |1>) = |y>$. The target register is $|y>$.

We have at the input of the oracle the qubit state denoted by (in two-qubit language)

$$|x> \otimes|y> = \frac{1}{\sqrt{2}}(|0> +|1>) \otimes \frac{1}{\sqrt{2}}(|0> - |1>)$$

$$= \frac{1}{2}(|00> +|10> - |01> - |11>) \tag{8.1}$$

The oracle applies a unitary transformation U_f on $|x> \otimes|y>$. The oracle keeps the input qubit register (the x bit, i.e. $|x>$) undisturbed but changes the target qubit register (the y bit, i.e. $|y>$) to $|y \oplus f(x) >$, i.e.

$$U_f|x> \otimes|y> = |x> \otimes|y \oplus f(x) >.$$

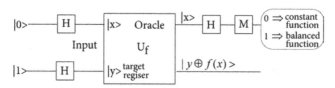

Figure 8.4. Circuit for Deutsch algorithm.

In the two-qubit language we have oracle output (using equation (8.1))

$$U_f|x> \otimes|y> = |x> \otimes|y \oplus f(x)>$$

$$= U_f \frac{1}{2}(|00> +|10> - |01> - |11 >)$$

$$= |x, y \oplus f(x)>$$

$$= \frac{1}{2}[|0,0 \oplus f(0) > +|1,0 \oplus f(1) > - |0,1 \oplus f(0) > - |1,1 \oplus f(1) >]$$

The operation of $\oplus$ that represents XOR or addition modulo 2 gives

$$0 \oplus A = A, \quad 1 + A = \bar{A}$$

(since $0\oplus0=0,0\oplus1=1$ and $1\oplus0=1 = \bar{0}$, $1\oplus1=0 = \bar{1}$) and so we have

$$U_f|x> \otimes|y > = \frac{1}{2}[|0, f(0) > +|1, f(1) > - |0, \bar{f}(0) > - |1, \bar{f}(1) >] \qquad (8.2)$$

(since $0 \oplus f(0) = f(0)$, $0 \oplus f(1) = f(1)$; $1 \oplus f(0) = \bar{f}(0)$, $1 \oplus f(1) = \bar{f}(1)$).

Case 1: Suppose $f(x)$ is constant function, i.e. $f(0) = f(1)$. Oracle output is

$$U_f|x> \otimes|y > = \frac{1}{2}[|0, f(0) > +|1, f(0) > - |0, \bar{f}(0) > - |1, \bar{f}(0) >]$$

$$= \frac{1}{2}[(|0 > +|1 >) \otimes (f(0) - \bar{f}(0))]$$

Since $f(0)$ takes values 0,1 it follows that $f(0) - \bar{f}(0) = 0-1$ or $1-0$. So the oracle output might have an overall global phase factor of $\pm = e^{\pm i\pi}$. Thus

$$f(0) - \bar{f}(0) \rightarrow \pm(|0 > - |1>). \qquad (8.3)$$

Hence the oracle output is

$$U_f|x> \otimes|y > = \frac{1}{\sqrt{2}}(|0 > +|1 >) \otimes \pm\frac{1}{\sqrt{2}}(|0 > - |1 >) \qquad (8.4)$$

The first qubit $\frac{1}{\sqrt{2}}(|0 > +|1 >)$ is passed through Hadamard Gate (figure 8.4)

$$\frac{1}{\sqrt{2}}(|0 > +|1 >) \xrightarrow{H} |0>$$

The final output thus becomes (from equation (8.4))

$$|output > = |0 > \otimes\pm\frac{1}{\sqrt{2}}(|0 > - |1 >)$$

8-10

Clearly the first qubit is |0> for $f(x)$= constant function.

Hence if measurement M (figure 8.4) on first qubit gives 0 it corresponds to a constant function.

Case 2: Suppose $f(x)$ is balanced function, i.e. $f(0) \neq f(1)$, i.e. $f(0) = \bar{f}(1)$, $f(1) = \bar{f}(0)$. Oracle output is (from equation (8.2))

$$U_f|x> \otimes |y> = \frac{1}{2}[|0, f(0)> + |1, f(1)> - |0, \bar{f}(0)> - |1, \bar{f}(1)>]$$

$$= \frac{1}{2}[|0, f(0)> + |1, \bar{f}(0)> - |0, \bar{f}(0)> - |1, f(0)>]$$

$$= \frac{1}{2}[(|0> - |1>) \otimes (f(0) - \bar{f}(0))]$$

As in case 1, here too we write: $f(0) - \bar{f}(0) \rightarrow \pm(|0> - |1>)$ (equation (8.3)). Hence the oracle output is

$$U_f|x> \otimes |y> = \frac{1}{\sqrt{2}}(|0> - |1>) \otimes \pm\frac{1}{\sqrt{2}}(|0> - |1>) \qquad (8.5)$$

The first qubit $\frac{1}{\sqrt{2}}(|0> - |1>)$ is passed through Hadamard Gate (figure 8.4)

$$|x> = \frac{1}{\sqrt{2}}(|0> - |1>) \xrightarrow{H} |1>$$

The final output thus becomes (from equation (8.5))

$$|output> = |1> \otimes \pm\frac{1}{\sqrt{2}}(|0> - |1>)$$

Clearly the first qubit is |1> for $f(x)$= balanced function.

Hence if a single measurement M (figure 8.4) on first qubit gives 1 then it corresponds to a balanced function.

It is evident from the above discussion that if a single measurement M (figure 8.4) on first qubit gives 0 it corresponds to a constant function and if it gives 1 then it corresponds to a balanced function. In other words a single measurement on the first qubit, i.e. a single query solves the problem of determining the nature of the function.

The Deutsch algorithm thus shows that classical computation requires two queries while quantum computation requires a single query. There is thus a 50% reduction in the computing resources.

In classical computation we could, through two queries, know the values $f(0), f(1)$). In quantum computation we do not require the values of $f(0), f(1)$ as our primary aim was to find whether function $f(x)$ is constant or balanced.

8.6 Deutsch–Jozsa (DJ) algorithm

We extend or generalize the Deutsch algorithm to multivariate function $f(x)$. Consider n qubit input (i.e. 2^n input qubit combinations of x) but 1 qubit output.

(The Deutsch algorithm involved 1 qubit input (i.e. $2^1 = 2$ values of x) and 1 qubit output represented as $f(x)$: $\{0,1\} \rightarrow \{0,1\}$ or $\{0,1\}^{\otimes 1} \rightarrow \{0,1\}$.)

Assume that a black box or an oracle is given to compute the value of the output function $f(x)$ for a given input x, i.e. to determine whether the output function $f(x)$ is constant or balanced. This problem can be symbolically represented as: $f(x)$: {inputs} $\rightarrow$ {outputs}, i.e. $f(x)$: $\{0,1\}^{\otimes n} \rightarrow \{0,1\}$.

Corresponding to 2^n input qubit combinations of x the task is to find whether function $f(x)$ is constant (in which all inputs are mapped to either 0 or 1) or balanced (in which half the inputs go to 0 and the other half go to 1). No other possibility is considered here.

Classically we need $2^{n-1} + 1$ queries to solve the problem (i.e. to determine whether function $f(x)$ is constant or balanced). (Classically, to solve the $n = 1$ qubit input problem we needed $2^{n-1} + 1 = 2^{1-1} + 1 = 2$ queries. But the same was solved with 1 query in the Deutsch algorithm.)

Illustration

x: Input ($n = 2$ qubits)	$f(x)$: Output (1 qubit)			
	Constant function (possible values)		Balanced function (possible values)	
00	$f(00) = 0$	$f(00) = 1$	$f(00) = 0$	$f(00) = 0$
01	$f(01) = 0$	$f(01) = 1$	$f(01) = 1$	$f(01) = 0$
10	$f(10) = 0$	$f(10) = 1$	$f(10) = 0$	$f(10) = 1$
11	$f(11) = 0$	$f(11) = 1$	$f(11) = 1$	$f(11) = 1$

In this problem $f(x)$ is either a constant or a balanced function. Suppose the outputs are $f(00) = 0$, $f(01) = 0$.

Classical computation would now need the third query $(2^{n-1} + 1 = 2^{2-1} + 1 = 3)$ viz. $f(10) = ?$ to determine whether $f(x)$ is constant or balanced. If $f(10) = 1$ (last column) then $f(x)$ is a balanced function since then $f(11) = 1$ is a forgone conclusion. If $f(10) = 0$ then $f(x)$ is a constant function since then $f(11) = 0$ is a foregone conclusion.

In quantum computation we need a single query to solve this problem through the DJ algorithm.

The circuit for implementing the DJ algorithm is shown in figure 8.5.

Start with $n + 1$ qubits, i.e. n qubits of $|0>$ and 1 qubit of $|1>$.

$|0>$ in each of n input registers is passed through n Hadamard gates. Each qubit $|0>$ is transformed by the Hadamard Gate as

$$|0> \xrightarrow{H} \frac{1}{\sqrt{2}}(|0> + |1>) = |x>$$

So when n qubits $|0>$ are passed through n Hadamard Gates what we get can be written as

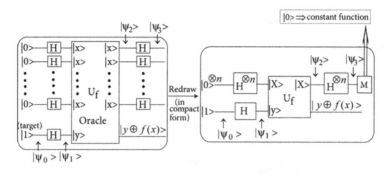

Figure 8.5. Circuit for Deutsch–Jozsa (DJ) algorithm.

$$|0>^{\otimes n} \xrightarrow{H^{\otimes n}} \frac{|0> + |1>}{\sqrt{2}} \otimes \frac{|0> + |1>}{\sqrt{2}} \otimes \frac{|0> + |1>}{\sqrt{2}} \cdots \cdots$$

$$H^{\otimes n} \ |0>^{\otimes n} = \frac{1}{\sqrt{2^n}}(\ |0000... > + |0100... > + |0010... > + \cdots \cdots |1111.... >)$$

$$= \frac{1}{\sqrt{2^n}} \sum_{x=0}^{2^n-1} |x> = |X> \text{ (let us use this short hand notation)}$$

where $|x>$ represents the n qubit basis. What we get at the input to the oracle is a linear combination of n qubit basis states of equal strength or a uniform linear combination of n qubit basis states. For instance if $n=2$ then we would get

$$|0>^{\otimes 2} \xrightarrow{H^{\otimes 2}} \frac{|0> + |1>}{\sqrt{2}} \otimes \frac{|0> + |1>}{\sqrt{2}} = \frac{1}{\sqrt{2^2}}(|00> + |01> + |10> + |11>)$$

The target qubit is set at $|1>$. So

$$|\psi_0> = |0>^{\otimes n} \otimes |1>$$

$|1>$ is passed through Hadamard gate which is transformed as

$|1> \xrightarrow{H} \frac{1}{\sqrt{2}}(|0> - |1>) = |y>$. The target register is $|y>$. Alternatively, it is written as

$$H|1> = \frac{1}{\sqrt{2}}(|0> - |1>) = |y>$$

We have at the input of the oracle the qubit state denoted by

$$|\psi_1> = H^{\otimes n}|0>^{\otimes n} \quad \otimes \quad H|1>$$

$$= |X> \otimes \frac{1}{\sqrt{2}}(|0> - |1>) = |X> \otimes |y>$$

Oracle applies a unitary transformation U_f on $|\psi_1>$, i.e. the oracle changes the target qubit register (the y bit, i.e. $|y>$) to $|y \oplus f(x)>$

$$|\psi_2> = U_f|\psi_1> = U_f|X> \otimes |y> = |X> \otimes |y \oplus f(x)> \qquad (8.6)$$

For $y = 0$, from equation (8.6)

$$U_f|X> \otimes|0> \quad =|X> \otimes|0 \oplus f(x)> =|X> \otimes|f(x)> \quad (\text{as } 0 \oplus f(x) = f(x))$$

For $y - 1$, from equation (8.6)

$$U_f|X> \otimes|1> \quad =|X> \otimes|1 \oplus f(x)> =|X> \otimes|\bar{f}(x)> \quad (1 \oplus f(x) = \bar{f}(x))$$

Hence subtracting the two equations and multiplying by $\frac{1}{\sqrt{2}}$ we have

$$U_f|X> \otimes\frac{1}{\sqrt{2}}(|0> - |1>) = |X> \otimes\frac{1}{\sqrt{2}}[|f(x)> - |\bar{f}(x)>]$$

$$U_f|X> \otimes|y> \ = \ |X> \otimes\frac{1}{\sqrt{2}}[|f(x)> - |\bar{f}(x)>] \tag{8.7}$$

If $f(x) = 0$, equation (8.7) gives

$$U_f|X> \otimes|y> \ = \ |X> \otimes\frac{1}{\sqrt{2}}(|0> - |1>) \tag{8.8}$$

If $f(x) = 1$ equation (8.7) gives

$$U_f|X> \otimes|y> \ = \ |X> \otimes\frac{1}{\sqrt{2}}(|1> - |0>)$$

$$=|X> \otimes(-1)\frac{1}{\sqrt{2}}(|0> - |1>) \tag{8.9}$$

Combining equation (8.8) and equation (8.9) we write for equation (8.7)

$$U_f|X> \otimes|y> \ =|X> \otimes(-1)^{f(x)}\frac{1}{\sqrt{2}}(|0> - |1>).$$

So the oracle output equation (8.6) is

$$|\psi_2> = U_f|\psi_1> \ =U_f|X> \otimes|y>$$

$$= \frac{1}{\sqrt{2^n}}\sum_{x=0}^{2^n-1}|x> \otimes(-1)^{f(x)}\frac{1}{\sqrt{2}}(|0> - |1>)$$

$$= \frac{1}{\sqrt{2^n}}\sum_{x=0}^{2^n-1}(-1)^{f(x)}|x> \otimes\frac{1}{\sqrt{2}}(|0> - |1>) \tag{8.10}$$

Apply n Hadamard gates in parallel to the first n qubits $|x> \ =|x_0x_1x_2...x_{n-1}>$ emerging from the oracle output which we denote as follows

$$|\psi_3> = H^{\otimes n}|\psi_2> = H^{\otimes n}\frac{1}{\sqrt{2^n}}\sum_{x=0}^{2^n-1}(-1)^{f(x)}|x>\otimes\frac{1}{\sqrt{2}}(|0>-|1>)$$

$$= \frac{1}{\sqrt{2^n}}\sum_{x=0}^{2^n-1}(-1)^{f(x)}\ (H^{\otimes n}\ |x>)\ \otimes\frac{1}{\sqrt{2}}(|0>-|1>)$$

(8.11)

Again,

$$H^{\otimes n}|x> = H^{\otimes n}|x_0 x_1 x_2 ... x_{n-1}>$$
$$= H|x_0>H|x_1>H|x_2>.....H|x_{n-1}>$$

(8.12)

Now $H|0> = \frac{1}{\sqrt{2}}[|0>+|1>]$ and $H|1> = \frac{1}{\sqrt{2}}[|0>-|1>]$. Combining, we write

$$H|x> = \frac{1}{\sqrt{2}}[|0>+(-1)^x|1>] = \frac{1}{\sqrt{2}}[(-1)^{x.0}|0>+(-1)^{x.1}|1>]$$

$$= \frac{1}{\sqrt{2}}\sum_{k=0}^{1}(-1)^{x.k}|k>.$$

Then from equation (8.12) we can write

$$H^{\otimes n}|x> = \sum_{k_0=0}^{1}\sum_{k_1=0}^{1}.....\sum_{k_{n-1}=0}^{1}\frac{1}{\sqrt{2^n}}(-1)^{x_0 k_0 + x_1 k_1 + ...+x_{n-1}k_{n-1}}|k_0 k_1......k_{n-1}>$$

$$= \sum_{k}\frac{1}{\sqrt{2^n}}(-1)^{x.k}|k>\quad \text{(in a compact notation)}$$

(8.13)

where $x.\ k = x_0 k_0 + x_1 k_1 + ... + x_{n-1}k_{n-1}$ = sum of bitwise product and

$$|k> = |k_0 k_1 ... k_{n-1}>.$$

Clearly putting equation (8.13) into equation (8.11) we write the output as

$$|\psi_3> = H^{\otimes n}|\psi_2>$$

$$= \frac{1}{\sqrt{2^n}}\sum_{x=0}^{2^n-1}(-1)^{f(x)}\left(\sum_{k}\frac{1}{\sqrt{2^n}}(-1)^{x.k}|k>\right)\otimes\frac{1}{\sqrt{2}}(|0>-|1>)$$

(8.14)

$$|\psi_3> = \sum_{x}\sum_{k}\frac{(-1)^{f(x)+x.k}}{2^n}|k>\otimes\frac{1}{\sqrt{2}}(|0>-|1>)$$

Let us focus attention on the coefficient of $|k>\ \sim\ |0>^{\otimes n}$ in $|\psi_3>$ which is

$$\sim\sum_{x}(-1)^{f(x)}.$$

Case 1: $f(x) =$ constant function

For $f(x) = 0$, $(-1)^{f(x)} = (-1)^0 = 1$. So the probability of getting $|k > \sim |0>^{\otimes n}$ is unity in a measurement. And for $f(x) = 1$, $(-1)^{f(x)} = (-1)^1 = -1$. So the probability of getting $|k > \sim |0>^{\otimes n}$ is unity in a measurement here too. So we will surely get $|0>$ in a measurement for a constant function.

Case 2: $f(x) =$ Balanced function

For half of the cases $f(x) = 1 \Rightarrow (-1)^{f(x)} = (-1)^1 = -1$ while for other half of the cases $f(x) = 0 \Rightarrow (-1)^{f(x)} = (-1)^0 = 1$. Clearly when summed we get $\sum_x (-1)^{f(x)} = 0$.

So $|\psi_3 > = 0$ for $|k > \sim |0>^{\otimes n}$. This means that we won't get 0 in a measurement for a balanced function.

Hence when we measure the first register ($\Rightarrow$ single query/ measurement denoted by M in figure 8.5) a result $|0>$ implies constant function, otherwise balanced function. A single query determines the nature of the function.

Though this algorithm does not have any practical application, it shows the importance and power of quantum algorithms.

8.7 Bernstein–Vazirani algorithm

This algorithm is an extension of the Deutsch–Josza algorithm where instead of distinguishing between constant and balanced functions it tries to identify an unknown string $|a > = |a_0 a_1 a_2 .... a_{n-1}>$ in an encoded function.

Suppose there is a function $f(x) = a . x$ ($\Rightarrow$ the promise) with $f: \{0,1\}^n \rightarrow \{0,1\}$. Here $|x > = |x_0 x_1 ..... x_{n-1}>$ represents n qubit input string and $|a > = |a_0 a_1 ..... a_{n-1}>$ represents unknown n qubit string ($\Rightarrow$ secret string). $f(x) = a . x = a_1 x_1 \oplus a_2 x_2 \oplus .... \oplus a_{n-1} x_{n-1}$ represents a bitwise product (addition modulo 2).

Classically, one requires n queries to find $|a >$, i.e. one has to perform n measurements of $f(100...0), f(010...0), ... f(000...1)$ to solve a system of linear equations for $|a>$. Using a quantum algorithm $|a>$ can be determined in a single query.

Using Deutsch–Josza algorithm (figure 8.5) the output is, with $f(x) = a . x$ (equation (8.14))

$$|\psi_3> = \sum_x \sum_k \frac{(-1)^{f(x)+x.k}}{2^n} |k > \otimes \frac{1}{\sqrt{2}}(|0 > - |1 >)$$

$$= \sum_x \sum_k \frac{(-1)^{a.x+x.k}}{2^n} |k > \otimes \frac{1}{\sqrt{2}}(|0 > - |1 >)$$

Performing sum over x we get

$$\sum_{x=0}^{2^n-1} (-1)^{a.x+k.x} = \prod_{i=1}^{n} \sum_{x_i=0}^{1} (-1)^{x_i(a_i+k_i)}$$

$$= \prod_{i=1}^{n} [1 + (-1)^{(a_i+k_i)}] \ (1 \text{ for } x_i = 0 \text{ and } (-1)^{(a_i+k_i)} \text{ for } x_i = 1)$$

$$= 2^n \delta_{ak} \text{ (survives only for } a = k=0 \ or \ 1. \text{ For } a \neq k \text{ contribution is zero)}$$

$$|\psi_3> = \sum_k \delta_{ak}|k> \otimes \frac{1}{\sqrt{2}}(|0> - |1>) = |a> \otimes \frac{1}{\sqrt{2}}(|0> - |1>)$$

So the measured qubits at the output (single query) leads to $|a> = |a_0 a_1 a_2 .... a_{n-1}>$

8.8 Simon algorithm

This algorithm is an extension of the Bernstein–Vazirani algorithm to the case of n qubit output. Given an n qubit input string $|x> = |x_0 x_1 ..... x_{n-1}>$ and a function $f: \{0,1\}^n \rightarrow \{0,1\}^n (2^n$ inputs).

For a pair of inputs x, y the promise is that $f(x) = f(y)$ if and only if $y = x \oplus \xi$ i.e. $x \oplus y = \xi$ where ξ is an unknown string or a secret message. [ξ is the bitwise XOR of the first and second inputs]. The task is to find ξ. Here $f(x)$ is a two to one function.

NOTE

✓ A function is said to be one to one if whenever $f(x) = f(y)$, $x = y$ so that for each y value there is only one x value.

✓ A two-to one-function is one in which there are two inputs x, y and for the pair x, y we get $f(x) = f(y)$ if and only if $y = x \oplus \xi$. We give example of such a two-to-one function in table 8.7 and explain it.

Inputs are denoted by x or y and the outputs are denoted by $f(x)$ for x and $f(y)$ for y. The function referred to in table 8.7 is a two-to-one function. So we will be able to find pairs of inputs that give identical outputs as illustrated in the table.

We find $x = 000$ and $y = 011$ such that $f(x = 000) = f(y = 011) = 011$. A bitwise XOR of $x = 000$ and $y = 011$ gives 011, i.e. $x \oplus y=000\oplus011=011=\xi$(since $000\oplus011 = 0\oplus0 \ 0\oplus1 \ 0\oplus1=0 \ 1 \ 1$).

We also find $x=001$ and $y=010$ such that f(x=001)=f(y=010)=010. A bitwise XOR of $x = 001$ and $y = 010$ gives 011, i.e. $x \oplus y=001\oplus010=011=\xi$.

Table 8.7. Example of a two-to-one function. This is the $n = 3$ bit case.

Inputs x, y	Observed outputs $f(x), f(y)$	$f(x) = f(y)$ occurs for	$x \oplus y = \xi$ (bitwise XOR)
000	011		
001	010	f(000)=f(011)=011	000 $\oplus$ 011=011
010	010	f(001)=f(010)=010	001 $\oplus$ 010=011
011	011		
100	111	f(100)=f(111)=111	100 $\oplus$ 111=011
101	110	f(101)=f(110)=110	101 $\oplus$ 110=011
110	110		
111	111		

Similarly, we get $\xi=011$ starting with the pair $x = 100$, $y = 111$ and also with the pair $x = 101$, $y = 110$ (since $100\oplus111 = 0\ 1\ 1$ and $101\oplus110=011$).

This ξ is called the secret message string that we desire to find out through a suitable algorithm.

- **Problem (solving with classical algorithm)**

(i) For three-bit input we get three-bit output, as shown in table 8.8. Identify the secret message string ξ if the function connecting them is a two-to-one function.

(ii) Identify all the pairs that give identical outputs.

(iii) Comment on run time of the algorithm.

Ans.

(i) Number of input bits $n = 3$ (denoted by x). Number of output bits $n = 3$. Observation sequence is as follows:

For the first four inputs 000 (output 111), 001 (output 000), 010 (output 110), 011 (output 010) the outputs do not match or coincide. For the fifth input 100 the output is 000 which coincides with the output corresponding to the second input, namely 001. The secret message is thus the bitwise XOR of these inputs, namely

$$001\oplus100=101 = \xi.$$

(ii) Let us find the pair of the first input 000 that gives identical output:

$000 \oplus\xi= 000 \oplus 101= 101$. So the first input 000 and the sixth input 101 will have the same output (see table 8.8, the output is 111).

Let us find the pair of second input 001 that gives identical output:

$001 \oplus \xi = 001 \oplus 101 = 100$. So the second input 001 and the fifth input 100 will have the same output (see table 8.8, the output is 000).

Let us find the pair of third input 010 that gives identical output:

$010 \oplus \xi = 010 \oplus 101 = 111$. So the third input 010 and the eighth input 111 will have the same output (see table 8.8, the output is 110).

Table 8.8. Problem of finding secret message string in a three-bit input output problem with a two-to-one function

Problem		Solution	
3 bit Input x	3 bit Output $f(x)$	3 bit Input x	3 bit Output observed $f(x)$
000	111	000	111
001	000	001	000
010	110	010	110
011	010	011	010
100	000	100	000
		101	111
Find secret message ξ		Secret message $\xi = 101$	110
		110	010
		111	110
		(8 strings)	

Let us find the pair of fourth input 011 that gives identical output:
011 $\oplus$ ξ = 011 $\oplus$ 101 = 110. So the fourth input 011 and the seventh
input 110 will have the same output (see table 8.8, the output is 010).

We thus have identified all the input pairs that give the same output

$$(000,101), \quad (001,100), \quad (010,111), \quad (011,110).$$

We note the following:

It is a problem with $n = 3$ number of input bits (denoted by x) and $n = 3$
number of output bits. The domain size is $2^n = 2^3 = 8$ strings. As the
function is two-to-one, $2^{n-1} = 2^{3-1} = 4$ inputs (i.e. half the inputs) will
produce unique outputs. At most we have to try $2^{n-1} + 1 = 2^{3-1} + 1 = 5$
inputs (i.e. one extra input than half) to obtain the first pair of inputs that
gives identical output. (In this problem we observed the inputs
000, 001, 010, 011, 100, i.e. five inputs to identify the first pair (001,100)
that gives identical output.

(iii) Clearly the run time of classical algorithm is $2^{n-1} + 1 \sim 2^n$, i.e. exponential. We can do better if we use quantum algorithm to obtain a linear run
time $\sim n$ (instead of the exponential run time of 2^n).

- **Randomised algorithm**

The classical computation involves computation of $f(x)$ for various x
sequentially and compare whether there is a coincidence or match with any
previous computation. In the case of match $f(x) = f(y)$ for any y we
compute $x \oplus y$ to evaluate ξ.

For a large number of inputs, finding a match may be difficult and we can
proceed using a random choice of inputs.

Suppose we randomly choose x and find $f(x)$. Then choose y and find
$f(y)$. If $f(x) = f(y)$ then the unknown string is given by $x \oplus y = \xi$ and the
problem is solved. But if $f(x) \neq f(y)$ the problem is not solved and we have
to proceed further. The classical probability that we get $f(x) = f(y)$ is given
by $\dfrac{1}{2^n - 1}$ (which is very small for n large) since $f(x)$ could have matched
(apart from itself) with $2^n - 1$ possible results or strings.

Suppose the first k strings did not yield a match. So the number of strings
checked for coincidence, taking two at a time, is $^kC_2 = \dfrac{k(k-1)}{2}$ which are to be
ruled out as they did not produce a match. Let us now pick up the $(k+1)$th
string and find the probability P_k that it matches any of the previous k strings.

$$P_k = \frac{k}{(2^n - 1) - {}^kC_2} = \frac{k}{(2^n - 1) - \dfrac{k(k-1)}{2}}$$

$$= \frac{2k}{2^{n+1} - 2 - k^2 + k} \leqslant \frac{2k}{2^{n+1} - k^2} \quad \begin{pmatrix} \text{reducing denominator} \\ \text{makes RHS larger} \end{pmatrix}$$

This holds for every attempt. Probability P that there will be a match in the first m attempts (i.e. success in m attempts) is obtained by summing the probabilities of success from $k = 2$ to $k = m$. Hence

$$P = \sum_{k=2}^{m} \frac{2k}{2^{n+1} - k^2} \leqslant \sum_{k=2}^{m} \frac{2m}{2^{n+1} - k^2} \leqslant m \frac{2m}{2^{n+1} - m^2} = \frac{2m^2}{2^{n+1} - m^2}$$

A reasonable estimate for a chance of success is to consider

$$\frac{2m^2}{2^{n+1} - m^2} \geqslant \frac{3}{4} \text{ or } m \geqslant \sqrt{\frac{6}{11} 2^n} \sim 2^{n/2} (\Leftarrow \text{exponential run time})$$

So number of attempts is very large, implying huge complexity for the classical problem. In other words, the classical algorithm requires an exponential number of queries.

- Let us proceed with the quantum algorithm called Simon algorithm accomplished by the circuit shown in figure 8.6.

Start with two quantum registers. The first register contains n qubits denoted by $|0>^{\otimes n}$. The second register is also a null register of n qubits denoted by $|0>^{\otimes n} = |y>$. So the input state is

$$|\psi_0> = |0>^{\otimes n} \otimes |0>^{\otimes n} = |0>^{\otimes n} \otimes |y> \tag{8.15}$$

After passing the n qubits of the first register through n Hadamard gates we get

$$|\psi_1> = [H^{\otimes n} |0>^{\otimes n}] \otimes |y>$$

where

$$H^{\otimes n} |0>^{\otimes n} = \frac{1}{\sqrt{2^n}} \sum_{x=0}^{2^n-1} |x> = |X> \tag{8.16}$$

The input to the oracle is therefore

$$|\psi_1> = |X> \otimes |0>^{\otimes n} = |X> \otimes |y> \tag{8.17}$$

The oracle U_f acts on $|\psi_1>$ leading to $|\psi_2>$.

$$|\psi_2> = U_f |\psi_1> = U_f |X> \otimes |y> \tag{8.18}$$

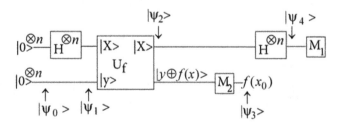

Figure 8.6. Circuit for Simon algorithm.

As $|y> =|0>^{\otimes n}$ is a target register of null strings, U_f operating on it gives

$$U_f|y> =|y \oplus f(x) > =|0 \oplus f(x) > =|f(x)>$$ (8.19)

since $0 \oplus f(x) = f(x)$

From equation (8.18) we have

$$|\psi_2 > =U_f|X > \otimes|y > =|X > \otimes U_f|y > =|X > \otimes |f(x)>$$ (8.20)

At this stage suppose a query or a measurement of second register denoted by M_2 (figure 8.6) yields a particular value $f(x_0)$. (So after measurement the second register will not be in superposition but it is in a pure state.)

If the function had been one-to-one fuction, each value of $f(x_0)$ would have been measured with probability $\frac{1}{2^n}$.

If the function is a two-to-one function, the probability of measuring a particular value of $f(x_0)$ is $\frac{1}{2^n-1}$, as for a pair of values, say x_0 and $x_0 + \xi$ the measured value is the same $f(x_0)$ in the second register.

Since our promise was $f(x_0) = f(x_0 \oplus \xi)$ it follows that the first register will not have superposition of all the possible inputs but will have superposition or linear combination of inputs that can produce $f(x_0)$. So the first register would contain a linear combination of $|x_0>$ and $|x_0 \oplus \xi>$ (x_0 being an input) viz. $\frac{1}{\sqrt{2}}(|x_0 > +|x_0 \oplus \xi>)$. ($\frac{1}{\sqrt{2}}$ is the normalization factor, implying that on measuring the first register $|x_0>$ and $|x_0 + \xi>$ are obtained with equal probability). Hence we have

$$|\psi_3 > =\frac{1}{\sqrt{2}}(|x_0 > +|x_0 + \xi>) \otimes |f(x_0)>$$ (8.21)

Now Hadamard transformation is applied to the first register (without affecting the second register)

$$|\psi_4 > =[H^{\otimes n}\frac{1}{\sqrt{2}}(|x_0 > +|x_0 + \xi>)] \otimes |f(x_0)>$$ (8.22)

Again,

$$H^{\otimes n}\frac{1}{\sqrt{2}}(|x_0 > +|x_0 + \xi>) = \frac{1}{\sqrt{2}}H^{\otimes n}|x_0 > +\frac{1}{\sqrt{2}}H^{\otimes n}|x_0 + \xi>$$

Using $H^{\otimes n}|x > =\sum_k \frac{1}{\sqrt{2^n}}(-1)^{x.k}|k>$ (equation (8.13)) we have

$$H^{\otimes n}\frac{1}{\sqrt{2}}(|x_0 > +|x_0 + \xi>)$$

$$=\frac{1}{\sqrt{2}}\sum_k \frac{1}{\sqrt{2^n}}(-1)^{x_0.k}|k > +\frac{1}{\sqrt{2}}\sum_k \frac{1}{\sqrt{2^n}}(-1)^{(x_0+\xi).k}|k>$$ (8.23)

$$=\frac{1}{\sqrt{2^{n+1}}}\sum_k(-1)^{x_0.k}[1 + (-1)^{\xi.k}]|k>$$

where $\xi. k = \xi_0 k_0 \oplus \xi_1 k_1 \oplus \ldots \oplus \xi_{n-1} k_{n-1} =$ sum of bitwise product

The output state is thus obtained from equation (8.22) using equation (8.23) to be

$$|\psi_4> = \frac{1}{\sqrt{2^{n+1}}} \sum_k (-1)^{x_0.k}[1 + (-1)^{\xi.k}]|k> \otimes |f(x_0)> \qquad (8.24)$$

Two cases arise.

Case (i): we note that for $\xi. k = 1, [1 + (-1)^{\xi.k}] = 1-1=0$. So amplitude of $|k>$ is zero. The probability that $\xi. k = 1$ is zero. Case (i) cannot occur.

Case (ii): we note that for $\xi. k = 0, [1 + (-1)^{\xi.k}] = 1+1=2 \neq 0$. This can occur.

As Case 2, corresponding to $\xi. k = 0$ occurs at the end of the circuit and this means that the string ξ and the strings k are perpendicular to each other. At the end of the circuit we get equal superposition of all (2^{n-1}) possible values of k.

We run the circuit multiple times and then collect $n - 1$ linearly independent values of k. Using the system of linear equations we can solve $\xi. k = 0$ for ξ. In other words n queries or measurements denoted by M_1 (figure 8.6) are needed that yields $k_0, k_1, , , , k_n$ satisfying $\xi_i. k_i = 0$ with $i = 0,1,\ldots n$ that leads to the unknown string ξ.

- **Example.**

Referring to table 8.9 we give an outline of solving a four-bit problem using the Simon algorithm.

We have two registers of bit size $n = 4$. Their tensor product gives $|\psi_0>$ (from equation (8.15)) (figure 8.6)

$$|\psi_0> = |0>^{\otimes n} \otimes |0>^{\otimes n} = |0>^{\otimes 4} \otimes |0>^{\otimes 4}$$

$$= |0000> \otimes |0000> = |0000> \otimes |y>$$

The Hadamard gate on first register (keeping second register intact) creates equal superposition of all possible strings of length 4 in the first register. So from equation (8.17)

$$|\psi_1> = [H^{\otimes 4} \otimes |0000>] \otimes |0000> = |X> \otimes |0000> = |X> \otimes |y>$$

where we used equation (8.16) viz.

Table 8.9. Application of Simon algorithm to $n = 4$ bit problem

Inputs x		Outputs $f(x)$
0000 , 1001		1111
0001 , 1000		0001
0010 , 1011		1110
0011 , 1010	Find ξ	1101
0100 , 1101		0000
0101 , 1100		0101
0110 , 1111		1010
0111 , 1110		1001

$$H^{\otimes 4} \otimes |0000> = \frac{1}{\sqrt{2^4}} \sum_{x=0}^{2^4-1} |x> = |X>$$

$$= \frac{1}{4}(|0000> +|0001> +|0010> + \ldots\ldots\ldots + |1110> +|1111 >)$$

($2^4 = 16$ input strings)

The oracle U_f acts on $|\psi_1>$ (equation (8.18)) to give

$$|\psi_2 > = U_f|\psi_1 > = U_f|X> \otimes|y > = |X> \otimes \ U_f|y>$$

Using equation (8.19)

$$U_f|y > = |y \oplus f(x) > = |0 \oplus f(x) > = |f(x)>$$

Hence we have (equation (8.20))

$$|\psi_2 > = |X> \otimes \ U_f|y > = |X> \otimes|f(x)>$$

From table 8.9 we expand and write, noting the content of first register x and second register $f(x)$, e.g. for $x = 0000$, $f(x) = 1111$ $x = 0001, f(x) = 0001$ etc. So

$$|\psi_2 > = \frac{1}{4}[|0000 > |1111 > +|0001 > |0001 >$$

$$+|0010 > |1110 > +\ldots. + |1111 > |1010 >]$$

Suppose measurement on second register yields a particular value $f(x_0) = 1010$. After measurement, the first register is no more in superposition of all inputs but it will have only the inputs that produce the output $f(x_0) = 1010$. From table 8.9 this output $|f(x_0) > = |1010>$ corresponds to the inputs 0110 and 1111. From equation (8.21) we thus have

$$|\psi_3 > = \frac{1}{\sqrt{2}}(|0110 > + |1111 >) \otimes |1010 >)$$

Now apply the Hadamard on the first register without affecting the second register (equation (8.22))

$$|\psi_4 > = [H^{\otimes 4}\frac{1}{\sqrt{2}}(|0110 > +|1111 >)] \ \otimes |1010>$$

$$= \left(\frac{1}{\sqrt{2}} \left[\frac{|0 > +|1>}{\sqrt{2}} \frac{|0 > - |1>}{\sqrt{2}} \frac{|0 > - |1>}{\sqrt{2}} \frac{|0 > +|1>}{\sqrt{2}} \right] \right.$$

$$\left. + \frac{1}{\sqrt{2}} \left[\frac{|0 > - |1>}{\sqrt{2}} \frac{|0 > - |1>}{\sqrt{2}} \frac{|0 > - |1>}{\sqrt{2}} \frac{|0 > - |1>}{\sqrt{2}} \right] \right) \otimes |1010>$$

Writing without ket symbol for convenience of calculation gives

$$|\psi_4> = \frac{1}{\sqrt{2^5}}[(00-01+10-11)(00+01-10-11) + (00-01-10+11)(00-01-10+11)]$$

$$= \frac{1}{\sqrt{2^5}}[\; (\overline{0000}+0001-\overline{0010}-0011)+(-\overline{0100}-0101+\overline{0110}+0111)$$
$$(1000+\overline{1001}-1010-\overline{1011})+(-1100-\overline{1101}+1110+\overline{1111})$$
$$+ \; (\overline{0000}-0001-\overline{0010}+0011)+(-\overline{0100}+0101+\overline{0110}-0111)$$
$$(-1000+\overline{1001}+1010-\overline{1011})+(1100-\overline{1101}-1110+\overline{1111}) \;]$$

$$= \frac{2}{\sqrt{2^5}}[(0000-0010-0100+0110+1001-1011-1101+1111) \;]$$

Writing within ket we have the result to be an equal superposition of $2^{4-1} = 8$ possible k values

$$|\psi_4> = \frac{1}{\sqrt{2^3}}[\; |0000> - |0010> - |0100> +|0110>$$

$$+|1001> - |1011> - |1101> +|1111>] \quad \otimes \quad |1010>$$

Each $|k>$ occurs with probability $\frac{1}{8}$. The unknown ξ string and the measured k string have to be perpendicular, i.e. $\xi_i k_i = 0$ (see note for verification).

The circuit is run multiple times to get $n - 1 = 4-1 = 3$ linearly independent values of k.

Suppose we measure $|k>$ to get, say $|k > \rightarrow |0000>$. We cannot work with it since a null vector is not linearly independent. So we run the circuit again and suppose we now measure $|k > \rightarrow |0010 >$, $|0100>$ and re-run the circuit to get the third $|k > \rightarrow |0110>$. We cannot proceed with $|0110>$ since it is linearly dependent on $|0010>$ and $|0100>$ as can be verified from $0010+0100=0110$. To get the third linearly independent $|k>$ we re-run the circuit and suppose we now get $|k > \rightarrow |1001>$. We thus have measured three linearly independent values of $|k>$ $\Rightarrow |0010 >$, $|0100 >$, $|1001>$ in our effort to find $|\xi>$ and now we have to construct the equation $\xi_i . k_i = 0$ or $.k_i \xi_i = 0$ using the bits and then solve.

Construct the matrix equation

$$.k_i \xi_i = 0 \Rightarrow \quad \begin{pmatrix} 0 & 0 & 1 & 0 \\ 0 & 1 & 0 & 0 \\ 1 & 0 & 0 & 1 \\ 0 & 0 & 0 & 0 \end{pmatrix} \begin{pmatrix} \xi_1 \\ \xi_2 \\ \xi_3 \\ \xi_4 \end{pmatrix} = 0$$

where the bits of $|k> \Rightarrow |0010 >$, $|0100 >$, $|1001>$ form the first three rows and the last row is the null ket $|0000>$. The column matrix is formed by $|\xi > = |\xi_1\xi_2\xi_3\xi_4>$. The solution gives values of $\xi_1, \xi_2, \xi_3, \xi_4$ leading to the secret message or string $|\xi > = |\xi_1\xi_2\xi_3\xi_4 > = |1001>$.

While the classical algorithm solves the problem in $2^{n/2}$ (= exponential run time) queries, the Simon's algorithm solves it in n queries (n being the number of qubits), i.e. in a polynomial run time.

- **NOTE**

Verify that $\xi_i k_i = 0$ holds for any $|k>$ in $|\psi_4>$.

✓ We got $|\xi> = |1001>$. Let us consider $|k> \rightarrow |0000>$ of $|\psi_4>$ and find the bitwise product $\xi_i k_i$ in mod 2.

$$\xi_i k_i \rightarrow |1001> |0000> = 1.0+0.0+0.0+1.0=0+0+0+0=0$$

✓ We got $|\xi> = |1001>$. Let us consider $|k> \rightarrow |1111>$ of $|\psi_4>$ and find the bitwise product $\xi_i k_i$ in mod 2.

$$\xi_i k_i \rightarrow |1001> |1111> = 1.1+0.1+0.1+1.1=1+0+0+1=2 \bmod 2=0$$

8.9 Grover's search algorithm

Consider an ordered list prepared with names of persons (aranged in some order) and their corresponding phone numbers appearing adjacent to their names. To search for a phone number of a known person from this list is easy but to search the name of an unknown person (if given the phone number) is extremely difficult. In other words searching something from a randomized (or unsorted) collection is difficult.

Grover's algorithm is intended to make a database search, i.e. to find a particular element in an unstructured or unsorted database of $N = 2^n$ elements, there being no order. For instance, finding the name of a person, whose telephone number is given from an alphabetically arranged telephone directory with N entries needs $O(N)$ searches or computational steps. Since success may be achieved in the first trial or at the worst in the last Nth trial we can say that such tasks require $O(\frac{N}{2})$ searches (O means 'of the order of'). Clearly for large N the search becomes increasingly difficult. So classical search requires $O(N)$ queries. Using a structured database, the number of queries is reduced. But drastic reduction in run time occurs with Grover's algorithm.

Grover's algorithm does the job in $O(\sqrt{N})$ searches. So the search speed rises quadratically. The circuit for Grover's algorithm is shown in figure 8.7.

Consider a $N = 2^n$ qubit data with n qubit inputs. $f: \{0,1\}^{\otimes n} \rightarrow \{0,1\}$. Applying $|0>$ in each of $|n>$ input registers and passing them through Hadamard gates we get $|X>$ as the input (n qubit states) of the oracle. The target register is $|y> = \frac{1}{\sqrt{2}}(|0> - |1>)$ (obtained after passing target qubit $|1>$ through a Hadamard gate).

Input to the oracle is $|X> \otimes |y>$

The oracle calculates a function $f_{w(x)}$ (with $0 \leqslant x \leqslant 2^n - 1$) for n qubit input strings and returns 1 when the input matches a particular string $x = w$ (called the marked string)—called success, and returns zero in all other cases $x \neq w$—called failure. So

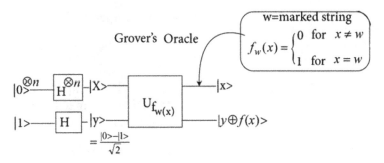

Figure 8.7. Circuit for Grover's algorithm.

$$f_x(x) = \begin{cases} 0 \text{ for } x \neq w \\ 1 \text{ for } x = w \end{cases}$$

The output of the oracle is

$$U_{f_{w(x)}}|X> \otimes |y> = |x> \otimes U_{f_{w(x)}}|y>$$

The oracle changes $|y>$ to $|y \oplus f_{w(x)}>$. So

$$U_{f_{w(x)}}|X> \otimes |y> = |x> \otimes |y \oplus f_{w(x)}> \tag{8.25}$$

For $f_{w(x)} = 0$, $|y \oplus 0> = |y> = \frac{1}{\sqrt{2}}(|0> - |1>)$

$$f_{w(x)} = 1, |y \oplus 1> = |\bar{y}> = \frac{1}{\sqrt{2}}(|1> - |0>) = (-)\frac{1}{\sqrt{2}}(|0> - |1>)$$

Combing the above we can write

$$|y \oplus f_{w(x)}> = (-)^{f_{w(x)}}\frac{1}{\sqrt{2}}(|0> - |1>)$$

Hence from equation (8.25) the oracle output is

$$U_{f_{w(x)}}|X> \otimes |y> = |x> \otimes (-)^{f_{w(x)}}\frac{1}{\sqrt{2}}(|0> - |1>)$$

$$= (-)^{f_{w(x)}}|x> \otimes \frac{1}{\sqrt{2}}(|0> - |1>) \tag{8.26}$$

From equation (8.26) it follows that the first register is flipped in sign (if $f_{w(x)} = 1$) while the second register is left unchanged. Clearly the sign of the state in the first register depends on the function $f_{w(x)}$. We thus focus attention to the state of the first register.

Define a unitary operator corresponding to $|w>$ (which is the state of the marked item $x = w$) which is a member of the orthonormalized input computational basis states $|x>$, i.e. orthogonal to remaining items in the database.

$$U_w = I - 2|w> <w| \quad [I \text{ is identity operator}]$$

$$U_w|w> = |w> - 2|w> <w|w> = |w> - 2|w> = -|w>.$$

So U_w flips sign of $|w>$.

Clearly $U_w = I - 2|w> <w|$ is a unitary operator that is capable of flipping the sign of the first register if $x = w$ but if input string does not match with the marked string w (i.e. $x \neq w$) then the sign of the first register is unaltered.

Define a reflection operator $U_s = 2|s> <s| - I$ where $|s> = \frac{1}{\sqrt{N}} \sum_{x=0}^{N-1} |x>$ is a standard state which is a linear combination of 2^n input computational basis states $|x>$ of which $|w>$ is also a member.

Clearly $<w|s> = <w| \frac{1}{\sqrt{N}} \sum_{x=0}^{N-1} |x> = \frac{1}{\sqrt{N}} \sum_{x=0}^{N-1} <w|x>$

$= \frac{1}{\sqrt{N}} \sum_{x=0}^{N-1} \delta_{xw} = \frac{1}{\sqrt{N}} =$ weight of each component state in the original linear combination (or equal superposition state).

Any state $|\psi>$ can be resolved into a component $|\psi_{\parallel}>$ along the standard state $|s>$ and $|\psi_{\perp}>$ perpendicular to the standard state $|s>$ as $|\psi> = |\psi_{\parallel}> + |\psi_{\perp}>$ (so that $<s|\psi_{\parallel}> \neq 0$, $<s|\psi_{\perp}> = 0$).

Now

$U_s|\psi_{\parallel}> = (2|s> <s| - I)|\psi_{\parallel}> = 2|s> <s|\psi_{\parallel}> - |\psi_{\parallel}>$ (this is along $|s>$)

$\Rightarrow$ component of $|\psi>$ parallel to $|s>$ remains unchanged under U_s.

$$U_s|\psi_{\perp}> = (2|s> <s| - I)|\psi_{\perp}> = 2|s> <s|\psi_{\perp}> - |\psi_{\perp}> = -|\psi_{\perp}>$$

since $<s|\psi_{\perp}> = 0$. Clearly U_s flips $|\psi_{\perp}>$

$\Rightarrow$ component of $|\psi>$ perpendicular to $|s>$ will change sign under U_s.

Grover rotation operator is defined as

$$R_G = U_s U_w.$$

Refer to figure 8.8 which gives a geometrical interpretation of the Grover operator R_G. It will be evident that rotation provides a basis for Grover's search.

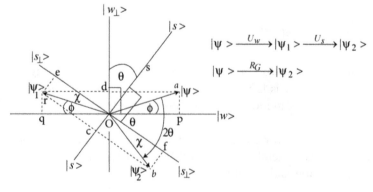

Figure 8.8. Geometrical interpretation of Grover operator R_G.

$|\psi>$ is an arbitrary ket (along Oa) making some angle ϕ with $|w>(\angle aOp = \phi)$, ($|w>$= marked state). Let us resolve $|\psi>$ along $|w>$ (i.e. along Op) and perpendicular to $|w>$, i.e. $|w_\perp>$ (i.e. along Od), $|\psi>$ being the resultant of the parallelogram $Opad$ with sides Op and Od—these sides being at right angles.

Now $U_w|w> = -|w>$, i.e. U_w flips sign of $|w>$, i.e. Op is flipped to Oq and $|\psi>$ is transformed to $|\psi_1>$ (along Or and so $\angle rOq = \angle aOp = \phi$). We thus have $U_w|\psi> = |\psi_1>$.

Let $|s>$ be the standard state along Os and $\theta = \angle sOd$ be the angle between $|w_\perp>$ and $|s>$. Resolve $|\psi_1>$ along $|s>$ (the projection being Oc) and perpendicular to $|s>$ (i.e. along $|s_\perp>$, the projection being Oe). So $|\psi_1>$ is the resultant of the parallelogram $Oerc$ with sides Oc and Oe, these sides being at right angles.

Since U_s flips sign of perpendicular component of $|\psi>$ (as per the relation $U_s|\psi_\perp> = -|\psi_\perp>$) Oe is flipped to Of along the $|s_\perp>$ line. This generates $|\psi_2>$ (along Ob and so $\angle fOb = \angle eOr = \chi$ say) which is the resultant of the parallelogram $Ofbc$ with sides Oc and Of. We thus have

$$R_G|\psi> = U_s U_w|\psi> = |\psi_2>. \qquad (|\psi> \xrightarrow{U_w} |\psi_1> \xrightarrow{U_s} |\psi_2>)$$

In other words Grover's operator R_G has rotated $|\psi>$ by angle $\angle aOb$. Let us evaluate $\angle aOb$.

$$\angle aOb = \angle aOp + \angle pOf + \angle fOb = \phi + \angle pOf + \chi$$

As $\angle dOp = \angle sOf = 90° \Rightarrow \angle pOf = \angle sOd = \theta$
Again $\angle pOf = \angle qOe = \chi + \phi \Rightarrow \chi + \phi = \theta$.
Hence $\angle aOb = \phi + \angle pOf + \chi = \phi + \chi + \angle pOf = \theta + \theta = 2\theta$. Clearly Grover operator R_G rotates an arbitrary state $|\psi>$ by 2θ where the angle between $|w>$ and $|s>$ is $\frac{\pi}{2} - \theta = \angle sOp$. This Grover rotation (of 2θ) is identified with initial probability amplitude as evident from the following example. Through Grover rotations the amplitude of $|w>$ is selectively amplified.

Consider the case $N = 4 = 2^2$. The initial equal superposition state is

$$|\psi> = \frac{1}{\sqrt{4}}[|00> + |01> + |10> + |11>]$$

Suppose there is one search string in the four states, i.e. one state is to be found out from an unsorted database of four elements. So there are three $|x \neq w>$ states and one $|x = w>$ marked state and we rewrite

$$|\psi> = \sqrt{\frac{3}{4}} \sum_{x \neq w} |x> + \sqrt{\frac{1}{4}} |x>|_{x=w} = \cos\theta \sum_{x \neq w} |x> + \sin\theta|x>|_{x=w}$$

where we have put $\sin\theta = \sqrt{\frac{1}{4}} = \frac{1}{2} \Rightarrow \theta = 30°$. So the angle between $|s>$ and $|w>$ is $\frac{\pi}{2} - \theta = 90° - 30° = 60°$.

Single rotation will rotate $|s>$ by $2\theta = 2(30°) = 60°$ and $|s>$ aligns with $|w>$. Obviously one Grover rotation or query (and one is less than $\sqrt{N} = \sqrt{2}$) is needed to search and align with $|w>$ (the marked state) starting from a standard state. (Classically, the search would need three queries.)

For the case $N = 8 = 2^3$ we have $\sin\theta = \sqrt{\frac{1}{8}} = \frac{1}{2\sqrt{2}} \Rightarrow \theta = 20.7°$ and the angle between $|s>$ and $|w>$ is $\frac{\pi}{2} - \theta = 90° - 20.7° = 69.3°$. The first Grover rotation by $2\theta = 2(20.7)=41.4°$ makes the angle between $|s>$ and $|w>$ reduce to $69.3° - 2(20.7°) = 27.9°$. A second rotation makes this angle $69.3° - 2(41.4°) = -13.5°$ which is closer to the marked state $|w>$. A third iteration takes the search to $69.3° - 3(41.4°) = -54.9°$ which is further away from the intended marked state $|w>$. It seems that we should *a priori* know the number of rotations needed for the purpose. Actually, an increase of N is associated with smaller rotation angles and so we have a greater control in the process of approaching the marked state. So for large N we can have a better and much more efficient control. It is clear that every Grover rotation will make the arbitrary state come closer to marked state $|w>$ by 2θ but, in the process if we cross over $|w>$ then rotations will take us further and further away.

✓ Number of iterations needed

Let us consider an arbitrary state $|\psi> = \sum_x a_x|x>$ expanded in the computational basis having an equal linear combination (a_x= amplitude in the state $|x>$) while the standard state is $|s> = \frac{1}{\sqrt{N}}\sum_x |x>$. Then

$$< s|\psi> = \frac{1}{\sqrt{N}}\sum_x <x|\sum_{x'} a_{x'}|x'> = \frac{1}{\sqrt{N}}\sum_x \sum_{x'} a_{x'} < x|x'>$$

$$= \frac{1}{\sqrt{N}}\sum_x \sum_{x'} a_{x'}\delta_{xx'} = \frac{1}{\sqrt{N}}\sum_x a_x$$

Defining $\bar{a} = \frac{1}{N}\sum_x a_x$ i.e. $\sum_x a_x = N\bar{a}$ i.e. we have

$$<s|\psi> = \frac{1}{\sqrt{N}}N\bar{a} = \sqrt{N}\bar{a}$$

where $\bar{a}$ is the mean amplitude of $|\psi>$ in the computational basis.

Let us find the effect of U_s on $|\psi>$.

$$U_s|\psi> = (2|s> <s| - I)|\psi> = 2|s> <s|\psi> - |\psi>$$

$$= 2|s> (\sqrt{N}\bar{a}) - |\psi>$$

$$U_s|\psi> = 2\sqrt{N}\bar{a}\frac{1}{\sqrt{N}}\sum_x |x> - \sum_x a_x|x> = \sum_x (2\bar{a} - a_x)|x>$$

$|\psi> = \sum_x a_x|x>$ thus becomes $U_s|\psi> = \sum_x (2\bar{a} - a_x)|x>$, i.e. amplitude changes from a_x to $2\bar{a} - a_x$ on application of U_s.

Suppose we measure the amplitude w.r.t the mean. Initial amplitude w.r.t the mean is $(a_x - \bar{a})$. The final amplitude w.r.t the mean is $(2\bar{a} - a_x) - a = a - a_x$. It is clear that the amplitude $(a_x - \bar{a})$, on application of U_s, becomes $(\bar{a} - a_x)$, i.e. the amplitude w.r.t mean gets inverted or flipped.

For illustration purpose, let us suppose $N = 8$. In $|s>$ each base state has amplitude $\frac{1}{\sqrt{8}}$ and $|w>$ is one of them also having amplitude $\frac{1}{\sqrt{8}}$. Now U_w inverts the amplitude of $|w>$ only (as per the relation $U_w|w> = -|w>$). So on application by U_w the amplitude of $|w>$ becomes $-\frac{1}{\sqrt{8}}$ (i.e. flipped). Hence mean of the eight kets becomes $\frac{1}{8}[7\frac{1}{\sqrt{8}} - 1.\frac{1}{\sqrt{8}}] = \frac{1}{8}6\frac{1}{\sqrt{8}} = \frac{3}{8\sqrt{2}}$.

Under application of U_s amplitude becomes $(2\bar{a} - a_x)$. So the mean becomes as follows:

For $|x \neq w>$ (unmarked states): $2\bar{a} - a_x = 2.\frac{3}{8\sqrt{2}} - (\frac{1}{\sqrt{8}}) = \frac{3}{4\sqrt{2}} - \frac{1}{2\sqrt{2}} = \frac{1}{4\sqrt{2}}$

For $|x = w>$ (marked state) : $2\bar{a} - a_x = 2.\frac{3}{8\sqrt{2}} - (-\frac{1}{\sqrt{8}}) = \frac{3}{4\sqrt{2}} + \frac{1}{2\sqrt{2}} = \frac{5}{4\sqrt{2}} = 5\frac{1}{4\sqrt{2}}$

Obviously, marked state amplitude is increased fivefold and so the probability is enhanced 25 times w.r.t each of the unmarked states (figure 8.9). This is called selective amplification of the amplitude of $|w>$.

Number of iterations or Grover rotations (which is of amount 2θ) needed (to move from $|s>$ to $|w>$, i.e. to move an angle $\frac{\pi}{2} - \theta$ is say m. Then

$$m(2\theta) \approx \frac{\pi}{2} - \theta \Rightarrow m = \frac{\pi}{4\theta} - \frac{1}{2}$$

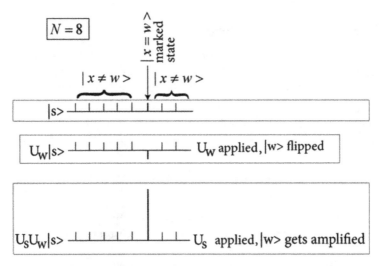

Figure 8.9. Selective amplification of marked state $|w>$.

We can get an estimate of m for large N by setting $\sin\theta = \frac{1}{\sqrt{N}}$

For large N, $\sin\theta$ is small and we approximate $\sin\theta \approx \theta$, so that $\theta = \frac{1}{\sqrt{N}}$. Hence

$$m = \frac{\pi}{4\theta} - \frac{1}{2} \approx \frac{\pi}{4}\sqrt{N} - \frac{1}{2} \approx \frac{\pi}{4}\sqrt{N} \sim O(\sqrt{N})$$

The number of iterations after which the standard state will come to the search string marked state $|w\rangle$ as close as possible is $O(\sqrt{N})$ for very large N. This shows that the number of queries is $O(\sqrt{N})$. This result is referred to as showing quadratic acceleration.

✓ If we make m iterations the angle between $|s\rangle$ to $|w\rangle$ will be

$$(\frac{\pi}{2} - \theta) - m(2\theta) = \frac{\pi}{2} - (2m + 1)\theta.$$

The amplitude of the state $|w\rangle$ in $|s\rangle$ is

$$(\text{recall: } |\psi\rangle \sim \cos(2m + 1)\theta \sum_{x \neq w} |x\rangle + \sin(2m + 1)\theta|x = w\rangle)$$

$|\sin(2m + 1)\theta| = |\sin(2.\frac{\pi\sqrt{N}}{4} + 1)\frac{1}{\sqrt{N}}|$

(putting $m \approx \frac{\pi}{4}\sqrt{N}, \theta = \frac{1}{\sqrt{N}}$, N=large)

$= |\sin(\frac{\pi}{2} + \frac{1}{\sqrt{N}})| = \cos\frac{1}{\sqrt{N}} \approx 1 - \frac{(\frac{1}{\sqrt{N}})^2}{2!} = 1 - \frac{1}{2N}$ for large N.

The amplitude is very close to 1. This shows the quadratic acceleration of the algorithm.

Probability of success of Grover's algorithm (to identify the marked state, i.e. to locate the search string) after optimal number (m) of iterations, for large (N) is given by

$|1 - \frac{1}{2N}|^2 = 1 - 2.1.\frac{1}{2N} + (\frac{1}{2N})^2 \approx 1 - \frac{1}{N}$ (as N is large)

8.10 Discrete integral transform

Integral transforms such as Laplace transform, Fourier transform are defined over continuous variable. These are used to simplify complicated mathematical problems.

In quantum computing we will rely on discrete variables. So we will deal with discrete integral transform that deals with a discrete function.

Let n belong to a set of natural numbers. Consider a set of $N = 2^n$ integers viz. $0,1,2, \ldots 2^n - 1$. Define the discrete integral transform $F(y)$ of the function $f(x)$ as

$$F(y) = \sum_{x=0}^{N-1} K(y, x)f(x) \tag{8.27}$$

where $K(y, x)$ is the Kernel, which is a bivariate function in general complex. Clearly x, y can assume any of the $0,1,2, \ldots 2^n - 1$ discrete values but $K(x, y)$ can be a continuous function.

As x, y are discrete, we can think of equation (8.27) as a matrix equation with F, f as $N \times 1$ column vectors and $K(y, x)$ representing $N \times N$ matrix.

Demanding that the Kernel $K(y, x)$ is invertible, in particular unitary $K^\dagger = K^{-1}$ an inverse transformation exists. Let us construct

$$\sum_{y=0}^{N-1} K^\dagger(x, y)F(y)$$

$$= \sum_{y=0}^{N-1} K^\dagger(x, y)[\sum_{z=0}^{N-1} K(y, z)f(z)] \text{using (8.27)}$$

$$= \sum_{z=0}^{N-1}\sum_{y=0}^{N-1} K^\dagger(x, y)K(y, z)f(z) = \sum_{z=0}^{N-1} \delta_{xz}f(z) = f(x)$$

Hence the inverse transform (to equation 8.27) is

$$f(x) = \sum_{y=0}^{N-1} K^\dagger(x, y)F(y) \tag{8.28}$$

We now extend this concept from a set of numbers to n qubit Hilbert space. Construct the n qubit ket vector

$|x> = |x_{n-1}x_{n-2}.....x_0>$ where $x_i = 0,1$.

So we have for the unitary operator U

$$U|x> = I \; U|x> = \sum_{y=0}^{N-1} |y> <y| \; U|x> = \sum_{y=0}^{N-1} U(y, x)|y> \tag{8.29}$$

using the completeness relation given by $I = \sum_{y=0}^{N-1} |y> <y|$

$U(y, x) = <y|U|x>$ is the matrix element of the unitary operator U.

Comparing $U|x> = \sum_{y=0}^{N-1} U(y, x)|y>$ (equation (8.29)) with $F(y) = \sum_{x=0}^{N-1} K(y, x)f(x)$

(equation (8.27)) let us construct a unitary matrix U such that

$$U|x> = \sum_{y=0}^{N-1} K(y, x)|y> .$$

We then can say that the unitary operator U computes the discrete integral transform.

Consider a superposition state $\sum_{x=0}^{N-1} f(x)|x>$. Action of U on this superposition state is as follows

$$U\sum_{x=0}^{N-1} f(x)|x> = \sum_{x=0}^{N-1} f(x)U|x> = \sum_{x=0}^{N-1} f(x)\sum_{y=0}^{N-1} K(y, x)|y>$$

$$= \sum_{y=0}^{N-1}\sum_{x=0}^{N-1} K(y, x)f(x)|y> = \sum_{y=0}^{N-1} F(y)|y>$$

where $F(y)$ is the discrete integral transform of $f(x)$. In other words U maps a state with probability amplitude $f(x)$ to another state with probability amplitude $F(y)$ that is related to $f(x)$ through the Kernel $K(y, x)$. The unitary matrix U computes the discrete integral transform $F(y)$ for all variables y by a single operation if it acts on the superposition (i.e. linear combination) of states $\sum_{x=0}^{N-1} f(x)|x>$. There are exponentially large numbers 2^n of y for a n qubit register and this provides a quantum computer exponentially fast computing power compared to classical computing.

8.11 Quantum Fourier transform

The Nth primitive root of unity is $w_n = e^{2\pi i/N}$, where $N = 2^n$. With it we define a Kernel

$$K(x, y) = \frac{1}{\sqrt{N}} e^{2\pi i xy/N} = \frac{1}{\sqrt{N}} (e^{2\pi i/N})^{xy} = \frac{1}{\sqrt{N}} w_n^{xy}$$

where x, y are usual decimal numbers and xy is usual decimal multiplication (not bitwise multiplication).

The discrete integral transform with the Kernel K is

$$F(y) = \sum_{x=0}^{N-1} K(y, x) f(x) = \sum_{x=0}^{N-1} \frac{1}{\sqrt{N}} w_n^{xy} f(x) = \frac{1}{\sqrt{N}} \sum_{x=0}^{N-1} w_n^{xy} f(x) \tag{8.30}$$

called discrete Fourier transform. The quantum integral transform with this kernel is called quantum Fourier transform (QFT).

Let us check the unitarity of the Kernel. Consider

$$<x|K^\dagger K|y> = <x|K^\dagger \sum_{z=0}^{N-1} |z> <z|K|y> = \sum_{z=0}^{N-1} <x|K^\dagger|z> <z|K|y>$$

since $\sum_{z=0}^{N-1} |z> <z| = I$ as per the completeness condition.

$$<x|K^\dagger K|y> = \sum_{z=0}^{N-1} K^\dagger(x, z) K(z, y) = \sum_{z=0}^{N-1} \frac{1}{\sqrt{N}} w_n^{-xz} \frac{1}{\sqrt{N}} w_n^{zy}$$

since

$$K^\dagger(x, z) = \left(\frac{1}{\sqrt{N}} w_n^{xz}\right)^\dagger = \frac{1}{\sqrt{N}} (w_n^\dagger)^{xz} = \frac{1}{\sqrt{N}} [(e^{2\pi i/N})^\dagger]^{xz} = \frac{1}{\sqrt{N}} (e^{-2\pi i/N})^{xz}$$

$$= \frac{1}{\sqrt{N}} (e^{2\pi i/N})^{-xz} = \frac{1}{\sqrt{N}} w_n^{-xz}$$

$$<x|K^\dagger K|y> = \frac{1}{N} \sum_{z=0}^{N-1} w_n^{z(y-x)} = \frac{1}{N} \sum_{z=0}^{N-1} (e^{2\pi i/N})^{z(y-x)}$$

$$= \frac{1}{N} \sum_{z=0}^{N-1} e^{2\pi i z(y-x)/N}$$

This is a finite geometric sum with common ratio $e^{2\pi i(y-x)/N}$ with $x \neq y$. Hence

$$<x|K^\dagger K|y> = \frac{1}{N}\frac{[e^{2\pi i(y-x)/N}]^N - 1}{e^{2\pi i(y-x)/N} - 1} = \frac{1}{N}\frac{e^{2\pi i(y-x)} - 1}{e^{2\pi i(y-x)/N} - 1}.$$

As the denominator $e^{2\pi i(y-x)/N} - 1 \neq 0$, but numerator $e^{2\pi i(y-x)} - 1 = 0$ we have for $x \neq y$, $<x|K^\dagger K|y> = 0$.

For $x = y$, $<x|K^\dagger K|y> = \frac{1}{N}\sum_{z=0}^{N-1} w_n^{z(y-x)} = \frac{1}{N}\sum_{z=0}^{N-1} 1 = \frac{1}{N}N = 1$.

Thus

$$<x|K^\dagger K|y> = \delta_{xy} = <x|I|y> \delta_{xy} \Rightarrow K^\dagger K = I$$

Example: $n = 1$, $N = 2^1 = 2$, set is (0,1)
$w_n = e^{2\pi i/N}$ becomes $w_1 = e^{2\pi i/2} = e^{\pi i} = -1$
$K(x, y) = \frac{1}{\sqrt{N}}w_n^{xy}$ becomes $K(x, y) = \frac{1}{\sqrt{2}}(e^{\pi i})^{xy} = \frac{1}{\sqrt{2}}(-1)^{xy}$

$$K = \begin{pmatrix} K(0,0) & K(0,1) \\ K(1,0) & K(1,1) \end{pmatrix} = \begin{pmatrix} \frac{1}{\sqrt{2}}(-1)^{0.0} & \frac{1}{\sqrt{2}}(-1)^{0.1} \\ \frac{1}{\sqrt{2}}(-1)^{1.0} & \frac{1}{\sqrt{2}}(-1)^{1.1} \end{pmatrix}$$

$$= \frac{1}{\sqrt{2}}\begin{pmatrix} 1 & 1 \\ 1 & -1 \end{pmatrix} = H = \text{Hadamard matrix}$$

In other words, the Hadamard gate implements QFT in Hilbert space which we can also call a Hadamard transform. So QFT in 2D Hilbert space is equivalent to doing a Hadamard gate operation.

Hence the QFT of $\alpha|0> + \beta|1>$ will be $\alpha\frac{|0>+|1>}{\sqrt{2}} + \beta\frac{|0>-|1>}{\sqrt{2}}$ since $H|0> = \frac{|0>+|1>}{\sqrt{2}}$, $H|1> = \frac{|0>-|1>}{\sqrt{2}}$. Hence $\alpha|0> + \beta|1> \xrightarrow{QFT\equiv H} \alpha\frac{|0>+|1>}{\sqrt{2}} + \beta\frac{|0>-|1>}{\sqrt{2}}$.

Example: $n = 2$, $N = 2^2 = 4$, set is (0,1,2,3)
$w_n = e^{2\pi i/N}$ becomes $w_2 = e^{2\pi i/4} = e^{\pi i/2}$. $K(x, y) = \frac{1}{\sqrt{N}}w_n^{xy}$ becomes

$$K(x, y) = \frac{1}{\sqrt{4}}(e^{\pi i/2})^{xy} = \frac{1}{2}i^{xy}$$

$$= \frac{1}{2}\begin{pmatrix} i^{0.0} & i^{0.1} & i^{0.2} & i^{0.3} \\ i^{1.0} & i^{1.1} & i^{1.2} & i^{1.3} \\ i^{2.0} & i^{2.1} & i^{2.2} & i^{2.3} \\ i^{3.0} & i^{3.1} & i^{3.2} & i^{3.3} \end{pmatrix} = \frac{1}{2}\begin{pmatrix} 1 & 1 & 1 & 1 \\ 1 & i & -1 & -i \\ 1 & -1 & 1 & -1 \\ 1 & -i & -1 & i \end{pmatrix}$$

8.12 Finding period using QFT

Definition of periodicity is

$f(x + P) = f(x)$ where P= period which is a discrete number.

The oracle calculates the periodic function corresponding to every x in the input and sends the linear combination to output target register.

Let us illustrate the process of finding period for a three-qubit input case (set contains integers 0, 1, 2, ..., $2^3 - 1 = 0,1,2,3,4,5,6,7$)

Suppose the input or first register (which is a three-qubit system) contains a uniform linear combination of three qubit base states

$$\frac{1}{\sqrt{2^3}}(|000> +|001> +|010> +|011> +|100> +|101> +|110> +|111 >)$$

$$=\frac{1}{\sqrt{8}}(|0> +|1> +|2> +|3> +|4> +|5> +|6> +|7 >)$$

The target register or the second register is set to $|000> \equiv |0>$.

Start with the $|000>$ qubit state (of first register) and pass each qubit through Hadamard gate so that we get a linear combination of various states in the three-qubit basis. Oracle U_f computes the function $f(x)$ and sends to output register. The output $|\psi>$ is entangled and is given by (where U_f denotes the oracle that computes the function $f(x)$)

$$|\psi> =U_f\frac{1}{\sqrt{8}}\sum_{x=0}^{8-1}|x> |0> = \frac{1}{\sqrt{8}}\sum_{x=0}^{7}|x> |f(x)>$$

$$=\frac{1}{\sqrt{8}}[|0, f(0) > +|1, f(1) > +|2, f(2) > +|3, f(3) > +|4, f(4) >$$

$$+|5, f(5) > +|6, f(6) > +|7, f(7) >]$$

where we have punched the states together and written in a single ket. The coefficient of each $|x> |f(x)>$ is $\frac{1}{\sqrt{8}}$ but it is zero for $|y> |f(x)>$ if $y \neq x$.

Apply QFT (equation (8.30)) on the first register keeping the second register unaltered.

$$QFT|\psi> =|\psi'> =\frac{1}{\sqrt{8}}\sum_{y=0}^{7}w_8^{xy}\frac{1}{\sqrt{8}}\sum_{x=0}^{7}|y, f(x)> =\frac{1}{8}\sum_{x=0}^{7}\sum_{y=0}^{7}e^{2\pi ixy/8}|y, f(x)>$$

$$=\frac{1}{8}\sum_{x=0}^{7}e^{2\pi ix(0)/8}|y = 0, f(x) > +\frac{1}{8}\sum_{x=0}^{7}e^{2\pi ix(1)/8}|y = 1, f(x) > +...$$

$$......... .... + \frac{1}{8}\sum_{x=0}^{7}e^{2\pi ix(7)/8}|y = 7, f(x)>$$

Clearly there will be 64 terms. We show some of them.

$$QFT|\psi> =|\psi'> =\frac{1}{8}|0> [|f(0)> +|f(1)> +.... + |f(7) >]$$

$$+\frac{1}{8}|1> [|f(0)> +e^{2\pi i(1)(1)/8}|f(1)> +e^{2\pi i(2)(1)/8}|f(2)> +....... + e^{2\pi i(7)(1)/8}|f(7) >]$$

....

$$+\frac{1}{8}|4> [|f(0)> +e^{2\pi i(1)(4)/8}|f(1)> +e^{2\pi i(2)(4)/8}|f(2)> +....... + e^{2\pi i(7)(4)/8}|f(7) >]$$

....

$$+\frac{1}{8}|7> [|f(0)> +e^{2\pi i(1)(7)/8}|f(1)> +e^{2\pi i(2)(7)/8}|f(2)> +....... + e^{2\pi i(7)(7)/8}|f(7) >]$$

$$QFT|\psi> =|\psi'> =\frac{1}{8}|0> [|f(0)> +|f(1)> +.... + |f(7) >]$$

$$+\frac{1}{8}|1> [|f(0)> +e^{2\pi i/8}|f(1)> +e^{4\pi i/8}|f(2)> +........ + e^{14\pi i/8}|f(7) >]$$

....

$$+\frac{1}{8}|4> [|f(0)> +e^{8\pi i/8}|f(1)> +e^{16\pi i/8}|f(2)> +....... + e^{56\pi i/8}|f(7) >]$$

....

$$+\frac{1}{8}|7> [|f(0)> +e^{14\pi i/8}|f(1)> +e^{28\pi i/8}|f(2)> +....... + e^{98\pi i/8}|f(7) >]$$

Assume a period P of the function $f(x)$: $f(x) = f(x + P)$. Supposing $P = 2$ let

$$f(0) = f(2) = f(4) = f(6) = a,$$

$$f(1) = f(3) = f(5) = f(7) = b \quad (b \neq a).$$

Then QFT$|\psi> =|\psi'>$ becomes

$$|\psi'> =\frac{1}{8}|0> [4|a> +4|b >]$$

$$+\frac{1}{8}|1> [|a> (1 + e^{4\pi i/8} + e^{8\pi i/8} + e^{12\pi i/8}) + |b> (e^{2\pi i/8} + e^{6\pi i/8} + e^{10\pi i/8}. +e^{14\pi i/8})]$$

....

$$+\frac{1}{8}|4> [|a> (1 + e^{16\pi i/8} + e^{32\pi i/8} + e^{48\pi i/8}) + |b> (e^{8\pi i/8} + e^{24\pi i/8} + e^{40\pi i/8} + e^{56\pi i/8})]$$

$+$............................

$$|\psi'> = \frac{1}{2}|0> (|a> +|b>)$$

$$+\frac{1}{8}|1> \left[|a> \{1 + i + (-1) + (-i)\} + |b> \left\{ \frac{1+i}{\sqrt{2}} + \frac{-1+i}{\sqrt{2}} + \frac{-1-i}{\sqrt{2}} + \frac{1-i}{\sqrt{2}} \right\} \right] + \ldots$$

$$+\frac{1}{8}|4> [|a> \{1+1+1+1\} + |b> \{(-1) + (-1) + (-1) + (-1)\}] + \ldots$$

$$|\psi'> = \frac{1}{2}|0> (|a> +|b>) + \frac{1}{8}|1> [|a> (0) + |b> (0)] + \ldots$$

$$+\frac{1}{8}|4> [4|a> - 4|b>] + \ldots$$

(all terms other than those corresponding to $|y = 0>$ and $|y = 4>$ vanish).

$$|\psi'> = = |0> \frac{1}{2}(|a> +|b>) + |4> \frac{1}{2}(|a> +|b>)$$

$$|\psi'> = \frac{1}{2}(|0 \; a> +|0 \; b>) + |4 \; a> +|4 \; b>$$

If we measure the first register the result is either 0 or 4 which is a direct consequence of periodicity of $f(x)$ viz. $P = 2$. So periodicity determines the non-vanishing states in the first register, i.e. the possible result of measurement of the first register.

Let us try to find a unitary transformation that will act similarly.

8.13 Implementation of QFT

Suppose the unitary operator U maps the state $\sum_x f(x)|x>$ to the state $\sum_y F(y)|y>$ as

$$U\sum_x f(x)|x> = \sum_y F(y)|y> \tag{8.31}$$

where $F(y) = \frac{1}{\sqrt{N}} \sum_{x=0}^{N-1} w_n^{xy} f(x)$ and $w_n = e^{2\pi i/N}$.

For $f(x') = \delta_{xx'}$ we get

$$F(y) = \frac{1}{\sqrt{N}} \sum_{x'=0}^{N-1} w_n^{x'y} f(x') = \frac{1}{\sqrt{N}} \sum_{x'=0}^{N-1} w_n^{x'y} \delta_{xx'} = \frac{1}{\sqrt{N}} w_n^{xy} = \frac{1}{\sqrt{N}} e^{2\pi ixy/N}$$

So equation (8.31) leads to

$$U\sum_{x'=0}^{N-1} \delta_{xx'}|x'> = \sum_{y=0}^{N-1} \frac{1}{\sqrt{N}} e^{2\pi ixy/N}|y>$$

$$U|x> = \frac{1}{\sqrt{N}} \sum_{y=0}^{N-1} e^{2\pi ixy/N}|y>$$

$$\Rightarrow U|x> <x| = \frac{1}{\sqrt{N}} \sum_{y=0}^{N-1} e^{2\pi ixy/N}|y> <x|$$

$$U \sum_{x=0}^{N-1}|x> <x| = \frac{1}{\sqrt{N}} \sum_{x=0}^{N-1}\sum_{y=0}^{N-1} e^{2\pi ixy/N}|y> <x|$$

$$\Rightarrow U = \frac{1}{\sqrt{N}} \sum_{x=0}^{N-1}\sum_{y=0}^{N-1} e^{2\pi ixy/N}|y> <x|$$

The structure of U being determined lets us build a new Fourier basis $|\tilde{x}>$ from the standard computational basis $|x>$ as

$$|\text{Computational basis} > \xrightarrow{\text{QFT}} |\text{Fourier basis}>$$

$$|\tilde{x}>=U|x> = \frac{1}{\sqrt{N}} \sum_{y} e^{2\pi ixy/N}|y>$$

Case: $n = 1$, $N = 2$(single-qubit case)
Let $|x>$ be a one qubit state. The Fourier transform is given by

$$|\tilde{x}>=U|x> = \frac{1}{\sqrt{N}} \sum_{y} e^{2\pi ixy/N}|y>$$

$$= \frac{1}{\sqrt{2}} \sum_{y=0,1} e^{2\pi ixy/2}|y> = \frac{1}{\sqrt{2}}[|0> + e^{2\pi ix/2}|1>]$$

In the fractional binary notation $\frac{x}{2} = 0. \ x$ (here x is in the base 2, dot represents a binary point) we can write

$$|\tilde{x}>=\frac{1}{\sqrt{2}}[|0> + e^{2\pi i(0. \ x)}|1>]$$

Also, $|\tilde{x}>=U|x> = \frac{1}{\sqrt{2}}[|0> + e^{2\pi ix/2}|1>]$

$$= \frac{1}{\sqrt{2}}[|0> + (e^{\pi i})^x|1>] = \frac{1}{\sqrt{2}}[|0> + (-1)^x|1>]$$

We note that
For $x = 0$: $|\tilde{x}>=U|0> = \frac{1}{\sqrt{2}}[|0> + |1>]$ and
For $x = 1$: $|\tilde{x}>=U|1> = \frac{1}{\sqrt{2}}[|0> - |1>]$

This is a Hadamard transformation. So QFT for $n = 1$, $N = 2$ can be implemented by a normal Hadamard transformation.

- **Example:**

Let $|\psi> = f(0)|0> + f(1)|1>$ be any one qubit data. The QFT will be

$$|\tilde{x}> = U|\psi> = U\,[f(0)|0> + f(1)|1>] = f(0)U|0> + f(1)U|1>$$

$$= f(0)\frac{1}{\sqrt{2}}(|0> + |1>) + f(1)\frac{1}{\sqrt{2}}(|0> - |1>)$$

which is a Hadamard gate operation.

Case: $n = 2$, $N = 2^2 = 4$ **(two-qubit case)**

Let $|x>$ be a two-qubit state. The Fourier transform is given by

$$|\tilde{x}> = U|x> = \frac{1}{\sqrt{N}}\sum_{y} e^{2\pi i x y/N}|y> = \frac{1}{\sqrt{4}}\sum_{y=0,1} e^{2\pi i x y/4}|y>$$

Let us write x in the binary form $|x_1 x_0>$ where $x = 2^1 x_1 + 2^0 x_0 = 2x_1 + x_0$ and y in the binary form $|y_1 y_0>$ where $y = 2^1 y_1 + 2^0 y_0 = 2y_1 + y_0$.

$$|\tilde{x}> = \frac{1}{2}\sum_{y_0 y_1=0,1} e^{2\pi i x(2y_1+y_0)/4}|y_1 y_0>$$

$$= \frac{1}{\sqrt{2}}\sum_{y_1=0,1} e^{2\pi i x\,2y_1/4}|y_1> \otimes \frac{1}{\sqrt{2}}\sum_{y_0=0,1} e^{2\pi i x y_0/4}|y_0>$$

$$= \frac{1}{\sqrt{2}}(|0> + e^{2\pi i x/2}|1>) \otimes \frac{1}{\sqrt{2}}(|0> + e^{2\pi i x/4}|1>)$$

$$|\tilde{x}> = \frac{1}{\sqrt{2}}(|0> + e^{2\pi i (2x_1+x_0)/2}|1>) \otimes \frac{1}{\sqrt{2}}(|0> + e^{2\pi i (2x_1+x_0)/4}|1>)$$

$$|\tilde{x}> = \frac{1}{\sqrt{2}}(|0> + e^{2\pi i x_1}e^{2\pi i x_0/2}|1>) \otimes \frac{1}{\sqrt{2}}(|0> + e^{2\pi i x_1/2}e^{2\pi i x_0/4}|1>)$$

$$= \frac{1}{\sqrt{2}}(|0> + e^{2\pi i (x_0/2)}|1>) \otimes \frac{1}{\sqrt{2}}(|0> + e^{2\pi i (x_1/2+x_0/4)}|1>)$$

(as $x_1 \to 0, 1 \Rightarrow e^{2\pi i x_1} = 1$)

In the fractional binary notation $\frac{x_0}{2} = 0.\,x_0$ and $\frac{x_1}{2} + \frac{x_0}{4} = \frac{x_1}{2} + \frac{x_0}{2^2} = 0.\,x_1 x_0$

$$|\tilde{x}> = \frac{1}{\sqrt{2}}(|0> + e^{2\pi i (0.x_0)}|1>) \otimes \frac{1}{\sqrt{2}}(|0> + e^{2\pi i (0.\,x_1 x_0)}|1>)$$

We note that $e^{2\pi i(0.x_0)} = e^{2\pi i x_0/2} = e^{\pi i x_0} = (e^{\pi i})^{x_0} = (-1)^{x_0}$ (in the first qubit) and $e^{2\pi i(0.x_1 x_0)} = e^{\pi i x_1}e^{\pi i x_0/2} = (-1)^{x_1}e^{\pi i x_0/2}$ (in the second qubit). We can also write

$$|\tilde{x}> = \frac{1}{\sqrt{2}}(|0> + (-1)^{x_0}|1>) \otimes \frac{1}{\sqrt{2}}(|0> + (-1)^{x_1}e^{\pi i x_0/2}|1>)$$

$$|\tilde{x}> = \frac{1}{\sqrt{2}}(|0> + (-1)^{x_0}|1>) \otimes \frac{1}{\sqrt{2}}(|0> + (-1)^{x_1}e^{2\pi i x_0/4}|1>) \qquad (8.32)$$

The first term is the ordinary Hadamard transform as it gives $|0> \pm |1>$ for x_0 taking value 0 or 1. The second term is a Hadamard transform followed by $2\pi x_0/4$. So only for $x_0 = 1$, there is a rotation of the state $|1>$ by $\frac{2\pi}{4}$.

Let us define a controlled B_{jk} gate through the matrix

$$B = \begin{pmatrix} 1 & 0 \\ 0 & \exp\left(\dfrac{2\pi i}{2^{k-j+1}}\right) \end{pmatrix}$$

where $j, k \to 0, 1, 2, \ldots.$ and $k \geqslant j$. It gives a rotation of the state $|1>$ only if the control bit is 1. The effect of controlled B_{jk} gate on state $|x, y> (x, y \to 0, 1)$ is given by

$$B_{jk}|x, y> = \exp\left(\frac{2\pi i}{2^{k-j+1}}xy\right)|x, y>$$

The first state is the control bit and the second bit is the target bit. If $x = 0$ the action of the gate is an identity operation. If $x = 1$ a phase acts on $|y>$. This gives

$$B_{jk}|x, y> = \begin{cases} |x, y> & \text{if } x = 0 \\ \exp\left(\dfrac{2\pi i}{2^{k-j+i}}\right)|x, y> & \text{if } x = 1 \end{cases}$$

$$B_{jk}|y> = \begin{cases} |y> & \text{for } y = 0 \\ \exp\left(\dfrac{2\pi i}{2^{k-j+i}}\right)|y> & \text{for } y = 1 \end{cases}$$

With this the Fourier basis state (for $n = 2$) equation (8.32) can be written as (x_0= control bit, x_1= target bit)

$$|\tilde{x}> = \frac{1}{\sqrt{2}}(|0> + (-1)^{x_0}|1>) \otimes \frac{1}{\sqrt{2}}\left(|0> + (-1)^{x_1}[\exp(\frac{2\pi i}{2^{2-1+1}})]^{x_0}|1>\right) \qquad (8.33)$$

$$= \frac{1}{\sqrt{2}}(|0> + (-1)^{x_0}|1>) \otimes \frac{1}{\sqrt{2}}(|0> + (-1)^{x_1}[B_{12}]^{x_0}|1>)$$

$$|\tilde{x}> = \frac{1}{\sqrt{2}}(|0> + (-1)^{x_0}|1>) \otimes \frac{1}{\sqrt{2}}B_{12}^{x_0}(|0> + (-1)^{x_1}|1>) \qquad (8.34)$$

(as $B_{12}^{x_0}|0> = |0>$)

$[B_{12}]^{x_0}$ represents a rotation by $\frac{2\pi}{2^{2-1+1}}=\frac{2\pi}{4}$ with x_0 as control. The states are entangled because the first qubit carries $(-1)^{x_0}$ while the second qubit carries $(-1)^{x_1}$ coupled with a phase factor $e^{2\pi i x_0/4}$(which is there for $x_0 = 1$). For the input state $|x_1 x_0\rangle$ the output is reversed.

The factors $\frac{1}{\sqrt{2}}(|0\rangle +(-1)^{x_0}|1\rangle)$, $\frac{1}{\sqrt{2}}(|0\rangle +(-1)^{x_1}|1\rangle)$ are Hadamard transformation denoted by U_H. The first term is a Hadamard transformation. The second term is a Hadamard transformation followed by a rotation of $\frac{2\pi x_0}{4}$(i.e. only if $x_0 = 1$ there is a phase rotation by $\frac{2\pi}{4}$ and hence it is called controlled phase rotation). Since we need x_0 to provide a control we cannot apply the gate on x_0 beforehand as it gets reversed at the output. So the circuit requires a reversal of x_0 and x_1 before various gates are applied. We thus need a swap of qubits (figure 8.10). So we have from equation (8.34)

$$|\tilde{x}\rangle = [U_H|x_0\rangle] \otimes B_{12}^{x_0}U_H|x_1\rangle$$
$$= [(U_H \otimes I)B_{12}^{x_0}(I \otimes U_H)]|x_0 x_1\rangle$$
$$= [(U_H \otimes I)B_{12}^{x_0}(I \otimes U_H)]\text{Swap}|x_1 x_0\rangle$$

Fourier transform requires swapping of the order of bits prior to application of Hadamard gate and controlled B_{jk} gates.

QFT for three qubits

Equation (8.33) was obtained by applying Hadamard on the second qubit followed by Hadamard on the first qubit along with a controlled phase rotation. The value of the second qubit can be changed only after its value has been used for the purpose of controlling the operations on the first qubit. This is achieved by interchanging x_1 and x_0 and then applying the Hadamard gate. Execution of Fourier transform requires swapping of the order of bits before application of the Hadamard and controlled B_{jk} gates.

The QFT for $|x\rangle=|x_2 x_1 x_0\rangle$ will be (binary representation now)

$$|\tilde{x}\rangle=U|x\rangle = \frac{1}{\sqrt{N}}\sum_y e^{2\pi i xy/N}|y\rangle =\frac{1}{\sqrt{8}}\sum_{y=0,1} e^{2\pi i xy/8}|y\rangle$$

Let us write x in the binary form $|x_2 x_1 x_0\rangle$ where $x = 2^2 x_2 + 2^1 x_1 + 2^0 x_0 = 4x_2 + 2x_1 + x_0$ and y in the binary form $|y_2 y_1 y_0\rangle$ where $y = 2^2 y_2 + 2^1 y_1 + 2^0 y_0 = 4y_2 + 2y_1 + y_0$.

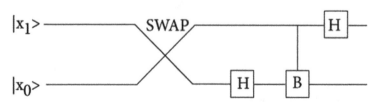

Figure 8.10. QFT implementation for $n = 2$.

$$|\tilde{x}> = \frac{1}{\sqrt{8}} \sum_{y_0 y_1 y_2 = 0,1} e^{2\pi i x (4y_2 + 2y_1 + y_0)/8} |y_2 y_1 y_0>$$

$$= \frac{1}{\sqrt{2}} \sum_{y_2 = 0,1} e^{2\pi i x 4 y_2/8} |y_2> \otimes \frac{1}{\sqrt{2}} \sum_{y_1 = 0,1} e^{2\pi i x 2 y_1/8} |y_2> \otimes \frac{1}{\sqrt{2}} \sum_{y_0 = 0,1} e^{2\pi i x y_0/8} |y_0>$$

$$= \frac{1}{\sqrt{2}}[|0> + e^{2\pi i x/2}|1>] \otimes \frac{1}{\sqrt{2}}[|0> + e^{2\pi i x/4}|1>] \otimes \frac{1}{\sqrt{2}}[|0> + e^{2\pi i x/8}|1>]$$

$$|\tilde{x}> = \frac{1}{\sqrt{2}}[|0> + e^{\pi i x}|1>] \otimes \frac{1}{\sqrt{2}}[|0> + e^{\pi i x/2}|1>] \otimes \frac{1}{\sqrt{2}}[|0> + e^{\pi i x/4}|1>]$$

$$= \frac{1}{\sqrt{2}}[|0> + e^{\pi i (4x_2 + 2x_1 + x_0)}|1>] \otimes \frac{1}{\sqrt{2}}[|0> + e^{\pi i (4x_2 + 2x_1 + x_0)/2}|1>]$$

$$\otimes \frac{1}{\sqrt{2}}[|0> + e^{\pi i (4x_2 + 2x_1 + x_0)/4}|1>]$$

$$= \frac{1}{\sqrt{2}}[|0> + e^{4\pi i x_2} e^{2\pi i x_1} e^{\pi i x_0}|1>] \otimes \frac{1}{\sqrt{2}}[|0> + e^{2\pi i x_2} e^{\pi i x_1} e^{\pi i x_0/2}|1>]$$

$$\otimes \frac{1}{\sqrt{2}}[|0> + e^{\pi i x_2} e^{\pi i x_1/2} e^{\pi i x_0/4}|1>]$$

$$|\tilde{x}> = \frac{1}{\sqrt{2}}[|0> + (+1)^{x_2}(+1)^{x_1}(-1)^{x_0}|1>]$$

$$\otimes \frac{1}{\sqrt{2}}[|0> + (+1)^{x_2}(-1)^{x_1}[\exp(\frac{2\pi i}{2^{1-0+1}})]^{x_0} |1>]$$

$$\otimes \frac{1}{\sqrt{2}}[|0> + (-1)^{x_2}[\exp(\frac{2\pi i}{2^{2-1+1}})]^{x_1}[\exp(\frac{2\pi i}{2^{2-0+1}})]^{x_0} |1>]$$

$$= \frac{1}{\sqrt{2}}[|0> + (-1)^{x_0}|1>] \otimes \frac{1}{\sqrt{2}}(|0> + (-1)^{x_1}[B_{01}]^{x_0}|1>)$$

$$\otimes \frac{1}{\sqrt{2}}[|0> + (-1)^{x_2}[B_{12}]^{x_1}[B_{02}]^{x_0}|1>]$$

$$|\tilde{x}> = \frac{1}{\sqrt{2}}[|0> + (-1)^{x_0}|1>] \oplus B_{01}^{x_0} \frac{1}{\sqrt{2}}[|0> + (-1)^{x_1}|1>]$$

$$\otimes \frac{1}{\sqrt{2}} B_{12}^{x_1} B_{02}^{x_0}[|0> + (-1)^{x_2}|1>]$$

The three terms in the product involve

(i) $\frac{1}{\sqrt{2}}[|0> + (-1)^{x_0}|1>]$ which is implemented by Hadamard gate H on third qubit

(ii) $B_{01}^{x_0} \frac{1}{\sqrt{2}}[|0> + (-1)^{x_1}|1>]$ which is implemented by Hadamard gate H on second qubit followed by a controlled gate B_{01}

(iii) $\frac{1}{\sqrt{2}} B_{12}^{x_1} B_{02}^{x_0}[|0> + (-1)^{x_2}|1>]$ which is implemented by Hadamard gate H on the first qubit followed by two controlled B_{jk} operations viz. $B_{12} B_{02}$.

For the three-qubit case it is necessary to interchange the most significant qubit (x_2) with the least significant qubit (x_0). QFT implementation for $n = 3$ case has been shown in figure 8.11.

We can generalize the result for n qubits.

8.14 Some definitions and GCD evaluation

Prime number and composite number

Prime number is a whole number $a \geqslant 2$, whose only factors are 1 and itself, i.e. a. Otherwise, the number is composite. As factors of 11 are 1 and 11 so it is a prime number. Other examples: 2, 3, 5, 7, 11 etc. But as $6 = 2 \times 3$ it is not a prime number.

GCD=Greatest common divisor

The greatest common divisor of two integers a and b is the greatest positive integer d denoted as $d = \mathrm{GCD}(a, b)$ that divides both a and b.

Co-prime (or relatively prime) numbers

If $\mathrm{GCD}(a, b) = 1$ we say that a and b are co-prime numbers. In other words co-prime numbers have no common factor. For example 100 and 19 are co-prime numbers since $\mathrm{GCD}(100, 19) = 1$, i.e. the greatest positive integer that divides both 100 and 19 is 1.

Finding GCD by brute force method

We explain the method through examples.

✓ Find GCD(4,8)

The divisors are $4 = \{1,2,4\}$ and $8 = \{1,2,4,8\}$.

The common divisors are $\{1,2,4\}$ and the greatest common divisor is 4. This is the largest number with which we can divide both 4 and 8. So GCD=4.

✓ Find GCD(10,15)

The divisors are: $10 = \{1,2,5,10\}$ and $15 = \{1,3,5,15\}$.

The common divisors are $\{1,5\}$ and the greatest common divisor is 5. This is the largest number with which we can divide both 10 and 15. So GCD = 5.

Finding GCD by Euclid algorithm

We can use Euclid algorithm to find GCD of two numbers a and b.

$\mathrm{GCD}(a, b) = \mathrm{GCD}(b, a \bmod b)$ where $a > b$, and we have to continue until $a \bmod b = 0$, i.e. until we have the form $\mathrm{GCD}(p, 0)$. Then GCD is p.

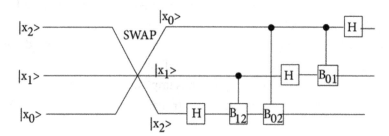

Figure 8.11. QFT Implementation for $n = 3$.

We explain through some illustrations as follows. The steps are shown in table 8.10, table 8.11 and table 8.12 which are self- explanatory.

8.15 Inverse modulo

Given two integers a and n that are co-prime. It can be shown that there exists a unique integer d such that $ad = 1 \mod n$. The integer d is called inverse modulo n of a and denoted by a^{-1}.

In other words as $ad = 1 \pmod n$, $d = a^{-1}$.

By Euclid's algorithm we can find a relation of the type

$$ad - nq = 1$$

(where q = some integer). Then if we take modulo n we get

$$ad = 1 \mod n \quad \text{since } nq \mod n = 0.$$

Things will be clear if we go through the illustrations, which, however, are based on the easier *trial and error* method.

- **Illustration**

(a) **Find** $3^{-1} \mod 26$

Ans. To find GCD(26,3)

GCD(26,3) = 1

This means 3 and 26 are co primes.

```
3 ) 26 ( 8
     24
     2 ) 3 ( 1
         2
         1 ) 2 ( 2
GCD          2
             0
```

Table 8.10. GCD (a,b) by Euclid algorithm.

$r_1 = a - bq_1$

$r_2 = b - r_1 q_2$

$r_3 = r_1 - r_2 q_3$

$r_4 = r_2 - r_3 q_4$

....

$r_n = r_{n-2} - r_{n-1} q_n$

$0 = r_{n-1} - r_n q_{n+1}$

$GCD = r_n$

Table 8.11. GCD (100, 19) by Euclid algorithm.

$b = 19$) $a = 100$ ($q_1 = 5$ $\quad\quad x_1 = 95$ $\quad\quad$ ───── $\quad\quad r_1 = 5$) $b = 19$ ($q_2 = 3$ $\quad\quad\quad\quad x_2 = 15$ $\quad\quad\quad\quad$ ───── $\quad\quad\quad\quad r_2 = 4$) $r_1 = 5$ ($q_3 = 1$ $\quad\quad\quad\quad\quad x_3 = 4$ $\quad\quad\quad\quad\quad$ ───── $\quad\quad\quad\quad\quad r_3 = 1$) $r_2 = 4$ ($q_4 = 4$ $GCD \nearrow \quad\quad\quad x_4 = 4$ $\quad\quad\quad\quad\quad\quad$ ───── $\quad\quad\quad\quad\quad\quad\quad 0$	GCD(100,19) =GCD(19, 100 mod 19) =GCD(19,5) =GCD(5, 19 mod 5) =GCD(5,4) =GCD(4, 5 mod 4) =GCD(4,1) =GCD(1, 4 mod 1) =GCD(1,0) =1
	Euclid algorithm $5 = 100 - (19)(5)$ $4 = 19 - (5)(3)$ $1 = 5 - (4)(1)$ $0 = 4 - (1)(4)$ $\quad\quad GCD = 1$

Table 8.12. Evaluation of GCD (10, 15) by Euclid algorithm

Example: Find GCD(10,15) by Euclid algorithm.	
Ans. GCD(10,15) = GCD(15,10) $\quad\quad$ (rearranging properly such that larger $\quad\quad$ number is put first) $\quad\quad\quad\quad$ =GCD(10, 15 mod 10) $\quad\quad\quad\quad$ =GCD(10,5) $\quad\quad\quad\quad$ =GCD(5, 10 mod 5) $\quad\quad\quad\quad$ =GCD(5,0) $\quad\quad\quad\quad$ =5	10) 15 (1 $\quad\quad 10$ $\quad\quad$ ───── $GCD \nearrow 5$) 10 (2 $\quad\quad\quad 10$ $\quad\quad\quad$ ───── $\quad\quad\quad 0$

We write: $3x \bmod 26 = 1$
On inspection we find that for $x = 9$:

$$3(9) \bmod 26 = 27 \bmod 26 = 1$$

As $3.9 = 1 \pmod{26}$, $9 = 3^{-1}$ (Here $a = 3$, $d = 9$, $n = 26$)
(b) **Find** $27^{-1} \bmod 392$.

Ans. To find GCD (392,27)
  GCD(392,27) = 1
This means 27 and 392 are co primes.

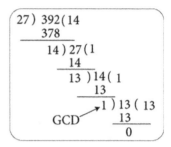

We write: $27x \bmod 392 = 1$
We inspect that: $392(1) = 392$ and $27(14) = 378$

$$392(2) = 784 \text{ and } 27(29) = 783$$

Clearly $27.29 = 783$ *falls short* of $392.2 = 784$ by 1. Since w.r.t mod 392, the number -29 and $392 - 29 = 363$ are equivalent let us check with 363 as follows. $(-d \bmod n = (n - d)\bmod n)$

$27(363)\bmod 392 = 9801 \bmod 392 = 1$
As $27.363 = 1 \pmod{392}$, $363 = 27^{-1}$,
(here $a = 27$, $d = 363$, $n = 392$).

(c) **Find** $20^{-1} \bmod 3$

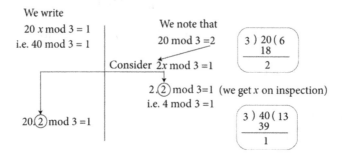

As $20.2=1 \pmod 3$, $2 = 20^{-1}$(Here

$$a = 20, d = 2, n = 3)$$

Table 8.13 shows some more examples of finding inverse modulo.

8.16 Shor's algorithm

Consider a number $N = pq$ which is the product of two large prime numbers.

There are classical algorithms which compute p and q but they are not fast such as Euclid algorithm for greatest common divisor. It takes $\sqrt{N}$ steps to solve for p and q.

It is easy to understand that factorization is a difficult job.

Finding 229 and 127 from 29083 is difficult but finding 29083 from 229 and 127 is easy (just multiply 229 and 127 and we get 29083).

Multiplication of two numbers can be done in polynomial time though factorization is not.

The Euclid algorithm serves no purpose if the numbers involved are very large.

Period finding

Let us first solve an equivalent problem called period finding. We discussed finding period with QFT earlier.

Given a number N. Choose a random number $m < N$ which is co-prime with N, i.e. $GCD(m, N) = 1$. This can be ensured by checking with the Euclid algorithm.

We will try to find out various powers of m say m^a and define a function $F_N(a) = m^a \bmod N$.

The smallest value of a, say p for which

$$m^p \bmod N = 1 \text{[or equivalently } m^p = 1 \pmod N)]$$

is called period of the function.

Period finding requires a quantum computer.

Example

Given a number $N = 21$, which is a composite number. A number $m < N$ and co-prime with $N = 21$ has to be chosen. This gives $m = 2,4,5,8,10,11,13,16,17,19,20$ as

Table 8.13. Simple problems on inverse modulo.

$5^{-1}\mathrm{mod}\ 14$	$3^{-1}\mathrm{mod}\ 7$	$3^{-1}\mathrm{mod}\ 17$	$7^{-1}\mathrm{mod}\ 24$
Write	Write	Write	Write
$5x\ \mathrm{mod}\ 14 =1$	$3x\ \mathrm{mod}\ 7 =1$	$3x\ \mathrm{mod}\ 17$	$7x\ \mathrm{mod}\ 24 =1$
$= 5.\text{③}\mathrm{mod}\ 14 =1$	$= 3.\text{⑤}\mathrm{mod}\ 7 =1$	$= 3.\text{⑥}\mathrm{mod}\ 17 =1$	$= 7.\text{⑦}\mathrm{mod}\ 24 =1$
Ans $3=5^{-1}$	Ans $5=3^{-1}$	Ans $6=3^{-1}$	Ans $7=7^{-1}$

$17^{-1}\mathrm{mod}\ 3120$	$3^{-1}\mathrm{mod}\ 7$
Write $17x\ \mathrm{mod}\ 3120=1$	Write $3x\ \mathrm{mod}\ 7=1$
We inspect that	We inspect that
$3120.1=3120$, $17.183=3111$	$7.1=7$, $3.1=3$
$3120.2=6240$, $17.\boxed{367}=6239$	$3.\boxed{2}=6$
$17.367 = 6239$ *falls short* of	$3.2= 6$ *falls short* of $7.1=7$ by 1.
$3920.2=6240$ by 1.	Since w.r.t mod 7, $-\boxed{2}$ and $7-\boxed{2}=\text{⑤}$
Since w.r.t mod 3120, $-\boxed{367}$ and	are equivalent we check with 5.
$3120-\boxed{367}=\text{②753}$ are equivalent	$3.\text{⑤}\mathrm{mod}\ 7 =15\ \mathrm{mod}\ 7 = 1$
we check with 2753.	$\Rightarrow 5 = 3^{-1}$
$17.\text{②753}\ \mathrm{mod}\ 3120$ $\quad 3120)\overline{46801}(15$	$\qquad 7)\overline{15}(2$
$=46801\mathrm{mod}\ 3120 =1 \quad\quad \underline{46800}$	$\qquad \underline{14}$
$\Rightarrow 2753 = 17^{-1} \qquad\qquad\qquad 1$	$\qquad\qquad 1$

$15^{-1}\mathrm{mod}\ 11$	$35^{-1}\mathrm{mod}\ 8$	$40^{-1}\mathrm{mod}\ 7$	$19^{-1}\mathrm{mod}\ 5$
Write	Write	Write	Write
$15x\ \mathrm{mod}\ 11=1$	$35x\ \mathrm{mod}\ 8=1$	$40x\ \mathrm{mod}\ 7=1$	$19x\ \mathrm{mod}\ 5=1$
We note that	We note that	We note that	We note that
$15\ \mathrm{mod}\ 11=4$)	$35\ \mathrm{mod}\ 8=3$)	$40\ \mathrm{mod}\ 7=5$)	$19\ \mathrm{mod}\ 5=4$)
$4x\ \mathrm{mod}\ 11 =1$	$3x\ \mathrm{mod}\ 8 =1$	$5x\ \mathrm{mod}\ 7 =1$	$4x\ \mathrm{mod}\ 5 =1$
$4.\text{③}\mathrm{mod}\ 11 =1$	$3.\text{③}\mathrm{mod}\ 8 =1$	$5.\text{③}\mathrm{mod}\ 7 =1$	$4.\text{④}\mathrm{mod}\ 5 =1$
Check	Check	Check	Check
$15.\text{③}\mathrm{mod}\ 11$	$35.\text{③}\mathrm{mod}\ 8$	$40.\text{③}\mathrm{mod}\ 7$	$19.\text{④}\mathrm{mod}\ 5$
$=45\ \mathrm{mod}\ 11=1$	$=105\ \mathrm{mod}\ 8=1$	$=120\ \mathrm{mod}\ 7=1$	$=76\ \mathrm{mod}\ 5=1$
Ans $3=15^{-1}$	Ans $3=35^{-1}$	Ans $3=40^{-1}$	Ans $4=19^{-1}$

can be checked using $\mathrm{GCD}(m,\ 21) = 1$ e.g. $\mathrm{GCD}(2,\ 21) = 1, \mathrm{GCD}(10,\ 21) = 1$ etc. (We did not choose 3,6,7,9 etc as they are not co-prime with $N = 21$ e.g. $\mathrm{GCD}(3,21)$ $=3 \neq 1$.)

✓ Now let us choose a random number $m < N$ say $m = 2$ and find its various powers 2^a viz. $2^1 = 2, 2^2 = 4, 2^3 = 8,\ 2^4 = 16,\ 2^5 = 32,\ 2^6 = 64$ and define the function $F_{21}(a) = 2^a\ \mathrm{mod}\ 21$ e.g.

$$F_{21}(5) = 2^5\ \mathrm{mod}\ 21=32\ \mathrm{mod}\ 21=11\neq1,$$

$$F_{21}(6) = 2^6\ \mathrm{mod}\ 21=64\ \mathrm{mod}\ 21=1\ (\text{or equivalently } 2^6 = 64=1\ (\mathrm{mod}\ 21)).$$

For $m = 2$, we thus have got the smallest value of a, i.e. $p = 6$, for which $m^p \bmod N = 2^6 \bmod 21 = 1$. So period is 6.

✓ What if $m = 4$?

With $m = 4$ we consider $4^1 = 4$, $4^2 = 16$, $4^3 = 64$ and define the function $F_{21}(a) = 4^a \bmod 21$, e.g. $F_{21}(3) = 4^3 \bmod 21 = 64 \bmod 21 = 1$. So period is 3.

✓ What if $m = 5$?

With $m = 5$ we consider $5^1 = 5$, $5^2 = 25$, $5^3 = 125$, $5^4 = 625$, $5^5 = 3125$, $5^6 = 15625$ and define the function $F_{21}(a) = 5^a \bmod 21$, e.g.

$$F_{21}(2) = 5^2 \bmod 21 = 25 \bmod 21 = 4 \neq 1$$

$$F_{21}(3) = 5^3 \bmod 21 = 125 \bmod 21 = 20 \neq 1$$

$$F_{21}(4) = 5^4 \bmod 21 = 625 \bmod 21 = 16 \neq 1$$

$$F_{21}(5) = 5^5 \bmod 21 = 3125 \bmod 21 = 17 \neq 1,$$

$$F_{21}(6) = 5^6 \bmod 21 = 15625 \bmod 21 = 1.$$

So period is 6.

✓ What if $m = 8$?

With $m = 8$ we consider $8^1 = 8$, $8^2 = 64$ and define the function $F_{21}(a) = 8^a \bmod 21$, e.g. $F_{21}(2) = 8^2 \bmod 21 = 64 \bmod 21 = 1$. So period is 2.

Connection of period calculation and factorization

Consider a quadratic equation: $x^2 = 1 \bmod N$.

If N is an odd prime then we have a trivial solution $x = \pm 1$.

If N is a composite number then in addition to the trivial solution there are also non-trivial solutions like $x = \pm a \bmod N$.

Example:

✓ Consider $x^2 = 1 \bmod 41$ has trivial solution $x = \pm 1 (N = 41$ is a prime number).

✓ Consider $x^2 = 1 \bmod 55 (N=55$ is a composite number). So, in addition to the trivial solution $x = \pm 1$ there will be a pair of non-trivial solutions that can be shown to be $x = \pm 21$. Since $(\pm 21)^2 = 441 = 440 + 1 \Rightarrow (\pm 21)^2 = 8(55) + 1$ we have $(\pm 21)^2 = 1 \pmod{55}$

As we have chosen $m^p = 1 \pmod{N}$ we take $x = m^{p/2}$ so that

$$x^2 = (m^{p/2})^2 = m^p = 1 \pmod{N}$$

is obtained. To achieve this one should ensure that the period of random number

$$p = \text{even}.$$

One of the requirements of Shor's algorithm is that p should be even.

If p is odd the algorithm fails and we have to use a different m.
If p = even we write the equation

$$x^2 = 1 \Rightarrow (m^{p/2})^2 = 1 \Rightarrow (m^{p/2} + 1)(m^{p/2} - 1) = 0 (\text{mod } N)$$

Now $(m^{p/2} - 1) = 0$ would lead to $m^{p/2} = 1$ which is not acceptable since we defined period p as the smallest integer for which $m^p = 1$ (mod N) and since we cannot have both $m^p = 1$ as well as $m^{p/2} = 1$. So $m^{p/2} = 1$ is ruled out. We are thus left with the possibility
$$m^{p/2} + 1 = 0 \ (\text{mod } N) \quad \Rightarrow \quad m^{p/2} + 1 = q . N + 0, \text{ i.e.}$$

$$m^{p/2} + 1 = q . N$$

But this equation may lead to a trivial solution and for a non-trivial solution we should find m such that

$$m^{p/2} + 1 \neq 0 \ (\text{mod } N).$$

If the two conditions are satisfied viz. $p=$ even and $m^{p/2} + 1 \neq 0$ (mod N) then the equation

$$(m^{p/2} + 1)(m^{p/2} - 1) = 0 \ (\text{mod } N)$$

will be satisfied only if $m^{p/2} + 1$ and $m^{p/2} - 1$ contain factors of N.

If N is large, a classical computer may require evaluation of $O(N)$ powers of m. In a quantum computer all the powers of m would be simultaneously determined. The task is to search for p. Hence the problem of factorization of a large number is essentially the problem of finding the period.

Illustration

To find the factors of 21 through Shor's algorithm

$N = 21$. Let us choose $m = 2$ ($< N = 21$) that is coprime with $N = 21$; GCD(2,21) =1.

As 2^6 mod 21= 64 mod 21=1, i.e. 2^6=64 =1 mod 21 the period $p = 6$ =even. So first condition of $p =$ even is satisfied.

Also $m^{p/2} + 1 = 2^{6/2} + 1 = 9$
and since $9=0(21)+9$ so 9 (mod 21) =9.
So $m^{p/2} + 1 = 2^{6/2} + 1 = 9 \neq 0$ (mod $N = 21$) which is the second condition is also satisfied.

$$21 \,) \, 9 \, (0$$
$$\underline{0}$$
$$9 = \text{Remainder}$$

Clearly for this choice of $m = 2$ both the conditions for Shor's algorithm are satisfied.

Then in the equation $(m^{p/2} + 1)(m^{p/2} - 1) = 0 \,(\text{mod } N)$ the factors $m^{p/2} + 1$ and $m^{p/2} - 1$ contain the non-trivial factors of N. Obviously $2^{6/2} + 1 = 9$ and $2^{6/2} - 1 = 7$ contain the non-trivial factors of 21. We see that:

$$\text{GCD}\,(21, 2^{6/2}+1) = \text{GCD}\,(21, 9) = 3$$

$$\text{GCD}\,(21, 2^{6/2}-1) = \text{GCD}\,(21, 7) = 7$$

$$
\begin{array}{l}
9\ )\ 21\ (\,2 \\
\quad \underline{18} \\
\text{GCD} \quad \nearrow 3\ )\ 9\ (\,3 \\
\qquad\quad \underline{9} \\
\qquad\quad 0
\end{array}
\qquad
\begin{array}{l}
\nearrow 7\ )\ 21\ (\,3 \\
\quad \underline{21} \\
\text{GCD}\quad 0
\end{array}
$$

Indeed 3×7=21.

✓ Can we test with $m = 4$?

We got $N = 21$, $m = 4$, GCD(4,21)=1

Also $4^3 \bmod 21 = 64 \bmod 21 = 1$ but the period $p = 3 =$ odd. We cannot proceed.

✓ Can we test with $m = 5$?

We got $N = 21$, $m = 5$, GCD(5,21)=1.

Also $5^6 \bmod 21 = 15\,625 \bmod 21 = 1$ and $p = 6 =$ even.

$$
\begin{array}{l}
21\ )\ 126\ (\,6 \\
\quad\ \underline{126} \\
\qquad\ 0 = \text{Remainder}
\end{array}
$$

Now $m^{p/2} + 1 = 5^{6/2} + 1 = 126$ and since $126 = 6(21) + 0$, so $126 = 0 \,(\text{mod } 21)$ and so the second condition $m^{p/2} + 1 \neq 0 (\text{mod } N)$ is not satisfied and so we cannot proceed.

✓ Can we test with $m = 8$?

We got $N = 21$, $m = 8$, GCD(8,21)=1.

Also $6^2 \bmod 21 = 64 \bmod 21 = 1$ and the period $p = 2 =$ even. So first condition of $p =$ even is satisfied.

Also $m^{p/2} + 1 = 8^{2/2} + 1 = 9$

and since $9 = 0(21) + 9$ so $9 \,(\text{mod }21) = 9 \neq 0$.

So the second condition $m^{p/2} + 1 \neq 0(\text{mod } N)$ is also satisfied.

$$
\begin{array}{l}
21\ )\ 9\ (\,0 \\
\quad\ \underline{0} \\
\qquad 9 = \text{Remainder}
\end{array}
$$

Clearly for this choice of $m = 8$ both the conditions for Shor's algorithm are satisfied.

Then in the equation $(m^{p/2} + 1)(m^{p/2} - 1) = 0 \pmod{N}$ the factors $m^{p/2} + 1$ and $m^{p/2} - 1$ contain the non-trivial factors of N. Obviously $8^{2/2} + 1 = 9$ and $8^{2/2} - 1 = 7$ contain the non-trivial factors of 21. We see that

$$\text{GCD}(21, 8^{2/2}+1) = \text{GCD}(21, 9) = 3$$
$$\text{GCD}(21, 8^{2/2}-1) = \text{GCD}(21, 7) = 7$$

```
          9 ) 21 ( 2              7 ) 21 ( 3
              18                      21
         3 ) 9 ( 3    GCD           0
    GCD      9
             0
```

Indeed 3×7=21.

Find factors of 35 through Shor's algorithm.

Given a number $N = 35 = $ composite number.

Choose a number $m < N$ and co-prime with $N = 35$ say $m = 13$ as can be checked using GCD(13,35) = 1.

```
13 ) 35 ( 2
     26
      9 ) 13 ( 1
          9
          4 ) 9 ( 2
              8
              1 ) 4 ( 4
        GCD       4
                  0
```

Now find various powers 13^a viz.

$13^1 = 13$, $13^2 = 169$, $13^3 = 2197$, $13^4 = 28561$ and define the function $F_{21}(a) = 13^a \mod 35$, e.g.

$$F_{35}(2) = 13^2 \mod 35 = 169 \mod 35 = 29 \neq 1,$$

$$F_{35}(3) = 13^3 \mod 35 = 2197 \mod 35 = 27 \neq 1,$$

$$F_{35}(4) = 13^4 \mod 35 = 28561 \mod 35 = 1.$$

For $m = 13$, we thus have got the smallest value of a, i.e. $p = 4$, for which $m^p \mod N = 13^4 \mod 35 = 1$. So period $p = 4 = $ even. So first condition of $p = $ even is satisfied.

Also $m^{p/2} + 1 = 13^{4/2} + 1 = 170$ and since $170 = 4(35) + 30$ so $170 \pmod{35} = 30$.

So $m^{p/2} + 1 = 13^{4/2} + 1 = 170 \mod 35 = 30 \neq 0 \pmod{N = 35}$ which is the second condition is also satisfied.

Clearly for this choice of $m = 13$ both the conditions for Shor's algorithm are satisfied.

Then in the equation $(m^{p/2} + 1)(m^{p/2} - 1) = 0 \pmod{N}$ the factors $m^{p/2} + 1$ and $m^{p/2} - 1$ contain the non-trivial factors of N. Obviously $13^{4/2} + 1 = 170$ and $13^{4/2} - 1 = 168$ contain the non-trivial factors of 35. We see that:

$$\text{GCD}(35, 13^{4/2}+1) = \text{GCD}(35, 170) = 5$$
$$\text{GCD}(35, 13^{4/2}-1) = \text{GCD}(35, 168) = 7$$

```
35 ) 170 ( 4        35 ) 168 ( 4
     140                 140
  30 ) 35 ( 1         28 ) 35 ( 1
       30                  28
    5 ) 30 ( 6          7 ) 28 ( 4
GCD    30          GCD      28
        0                    0
```

Indeed 5×7=35.

In illustrations we have chosen small numbers for quick elucidation of the principle.

Shor's algorithm—implementation

Consider an l qubit register. Choose l such that $N^2 < 2^l < 2N^2$. For performing QFT smoothly let us choose $Q = 2^l$.

Initialise two l qubit registers to null state.

$$|\psi_0> = |0> \otimes |0> .$$

Apply QFT on first register (Hadamard gates) to get a uniform linear combination of computational basis states. So state of the system is

$$|\psi_1> = \frac{1}{\sqrt{Q}} \sum_{x=0}^{Q-1} |x>|0>$$

Choose a random number $m < N$ and co-prime with N. Apply the oracle to compute $m^a \bmod N$ and store the result in the second register.

As an illustration take $N=55$, $m = 13$. Let us try to find the period through the equation $m^p = 1 \pmod{N}$, i.e. $13^p = 1 \pmod{55}$ (table 8.14).

$$13^p = 13^{20} = 1 \pmod{55} \Rightarrow p = 20.$$

$$N = 55, \quad N^2 = 55^2 = 3025, \quad 2N^2 = 6050$$

As $N^2 < 2^l < 2N^2$ means $3025 < 2^l < 6050$ this implies that $l = 12$ since $2^{12} = 2048$, $2^{12} = 4096$, $2^{13} = 8192$.

So we choose a 12-qubit register.

$|\psi_1>$ obtained by passing the null state of the first register through the Hadamard gate is given by

$$|\psi_1> = \frac{1}{\sqrt{2^{12}}}[|0,0> + |1,0> + \ldots\ldots + |4095,0>]$$

$$= \frac{1}{\sqrt{4096}}[|0,0> + |1,0> + \ldots\ldots + |4095,0>]$$

Table 8.14. Finding p in $13^p = 1 \pmod{55}$.

$13^1 = \boxed{13} \pmod{55}$	$13^{12} = 13^8 \, 13^4 \pmod{55}$
$13^2 = 169 \pmod{55} = \boxed{4} \pmod{55}$	$= 13^8 \pmod{55} \, 13^4 \pmod{55}$
$13^3 = 13^2 \, 13^1 \pmod{55}$	$= 36.16 \pmod{55}$
$\quad = 13^2 \pmod{55} \, 13^1 \pmod{55}$	$= 576 \pmod{55} = \boxed{26} \pmod{55}$
$\quad = 4.13 \pmod{55} = \boxed{52} \pmod{55}$	$13^{13} = 13^8 \, 13^5 \pmod{55}$
$13^4 = (13^2)^2 = 4^2 \pmod{55} = \boxed{16} \pmod{55}$	$\quad = 13^8 \pmod{55} \, 13^5 \pmod{55}$
$13^5 = 13^4 \, 13^1 \pmod{55}$	$\quad = 36.43 \pmod{55}$
$\quad = 13^4 \pmod{55} \, 13^1 \pmod{55}$	$\quad = 1548 \pmod{55} = \boxed{8} \pmod{55}$
$\quad = 16.13 \pmod{55}$	$13^{14} = 13^8 \, 13^6 \pmod{55}$
$\quad = 208 \pmod{55} = \boxed{43} \pmod{55}$	$\quad = 13^8 \pmod{55} \, 13^6 \pmod{55}$
$13^6 = 13^4 \, 13^2 \pmod{55}$	$\quad = 36.9 \pmod{55}$
$\quad = 13^4 \pmod{55} \, 13^2 \pmod{55}$	$\quad = 324 \pmod{55} = \boxed{49} \pmod{55}$
$\quad = 16.4 \pmod{55}$	$13^{15} = 13^8 \, 13^7 \pmod{55}$
$\quad = 64 \pmod{55} = \boxed{9} \pmod{55}$	$\quad = 13^8 \pmod{55} \, 13^7 \pmod{55}$
$13^7 = 13^4 \, 13^3 \pmod{55}$	$\quad = 36.7 \pmod{55}$
$\quad = 13^4 \pmod{55} \, 13^3 \pmod{55}$	$\quad = 252 \pmod{55} = \boxed{32} \pmod{55}$
$\quad = 16.52 \pmod{55}$	$13^{16} = (13^8)^2 = 36^2 \pmod{55}$
$\quad = 832 \pmod{55} = \boxed{7} \pmod{55}$	$\quad = 1296 \pmod{55} = \boxed{31} \pmod{55}$
$13^8 = (13^4)^2 = 16^2 \pmod{55} = 256 \pmod{55}$	$13^{17} = 13^{16} 13^1 \pmod{55}$
$\quad = \boxed{36} \pmod{55}$	$\quad = 13^{16} \pmod{55} 13^1 \pmod{55}$
$13^9 = 13^8 \, 13^1 \pmod{55}$	$\quad = 31.13 \pmod{55}$
$\quad = 13^8 \pmod{55} \, 13^1 \pmod{55}$	$\quad = 403 \pmod{55} = \boxed{18} \pmod{55}$
$\quad = 36.13 \pmod{55}$	$13^{18} = 13^{16} 13^2 \pmod{55}$
$\quad = 468 \pmod{55} = \boxed{28} \pmod{55}$	$\quad = 13^{16} \pmod{55} 13^2 \pmod{55}$
$13^{10} = 13^8 \, 13^2 \pmod{55}$	$\quad = 31.4 \pmod{55}$
$\quad = 13^8 \pmod{55} \, 13^2 \pmod{55}$	$\quad = 124 \pmod{55} = \boxed{14} \pmod{55}$
$\quad = 36.4 \pmod{55}$	$13^{19} = 13^{16} 13^3 \pmod{55}$
$\quad = 144 \pmod{55} = \boxed{34} \pmod{55}$	$\quad = 13^{16} \pmod{55} 13^3 \pmod{55}$
$13^{11} = 13^8 \, 13^3 \pmod{55}$	$\quad = 31.52 \pmod{55}$
$\quad = 13^8 \pmod{55} \, 13^3 \pmod{55}$	$\quad = 1612 \pmod{55} = \boxed{17} \pmod{55}$
$\quad = 36.52 \pmod{55}$	
$\quad = 1872 \pmod{55} = \boxed{2} \pmod{55}$	

$$13^{20} = 13^{16} 13^4 \pmod{55} = 13^{16} \pmod{55} \, 13^4 \pmod{55} = 31.16 \pmod{55}$$
$$= 496 \pmod{55} = \boxed{1} \pmod{55} \quad \Rightarrow \; p = 20$$

Apply the oracle to compute $m^a = 13^a \bmod N$ and store the result in the second (target) register. The second register had 0 and so the oracle will give the value of the function without destroying the first register value.

$$|\psi_2 \rangle \; =$$

$$\frac{1}{\sqrt{4096}} [\, |0, \, 13^0 (\bmod \; 55) \rangle + |1, \, 13^1 (\bmod \; 55) \rangle + |2, \, 13^2 (\bmod \; 55) \rangle$$

$$+ \ldots\ldots + |20, \, 13^{20} (\bmod \; 55) \rangle + \ldots\ldots + |4095, \, 13^{4095} (\bmod \; 55) \rangle \,]$$

$$|\psi_2> =$$

$$\frac{1}{\sqrt{4096}}[|0,1(\text{mod } 55) > +|1,13(\text{mod } 55) > +|2,4(\text{mod } 55) >$$

$$+.......... + |20,1(\text{mod } 55) > +........ + |4095,32(\text{mod } 55) >]$$

(since $13^{4095} = 13^{4080+15} = 13^{4080}13^{15} = (13^{20})^{204}13^{15}$hence

$$13^{4095} \text{ mod } 55 = (13^{20})^{204} \text{ mod } 55.13^{15} \text{ mod } 55$$

$$=(13^{20} \text{ mod } 55)^{204}.13^{15} \text{ mod } 55=1 \text{ mod } 55.32 \text{ mod } 55$$

$=32 \text{ mod } 55)$ (table 8.13)

If we measure the second register we would get a random value, i.e. any of 13^a (mod 55)= 13,4,52,16,43,9,7,36,28,34,2,26,8,49,32,31,18,14,17,1. (table 8.13)

Let it be 9 (recall that $13^6 = 9(\text{mod } 55)$) and 9 appears 205 times in between 4096 bits (starting from 0 to 4095; periodic recurrence being at

$$6,6+1.20=26,6+2.20=46,..................,6+204.20=4086$$

and 9 occurs once more in the rest up to 4095).

So there are 205 values in the first register corresponding to the second register having value 9 with first register starting at 6 and ending at 4086. Thus we can write

$$|\psi_3 > =\frac{1}{\sqrt{205}}[|6,9 > +|26,9 > +..........|4086,6 >]$$

$$=\frac{1}{\sqrt{205}} \sum_{d=0}^{205-1} |6 + d.20,>$$

If M = number of states in the first register corresponding to a given value in the second register (= 205 in the example shown corresponding to second register value 9), k = measurement in the second register (=9 in the example), x_0=the starting value (= 6 in the example), p= period (=20 in the example) then we can represent the general structure as (run index d = position in the period)

$$|\psi_3 > =\frac{1}{\sqrt{M}} \sum_{d=0}^{M-1} |x_0 + d. p, k>$$

QFT on the first register which has Q number of bits gives

$$|\psi_4 > =\frac{1}{\sqrt{Q}} \sum_{y=0}^{Q-1} \frac{1}{\sqrt{M}} \sum_{d=0}^{M-1} e^{2\pi i y(x_0+dp)/Q}|y \ k>$$

$$=\frac{1}{\sqrt{QM}} \sum_{y=0}^{Q-1} e^{2\pi i y x_0/Q} \sum_{d=0}^{M-1} e^{2\pi i y dp/Q}|y \ k>$$

$$= \frac{1}{\sqrt{QM}} \sum_{y=0}^{Q-1} e^{2\pi i y x_0/Q} \left(\sum_{d=0}^{M-1} z^d \right) |y \ k\rangle \ \text{where} \ z = e^{2\pi i y p/Q}$$

If we measure the first register we would get a particular value of $|y\rangle$ with a probability given by

$$P(y) = \left| \frac{1}{\sqrt{QM}} e^{2\pi i y x_0/Q} \left(\sum_{d=0}^{M-1} z^d \right) \right|^2 = \frac{1}{QM} \left| \sum_{d=0}^{M-1} z^d \right|^2$$

We note that $\displaystyle\sum_{d=0}^{M-1} z^d =$ geometric series with common ratio z and is a finite sum

$$= \frac{1 - z^M}{1 - z} = \frac{z^{M/2}(z^{-M/2} - z^{M/2})}{z^{1/2}(z^{-1/2} - z^{1/2})}$$

$$\left| \sum_{d=0}^{M-1} z^d \right| = \left| \frac{z^{M/2}(z^{-M/2} - z^{M/2})}{z^{1/2}(z^{-1/2} - z^{1/2})} \right|$$

$$= \frac{|z^{M/2}|}{|z^{1/2}|} \cdot \left| \frac{z^{-M/2} - z^{M/2}}{z^{-1/2} - z^{1/2}} \right| = \left| \frac{z^{M/2} - z^{-M/2}}{z^{1/2} - z^{-1/2}} \right|$$

since $|z| = |e^{2\pi i y p/Q}| = 1., |z^{M/2}| = 1, |z^{1/2}| = 1$

$$\left| \sum_{d=0}^{M-1} z^d \right| = \left| \frac{e^{i\pi y p M/Q} - e^{-i\pi y p M/Q}}{e^{i\pi y p/Q} - e^{-i\pi y p/Q}} \right| = \left| \frac{\sin \dfrac{\pi y p M}{Q}}{\sin \dfrac{\pi y p}{Q}} \right| \quad (\text{as } z = e^{2\pi i y p/Q})$$

Measurement of the first register will give $|y\rangle$ with the probability

$$P(y) = \frac{1}{QM} \left| \left(\sum_{d=0}^{M-1} z^d \right) \right|^2 = \frac{1}{QM} \left| \frac{\sin \dfrac{\pi y p M}{Q}}{\sin \dfrac{\pi y p}{Q}} \right|^2 = \frac{1}{QM} \left| \frac{\sin \xi M}{\sin \xi} \right|^2 \quad \left(\text{where } \xi = \frac{\pi y p}{Q} \right)$$

Now

$P(y)=$ maximum (interferes constructively), when denominator $= 0$, i.e. when $\sin \frac{\pi y p}{Q} = 0 = \sin n \ \pi$, $n=$ integer $\Rightarrow \frac{y p}{Q} = n =$ integer. The probability then is

$$P(y) = \frac{1}{QM} \left| \frac{\sin \xi M}{\sin \xi} \right|^2 = \frac{1}{QM} M^2 = \frac{M}{Q}$$

since for $\xi = \frac{\pi y p}{Q} \to 0, \ \underset{\xi \to 0}{Lt} |\frac{\sin \xi M}{\sin \xi}|^2 = M^2$

$$P(y) = \frac{M}{Q} = \frac{205}{4096} \approx 0.05 \to 5\%.$$

For $\frac{yp}{Q}$ = non − integral the terms add incoherently (destructively interfere).

Measurement of first register will project out those values (with significant probability) for which y is a multiple of 205 say 410 or a close value of it say 408. Moving away means probability falls.

The task is now to find the period p from probability and we can do it by following the method of continued fraction on $\frac{y}{Q} = \frac{n}{p}$. (table 8.15)

We can express $\frac{y}{Q}$ as a continued fraction and look at various convergences of it. We shall approximate (i.e. stop) when we get a denominator larger than $N = 55$ as illustrated in the following. The period will be the approximated denominator or its multiple.

Suppose measured value of y is 408 and $Q = 4096$. Let us express $\frac{408}{4096}$ as a continued fraction.

$$\frac{408}{4096} = \frac{1}{\frac{408}{4096}} = \frac{1}{10 + \frac{16}{408}} = \frac{1}{10 + \frac{1}{\frac{408}{16}}} = \frac{1}{10 + \frac{1}{25 + \frac{8}{16}}} = \frac{1}{10 + \frac{1}{25 + \frac{1}{2}}}$$

$$= [0,10,25,2]$$

$$\xrightarrow{\text{1st convergence}} \frac{1}{10} \quad \text{(approximate)}$$

$$\xrightarrow{\text{2nd convergence}} \frac{1}{10 + \frac{1}{25}} = \frac{25}{251} \quad \text{(we will not take it as } 251 > N = 55\text{)}$$

Table 8.15. Method of continued fraction.

We can express a real number as follows:

$$R = a_0 + \frac{1}{a_1 + \frac{1}{\cdots\cdots + \frac{1}{a_n}}} = [a_0, a_1, \ldots\ldots\ldots, a_j] \qquad \text{(convergence upto jth order)}$$

Example

$$R = \frac{15}{4} = 3 + \frac{3}{4} = 3 + \frac{1}{\frac{4}{3}} = 3 + \frac{1}{1 + \frac{1}{3}} = [3,1,3]$$

$$R = \frac{4}{15} = 0 + \frac{1}{\frac{15}{4}} = 0 + \frac{1}{3 + \frac{3}{4}} = 0 + \frac{1}{3 + \frac{1}{\frac{4}{3}}} = 0 + \frac{1}{3 + \frac{1}{1 + \frac{1}{3}}} = [0,3,1,3]$$

Upon approximation we get

$$\frac{y}{Q} = \frac{n}{p} = \frac{1}{10}.$$

So the possibilities are
$p = n(10) = 10,20,30,40,50$ (we cannot take $p = 60$ as $N = 55$ is exceeded).
Now we will find 13^p:
$13^{10} = 34(\mod 55)$, $13^{20} = 1(\mod 55)$.(table 8.14)
So $p = 20$. Period is determined.
Clearly because of the efficiency of the QFT, Shor's algorithm is able to execute integer factorization in polynomial time (i.e. time taken is polynomial in $\log N$, where N is the integer to be factored). This is exponentially faster than classical factoring algorithms.

Shor's algorithm is probabilistic, i.e. provides answers with probability. And probability of failure decreases by repeating the algorithm.

8.17 Further reading

[1] Ghosh D 2016 Online course on quantum information and computing (Bombay: Indian Institute of Technology)
[2] Nielsen M A and Chuang I L 2019 *Quantum Computation and Quantum Information* (Cambridge: Cambridge University Press)
[3] Nakahara M and Ohmi T 2017 *Quantum Computing* (Boca Raton, FL: CRC Press)
[4] Pathak A 2016 *Elements of Quantum Computing and Quantum Communication* (Boca Raton, FL: CRC Press)

8.18 Problems

1. *Find the continued fraction representation of* $\frac{61}{45}$.

   Ans. $\frac{61}{45} = 1 + \frac{16}{45} = 1 + \frac{1}{\frac{45}{16}} = 1 + \frac{1}{2+\frac{13}{16}}$

   $= 1 + \cfrac{1}{2 + \cfrac{1}{\frac{16}{13}}} = 1 + \cfrac{1}{2 + \cfrac{1}{1 + \frac{3}{13}}} = 1 + \cfrac{1}{2 + \cfrac{1}{1 + \cfrac{1}{\frac{13}{3}}}} = 1 + \cfrac{1}{2 + \cfrac{1}{1 + \cfrac{1}{4 + \frac{1}{3}}}}$

   $= [1,2,1,4,3]$

2. *In the Deutsch algorithm* $|x> = \frac{|0> + |1>}{\sqrt{2}}$, $|y> = \frac{|0> - |1>}{\sqrt{2}}$ *are the first and second input registers, respectively. If the oracle calculates the function for the case* $f(0) = 0, f(1) = 1$ *find the operator* $\hat{U}_f$ *that can be a substitute for the oracle.*

   Ans. Input $= |x> \otimes |y> \xrightarrow{\hat{U}_f} |x> \otimes |y \oplus f(x)> =$ Output

$$\text{Input} = |x> \otimes |y> = \frac{|0> + |1>}{\sqrt{2}} \otimes \frac{|0> - |1>}{\sqrt{2}}$$

$$= \frac{1}{2}(|00> + |10> - |01> - |11>)$$

$$\text{Output} = |x> \otimes |y \oplus f(x)> = |x, \ y \oplus f(x)>$$

$$= \frac{1}{2}(|0, \ 0 \oplus f(0)> + |1, \ 0 \oplus f(1)> - |0, \ 1 \oplus f(0)> - |1, \ 1 \oplus f(1)>)$$

Given: $f(0) = 0, f(1) = 1$

$$\text{Output} = \frac{1}{2}(|0, \ 0 \oplus 0> + |1, \ 0 \oplus 1> - |0, \ 1 \oplus 0> - |1, \ 1 \oplus 1>)$$

$$= \frac{1}{2}(|00> + |11> - |01> - |10>)$$

Clearly the input and output are related by a CNOT with control = first qubit and target = second qubit, i.e.

$$\frac{1}{2}(|00> + |10> - |01> - |11>) \overset{\hat{U}_f = CNOT}{\longrightarrow} \frac{1}{2}(|00> + |11> - |01> - |10>)$$

3. *In using Shor's algorithm for N = 21 check which of the numbers 2,3,4,5 can be used to determine the period of m^a?*
   Ans. GCD(m, N) = 1
   Clearly the numbers 2,4,5 are co-prime with 21 (3 is a factor of 21). So 3 cannot be used as m^a but 2,4,5 can be used.
4. *What is the period of 11 mod 15?*
   Ans. Period of any positive integer m mod any natural number n i.e. m mod n is the smallest integer $p > 0$ that satisfies m^p mod $n=1$. Now

$$11^0 \text{ mod } 15=1$$
$$11^1 \text{ mod } 15=11$$
$$11^2 \text{ mod } 15=121 \text{ mod } 15=1.$$

So period $p = 2$
5. *Write down the order in which gates are applied in implementing QFT on a four-qubit register (after the SWAP operation). Use the notation in which B_{jk} represents the gate applied on the j th qubit with k th qubit as control and H_i is the Hadamard gate applied on the i th qubit.*

Find the number of B_{jk} gates that are needed to implement QFT on a four-qubit register.

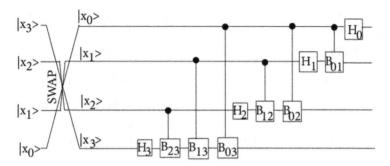

Figure 8.12. QFT implementation for $n = 4$.

Find the number of B_{jk} gates that are needed to implement QFT on a l-qubit register.

Ans. Refer to the circuit diagram of figure 8.12.

The sequence of gates as evident from the QFT implementation circuit of figure 8.12 is

$$\hat{U}_{QFT} \longrightarrow (\hat{H}_0)(\hat{B}_{01}\hat{H}_1)(\hat{B}_{02}\hat{B}_{12}\hat{H}_2)(\hat{B}_{03}\hat{B}_{13}\hat{B}_{23}\hat{H}_3)$$

- The number of B_{jk} gates that are needed to implement QFT on the four-qubit register is

  $0+1(\hat{B}_{01})+2(\hat{B}_{12}\hat{B}_{02})+3(\hat{B}_{23}\hat{B}_{13}\hat{B}_{03})$, i.e. $0+1+2 + (4-1) = 6$ for four-qubit register.

  (For a 3-qubit register the number will be $0+1 + (3-1) = 3$).

- For l-qubit register the number will be given by

$0+1+2+3+.. + (l - 1)$

$$=1+2+3+....(l - 1) = \frac{(l - 1)(l - 1+1)}{2} = \frac{l(l - 1)}{2}$$

6. *Suppose the target input bit of the Deutsch algorithm is |1> and measurement of the first register gives the state |1> then comment on the nature of function (constant or balanced).*

    Ans. Measurement of the first register gives 1. So it is a balanced function (details done in Deutsch algorithm, section 8.5).

7. *In the Deutsch–Jozsa algorithm, suppose the input is a uniform linear combination of two-qubit basis states $|00>$, $|01>$, $|10>$, $|11>$. The function has the property*
    (a) $f(00) = 0, f(01) = f(10) = f(11) = 1$
    (b) $f(00) = f(11) = 0, \cdot f(01) = f(10) = 1$

What will be the state of the first register after execution of the algorithm (before measurement of state)?

Ans. Inputs are $|x> = \frac{1}{\sqrt{2^2}}(|00> + |01> + |10> + |11>)$, $|y> = \frac{|0> - |1>}{\sqrt{2}}$

Input to the oracle $= |x> \otimes |y> = \frac{1}{2}(|00> + |01> + |10> + |11>) \otimes \frac{|0> - |1>}{\sqrt{2}}$

$= \frac{1}{2\sqrt{2}}(|000> + |010> + |100> + |110> - |001> - |011> - |101> - |111>)$

(First two qubits are of x and the third qubit is of y).

(a) Output of the oracle $= |x> \otimes |y \oplus f(x)> = |x \ y \oplus f(x)>$

$= \frac{1}{2\sqrt{2}}[|00 \ 0 \oplus f(00)> + |01 \ 0 \oplus f(01)> + |10 \ 0 \oplus f(10)> + |11 \ 0 \oplus f(11)>$

$- |00 \ 1 \oplus f(00)> - |01 \ 1 \oplus f(01)> - |10 \ 1 \oplus f(10)> - |11 \ 1 \oplus f(11)>]$

Given $f(00) = 0, f(01) = f(10) = f(11) = 1$

Output $= \frac{1}{2\sqrt{2}}[|00 \ 0 \oplus \ 0> + |01 \ 0 \oplus \ 1> + |10 \ 0 \oplus 1> + |11 \ 0 \oplus 1>$

$- |00 \ 1 \oplus \ 0> - |01 \ 1 \oplus \ 1> - |10 \ 1 \oplus 1> - |11 \ 1 \oplus 1>]$

Output $= \frac{1}{2\sqrt{2}}[|000> + |011> + |101> + |111> - |001> - |010>$

$- |100> - |110>]$

Rearranging,

Output $= \frac{1}{2\sqrt{2}}[(|000> - |001>) + (|011> - |010>) + (|101>$

$- |100>) + (|111> - |110>)]$

$= \frac{1}{2}[|00> \frac{|0> - |1>}{\sqrt{2}} - |01> \frac{|0> - |1>}{\sqrt{2}} - |10> \frac{|0> - |1>}{\sqrt{2}} - |11> \frac{|0> - |1>}{\sqrt{2}}]$

Output $= \frac{1}{2}[|00> - |01> - |10> - |11>]\frac{|0> - |1>}{\sqrt{2}}$

$= \frac{1}{2}\left[\begin{pmatrix} 1 \\ 0 \\ 0 \\ 0 \end{pmatrix} - \begin{pmatrix} 0 \\ 1 \\ 0 \\ 0 \end{pmatrix} - \begin{pmatrix} 0 \\ 0 \\ 1 \\ 0 \end{pmatrix} - \begin{pmatrix} 0 \\ 0 \\ 0 \\ 1 \end{pmatrix}\right]\frac{|0> - |1>}{\sqrt{2}} = \frac{1}{2}\begin{pmatrix} 1 \\ -1 \\ -1 \\ -1 \end{pmatrix}\frac{|0> - |1>}{\sqrt{2}}$

Hadamard is applied on first register of two qubits. The operator has the form

$$\hat{H} \otimes \hat{H} = \frac{1}{\sqrt{2}}\begin{pmatrix} 1 & 1 \\ 1 & -1 \end{pmatrix} \otimes \frac{1}{\sqrt{2}}\begin{pmatrix} 1 & 1 \\ 1 & -1 \end{pmatrix} = \frac{1}{2}\begin{pmatrix} 1\begin{pmatrix} 1 & 1 \\ 1 & -1 \end{pmatrix} & 1\begin{pmatrix} 1 & 1 \\ 1 & -1 \end{pmatrix} \\ 1\begin{pmatrix} 1 & 1 \\ 1 & -1 \end{pmatrix} & -1\begin{pmatrix} 1 & 1 \\ 1 & -1 \end{pmatrix} \end{pmatrix}$$

$$= \frac{1}{2}\begin{pmatrix} 1 & 1 & 1 & 1 \\ 1 & -1 & 1 & -1 \\ 1 & 1 & -1 & -1 \\ 1 & -1 & -1 & 1 \end{pmatrix}$$

This gives the output to be

$$\hat{H} \otimes \hat{H} \frac{1}{2}\begin{pmatrix} 1 \\ -1 \\ -1 \\ -1 \end{pmatrix}\frac{|0> - |1>}{\sqrt{2}} = \frac{1}{2}\begin{pmatrix} 1 & 1 & 1 & 1 \\ 1 & -1 & 1 & -1 \\ 1 & 1 & -1 & -1 \\ 1 & -1 & -1 & 1 \end{pmatrix}\frac{1}{2}\begin{pmatrix} 1 \\ -1 \\ -1 \\ -1 \end{pmatrix}\frac{|0> - |1>}{\sqrt{2}}$$

$$= \frac{1}{4}\begin{pmatrix} -2 \\ 2 \\ 2 \\ 2 \end{pmatrix}\frac{|0> - |1>}{\sqrt{2}}$$

$$\text{Output} = \frac{1}{2}\begin{pmatrix} -1 \\ 1 \\ 1 \\ 1 \end{pmatrix}\frac{|0> - |1>}{\sqrt{2}} = \frac{1}{2}\left[-\begin{pmatrix} 1 \\ 0 \\ 0 \\ 0 \end{pmatrix} + \begin{pmatrix} 0 \\ 1 \\ 0 \\ 0 \end{pmatrix} + \begin{pmatrix} 0 \\ 0 \\ 1 \\ 0 \end{pmatrix} + \begin{pmatrix} 0 \\ 0 \\ 0 \\ 1 \end{pmatrix}\right]\frac{|0> - |1>}{\sqrt{2}}$$

$$\text{Output} = \frac{1}{2}[-|00> + |01> + |10> + |11>]\frac{|0> - |1>}{\sqrt{2}}$$

The state of the first register after execution of the algorithm (before measurement of state) is $\frac{1}{2}[-|00> + |01> + |10> + |11>]$

(b) Output of the oracle $= |x> \otimes |y \oplus f(x)> = |x\ y \oplus f(x)>$

$$= \frac{1}{2\sqrt{2}}[|00\ 0 \oplus f(00)> + |01\ 0 \oplus f(01)> + |10\ 0 \oplus f(10)> + |11\ 0 \oplus f(11)>$$

$$- |00\ 1 \oplus f(00)> - |01\ 1 \oplus f(01)> - |10\ 1 \oplus f(10)> - |11\ 1 \oplus f(11)>]$$

Given $f(00) = f(11) = 0$, $f(01) = f(10) = 1$

$$\text{Output} = \frac{1}{2\sqrt{2}}[|00\ 0 \oplus\ 0> + |01\ 0 \oplus\ 1> + |10\ 0\oplus 1> + |11\ 0\oplus 0>$$

$$- |00\ 1 \oplus\ 0> - |01\ 1 \oplus\ 1> - |10\ 1\oplus 1> - |11\ 1\oplus 0>]$$

$$\text{Output} = \frac{1}{2\sqrt{2}}[|000> +|011> +|101> +|110> - |001> - |010>$$
$$- |100> - |111>]$$

Rearranging,

$$\text{Output} = \frac{1}{2\sqrt{2}}[(|000> - |001>) + (|011> - |010>) + (|101> - |100>)$$
$$+ (|110> - |111>)]$$
$$=\frac{1}{2}[|00> \frac{|0> - |1>}{\sqrt{2}} - |01> \frac{|0> - |1>}{\sqrt{2}} - |10> \frac{|0> - |1>}{\sqrt{2}}$$
$$+ |11> \frac{|0> - |1>}{\sqrt{2}}]$$
$$=\frac{1}{2}[|00> - |01> - |10> +|11>]\frac{|0> - |1>}{\sqrt{2}}$$
$$=\frac{1}{2}[(|00> - |10>) - (|01> - |11>)]\frac{|0> - |1>}{\sqrt{2}}$$
$$=\frac{1}{2}[(|0> - |1>)|0> - (|0> - |1>)|1>]\frac{|0> - |1>}{\sqrt{2}}$$
$$\text{Output} = \frac{|0> - |1>}{\sqrt{2}}\frac{|0> - |1>}{\sqrt{2}}\frac{|0> - |1>}{\sqrt{2}}$$

Applying Hadamard gate on the first two qubits we have

$$\text{Output} = (|1> \otimes|1>)\frac{|0> - |1>}{\sqrt{2}}$$

$$\text{Output} = 11 > \frac{|0> - |1>}{\sqrt{2}}$$

The state of the first register after execution of the algorithm (before measurement of state) is $|11>$

8. *How many iterations are needed to find a marked state among 64 states in Grover's algorithm?*

Ans. $N = 64$, $\sin \theta = \frac{1}{\sqrt{N}} = \frac{1}{\sqrt{64}} = \frac{1}{8} \Rightarrow \theta = 7.18°$.

Angle between $|s>$ and $|w>$ is $\frac{\pi}{2} - \theta = \frac{\pi}{2} - 7.18° = 82.81°$.

For every iteration angle traced $=2\theta = 2(7.18°)=14.36°$

Number of iterations needed to approach 82.81° as closely as possible is determined from the following observations:

$$82.81° - 5(14.36°) = 11.01°$$

$$82.81° − 6(14.36°) = −3.35°$$

So required number is 6.

---

9. *Consider the description of a two-to-one function* $f(x)$
corresponding to three-qubit input x. There exists a code ξ
such that $f(x) = f(y)$ *if and only if* $x + \xi = y$. *Find* ξ.
A Simon algorithm is executed and the second register is
measured that gives $|110\rangle$. *If the first register is passed*
through Hadamard gates then what will be the result of
measurement on the first register?

x	$f(x)$
000	101
001	010
010	000
011	110
100	000
101	110
110	101
111	010

Ans. From the table it follows that $f(x) = 101$ for both $x = 000$ and
$x = 110 = y$ say. So we have found x and y such that $f(x) = f(y)$. So using
$x + \xi = y \Rightarrow \xi = x \oplus y$
$\xi = 000 \oplus 110 = 110$. So $|\xi\rangle = |110\rangle$.
The circuit to implement Simon algorithm is shown in figure 8.6.
We have two registers of bit size $n = 3$. From equation (8.15)

$$|\psi_0\rangle = |0\rangle^{\otimes n} \otimes |0\rangle^{\otimes n} = |0\rangle^{\otimes 3} \otimes |0\rangle^{\otimes 3} = |000\rangle \otimes |000\rangle$$

Hadamard gate on the first register (keeping second register intact) creates equal
superposition of all possible strings of length three in the first register. From
equation (8.17)

$$|\psi_1\rangle = [H^{\otimes 3} \otimes |000\rangle] \otimes |000\rangle = |X\rangle \otimes |y\rangle$$

where we used equation (8.16) viz.

$$H^{\otimes 3} \otimes |000\rangle = \frac{1}{\sqrt{2^3}} \sum_{x=0}^{2^3-1} |x\rangle = |X\rangle \quad (2^3 = 8 \text{ strings})$$

$$= \frac{1}{4}(|000\rangle + |001\rangle + |010\rangle + .|011\rangle + |100\rangle + |101\rangle + |110\rangle + |111\rangle)$$

The oracle U_f acts on $|\psi_i>$ (equation (8.18)) to give

$$|\psi_2> \ = U_f|\psi_1> \ = U_f|X> \otimes |y> \ = |X> \otimes U_f|y>$$

Using equation (8.19): $U_f|y> \ = |y \oplus f(x)> \ = |0 \oplus f(x)> \ = |f(x)>$ we have (equation (8.20))

$$|\psi_2> \ = |X> \otimes U_f|y> = |X> \otimes |f(x)>$$

From the table we expand and write, noting the content of first register x and second register $f(x)$

$$|\psi_2> \ = \frac{1}{2\sqrt{2}}[|000> |101> +|001> |010> +|010> |000> +|011> |110>$$

$$+|100> |000> +|101> |110> +|110> |101> +|111> |010>]$$

Suppose measurement on the second register yields a particular value $f(x_0) = 110$. After measurement, the first register is no longer in superposition of all inputs but it will have only the inputs that produce the output $f(x_0) = 110$. From the table this output $|f(x_0)> \ = |110>$ corresponds to the inputs 011 and 101. From equation (8.21) we thus have

$$|\psi_3> \ = \frac{1}{\sqrt{2}}(|011> +|101>) \otimes |110>$$

Now apply Hadamard on the first register without affecting the second register (equation (8.22))

$$|\psi_4> \ = [H^{\otimes 3} \frac{1}{\sqrt{2}}(|011> +|101>)] \otimes |110>$$

$$= \frac{1}{\sqrt{2}}\left[\frac{|0> +|1>}{\sqrt{2}} \ \frac{|0> - |1>}{\sqrt{2}} \ \frac{|0> - |1>}{\sqrt{2}}\right]$$

$$+ \frac{1}{\sqrt{2}}\left[\frac{|0> - |1>}{\sqrt{2}} \ \frac{|0> +|1>}{\sqrt{2}} \ \frac{|0> - |1>}{\sqrt{2}}\right] \otimes |110>$$

$$= \frac{1}{4}[(|00> - |01> +|10> - |11>)(|0> - |1>)$$

$$+ (|00> +|01> - |10> - |11>)(|0> - |1>)] \otimes |110>$$

$$= \frac{1}{4}(|00> - |01> +|10> - |11> +|00> +|01> - |10> - |11>)(|0> - |1)>$$

$$= \frac{1}{4}[2(|00> - |11>)(|0> - |1>)] \otimes |110>$$

$$= \frac{1}{2}[|000> - |001> - |110> +|111>] \otimes |110>$$

We obtain an equal superposition of $2^{3-1} = 4$ possible k values. Each $|k>$ occurs with probability $|\frac{1}{2}|^2 = \frac{1}{4}$. The result of measurement of the first register is thus

$$\frac{1}{2}(|000> - |001> - |110> +|111 >)$$

10. *Obtain the probability of failure of Grover's algorithm after two application of Grover rotation for identifying a single marked state amidst eight items?*
    Ans. $N = 8$, Amplitude $= \frac{1}{\sqrt{8}} = \frac{1}{2\sqrt{2}}$

    Application of U_w on state $|s>$ inverts component parallel to $|w>$. So amplitude of marked state $a_w = -\frac{1}{2\sqrt{2}}$ (flipped) and for other states $a_x = \frac{1}{2\sqrt{2}}$.

    Mean amplitude $\bar{a} = \frac{1}{8}(7.\frac{1}{2\sqrt{2}} - 1.\frac{1}{2\sqrt{2}}) = \frac{3}{8\sqrt{2}}$.

    Applying U_s: $a_w \rightarrow 2\bar{a} - a_w = 2(\frac{3}{8\sqrt{2}}) - (-\frac{1}{2\sqrt{2}}) = \frac{5}{4\sqrt{2}}$

    $$a_x \rightarrow 2\bar{a} - a_x = 2(\frac{3}{8\sqrt{2}}) - (\frac{1}{2\sqrt{2}}) = \frac{1}{4\sqrt{2}}$$

    After one iteration amplitude of one marked state is five times the other states. So probability of marked state increases $5^2 = 25$ times.

    In the next rotation the same operations are repeated but with initial amplitude as $\frac{5}{4\sqrt{2}}$ for a_w and $\frac{1}{4\sqrt{2}}$ for a_x. After applying U_w again we have

    $a_w = -\frac{5}{4\sqrt{2}}$(flipped), $a_x = \frac{1}{4\sqrt{2}}$.

    The new mean is $\bar{a} = \frac{1}{8}(7.\frac{1}{4\sqrt{2}} - \frac{5}{4\sqrt{2}}) = \frac{1}{16\sqrt{2}}$.

    Apply U_s: $a_x \rightarrow 2\bar{a} - a_x$, $a_w \rightarrow 2\bar{a} - a_w = 2(\frac{1}{16\sqrt{2}}) - (-\frac{5}{4\sqrt{2}}) = \frac{11}{8\sqrt{2}}$.

    Probability of success after two Grover iterations is $|amplitude|^2 = |a_w|^2 = |\frac{11}{8\sqrt{2}}|^2 = \frac{121}{128}$.

    And probability of failure is $1 - \frac{121}{128} = \frac{7}{128}$.

11. *What is the probability of success of Grover's algorithm after optimal number of iterations, for large N?*

    (a) $1 - \frac{1}{2N}$,*(b)* $|1 - \frac{1}{2N}|^2$, *(c)* $\frac{1}{\sqrt{N}}$, *(d)* $1 - \frac{1}{\sqrt{N}}$.

    Ans. (b), (d). (section 8.9)

12. *Consider a database of 16 items, Suppose n iterations of Grover algorithm is done. Find the probability p that it will fail to identify the marked state corresponding to n = 1,2.*

    Ans. $N = 16$. Amplitude is $\frac{1}{\sqrt{16}} = \frac{1}{4}$(this is the coefficient or the normal-ization constant). Hence $a_w = \frac{1}{4}$, $a_x = \frac{1}{4}$.

    Applying U_w: $a_w = -\frac{1}{4}$ (flipped),$a_x = \frac{1}{4}$.

Mean $\bar{a} \to \frac{1}{16}(15.\frac{1}{4} - \frac{1}{4}) = \frac{7}{32}$.

Applying U_s: $a_w \to 2\bar{a} - a_w = 2.\frac{7}{32} - (-\frac{1}{4}) = \frac{11}{16}$

$$a_x \to 2\bar{a} - a_x = 2.\frac{7}{32} - \frac{1}{4} = \frac{3}{16}$$

After $n = 1$ iteration, probability of success is $|\text{amplitude}|^2 = |a_w|^2 = (\frac{11}{16})^2$, probability of failure $p = 1 - (\frac{11}{16})^2 = \frac{135}{256}$.

Second iteration:

Applying U_w: $a_w = -\frac{11}{16}$(flipped), $a_x = \frac{3}{16}$.

Mean $\bar{a} \to \frac{1}{16}(15.\frac{3}{16} - \frac{11}{16}) = \frac{7}{8(16)}$.

Applying U_s: $a_w \to 2\bar{a} - a_w = 2.\frac{17}{8(16)} - (-\frac{11}{16}) = \frac{17}{64} + \frac{11}{16} = \frac{61}{64}$,

after $n = 2$ iteration, probability of success is $|\text{amplitude}|^2 = |a_w|^2 = (\frac{61}{64})^2$, probability of failure $p = 1 - (\frac{61}{64})^2 = \frac{375}{4096}$

13. *Make a chart of the following functions for $x = 0,1,2,3,4,...$and mention whether they are constant function, balanced function or not:*

$$\sin \pi x, \ \cos \pi x, \ \sin \frac{\pi x}{2}, \ |\cos \pi x|, \ \left|\sin \frac{\pi x}{2}\right|, \ \left|\cos \frac{\pi x}{2}\right|$$

Ans.

x	$\sin \pi x$	$\cos \pi x$	$\sin \frac{\pi x}{2}$	$|\cos \pi x|$	$\left|\sin \frac{\pi x}{2}\right|$	$\left|\cos \frac{\pi x}{2}\right|$
0	0	1	0	1	0	1
1	0	−1	1	1	1	0
2	0	1	0	1	0	1
3	0	−1	−1	1	1	0
4	0	1	0	1	0	1
	constant	Balanced	not constant not Balanced	constant	Balanced	Balanced

14. *Outline in brief the discrete Fourier transform and the QFT. Find the QFT of the following*

(a) $\frac{|0> - |1>}{\sqrt{2}}$

(b) $\sqrt{\frac{2}{N}} \sum_{x=0}^{N-1} \cos \frac{2\pi x}{N}|x>$

(c) $\sqrt{\frac{2}{N}} \sum_{x=0}^{N-1} \sin \frac{2\pi x}{N}|x>$

(d) $\sqrt{\frac{2}{N}} \sum_{x=0}^{N-1} \cos \frac{4\pi x}{N}|x>$

Ans.

- Discrete Fourier transform

  Discrete Fourier transform transforms a discrete function f to another discrete function g There are N inputs of f and N outputs of g. For convenience of writing we can use the run index $i = 0,1,2,3,....., N - 1$ and denote the functions as $f(i) = a_i$ and $g(i) = b_i$. Let $N = 2^n$, n = natural number. In a nutshell

  $$a_i \xrightarrow{\text{DFT}} b_j \quad (i, j = 0,1,2,3,....., N - 1)$$

  Formula defining DFT is

  $$b_j = \frac{1}{\sqrt{N}} \sum_{i=0}^{N-1} e^{(2\pi i,j)i/N} a_i \ (i \text{ outside bracket is } \sqrt{-1}).$$

  Denoting $e^{2\pi i/N} = w$ we can write

  $$b_j = \frac{1}{\sqrt{N}} \sum_{i=0}^{N-1} w^{ij} a_i.$$

  In matrix representation

  $$\begin{pmatrix} b_0 \\ b_1 \\ b_{N-1} \end{pmatrix} = \frac{1}{\sqrt{N}} \begin{pmatrix} w^{0.0} & w^{0.1} & w^{0.2} & ...... & w^{0.(N-1)} \\ w^{1.0} & w^{1.1} & w^{1.2} & ...... & w^{1.(N-1)} \\ w^{(N-1).0} & w^{(N-1).1} & w^{(N-1).2} & ...... & w^{(N-1).(N-1)} \end{pmatrix} \begin{pmatrix} a_0 \\ a_1 \\ a_{N-1} \end{pmatrix}.$$

- Quantum Fourier transform

Consider state $|\psi\rangle$ that is a superposition of orthonormalized base states $|i\rangle$

$$|\psi\rangle = \sum_i a_i |i\rangle \quad i = 0,1,2,3,.....N - 1, \ N = 2^n$$

QFT on $|\psi\rangle$ is $|\phi\rangle = U_{\text{QFT}}|\psi\rangle = \sum_{i=0}^{N-1}\sum_{j=0}^{N-1} \frac{a_i}{\sqrt{N}} e^{(2\pi i,j)i/N}|j\rangle$

(i outside bracket is $\sqrt{-1}$)
QFT basically relates the coefficients of $|\psi\rangle$ and $|\phi\rangle$.

(a) $\frac{|0\rangle - |1\rangle}{\sqrt{2}}$ is a one qubit state. So $n = 1$, $N = 2^1 = 2$

The matrix is given by

$$\hat{U}_{\text{QFT}} = \frac{1}{\sqrt{2}} \begin{pmatrix} w^{0.0} & w^{0.1} \\ w^{1.0} & w^{1.1} \end{pmatrix} = \frac{1}{\sqrt{2}} \begin{pmatrix} w^0 & w^0 \\ w^0 & w^1 \end{pmatrix} = \frac{1}{\sqrt{2}} \begin{pmatrix} 1 & 1 \\ 1 & w \end{pmatrix},$$

where $w = e^{2\pi i/N} = e^{2\pi i/2} = e^{\pi i} = -1$

$$\hat{U}_{QFT} = \frac{1}{\sqrt{2}}\begin{pmatrix} 1 & 1 \\ 1 & -1 \end{pmatrix} = \hat{H} = \text{Hadamard gate}$$

The operator corresponding to QFT for one qubit state

$$\frac{|0> - |1>}{\sqrt{2}} = \frac{1}{\sqrt{2}}[\begin{pmatrix} 1 \\ 0 \end{pmatrix} - \begin{pmatrix} 0 \\ 1 \end{pmatrix}] = \frac{1}{\sqrt{2}}\begin{pmatrix} 1 \\ -1 \end{pmatrix}$$

is the same as the Hadamard gate. So QFT leads to state

$$\hat{U}_{QFT}|\psi> = \hat{H}|\psi> = \frac{1}{\sqrt{2}}\begin{pmatrix} 1 & 1 \\ 1 & -1 \end{pmatrix}\frac{1}{\sqrt{2}}\begin{pmatrix} 1 \\ -1 \end{pmatrix} = \frac{1}{2}\begin{pmatrix} 0 \\ 2 \end{pmatrix} = \begin{pmatrix} 0 \\ 1 \end{pmatrix} = |1>$$

(b) $|\psi> = \sqrt{\frac{2}{N}} \displaystyle\sum_{x=0}^{N-1} \cos\frac{2\pi x}{N}|x>$ (compare with $|\psi> = \displaystyle\sum_i a_i|i>$)

$$QFT|\psi> = \sqrt{\frac{2}{N}} \sum_{y=0}^{N-1}\sum_{x=0}^{N-1} \frac{\cos(\frac{2\pi x}{N})}{\sqrt{N}} e^{(2\pi xy)i/N}|y>$$

(replacing a_i by $\cos(\frac{2\pi x}{N})$ in $U_{QFT}|\psi> = \displaystyle\sum_{i=0}^{N-1}\sum_{j=0}^{N-1} \frac{a_i}{\sqrt{N}} e^{(2\pi i.j)i/N}|j>$)

Using $\cos x = \dfrac{e^{ix} + e^{-ix}}{2}$ we get

$$QFT|\psi> = \frac{\sqrt{2}}{N} \sum_{y=0}^{N-1}\sum_{x=0}^{N-1} \frac{e^{\frac{2\pi x}{N}i} + e^{-\frac{2\pi x}{N}i}}{2} e^{(2\pi xy)i/N}|y>$$

$$= \frac{\sqrt{2}}{N} \sum_{y=0}^{N-1}\sum_{x=0}^{N-1} \frac{e^{2\pi x(y+1)i/N} + e^{2\pi x(y-1)i/N}}{2} = \frac{\sqrt{2}}{N} \sum_{y=0}^{N-1} \frac{S_1 + S_2}{2}|y>$$

where

$$S_1 = \sum_{x=0}^{N-1} e^{2\pi x(y+1)i/N} \quad , \quad S_2 = \sum_{x=0}^{N-1} e^{2\pi x(y-1)i/N}$$

Now

$$S_1 = \sum_{x=0}^{N-1} e^{2\pi xi} = \sum_{x=0}^{N-1} 1 = N \quad (\text{for } y+1 = N \Rightarrow y = N-1)$$

$$S_1 = \sum_{x=0}^{N-1} e^{2\pi x(y+1)i/N} = \sum_{x=0}^{N-1} (e^{2\pi(y+1)i/N})^x (=\text{a geometric series for } y \neq N-1)$$

$$= \frac{1 - (e^{2\pi(y+1)i/N})^N}{1 - e^{2\pi(y+1)i/N}} = \frac{1 - e^{2\pi(y+1)i}}{1 - e^{2\pi(y+1)i/N}} = \frac{1-1}{1 - e^{2\pi(y+1)i/N}} = 0 \quad \text{for } y \neq N-1$$

Combining the results and using Kroneckar delta symbol

$$S_1 = N\delta_{y,N-1}$$

$$S_2 = \sum_{x=0}^{N-1} e^{2\pi x(y-1)i/N} = \sum_{x=0}^{N-1} 1 = N \quad \text{for } y = 1$$

$$S_2 = \sum_{x=0}^{N-1} e^{2\pi x(y-1)i/N} = \sum_{x=0}^{N-1} (e^{2\pi(y-1)i/N})^x (= \text{a geometric series for } y \neq 1)$$

$$= \frac{1 - (e^{2\pi(y-1)i/N})^N}{1 - e^{2\pi(y-1)i/N}} = \frac{1 - e^{2\pi(y-1)i}}{1 - e^{2\pi(y-1)i/N}} = \frac{1-1}{1 - e^{2\pi(y-1)i/N}} = 0 \quad \text{for } y \neq 1$$

Combining the results and using Kroneckar delta symbol

$$S_2 = N\delta_{y,1}$$

Thus

$$\text{QFT}|\psi> = \frac{\sqrt{2}}{N}\sum_{y=0}^{N-1} \frac{S_1 + S_2}{2}|y> = \frac{\sqrt{2}}{N}\sum_{y=0}^{N-1} \frac{1}{2}(N\delta_{y,N-1} + N\delta_{y,1})|y>$$

$$= \frac{1}{\sqrt{2}}(|N-1> + |1>) = \frac{1}{\sqrt{2}}(|1> + |N-1>)$$

(c) $|\psi> = \sqrt{\frac{2}{N}}\sum_{y=0}^{N-1} \sin\frac{2\pi x}{N}|x>$

$$\text{QFT}|\psi> = \sqrt{\frac{2}{N}}\sum_{y=0}^{N-1}\sum_{x=0}^{N-1} \frac{\sin(\frac{2\pi x}{N})}{\sqrt{N}} e^{(2\pi xy)i/N}|y>$$

Using $\sin x = \frac{e^{ix} - e^{ix}}{2i}$ we get

$$\text{QFT}|\psi> = \frac{\sqrt{2}}{N}\sum_{y=0}^{N-1}\sum_{x=0}^{N-1} \frac{e^{\frac{2\pi x}{N}i} - e^{-\frac{2\pi x}{N}i}}{2i} e^{(2\pi xy)i/N}|y>$$

$$= \frac{\sqrt{2}}{N}\sum_{y=0}^{N-1}\sum_{x=0}^{N-1} \frac{e^{2\pi x(y+1)i/N} - e^{2\pi x(y-1)i/N}}{2i}|y> = \frac{\sqrt{2}}{N}\sum_{y=0}^{N-1} \frac{S_1 - S_2}{2i}|y>$$

where

$$S_1 = \sum_{x=0}^{N-1} e^{2\pi x(y+1)i/N} \quad , \quad S_2 = \sum_{x=0}^{N-1} e^{2\pi x(y-1)i/N}$$

As before

$$S_1 = N\delta_{y,N-1}, \; S_2 = N\delta_{y,1}$$

Thus

$$\text{QFT}|\psi> = \frac{\sqrt{2}}{N}\sum_{y=0}^{N-1}\frac{S_1 - S_2}{2i}|y> = \frac{\sqrt{2}}{N}\sum_{y=0}^{N-1}\frac{1}{2i}(N\delta_{y,N-1} - N\delta_{y,1})|y>$$

$$= \frac{1}{i\sqrt{2}}(|N-1> - |1>) = \frac{i}{\sqrt{2}}(|1> - |N-1>)$$

(d) $|\psi> = \sqrt{\frac{2}{N}}\sum_{x=0}^{N-1}\cos\frac{4\pi x}{N}|x>$

$$\text{QFT}|\psi> = \sqrt{\frac{2}{N}}\sum_{y=0}^{N-1}\sum_{x=0}^{N-1}\frac{\cos(\frac{4\pi x}{N})}{\sqrt{N}}e^{(2\pi xy)i/N}|y>$$

Using $\cos x = \frac{e^{ix} + e^{-ix}}{2}$ we get

$$\text{QFT}|\psi> = \frac{\sqrt{2}}{N}\sum_{y=0}^{N-1}\sum_{x=0}^{N-1}\frac{e^{\frac{4\pi x}{N}i} + e^{-\frac{4\pi x}{N}i}}{2}e^{(2\pi xy)i/N}|y>$$

$$= \frac{\sqrt{2}}{N}\sum_{y=0}^{N-1}\sum_{x=0}^{N-1}\frac{e^{2\pi x(y+2)i/N} + e^{2\pi x(y-2)i/N}}{2}|y> = \frac{\sqrt{2}}{N}\sum_{y=0}^{N-1}\frac{S_1 + S_2}{2}|y>$$

where

$$S_1 = \sum_{x=0}^{N-1}e^{2\pi x(y+2)i/N} \quad, \; S_2 = \sum_{x=0}^{N-1}e^{2\pi x(y-2)i/N}$$

Now

$$S_1 = \sum_{x=0}^{N-1}e^{2\pi xi} = \sum_{x=0}^{N-1}1 = N \quad (\text{for } y+2 = N \Rightarrow y = N-2)$$

$$S_1 = \sum_{x=0}^{N-1}e^{2\pi x(y+2)i/N} = \sum_{x=0}^{N-1}(e^{2\pi(y+2)i/N})^x (=\text{a geometric series for } y \neq N-2)$$

$$= \frac{1 - (e^{2\pi(y+2)i/N})^N}{1 - e^{2\pi(y+2)i/N}} = \frac{1 - e^{2\pi(y+2)i}}{1 - e^{2\pi(y+2)i/N}} = \frac{1-1}{1 - e^{2\pi(y+2)i/N}} = 0 \quad \text{for } y \neq N-2$$

Combining the results and using Kroneckar delta symbol

$$S_1 = N\delta_{y, N-2}$$

$$S_2 = \sum_{x=0}^{N-1} e^{2\pi x(y-2)i/N} = \sum_{x=0}^{N-1} 1 = N \quad \text{for } y = 2$$

$$S_2 = \sum_{x=0}^{N-1} e^{2\pi x(y-2)i/N} = \sum_{x=0}^{N-1} (e^{2\pi(y-2)i/N})^x (= \text{a geometric series for } y \neq 2)$$

$$= \frac{1 - (e^{2\pi(y-1)i/N})^N}{1 - e^{2\pi(y-1)i/N}} = \frac{1 - e^{2\pi(y-2)i}}{1 - e^{2\pi(y-2)i/N}} = \frac{1-1}{1 - e^{2\pi(y-2)i/N}} = 0 \quad \text{for } y \neq 2$$

Combining the results and using Kroneckar delta symbol

$$S_2 = N\delta_{y,2}$$

Thus

$$\text{QFT}|\psi> = \frac{\sqrt{2}}{N} \sum_{y=0}^{N-1} \frac{S_1 + S_2}{2} |y> = \frac{\sqrt{2}}{N} \sum_{y=0}^{N-1} \frac{1}{2}(N\delta_{y, N-2} + N\delta_{y, 2})|y>$$

$$= \frac{1}{\sqrt{2}}(|N-2> + |2>) = \frac{1}{\sqrt{2}}(|2> + |N-2>)$$

15.

(a) *The state* $\frac{|00> + |01>}{\sqrt{2}}$ *is subjected to QFT. Now the first qubit is measured. Find the probability that state $|0>$ is obtained.*

(b) *Find QFT of the state* $\frac{1}{2}(|0> + |1> + |2> + |3>)$.

Ans.

(a)

$$|\psi> = \frac{|00> + |01>}{\sqrt{2}} = \frac{1}{\sqrt{2}}|0> + \frac{1}{\sqrt{2}}|1> = \sum_i a_i|i> \quad (a_0 = a_1 = \frac{1}{\sqrt{2}}, \ a_2 = a_3 = 0)$$

Apply QFT. Here $n = 2$, $N = 2^n = 2^2 = 4$, (two-qubit states). Coefficient of any state $|y>$ ($y \rightarrow 0,1,2,3$) is

$$c_y = \sum_{x=0}^{3} \frac{a_x}{\sqrt{N}} e^{2\pi xyi/N} = \sum_{x=0}^{3} \frac{a_x}{\sqrt{4}} e^{2\pi xyi/4}$$

$$c_0 = \sum_{x=0}^{3} \frac{a_x}{\sqrt{4}} e^{2\pi x0i/4} = \sum_{x=0}^{3} \frac{a_x}{2} = \frac{a_0 + a_1 + a_2 + a_3}{2} = \text{coefficient of state } |00> = |0>$$

$$= \frac{1}{2}\left(\frac{1}{\sqrt{2}} + \frac{1}{\sqrt{2}} + 0+0\right) = \frac{1}{\sqrt{2}}$$

To get the probability of state $|0>$ in the first qubit we deal with kets $|00>$(coefficient c_0) and $|01>$ (coefficient c_1) where 0 occurs as the first qubit. Now

$$c_1 = \sum_{x=0}^{3} \frac{a_x}{\sqrt{4}} e^{2\pi x 1 i/4} = \sum_{x=0}^{3} \frac{a_x}{2} e^{\pi x i/2} = \frac{a_0}{2} + \frac{a_1}{2} e^{\pi i/2} + 0+0 = \frac{1}{2}\frac{1}{\sqrt{2}} + \frac{1}{2}\frac{1}{\sqrt{2}}i$$

$$= \frac{1}{2\sqrt{2}}(1 + i)$$

Probability of first qubit 0 is

$$|c_{|00>}|^2 + |c_{|01>}|^2 = |c_0|^2 + |c_1|^2 = \left|\frac{1}{\sqrt{2}}\right|^2 + \left|\frac{1}{2\sqrt{2}}(1+i)\right|^2 = \frac{1}{2} + \frac{1-i^2}{8} = \frac{1}{2} + \frac{1}{4} = \frac{3}{4}$$

(b) Given state $|\psi> = \frac{1}{2}(|0> +|1> +|2> +|3>) = \sum_{x=0}^{3} a_i|i>$ $(a_i = \frac{1}{2}$ for all $i)$

Let $n = 2$, $N = 2^n = 2^2 = 4$, (2 qubit states). So we get states 0,1,2,3,

To get the QFT let us evaluate coefficients term by term, i.e. coefficient of each of the basis kets in the Fourier transform. So

$$c_y = \sum_{x=0}^{3} \frac{1}{\sqrt{N}} a_x e^{2\pi xyi/N}$$

$$c_0 = \sum_{x=0}^{3} \frac{1}{2} a_x e^{2\pi x 0i/4} = \sum_{x=0}^{3} \frac{1}{2} a_x = \frac{1}{2}(a_0 + a_1 + a_2 + a_3) = \frac{1}{2}\left(\frac{1}{2} + \frac{1}{2} + \frac{1}{2} + \frac{1}{2}\right) = 1$$

As $c_0 = 1$ coefficient of all other states is 0. So $c_1 = c_2 = c_3 = 0$ (for normalized Fourier transform). So QFT is $|0>$.

1. *Grover algorithm for $N = 8$ was used to identify one marked state. After some iterations the amplitude of the marked state was found to be $\sqrt{\frac{6}{7}}$ and that of each marked state $\frac{1}{7}$. Find the amplitude of the marked state after two more iterations.*

Ans. *N=8*

a_w= Amplitude of the marked state $=\sqrt{\frac{6}{7}}$.

a_x= Amplitude of the unmarked state $= \frac{1}{7}$.

We perform two iterations as follows:

Apply U_w: $a_w = -\sqrt{\frac{6}{7}}$ (flipped) ,$a_x = \frac{1}{7}$

Mean $\bar{a} \rightarrow \frac{1}{8}(7.\frac{1}{7} - 1\sqrt{\frac{6}{7}}) = \frac{1}{8}(1 - \sqrt{\frac{6}{7}})$

Applying U_s: $a_w \rightarrow 2\bar{a} - a_w = 2\frac{1}{8}(1 - \sqrt{\frac{6}{7}}) - (-\sqrt{\frac{6}{7}}) = \frac{1}{4} + \frac{3}{4}\sqrt{\frac{6}{7}}$,

$$a_x \rightarrow 2\bar{a} - a_x = 2\frac{1}{8}(1 - \sqrt{\frac{6}{7}}) - (\frac{1}{7}) = \frac{3}{28} - \frac{1}{4}\sqrt{\frac{6}{7}}$$

Apply U_w: $a_w = -(\frac{1}{4} + \frac{3}{4}\sqrt{\frac{6}{7}})$(flipped), $a_x = \frac{3}{28} - \frac{1}{4}\sqrt{\frac{6}{7}}$

Mean $\bar{a} \rightarrow \frac{1}{8}(7\frac{3}{28} - \frac{1}{4}\sqrt{\frac{6}{7}}) - (\frac{1}{4} + \frac{3}{4}\sqrt{\frac{6}{7}}) = \frac{1}{8}(\frac{1}{2} - \frac{5}{2}\sqrt{\frac{6}{7}})$

Applying U_s: $a_w \rightarrow 2\bar{a} - a_w = 2\frac{1}{8}(\frac{1}{2} - \frac{5}{2}\sqrt{\frac{6}{7}}) - [-(\frac{1}{4} + \frac{3}{4}\sqrt{\frac{6}{7}})]$

$$= \frac{1}{8}(1 - 5\sqrt{\frac{6}{7}} + 2 + 6\sqrt{\frac{6}{7}}]$$

$= \frac{1}{8}(3 + \sqrt{\frac{6}{7}})=$ amplitude of the marked state after two more iterations.

IOP Publishing

Quantum Optics and Quantum Computation
An introduction
Dipankar Bhattacharyya and Jyotirmoy Guha

Chapter 9

Quantum error correction

Apart from addressing computation problems, computers are used in communication, e-mail, social media etc in a big way. Let us explore the possibility of use of quantum computers for the purpose of communication.

Communication occurs between a sender and receiver. The sender sends a signal that is received by a receiver. The sending or receiving of information, i.e. signal, occurs via a wired connection (i.e. through a telephone). This model of communication has undergone many changes and up gradations. The intermediate wire is often not necessary, as evident in the microwave connectivity between two mobiles. We shall discuss the connectivity or communication pattern in the quantum computer regime.

In day-to-day communication if a sender misspells something the receiver can easily sense the mistake and communication occurs uninterrupted. This is primarily because of the human ability to assimilate the sense conveyed while communicating, and a spelling mistake or grammatical issues do not set up an insurmountable hindrance. Also, if the error appears to be too big to handle we request a re-sending of information.

We say that communication of data (through strings of bits or bytes or words) occurs through a noisy channel if errors occur due to the effects of random influences that are not related to or performed by the sender. We should have some mechanism in the protocol through which such errors that creep in, due to a noisy channel, get detected and corrected as far as possible.

When communication occurs between computers or in the computing language, we speak of fault tolerant computing or fault tolerant communication, where the communication protocol is so designed that it is able to withstand errors to a certain limit.

9.1 Error in classical computing

In classical computing, an error in communication involves a bit flip, i.e. the receiver may receive a 0 instead of 1 or may receive a 1 instead of 0. When several bits are being communicated several such flips might occur. If that happens we say that the data is being corrupted since the receiver receives data which is different from what was actually sent by the sender. Such corruption in data necessitates detection and correction. The reason for such corruption is attributed to transmission through a noisy (unfaithful or unreliable) channel since the channel fails to maintain or hold on to purity of the data that the sender sends through it.

We can speak of two types of errors, as depicted in figures 9.1 and 9.2.

Single-bit error

One bit gets flipped, i.e. corrupted (an original bit gets complemented due to the noisy channel). This is shown in figure 9.1.

Burst error or multiple-bit error

More than one bits get flipped, i.e. corrupted due to the noisy channel.

The corrupted bits may occur consecutively or may be non-adjacent.

The length of burst error is measured from the first corrupted bit to the last corrupted bit, even though some bits situated in between may not be corrupted.

This is shown in figure 9.2.

Error detection

There are various methods to detect errors and for communication purpose some protocols are maintained. For instance, the protocol might be that the words should have an even number of 1s, i.e. even parity words are only legitimate. We call it say, even parity protocol. This means that one has to manipulate each word by adding 0 or 1 to make the word have even parity. Then if the noisy channel changes (flips) one bit (say), parity would change and the error would be detected at once. In such a case, the additional bit (that makes the protocol of sending only even parity words) is referred to as parity bit or check bit.

Parity bit is only suitable for detecting errors—it cannot correct errors since there is no way to determine which particular bit has been corrupted by the noisy channel. The data that has been established to be erroneous is discarded and a request for re-transmission of the same data maintaining the protocol is made.

Suppose the sender wishes to send data 1101 = **0001101**. It has three 1s and so is of odd parity. Maintaining protocol, the sender includes a parity bit 1 to make it **00011011**. Now it has four 1s and so is of even parity.

On the other hand if the sender wishes to send data 1001 = **0001001**, it has two 1s and has even parity and so is suitable for data transmission under even parity protocol. Maintaining protocol, the sender includes a parity bit 0 to keep it **00010010**. The resultant word has two 1s and so is of even parity and is suitable for data transmission under even parity protocol. If the noisy channel changes one bit say **00011010**, then parity becomes odd (as there are three 1s now)

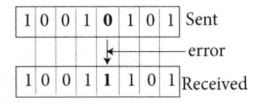

Figure 9.1. Single-bit error

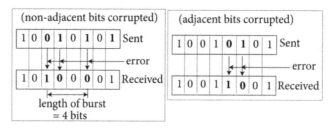

Figure 9.2. Burst error/multiple-bit error.

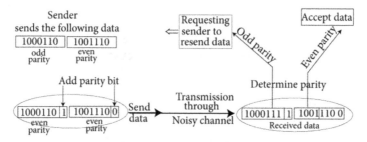

Figure 9.3. Flow chart for error detection in classical communication.

and receiver concludes that a single-bit error has occurred. However, change of two bits (that keeps the parity intact cannot be detected through this method).

Clearly this mechanism enables the detection of single-bit errors. The flow chart of figure 9.3 summarizes the procedure.

Redundancy

In this method the sender does not send a single bit but replicates (or duplicates) the bit (to be sent) odd number of times (say three times or five times). And the receiver detects or extracts the correct bit through a majority vote, as explained in the following.

Instead of 0 the sender will send either 000 or 00000 (defined as logical 0 denoted as 0_L) and similarly, instead of 1, the sender will send either 111 or 11111 (defined as logical 1 denoted as 1_L).

Suppose that the protocol is to send three replicas of a bit.

Suppose the sender is to transmit a bit 0 and so he transmits $0_L = 000$ obeying the protocol. Also assume that the noisy channel can inflict one bit flip.

Suppose data received is 010. The receiver now checks the received data through auto correct mechanism, according to which the data received is wrong as it consists of a mixture of 0 and 1 and the correct data would be 000 since as per majority vote the majority of the bits are 0 (two 0s and one 1), so all three should be 0s.

9.2 Errors in quantum computing/communication

Quantum computing or communication occurs from sender to receiver through a quantum channel and one sends qubits (instead of classical bits) through it. The qubits may be either 0 or 1 or a linear combination of the basis states.

The quantum channel may be a noisy channel in which the effects of random influences cause a certain amount of errors in the qubits (that are sent by the sender). Again, quantum states evolve continuously with time and so errors might creep in. Dealing with errors in quantum computing is different from the way errors in classical communication are handled because of the following reasons.

Due to the no-cloning theorem we cannot make replica of a qubit, i.e. we cannot duplicate a qubit. So we cannot send multiple logical qubits (as can be done in the classical case). In other words, multiplication or replication of qubits is not possible in quantum computing or communication.

Unlike the classical case, we cannot determine errors in a quantum state through the process of measurement because of the collapse associated in a measurement process, by which the entire quantum state gets destroyed. So monitoring qubits through measurement is not possible.

Quantum states are extremely fragile and the quantum state is likely to easily decohere (i.e. lose the phase relationship) due to interaction with the surroundings. Maintaining a coherent superposition of multi-qubit quantum states over a long period of time (during which a quantum algorithm can be executed) is extremely difficult.

In quantum computing as data is communicated through a quantum channel the errors that are introduced into the data are different than in the classical case. In classical computing and communication the error is only the bit flip. But in quantum computing the error may be of two types. The noisy channel or surroundings can inflict a qubit flip and/or phase flip (decohere).

9.3 The phase flip

A linear superposition of states has a certain phase relationship. Having an overall phase between different components of qubits is not an important issue but change of relative phase (and the error therein) in the process of communication is significant.

Consider the computational basis states

$$|0> = \begin{pmatrix} 1 \\ 0 \end{pmatrix}, |1> = \begin{pmatrix} 0 \\ 1 \end{pmatrix}$$

and in this basis we consider the state

$$|\psi> = \frac{|0> + |1>}{\sqrt{2}} = \frac{1}{\sqrt{2}}[\begin{pmatrix} 1 \\ 0 \end{pmatrix} + \begin{pmatrix} 0 \\ 1 \end{pmatrix}] = \frac{1}{\sqrt{2}}\begin{pmatrix} 1 \\ 1 \end{pmatrix} = \begin{pmatrix} 1/\sqrt{2} \\ 1/\sqrt{2} \end{pmatrix} = |+>.$$

Consider the Pauli Z gate $\hat{Z} = \begin{pmatrix} 1 & 0 \\ 0 & -1 \end{pmatrix}$ whose eigenstates are $|0>, |1>$ with eigenvalues ± 1 since $\hat{Z}|0> = |0>$, $\hat{Z}|1> = -|1>$. Clearly measurement of $\hat{Z}$ in the state $|\psi> = \frac{|0> + |1>}{\sqrt{2}}$ is $|0>$ with probability $|\frac{1}{\sqrt{2}}|^2 = \frac{1}{2}$ and $|1>$ with probability $|\frac{1}{\sqrt{2}}|^2 = \frac{1}{2}$.

Now suppose, due to noise or due to interaction with surroundings (i.e. due to a phase perturbation) $|\psi> = \frac{|0>+|1>}{\sqrt{2}}$ suffers a change to the qubit state

$$|\psi'> = \frac{|0>+e^{i\alpha}|1>}{\sqrt{2}} = \frac{1}{\sqrt{2}}\left[\begin{pmatrix}1\\0\end{pmatrix} + e^{i\alpha}\begin{pmatrix}0\\1\end{pmatrix}\right] = \frac{1}{\sqrt{2}}\begin{pmatrix}1\\e^{i\alpha}\end{pmatrix} = \begin{pmatrix}1/\sqrt{2}\\e^{i\alpha}/\sqrt{2}\end{pmatrix}.$$

Measurement of $\hat{Z}$ in the state $|\psi'>$ is $|0>$ with probability $|\frac{1}{\sqrt{2}}|^2 = \frac{1}{2}$ and $|1>$ with probability $|\frac{e^{i\alpha}}{\sqrt{2}}|^2 = \frac{e^{-i\alpha}}{\sqrt{2}}\frac{e^{i\alpha}}{\sqrt{2}} = \frac{1}{2}$. So the probability distribution does not change.

Let us consider Pauli X gate $\hat{\sigma}_x = \hat{X} = \begin{pmatrix}0&1\\1&0\end{pmatrix}$ whose eigenstates are not $|0>$, $|1>$ since $\hat{X}|0> = |1>$, $\hat{X}|1> = |0>$. $\hat{X}$ has eigenvalues $+1$ with eigenvector $\frac{1}{\sqrt{2}}\begin{pmatrix}1\\1\end{pmatrix} = |+>$ and -1 with eigenvector $\frac{1}{\sqrt{2}}\begin{pmatrix}1\\-1\end{pmatrix} = |->$ (section 1.27). Let us construct the projection operator for eigenvalue $+1$ as

$$\hat{P}_+ = |+><+| = \frac{1}{\sqrt{2}}\begin{pmatrix}1\\1\end{pmatrix}\frac{1}{\sqrt{2}}(1\ \ 1) = \frac{1}{2}\begin{pmatrix}1&1\\1&1\end{pmatrix}$$

$$\hat{P}_+|\psi> = \hat{P}_+|+> = |+><+|+> = |+>=1|+>$$

as $+1$ is eigenvalue it is projected out with probability 1 (100%) if we measure $\hat{X}$ in state $|\psi> = |+>$. Let us find with what probability $+1$ is projected out from the modified state

$$|\psi'> = \frac{|0>+e^{i\alpha}|1>}{\sqrt{2}} = \frac{1}{\sqrt{2}}\begin{pmatrix}1\\e^{i\alpha}\end{pmatrix} = \frac{1}{\sqrt{2}}e^{i\alpha/2}\begin{pmatrix}e^{-i\alpha/2}\\e^{i\alpha/2}\end{pmatrix}$$

$$\hat{P}_+|\psi'> = \frac{1}{2}\begin{pmatrix}1&1\\1&1\end{pmatrix}\frac{1}{\sqrt{2}}e^{i\alpha/2}\begin{pmatrix}e^{-i\alpha/2}\\e^{i\alpha/2}\end{pmatrix} = \frac{1}{2\sqrt{2}}e^{i\alpha/2}\begin{pmatrix}e^{i\alpha/2}+e^{-i\alpha/2}\\e^{i\alpha/2}+e^{-i\alpha/2}\end{pmatrix}$$

$$= \frac{1}{\sqrt{2}}e^{i\alpha/2}\frac{1}{2}(e^{i\alpha/2}+e^{-i\alpha/2})\begin{pmatrix}1\\1\end{pmatrix}$$

$$= e^{i\alpha/2}\cos\frac{\alpha}{2}\frac{1}{\sqrt{2}}\begin{pmatrix}1\\1\end{pmatrix} = e^{i\alpha/2}\cos\frac{\alpha}{2}|+>$$

So the probability with which eigenvalue $+1$ appears if we measure $\hat{X}$ in the phase modified state $|\psi'>$ is $|e^{i\alpha/2}\cos\frac{\alpha}{2}|^2 = \cos^2\frac{\alpha}{2}$.

It follows therefore that apart from the qubit flip the loss of coherence or decoherence of a state due to external phase perturbation (i.e. the change of phase) has to be considered as qubits are transmitted from sender to receiver.

9.4 Qubit transmission from Alice to Bob

Suppose Alice is the sender and Bob is the receiver and the protocol is that duplication is not allowed. Suppose Alice sends a quantum state to Bob.

$|\psi> = \alpha|0> + \beta|1>$, where $|\alpha|^2 + |\beta|^2 = 1$ (normalized state)

For the purpose of transmission of this quantum state Alice introduces two ancila qubits $|0>$, $|0>$ and sends them through two CNOT gates, as shown in figure 9.4.

$|\psi> = |$quantum state $> |$1st ancila qubit $> |$2nd ancila qubit$>$

$= (\alpha|0> + \beta|1>)|0> |0> = \alpha|000> + \beta|100>$ (written in this representation)

Also, we use C for control qubit and T for target qubit. Only C = 1 will flip the target. The operation of the circuit has been described in detail in figure 9.4. The quantum state at the senders end (i.e. Alice's end) is

$$|\psi''> = \alpha|000> + \beta|111>$$

Bob receives a quantum state that we denote by say $|\psi_1>$ through the quantum channel. This quantum state $|\psi_1>$ that Bob receives is, in general, different from the quantum state that Alice sent viz. $|\psi''> = \alpha|000> + \beta|111>$ due to infliction of error or modification by the noisy quantum channel, say due to qubit changes.

Let us discuss the states which Bob can receive and the corresponding probability.

Suppose the quantum channel is noisy and the probability that a qubit flips is p while the probability that a qubit does not flip is $1 - p$. The quantum state $|\psi_1>$ received by Bob can be of eight types. These are shown with the corresponding probabilities in table 9.1.

To decode the quantum state $|\psi_1>$ that Bob got from Alice through the noisy quantum channel, Bob adds two ancila qubits $|0> |0>$ and so the quantum state becomes

$$|\psi''> \xrightarrow{\text{Add ancila}|0>|0>} |\psi_1> |0> |0> = |\psi_1> |00>.$$

Bob now uses four CNOT gates, as depicted in figure 9.5.
Bob passes $|\psi_1> |00>$ through his circuit and gets $|\psi_2> |00>$ at the output.

$$|\psi_1> |00> \xrightarrow{\text{4 CNOTs in Bob's circuit (figure 9.5)}} |\psi_2> |\alpha\beta>$$

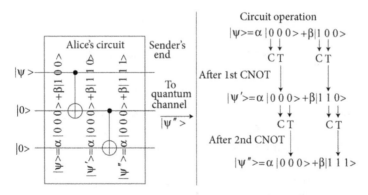

Figure 9.4. Alice's circuit for qubit transmission.

Table 9.1. Quantum states received by Bob from Alice and the corresponding probabilities due to modifications by the noisy quantum channel introducing errors in the process. (For one and two qubit flips probabilities are added as the events are mutually exclusive.)

Possible States $\|\psi_1>$ (Bob receives)	Probability (first qubit)(2nd qubit)(3rd qubit)	Total Probability
No qubit flip		
$\alpha\|000>+\beta\|111>$	$(1-p)(1-p)(1-p)=(1-p)^3$	$(1-p^3$
One qubit flip		
$\alpha\|100>+\beta\|011>$	$p(1-p)(1-p)=p(1-p)^2$	
$\alpha\|010>+\beta\|101>$	$(1-p)p(1-p)=p(1-p)^2$	$3p(1-p)^2$
$\alpha\|001>+\beta\|110>$	$(1-p)(1-p)p=p(1-p)^2$	
Two qubit flip		
$\alpha\|110>+\beta\|001>$	$pp(1-p)=p^2(1-p)$	
$\alpha\|101>+\beta\|010>$	$p(1-p)p=p^2(1-p)$	$3p^2(1-p)$
$\alpha\|011>+\beta\|100>$	$(1-p)pp=p^2(1-p)$	
All qubits flip		
$\alpha\|111>+\beta\|000>$	p^3	p^3

Note: left margin reads $\|$Alice's quantum state$>=\|\psi">=\alpha\|000>+\beta\|111>$

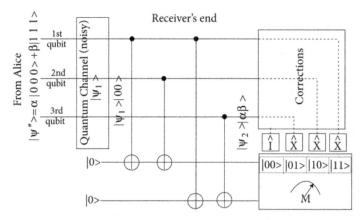

Figure 9.5. Bob's circuit (at the receiving end).

$$(\|\alpha\beta > \Rightarrow |00 > , |01 > , |10, |11 >)$$

As $|\psi_1>$ can assume eight values, $|\psi_2>$ also assumes eight values. These are shown in the following tables 9.2, 9.3 (four are worked out, the rest can be found out similarly). As before, we use C for control qubit and T for target qubit. Only C = 1 will flip the target.

Let us collect all the results. The following are all the possible results along with corresponding probability that Bob deciphers at the end of his circuit (M), i.e. the

Table 9.2. Operation within Bob's circuit for no qubit flip or one qubit flip upon Alice's data (refer to figure 9.5).

For no qubit flip	For one qubit flip
$\|\psi_1 > \| 00 > = (\alpha \| 000 > + \beta \| 111 >) \| 00 >$	$\|\psi_1 > \| 00 > = (\alpha \| 100 > + \beta \| 011 >) \| 00 >$
$= \alpha \| 000 > \| 00 > + \beta \| 111 > \| 00 >$	$= \alpha \| 100 > \| 00 > + \beta \| 011 > \| 00 >$
$\downarrow \quad \downarrow \quad \downarrow \quad \downarrow$	$\downarrow \quad \downarrow \quad \downarrow \quad \downarrow$
C $\quad$ T $\quad$ C $\quad$ T	C $\quad$ T $\quad$ C $\quad$ T
After 1st CNOT	After 1st CNOT
$= \alpha \| 000 > \| 00 > + \beta \| 111 > \| 10 >$	$= \alpha \| 100 > \| 10 > + \beta \| 011 > \| 00 >$
$\downarrow \quad \downarrow \quad \downarrow \quad \downarrow$	$\downarrow \quad \downarrow \quad \downarrow \quad \downarrow$
C $\quad$ T $\quad$ C $\quad$ T	C $\quad$ T $\quad$ C $\quad$ T
After 2nd CNOT	After 2nd CNOT
$= \alpha \| 000 > \| 00 > + \beta \| 111 > \| 00 >$	$= \alpha \| 100 > \| 10 > + \beta \| 011 > \| 10 >$
$\downarrow \quad \downarrow \quad \downarrow \quad \downarrow$	$\downarrow \quad \downarrow \quad \downarrow \quad \downarrow$
C $\quad$ T $\quad$ C $\quad$ T	C $\quad$ T $\quad$ C $\quad$ T
After 3rd CNOT	After 3rd CNOT
$= \alpha \| 000 > \| 00 > + \beta \| 111 > \| 01 >$	$= \alpha \| 100 > \| 11 > + \beta \| 011 > \| 10 >$
$\downarrow \quad \downarrow \quad \downarrow \quad \downarrow$	$\downarrow \quad \downarrow \quad \downarrow \quad \downarrow$
C $\quad$ T $\quad$ C $\quad$ T	C $\quad$ T $\quad$ C $\quad$ T
After 4th CNOT	After 4th CNOT
$= \alpha \| 000 > \| 00 > + \beta \| 111 > \| 00 >$	$= \alpha \| 100 > \| 11 > + \beta \| 011 > \| 11 >$
$= (\alpha \| 000 > + \beta \| 111 >) \| 00 > = \|\psi_2 > \| 00>$	$= (\alpha \| 100 > + \beta \| 011 >) \| 11 > = \|\psi_2 > \| 11>$

Table 9.3. Operation within Bob's circuit for two qubit flip and for all qubit flip upon Alice's data (refer to figure 9.5).

For two qubit flip	For all qubit flip
$\|\psi_1 > \| 00 > = (\alpha \| 110 > + \beta \| 001 >) \| 00 >$	$\|\psi_1 > \| 00 > = (\alpha \| 111 > + \beta \| 000 >) \| 00 >$
$= \alpha \| 110 > \| 00 > + \beta \| 001 > \| 00 >$	$= \alpha \| 111 > \| 00 > + \beta \| 000 > \| 00 >$
$\downarrow \quad \downarrow \quad \downarrow \quad \downarrow$	$\downarrow \quad \downarrow \quad \downarrow \quad \downarrow$
C $\quad$ T $\quad$ C $\quad$ T	C $\quad$ T $\quad$ C $\quad$ T
After 1st CNOT	After 1st CNOT
$= \alpha \| 110 > \| 10 > + \beta \| 001 > \| 00 >$	$= \alpha \| 111 > \| 10 > + \beta \| 000 > \| 00 >$
$\downarrow \quad \downarrow \quad \downarrow \quad \downarrow$	$\downarrow \quad \downarrow \quad \downarrow \quad \downarrow$
C $\quad$ T $\quad$ C $\quad$ T	C $\quad$ T $\quad$ C $\quad$ T
After 2nd CNOT	After 2nd CNOT
$= \alpha \| 110 > \| 00 > + \beta \| 001 > \| 00 >$	$= \alpha \| 111 > \| 00 > + \beta \| 000 > \| 00 >$
$\downarrow \quad \downarrow \quad \downarrow \quad \downarrow$	$\downarrow \quad \downarrow \quad \downarrow \quad \downarrow$
C $\quad$ T $\quad$ C $\quad$ T	C $\quad$ T $\quad$ C $\quad$ T
After 3rd CNOT	After 3rd CNOT
$= \alpha \| 110 > \| 01 > + \beta \| 001 > \| 00 >$	$= \alpha \| 111 > \| 01 > + \beta \| 000 > \| 00 >$
$\downarrow \quad \downarrow \quad \downarrow \quad \downarrow$	$\downarrow \quad \downarrow \quad \downarrow \quad \downarrow$
C $\quad$ T $\quad$ C $\quad$ T	C $\quad$ T $\quad$ C $\quad$ T
After 4th CNOT	After 4th CNOT
$= \alpha \| 110 > \| 01 > + \beta \| 001 > \| 01 >$	$= \alpha \| 111 > \| 00 > + \beta \| 000 > \| 00 >$
$= (\alpha \| 110 > + \beta \| 001 >) \| 01 > = \|\psi_2 > \| 01>$	$= (\alpha \| 111 > + \beta \| 000 >) \| 00 > = \|\psi_2 > \| 00>$

Table 9.4. Coupled states of Bob's circuit.

State received $|\psi_1>$	Bob's coupled states at the end	Probability
No qubit flip		
$\alpha\,|\,000> +\beta\,|\,111>$	$(\alpha|\,000> +\beta\,|\,111>)|00>$	$(1-p)^3$
One qubit flip		
$\alpha\,|\,100> +\beta\,|\,011>$	$(\alpha|\,100> +\beta\,|\,011>)|11>$	$p(1-p)^2$
$\alpha\,|\,010> +\beta\,|\,101>$	$(\alpha|\,010> +\beta\,|\,101>)|10>$	$p(1-p)^2$
$\alpha\,|\,001> +\beta\,|\,110>$	$(\alpha|\,001> +\beta\,|\,110>)|01>$	$p(1-p)^2$
Two qubit flip		
$\alpha\,|\,110> +\beta\,|\,001>$	$(\alpha|\,110> +\beta\,|\,001>)|01>$	$p^2(1-p)$
$\alpha\,|\,101> +\beta\,|\,010>$	$(\alpha|\,101> +\beta\,|\,010>)|10>$	$p^2(1-p)$
$\alpha\,|\,011> +\beta\,|\,100>$	$(\alpha|\,011> +\beta\,|\,100>)|11>$	$p^2(1-p)$
All qubits flip		
$\alpha\,|\,111> +\beta\,|\,000>$	$(\alpha|\,111> +\beta\,|\,000>)|00>$	p^3

state $|\psi_2>$ (along with ancillary bits $|\alpha\beta> \Rightarrow |00>$, $|01>$, $|10$, $|11>$). This is depicted in table 9.4 along with their probability of occurrence.

Alice sent $|\psi_1> = \alpha|000> +\beta|111>$ and Bob wishes to receive this state.

Bob does measurement (M) and depending on measurement he does some corrections (applies $\hat{X}$ on appropriate qubit to get the original state: $\hat{X}|0> =|1>$, $\hat{X}|1> =|0>$).

Bob started with ancila bits $|0> |0> =|00>$ but ends with any one of $|00>$, $|01>$, $|10>$, $|11>$. Let us analyze the results as shown in table 9.5.

So four out of eight cases turn out to have error of amount

$$\text{Error} = p^3 + p^2(1-p) + p^2(1-p) + p^2(1-p) = p^3 + 3p^2(1-p)$$

Generally, $p = 0.01$, error $\approx 3 \times 10^{-4} =$ small.

In this way a three-qubit code, sent by Alice, can be recovered by Bob.

In this example only qubit flips has been considered (not phase flip). We now discuss the case where both qubit flips and phase flips are taken care of.

9.5 Converting a phase flip error to qubit flip error

In a classical computer, flipped bits are the only kind of error. But in a quantum computer there can be another error called phase error. This occurs when the relative signs between |0> and |1> get inverted.

Table 9.5. Bob's corrective actions on his circuit.

Ancila qubits received by Bob $|\alpha\beta>$	Bob receives State $|\psi_2>$	Probability	status	Action by Bob on received state
$|00>$	$\alpha\,|000>+\beta\,|111>$	$(1-p)^3$	No qubit flip	No action as this is desired state $\alpha\,|000>+\beta\,|111>$
	$\alpha\,|111>+\beta\,|000>$	p^3	All qubits flip	error
$|01>$	$\alpha\,|001>+\beta\,|110>$	$p(1-p)^2$	One qubit flip	Applies $\hat{X}$ on third qubit $\alpha\,|000>+\beta\,|111>$
	$\alpha\,|110>+\beta\,|001>$	$p^2(1-p)$	Two qubit flip	error
$|10>$	$\alpha\,|010>+\beta\,|101>$	$p(1-p)^2$	One qubit flip	Applies $\hat{X}$ on second qubit $\alpha\,|000>+\beta\,|111>$
	$\alpha\,|101>+\beta\,|010>$	$p^2(1-p)$	Two qubit flip	error
$|11>$	$\alpha\,|100>+\beta\,|011>$	$p(1-p)^2$	One qubit flip	Applies $\hat{X}$ on first qubit $\alpha\,|000>+\beta\,|111>$
	$\alpha\,|011>+\beta\,|100>$	$p^2(1-p)$	Two qubit flip	error

Phase error is due to rotation of a qubit by an arbitrary angle ϕ so that a state

$$a|0> +b|1>$$

is converted to state

$$a|0> +e^{i\phi}b|1>$$

This ϕ can take any value and is continuous. So quantum errors are continuous contrary to the discrete nature of classical errors. For simplicity we consider rotation of amount $\phi = \pi$ (called a phase flip by π) as depicted in the following transformation:

$$a|0> +b|1> \quad \rightarrow \quad a|0> +e^{i\pi}b|1> = = a|0> -b|1>$$

This is equivalent to action of Pauli $\hat{Z}$ gate since $(\hat{Z}|0> =|0> , \hat{Z}|1> =-|1>)$

$$\hat{Z}\,(a|0> +b|1>)= a\hat{Z}|0> +b\hat{Z}|1> =a|0> -b|1> \text{ (defined as phase flip).}$$

Figure 9.6 demonstrates the bit flip and the phase flip over a Bloch sphere. In other words, ($\hat{Z}$ flips $|1>$)

$$\frac{1}{\sqrt{2}}(|0> +|1>)\xrightarrow{\text{phase flip}} \frac{1}{\sqrt{2}}(|0> -|1>) \tag{9.1}$$

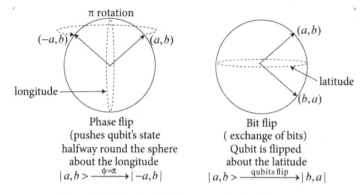

Figure 9.6. Showing bit flip error and phase flip error.

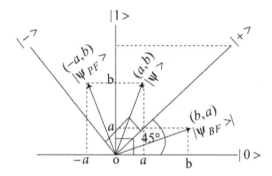

Figure 9.7. Phase flip error converting to qubit flip error.

The LHS and the RHS of equation (9.1) in the Hadamard basis are, respectively

$$\tfrac{1}{\sqrt{2}}(|0> +|1 >) = |+>, \quad \tfrac{1}{\sqrt{2}}(|0> -|1 >) = | - >$$

where $|+>, | - >$ are the qubits in the Hadamard representation. So equation (9.1) when rewritten in the Hadamard basis (or diagonal basis) becomes a bit flip

$$|+> \xrightarrow{\text{qubit flip}} | - > \tag{9.2}$$

The phase flip in computational basis is simply a qubit flip in the diagonal basis. So phase flip error can be handled by treating it as a bit flip error in the diagonal basis.

- A geometrical interpretation has been shown in figure 9.7 where computational basis $|0>$ is taken along Cartesian X-axis and $|1>$ is taken along Cartesian Y-axis. The diagonal basis makes angle of 45° with these axes.

Computational basis is $|0 > , |1>$. Diagonal basis is $|+>, | - >$
Consider an arbitrary state $|\psi>$

$$|\psi > =a|0 > +b|1>$$

(projections of $|\psi>$ are a along $|0>$ and b along $|1>$)

Make a qubit flip:

$$|0> \rightarrow |1> \quad |1> \rightarrow |0>$$

Then we arrive at $|\psi_{BF}>$ whose projections are b along $|0>$ and a along $|1>$) i.e.

$$|\psi> = a|0> +b|1> \xrightarrow{\text{qubit flip}} a|1> +b|0> = |\psi_{BF}>$$

But from the placement of $|\psi>$ and $|\psi_{BF}>$ it is clear that this is equivalent to reflection of the state $|\psi>$ about the direction $|+>$.

Also since

$$a|0> +b|1> \xrightarrow{\pi \ \text{phaseflip}} a|0> +e^{i\pi}b|1> = e^{i\pi}(e^{-i\pi}a|0> +b|1>)$$
$$= e^{i\pi}(-a|0> +b|1>)$$
$$\xrightarrow{\text{overlook global phase}\pi} (-a|0> +b|1>) = |\psi_{PF}>$$

$|\psi_{PF}> = -a|0> +b|1>$ which is π phase flip of state $|\psi>$ can be considered as a reflection of $|\psi>$ about $|1>$.

We have explained that a π phase flip in the computational basis is actually a bit flip in the diagonal basis. So by changing basis we can correct phase flip.

Phase flip errors

The diagonal basis is related to the computational basis as

$$|+>=\hat{H}|0> = \frac{|0>+|1>}{\sqrt{2}}, \quad |->=\hat{H}|1> = \frac{|0>-|1>}{\sqrt{2}}.$$

For correcting phase errors one should encode qubits as triplets in the diagonal basis in the form

$$|\psi> = a|+++>+b|--->.$$

In computational basis the state is

$$|\psi> = a(\frac{|0>+|1>}{\sqrt{2}}\frac{|0>+|1>}{\sqrt{2}}\frac{|0>+|1>}{\sqrt{2}}) + b(\frac{|0>-|1>}{\sqrt{2}}\frac{|0>-|1>}{\sqrt{2}}\frac{|0>-|1>}{\sqrt{2}})$$

$$= \frac{a}{2}[|00> +|01> +|10> +|11>]\frac{|0>+|1>}{\sqrt{2}}$$

$$+ \frac{b}{\sqrt{2}}[|00> -|01> -|10> +|11>]\frac{|0>-|1>}{\sqrt{2}}$$

$$= \frac{a}{2\sqrt{2}}(|000> +|010> +|100> +|110> +|001> +|011> +|101> +|111>)$$

$$+ \frac{b}{2\sqrt{2}}(|000> -|010> -|100> +|110> -|001> +|011> +|101> -|111>)$$

$$|\psi> = \frac{a+b}{2\sqrt{2}}(|000> +|110> +|011> +|101>)$$

$$+ \frac{a-b}{2\sqrt{2}}(|010> +|100> +|001> +|111>)$$

Measure any of the qubits (say first qubit) in a computational basis. The state collapses either to the state

$$[1^{st} \text{ qubit0}]\frac{a+b}{2\sqrt{2}}(|000> +|011 >) + \frac{a-b}{2\sqrt{2}}(|010 > +|001 >)$$

or to the state

$$[1^{st} \text{ qubit1}]\frac{a+b}{2\sqrt{2}}(|110 > +|101 >) + \frac{a-b}{2\sqrt{2}}(|100 > +|111 >)$$

Assume it collapsed to the first state

$$\frac{a+b}{2\sqrt{2}}(|000> +|011 >) + \frac{a-b}{2\sqrt{2}}(|010 > +|001 >)$$

We now have to do a parity check in the diagonal basis by introducing two ancila qubits and measuring the ancila qubits after CNOT operations as in the bit flip case.

- In the diagonal basis suppose the collapsed state is of the form

$$[a|+++>+b| - --->](\text{without error})[a|-++>+b| + --->] \text{ (one-qubit flip error)}$$

Such errors can be corrected after appropriate corrective action following ancila measurement.

9.6 Shor's nine-qubit error code

Any arbitrary quantum error can be looked upon as a combination of qubit flip error and phase flip error. We have to develop a code that can take care of both qubit flip error as well as phase flip error. We shall now discuss Shor's nine-qubit error code that takes care of both the flips.

We note that for the error code to take care of qubit flip error we can use

$$\hat{X} = \begin{pmatrix} 0 & 1 \\ 1 & 0 \end{pmatrix} \text{ since } \hat{X}\begin{pmatrix} a \\ b \end{pmatrix} = \begin{pmatrix} 0 & 1 \\ 1 & 0 \end{pmatrix}\begin{pmatrix} a \\ b \end{pmatrix} = \begin{pmatrix} b \\ a \end{pmatrix}$$

To take care of phase flip error we can use

$$\hat{Z} = \begin{pmatrix} 1 & 0 \\ 0 & -1 \end{pmatrix} \text{ since } \hat{Z}\begin{pmatrix} a \\ b \end{pmatrix} = \begin{pmatrix} 1 & 0 \\ 0 & -1 \end{pmatrix}\begin{pmatrix} a \\ b \end{pmatrix} = \begin{pmatrix} a \\ -b \end{pmatrix}.$$

The phase flip and qubit flip error can be taken care of by using

$$\hat{Y} = i\hat{X}\hat{Z} = i \begin{pmatrix} 0 & 1 \\ 1 & 0 \end{pmatrix}\begin{pmatrix} 1 & 0 \\ 0 & -1 \end{pmatrix} = \begin{pmatrix} 0 & -i \\ i & 0 \end{pmatrix}$$

since

$$\hat{Y}\begin{pmatrix} a \\ b \end{pmatrix} = \begin{pmatrix} 0 & -i \\ i & 0 \end{pmatrix}\begin{pmatrix} a \\ b \end{pmatrix} = i\begin{pmatrix} -b \\ a \end{pmatrix} = e^{i\pi/2}\begin{pmatrix} -b \\ a \end{pmatrix}$$

In Shor's 9-qubit error code we encode

$|0>$ as $|0 >_L = | + ++>$ and $|1>$ as $|1 >_L = | - --->$.

Let us be interested to send the state $a|0 > +b|1>$ through a quantum channel from Alice to Bob.

9.6.1 Encoding circuit

Suppose Alice is to send to Bob the quantum state $|\psi> = a|0> +b|1>$. Two ancilas $|0>$, $|0>$ are introduced and so we have the quantum state

$$|\psi> = [a|0> +b|1>]|0> |0> = a|000> +b|100> \text{'}$$

The circuit diagram given in figure 9.8 shows the encoding circuit that consists of two parts—the phase flip coding part and the qubit flip coding part.

We discuss first the passage through phase flip coding part of the circuit.

$|\psi>$ is passed through CNOT twice and we get $|\psi''> = a|000> +b|111>$.

(C is control and T is target).

Now we apply Hadamard gate (using the fact that $\hat{H}|0> = \frac{|0>+|1>}{\sqrt{2}}$, $\hat{H}|1> = \frac{|0>-|1>}{\sqrt{2}}$)

$$|\psi_1> = a[\frac{|0>+|1>}{\sqrt{2}}\frac{|0>+|1>}{\sqrt{2}}\frac{|0>+|1>}{\sqrt{2}}] + b[\frac{|0>-|1>}{\sqrt{2}}\frac{|0>-|1>}{\sqrt{2}}\frac{|0>-|1>}{\sqrt{2}}]$$

$$= \frac{a}{2\sqrt{2}}(|0>+|1>)(|0>+|1>)(|0>+|1>)]$$

$$+ \frac{b}{2\sqrt{2}}[(|0>-|1>)(|0>-|1>)(|0>-|1>)]$$

After Hadamard operation has been done, we introduce 2 more anciliary qubits with each of the Hadamard gates, as depicted in the qubit flip encoding part of the circuit.

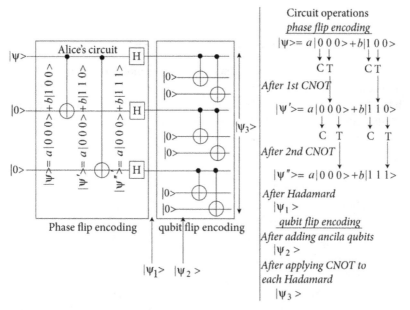

Figure 9.8. Encoding circuit. Shor's 9-qubit code: $|\psi>$, first 2 ancila qubits $|0>$, $|0>$, 6 ancila qubits to 3 Hadamard gates $|0>$, $|0>$;$|0>$, $|0>$;$|0>$, $|0>$.

9.6.2 Why is it called Shor's 9-qubit error code?

This is because Shor code utilizes 9 physical qubits to encode a single logical qubit (they being : $|\psi_1\rangle$, first 2 ancila qubits $|0\rangle$, $|0\rangle$; and 2 each ancilla qubits applied to 3 Hadamard gates, i.e. 6 ancilas qubits, and so $1 + 2 + 6 = 9$ codes are used.)

Let us study the passage through the qubit flip encoding part of the circuit. Introduction of 2 anciliary qubits gives us

$$|\psi_1\rangle \xrightarrow{\text{introduce 6 ancilary qubits}} |\psi_2\rangle$$

$$|\psi_2\rangle = \frac{a}{2\sqrt{2}}[(|0\rangle + |1\rangle)|00\rangle \ (|0\rangle + |1\rangle)|00\rangle \ (|0\rangle + |1\rangle)|00\rangle]$$

$$+ \frac{b}{2\sqrt{2}}(|0\rangle - |1\rangle)|00\rangle \ (|0\rangle - |1\rangle)|00\rangle \ (|0\rangle - |1\rangle)|00\rangle]$$

$$|\psi_2\rangle = \frac{a}{2\sqrt{2}}[(|000\rangle + |100\rangle)(|000\rangle + |100\rangle)(|000\rangle + |100\rangle)]$$

$$+ \frac{b}{2\sqrt{2}}(|000\rangle - |100\rangle)(|000\rangle - |100\rangle)(|000\rangle - |100\rangle)]$$

Again to each Hadamard gate, CNOT is applied, as depicted in the qubit flip coding part of the circuit. This gives the following as explained in figure 9.9.

$$|\psi_3\rangle = \frac{a}{2\sqrt{2}}[(|000\rangle + |111\rangle)(|000\rangle + |111\rangle)(|000\rangle + |111\rangle)]$$

$$+ \frac{b}{2\sqrt{2}}[(|000\rangle - |111\rangle)(|000\rangle - |111\rangle)(|000\rangle - |111\rangle)] \tag{9.3}$$

We can put this in a compact notation writing $\frac{1}{\sqrt{2}}(|000\rangle + |111\rangle)$ as $|+\rangle$ and $\frac{1}{\sqrt{2}}(|000\rangle - |111\rangle)$ as $|-\rangle$

$$|\psi_3\rangle = a|+++\rangle + b|---\rangle$$

In this notation $|+\rangle|$, $|-\rangle$ do not stand for diagonal basis state.

So the prescription is as follows.

We have to start with one qubit. Then we have to introduce two ancila qubits. Then through certain operations on the ancilas we have three qubits at the end of the

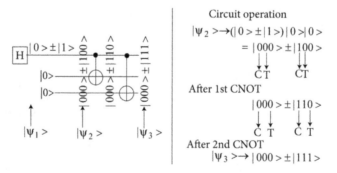

Figure 9.9. CNOT operation on the Hadamard in the qubit flip encoding part.

first group. Now with each of these modified groups we apply 2 more ancilas to finally end up with 9 qubits. This is the final coding received.

Assume that out of the 9 qubits passing through a noisy channel say at most one qubit gets affected and the nature of error in the qubit may be either a bit flip or phase flip or simultaneously a bit and phase flip. Let p be the probability that a single qubit is affected, e.g. a bit flip or a phase flip or a mixture of the two. So the probability that a single qubit is not affected is $1 - p$.

Clearly then the probability that no qubit is affected (i.e. probability of zero error) is $P = (1 - p)^9$.

Generally $p < 0.01$ (i.e. p is a small number less than 1%) and we can have the binomial series to be

$$P \approx 1 - 9p + \frac{9(9-1)}{2!}p^2 = 1 - 9p + 36p^2.$$

The probability that any of the nine qubits gets affected (i.e. probability of single error) is

$$P_1 = 9p(1 - p)^8 = 9p(1 - 8p + ...) \approx 9p - 72p^2$$

The probability of either zero error or single error is (as they are mutually exclusive events) is given by

Probability of zero error + Probability of single error

$$= P + P_1 = (1 - 9p + 36p^2) + (9p - 72p^2) = 1 - 36p^2$$

So the probability of having more than one error is

$$= 1 - (1 - 36p^2) = 36p^2 < < p \qquad (p = \text{small}, \ p^2 = \text{smaller})$$

Clearly we have described the encoding circuit of Shor's 9 qubit error code which tries to correct both bit flip and phase flip or their combination. Put in a nutshell, in this encoding circuit, we send the single qubit $|\psi>$ having the structure $|\psi> = a|0> + b|1>$ and to do that we generate a 9-qubit code (incorporating the phase flip coding plus the qubit flip coding)

$$(a|0> + b|1>)|0>|0> \rightarrow a|+++> + b|---> = a|0>_L + b|1>_L$$

where $|0>_L = |+++>$, $|1>_L = |--->$ (suffix L for logical: we encode the state $|0>$ as $|0>_L$ in a compact form and the state $|1>$ as $|1>_L$ in a compact form).

9.6.3 Decoding circuit

We now decode the encoded circuit. In the 9 qubits we have a single qubit error and suppose the error has occurred in the first qubit which is both a bit flip and a phase flip. Though we have assumed first qubit error, it is indistinguishable from the case in which error occurs instead in the second or third qubit because that would give a relative minus sign.

The decoding circuit has been depicted in figure 9.10.

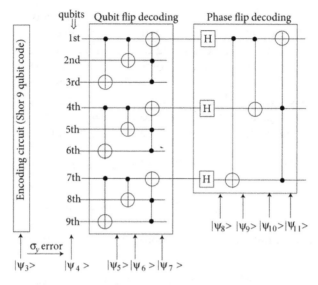

Figure 9.10. Decoding circuit.

Assuming first qubit is flipped with a σ_y error (that is, both bit flip and phase flip:

$\hat{Y}|0>\sim|1>,\hat{Y}|1>\sim-|0>)$ we get from eqn 9.3 [$|\psi_3> \xrightarrow{\sigma_y \text{ error}} |\psi_4>$]

$$|\psi_3> = \frac{a}{2\sqrt{2}}[(|000>+|111>)(|000>+|111>)(|000>+|111>)]$$
$$+ \frac{b}{2\sqrt{2}}[(|000>-|111>)(|000>-|111>)(|000>-|111>)]$$

$$|\psi_4> = \frac{a}{2\sqrt{2}}[(|100>-|011>)(|000>+|111>)(|000>+|111>)]$$
$$\frac{b}{2\sqrt{2}}[(|100>+|011>)(|000>-|111>)(|000>-|111>)]$$

.... (9.4)

Let us treat the 9 qubits as three blocks of three qubits each for convenience. The operation of the decoding circuit has been explained in detail in table 9.6.

$$|\psi_3> \xrightarrow{\sigma_y \text{ error}} |\psi_4> \xrightarrow{\text{1st CNOT}} |\psi_5> \xrightarrow{\text{2nd CNOT}} |\psi_6> \xrightarrow{\text{CCNOT}} |\psi_7>$$

Collecting the common qubits in $|\psi_7>$ (table 9.6) we can write

$$|\psi_7> = a[\frac{|0>-|1>}{\sqrt{2}}|11> \frac{|0>+|1>}{\sqrt{2}}|00> \frac{|0>+|1>}{\sqrt{2}}|00>]$$

$$+b[\frac{|0>+|1>}{\sqrt{2}}|11> \frac{|0>-|1>}{\sqrt{2}}|00> \frac{|0>-|1>}{\sqrt{2}}|00>]$$

We concentrate on the first, fourth and seventh qubits which now would pass through Hadamard gates as shown in the decoding circuit of figure 9.10. Using $\hat{H}\frac{|0>-|1>}{\sqrt{2}} = |1>, \hat{H}\frac{|0>+|1>}{\sqrt{2}} = |0>$ we rewrite $|\psi_7>$ as

Table 9.6. Detailed operation of qubit flip part of decoding circuit of figure 9.10.

From encoding circuit

$$|\psi_3> = \frac{a}{2\sqrt{2}}[(|000> + |111>)(|000> + |111>)(|000> + |111>)]$$

$$+ \frac{b}{2\sqrt{2}}[(|000> - |111>)(|000> - |111>)(|000> - |111>)] \quad ....(9.3)$$

After error in first qubit, both bit flip and phase flip

$$|\psi_4> = \frac{a}{2\sqrt{2}}[(|100> - |011>)(|000> + |111>)(|000> + |111>)]$$
$$\qquad\quad\ \ \text{C T}\quad\ \text{C T}\quad\ \text{C T}\quad\ \text{C T}\quad\ \text{C T}\quad\ \text{C T}$$

$$+ \frac{b}{2\sqrt{2}}[(|100> + |011>)(|000> - |111>)(|000> - |111>)] \quad ....(9.4)$$
$$\qquad\quad\ \ \text{C T}\quad\ \text{C T}\quad\ \text{C T}\quad\ \text{C T}\quad\ \text{C T}\quad\ \text{C T}$$

After 1st CNOT

$$|\psi_5> = \frac{a}{2\sqrt{2}}[(|101> - |011>)(|000> + |110>)(|000> + |110>)]$$
$$\qquad\quad\ \ \text{CT}\quad\ \ \text{CT}\quad\ \ \ \text{CT}\quad\ \ \text{CT}\quad\ \ \text{CT}\quad\ \ \text{CT}$$

$$+ \frac{b}{2\sqrt{2}}[(|101> + |011>)(|000> - |110>)(|000> - |110>)]$$
$$\qquad\quad\ \ \text{CT}\quad\ \ \text{CT}\quad\ \ \ \text{CT}\quad\ \ \text{CT}\quad\ \ \text{CT}\quad\ \ \text{CT}$$

After 2nd CNOT

$$|\psi_6> = \frac{a}{2\sqrt{2}}[(|111> - |011>)(|000> + |100>)(|000> + |100>)]$$
$$\qquad\quad\ \ \text{TCC}\quad\ \text{TCC}\quad\ \text{TCC}\quad\ \text{TCC}\quad\ \text{TCC}\quad\ \text{TCC}$$

$$+ \frac{b}{2\sqrt{2}}[(|111> + |011>)(|000> - |100>)(|000> - |100>)]$$
$$\qquad\quad\ \ \text{TCC}\quad\ \text{TCC}\quad\ \text{TCC}\quad\ \text{TCC}\quad\ \text{TCC}\quad\ \text{TCC}$$

After CCNOT

$$|\psi_7> = \frac{a}{2\sqrt{2}}[(|011> - |111>)(|000> + |100>)(|000> + |100>)]$$

$$+ \frac{b}{2\sqrt{2}}[(|011> + |111>)(|000> - |100>)(|000> - |100>)]$$

$$|\psi_7> \xrightarrow{\text{1st, 4th, 7th qubits through H}} |\psi_8>$$

$$|\psi_8> = a|1>_1 |11>_{2,3} |0>_4 |00>_{5,6} |0>_7 |00>_{8,9}$$
$$+ b|0>_1 |11>_{2,3} |1>_4 |00>_{5,6} |1>_7 |00>_{8,9}$$

Ignoring the common ancila qubits that have been projected out we rewrite $|\psi_8>$ as (showing only the first, fourth and seventh qubits)

$$|\psi_8> = a|1>_1 |0>_4 |0>_7 \quad + \quad b|0>_1 |1>_4 |1>_7 \qquad (9.5)$$

Now these qubits are passed through two CNOT and one CCNOT gates and the detailed operation has been shown in table 9.7.

$$|\psi_8> \xrightarrow{\text{First CNOT}} |\psi_9> \xrightarrow{\text{Second CNOT}} |\psi_{10}> \xrightarrow{\text{CCNOT}} |\psi_{11}>$$

Rewriting $|\psi_{11}>$ (table 9.7) with only the first qubit we have

$$|\psi>_{11} = a|0>_1 + b|1>_1 = a|0> + b|1> = |\psi>$$

This is the original quantum state that was sent and Bob has decoded it.

Table 9.7. Detailed operation of the phase flip part of the decoding circuit of figure 9.10.

$|\psi_8. \rangle = a\,[\,|1>_1|0>_4|0>_7\,] + b\,[\,|0>_1|1>_4|1>_7\,]$ ....(9.5)
  C   T   C   T

After first CNOT

$|\psi_9. \rangle = a\,[\,|1>_1|0>_4|1>_7\,] + b\,[\,|0>_1|1>_4|1>_7\,]$
  C   T   C   T

After second CNOT

$|\psi_{10}\rangle = a\,[\,|1>_1|1>_4|1>_7\,] + b\,[\,|0>_1|1>_4|1>_7\,]$
  T   C   C   T   C   C

After CCNOT

$|\psi_{11}\rangle = a\,[\,|0>_1|1>_4|1>_7\,] + b\,[\,|1>_1|1>_4|1>_7\,]$

$\qquad = [\,a\,|0>_1 + b\,|1>_1\,]\,|1>_4|1>_7$

Here we have analyzed the error in first qubit. A general approach would involve a parity check to find out the bit that suffered an error.

This is a short account of quantum error correction code which has far reaching consequences in communication problems. Purity of a quantum state is of utmost importance in dealing with any quantum algorithm.

9.7 Further reading

[1] Ghosh D 2016 Online course on quantum information and computing (Bombay, Indian Institute of Technology)
[2] Nielsen M A and Chuang I L 2019 *Quantum Computation and Quantum Information* (Cambridge: Cambridge University Press)
[3] Nakahara M and Ohmi T 2017 *Quantum Computing* (Boca Raton, FL: CRC Press)
[4] Pathak A 2016 *Elements of Quantum Computing and Quantum Communication* (Boca Raton, FL: CRC Press)

9.8 Problems

1. *A string of 18 bits viz. 000101001111000101 has been received. What does it represent in a classical 3 bit error code?*

      Ans. Dissolve the string into chunks of three bits and use the majority principle.

$$\underline{000}\ \ \underline{101}\ \ \underline{001}\ \ \underline{111}\ \ \underline{000}\ \ \underline{101}$$
$$\ \ 0\ \ \ \ \ 1\ \ \ \ \ 0\ \ \ \ \ 1\ \ \ \ \ 0\ \ \ \ \ 1$$

      Message = 010101

2. *Suppose we encode one qubit by n qubits. What is the minimum value of n that will tell us exactly what error has occurred in the qubit?*

      Ans. 9.

3. *Bits are sent individually in a 3-bit classical code. Suppose the probability of bit flip is 0.1. Find the probability of getting an error-free message if the protocol used employs majority vote.*

Ans. Bit flip probability $p = 0.1$. The possible cases (for majority vote case) are (let the correct bit be X, wrong bit Y) $X \xrightarrow{prob} 1 - p$, $Y \xrightarrow{prob} p$

3 bits are correct (no bit flip)	XXX	probability $= (1 - p)^3$
2 bits are correct (one bit flip)	XXY, XYX, YXX	probability $= 3p(1 - p)^2$.

Overall probability $= (1 - p)^3 + 3p(1 - p)^2 = (1-0.1)^3 + 3(0.1)(1-0.1)^2 = 0.972 \rightarrow 97.2\%$

4. *In three qubit code a message is sent as $\alpha|000> + \beta|111>$. Which of the following received messages do not require any corrective action: $\alpha|100> + \beta|011>$, $-\alpha|000> - \beta|111>$, $\alpha|110> + \beta|001>$, $\alpha|100> - \beta|011>$.*

Ans. Refer to table 9.5:

Received message	Corresponding ancila bits
$\alpha|100> + \beta|011>$	$|11>$
$-\alpha|000> - \beta|111>$	$|00>$
$\alpha|110> + \beta|001>$	$|01>$
$\alpha|100> - \beta|011>$	$|11>$

Corrective measures are taken except for the case when ancila bits are $|00>$. So the message $-\alpha|000> - \beta|111>$ does not need any corrective measure.

5. *In three-qubit error code assume that the probability of a single bit flip error is 0.05. The ancila bits are measured to be $|00>$. Find the probability that the measured state is error-free.*

Ans. Refer to table 9.5:

$p = 0.05$, We get $|00>$ ancila bits in the states $(\alpha|000> + \beta|111>)|00>$ with probability $(1 - p)^3$ and $(\alpha|111> + \beta|000>)|00>$ with probability p^3. So probability of occurrence of error free state $(\alpha|000> + \beta|111>)|00>$ is

$$\frac{(1-p)^3}{(1-p)^3 + p^3} = \frac{(1-0.05)^3}{(1-0.05)^3 + (0.05)^3} = \frac{(0.95)^3}{(0.95)^3 + (0.05)^3} = =0.999854.$$

6. *In three qubit error code assume that the probability of a single bit flip error is 0.05. The ancila bits are measured to be $|01>$. Find the probability that the measured state is error-free after Bob takes corrective measures.*

Ans. Refer to table 9.5:

$p = 0.05$, We get $|01>$ ancila bits in the states $(\alpha|001> + \beta|110>)|01>$ with probability $p(1 - p)^2$ (one error) and $(\alpha|110> + \beta|001>)|01>$ with probability $p^2(1 - p)$ (two errors). So probability of occurrence of $(\alpha|001> + \beta|110>)|01>$ state (that can be made error free by applying $\hat{X}$ on third qubit) is

$$\frac{p(1-p)^2}{p(1-p)^2 + p^2(1-p)} = \frac{1-p}{(1-p)+p} = 1 - p = 1 - 0.05 = 0.95$$

7. *Describe the process performed by the circuit of figure* 9.11.

Ans. Let $|\psi_{in}\rangle = a|0\rangle + b|1\rangle$, $|\psi_e\rangle = \sqrt{p}|0\rangle + \sqrt{1-p}|1\rangle$

Input $= |\psi_{in}\rangle \otimes |\psi_e\rangle = \sqrt{p}(a|00\rangle + b|10\rangle) + \sqrt{1-p}(a|01\rangle + b|11\rangle)$

Second qubit = Control, First qubit = Target. After passage through CNOT

$$|\psi_{output}\rangle = \sqrt{p}(a|00\rangle + b|10\rangle) + \sqrt{1-p}(a|11\rangle + b|01\rangle)$$

Measurement (M) is performed on second qubit.
If the second qubit is 1 the result obtained is

$a|0\rangle + b|1\rangle \longrightarrow a|11\rangle + b|01\rangle$ with probability $\left| \sqrt{1-p} \right|^2$ then it is evident that
the input bit (first qubit) gets flipped:

$$a|0\rangle + b|1\rangle \xrightarrow{\text{1st qubit}} a|1\rangle + b|0\rangle$$

If second qubit is 0 the result obtained is

$a|0\rangle + b|1\rangle \longrightarrow a|00\rangle + b|10\rangle$ with probability $\left| \sqrt{p} \right|^2$ then no change or no flip

occurs in the first qubit. $a|0\rangle + b|1\rangle \xrightarrow{\text{1st qubit}} a|0\rangle + b|1\rangle$

Hence

Probability of bit flip = Probability of second qubit to be 1

$$= \left| \sqrt{1-p}\, a \right|^2 + \left| \sqrt{1-p}\, b \right|^2$$

$= (1-p)(|a|^2 + |b|^2) = 1 - p$ since $|a|^2 + |b|^2 = 1$, $|\psi_{in}\rangle$ being normalized.

The process performed in the circuit is a bit flip with probability $1 - p$.

8. *Describe the process performed by the circuit of figure* 9.12.

Ans. Let $|\psi_{in}\rangle = a|0\rangle + b|1\rangle$, $|\psi_e\rangle = \sqrt{p}|0\rangle - \sqrt{1-p}|1\rangle$

Input $= |\psi_{in}\rangle \otimes |\psi_e\rangle = \sqrt{p}(a|00\rangle + b|10\rangle) - \sqrt{1-p}(a|01\rangle + b|11\rangle)$

Second qubit = Control, First qubit = Target. After passage through CNOT

$$|\psi_{output}\rangle = \sqrt{p}(a|00\rangle + b|10\rangle) - \sqrt{1-p}(a|11\rangle + b|01\rangle)$$

Measurement (M) is performed on the second qubit.
If the second qubit is 1 the result obtained is

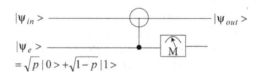

Figure 9.11. Circuit of problem 7.

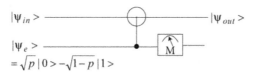

Figure 9.12. Circuit of problem 8.

$a|0 > +b|1 > \longrightarrow -(a|11 > +b|01 >)$ with probability $\left| \sqrt{1-p} \right|^2$ then it is evident that the input bit (first qubit) gets flipped and there is a phase flip too.

$$a|0 > +b|1 > \xrightarrow{\text{1st qubit}} - a|1 > +b|0>$$

If second qubit is 0 the result obtained is

$a|0 > +b|1 > \longrightarrow a|00 > +b|10>$ with probability $\left| \sqrt{p} \right|^2$ then no change occurs in first qubit.

$$a|0 > +b|1 > \xrightarrow{\text{1st qubit}} a|0 > +b|1>$$

Hence,

Probability of bit flip and phase flip = Probability of second qubit to be 1

$= \left| \sqrt{1-p}\, a \right|^2 + \left| \sqrt{1-p}\, b \right|^2 = 1 - p$ since $|a|^2 + |b|^2 = 1$, $|\psi_{in}>$ being normalized.

The process performed in the circuit is a bit flip and a phase flip with probability $1 - p$.

    9. *Describe the process performed by the circuit of figure 9.13. Take $|\psi > = |0>$ or $|1>$.*

       Ans.

$\|\psi > = \|0>$	$\|\psi > = \|1>$
Input $= \|000 > = \|0 > \|00>$	Input $= \|100 > = \|1 > \|00>$

$|\psi > = |0>$

Input $= |000 > = |0 > |00>$

$$\xrightarrow{H} \frac{|0 > + |1 >}{\sqrt{2}}|00>$$

$$\rightarrow \frac{|000 > + |100 >}{\sqrt{2}}$$

CNOT

(1st qubit = Control, 2nd qubit Target)

$$\rightarrow \frac{|000 > + |110 >}{\sqrt{2}}$$

Z gate on 2nd qubit

$$\rightarrow \frac{|000 > - |110 >}{\sqrt{2}}$$

CNOT

(1st qubit = Control, 3rd qubit Target)

$$\rightarrow \frac{|000 > - |111 >}{\sqrt{2}} = |->(\text{say})$$

$|\psi > = |1>$

Input $= |100 > = |1 > |00>$

$$\xrightarrow{H} \frac{|0 > - |1 >}{\sqrt{2}}|00>$$

$$\rightarrow \frac{|000 > - |100 >}{\sqrt{2}}$$

CNOT

(1st qubit = Control, 2nd qubit Target)

$$\rightarrow \frac{|000 > - |110 >}{\sqrt{2}}$$

Z gate on 2nd qubit.

$$\rightarrow \frac{|000 > + |110 >}{\sqrt{2}}$$

CNOT

(1st qubit = Control, 3rd qubit Target)

$$\rightarrow \frac{|000 > + |111 >}{\sqrt{2}} = |+>(\text{say})$$

Hence

$|000 > \longrightarrow | - >$

Hence

$|100 > \longrightarrow | + >$

---

10. *Consider a three-qubit code for correcting a bit flip.* $|\psi > = a|0 > +b|1>$ *is encoded as* $|\psi' > = a|000 > +b|111>$. *How will the encoded state appear to be if the third qubit is distorted by a 60° rotation about the X-axis ?*

Ans. $|\psi' > = a|000 > +b|111> = a|00 > |0 > +b|11 > |1>$

Third qubit suffers rotation. A rotation of $\theta = 60°$ occurs about the X-axis. This means the Bloch vector $|0>$ is in the YZ plane and its new position after rotation corresponds to $\theta = 60°$, $\phi = 90°$. Hence

$$|0 > =|r = 1, \theta = 0°, \phi = 0° > \xrightarrow{\theta = 60°, \phi = 90°} |0 > =|r = 1, \theta = 60°, \phi = 90°>$$

Again, the Bloch vector $|1>$ is also in the YZ plane and its new position after rotation corresponds to $\theta = 180° - 60° = 120°$, $\phi = 90°$. Hence

$$|1 > =|r = 1, \theta = 180°, \phi = 0° > \xrightarrow{\theta = 60°, \phi = 90°} |1 > =|r = 1, \theta = 120°, \phi = 90°>$$

We have shown $|0 > , |1>$ in the Bloch sphere in figure 9.14.

Again, $|0>$ and $|1>$ suffer rotation as per the equation $|\psi > = \cos \frac{\theta}{2}|0 > +e^{i\phi} \sin \frac{\theta}{2}|1>$

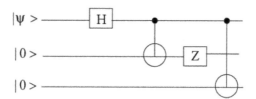

Figure 9.13. Circuit of problem 9.

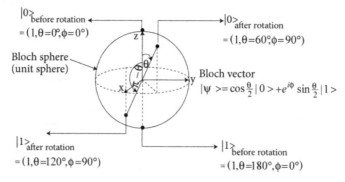

Figure 9.14. Bloch sphere representation of qubit.

$$|0> \xrightarrow{\theta=60°, \phi=90°} \cos\frac{60°}{2}|0> + e^{i\frac{\pi}{2}}\sin\frac{60°}{2}|1> = \frac{\sqrt{3}}{2}|0> + i\frac{1}{2}|1>$$

$$|1> \xrightarrow{\theta=120°, \phi=90°} \cos\frac{120°}{2}|0> + e^{i\frac{\pi}{2}}\sin\frac{120°}{2}|1> = \frac{1}{2}|0> + i\frac{\sqrt{3}}{2}|1>$$

Hence $|\psi> \xrightarrow{\text{rotation}} |\psi'>$ where

$$|\psi'> = a|00> (\frac{\sqrt{3}}{2}|0> + i\frac{1}{2}|1>) + b|11> (\frac{1}{2}|0> + i\frac{\sqrt{3}}{2}|1>)$$

$$|\psi'> = \frac{\sqrt{3}}{2}[a|000> + ib|111>] + \frac{1}{2}[ia|001> + b|110>].$$

Chapter 10

Quantum information

We will discuss various protocols in quantum communication.

10.1 Classical information theory

Let us define what we mean by the word information. Information is a piece of knowledge or facts (details or particulars) provided or learned about something or someone.

The questions that occur to our mind are as follows. Does every piece of information convey some knowledge? How can we compare two pieces of information, i.e. how can we decide which information is richer in content? And how does one quantify or measure information? In day-to-day usage we constantly deal with various pieces of information qualitatively. And we would like messages/statements/information that give us more clarity and more details and more precision.

Example 1 Consider the statement:
'I love to use pencils for sketching.'
This statement conveys some information. But the information that it carries is not clear or precise because details of pencils that I love to use are not specified with certainty.
Consider the modified statement:
'I love to use graphite pencil for sketching.'
This statement is more revealing, meaning that I prefer to use graphite pencils over charcoal pencils or coloured pencils or mechanical pencils. But the detail of which type of graphite pencils (H or B or F or HB) that I use is not at all clear.

Example 2 Consider the statement:
'I live in a cold place.'

This statement is not specific about what temperature is being referred to and one living in another place cannot realize whether his/her place is colder or hotter compared to the place I live in. So the statement carries some information but it is imprecise in comparison to our curiosity to know more.

The statement can be made richer in information if we try to put things quantitatively, say first defining what we mean by a cold place. Suppose we define a place to be cold if the temperature θ is 10 °C $\leqslant \theta \leqslant$ 15 °C. Having defined what we mean by cold, the statement 'I live in a cold place', means that the place I live in has a temperature lying in the range 10 °C $\leqslant \theta \leqslant$ 15 °C. The information level has definitely increased (w.r.t this definition of 'cold' place)—but there is uncertainty regarding whether the temperature is 10 °C, 11 °C, 12 °C, 13 °C, 14 °C or 15 °C and there is equal probability ($=\frac{1}{6}$) that my place has any of the aforesaid temperatures. Clearly, we have scaled down the level of uncertainty and we are in a position to make predictions (say taking temperature to be 12 °C) with a certain amount of precision level (assuming 12 °C means we are not widely off the mark but within a specified range). If I say that yesterday it was 12 °C and today it is colder, then the level of precision has immensely increased and the probability of correctness is 50% (since the possible temperatures are then 10 °C, 11 °C with probability $\frac{1}{2}$). So information has become more precise and sharp (as there are only two alternatives and we can choose either with a certain degree of accuracy).

The examples mentioned demonstrate that whenever we make a statement regarding some event/phenomena we get some facts and there are many facts which we do not get (and these facts that are not mentioned in the statement are the uncertainties involved in the statement). In other words, whenever we say something, there are things which we do not say. So information content in a statement is a measure of uncertainty associated with an event. The amount of uncertainty can be quantitatively defined (as we did in example 2) and as we reduce the uncertainty the more precise and sharp the measure of information is.

Another example

Suppose we see a light at a far distance which does not give much information. But if we see a flickering light source, being switched on and off with a particular frequency then we can interpret from this phenomenon that something is being communicated to someone and the flickering pattern might represent a code (while a uniformly glowing light would carry very little or no information).

Construction of a mathematical measure of information contained in a message requires specification of amount of uncertainty in a message as well as the uncertainty at the receiving end which affects the faithfulness or fidelity of transmission.

For content-neutral data transmission one should have a system capable of staying, in at least, two stable states. One we code as 0 and the other as 1 and define it as a bit (cbit for a classical digital system). For instance voltage levels with ground

(0) and +5 volt (1) or a spin-up atomic state (0) and a spin-down atomic state (1) etc. If we employ a system capable of staying in m different states we require more than a single bit to encode such a system. For instance a 4-state system requires 2 bits to describe it and we can represent the 4 states by 00, 01, 10, 11.

For a system capable of staying in m different states we need a minimum of n bits for its description where $2^n < m$.

Hartley's formula

The number of bits n that is required to send m different messages is given by

$$n = \log_2 m \text{ or } 2^n = m$$

One byte that is a string of eight bits ($n = 8$) can send $2^8 = 256$ different messages. There are two distinct approaches to define information content in a message.

(a) **Shannon's approach:**

This approach tells us the number of bits needed to be transmitted in order to select the correct answer from a list of previously agreed choices. This gives rise to a decision tree.

(b) **Kolmogorov and Chaitin's approach:**

This approach tells us the minimum number of bits needed to compress or store a given message. The information content or complexity of a message is measured by the length of its shortest description. In other words algorithmic capacity of a string (message) is the length of the shortest program describing it. This approach is termed as the algorithmic approach and the length of the shortest algorithm is its complexity. For instance a 64 bit message string

010 101.....01

can be written as '32 repetitions of 01.'

10.2 Decision tree

This tells us the number of bits that are needed to be transmitted for selecting the correct answer from a list of previously agreed upon choices. So we have to ask a set of questions and the answers have to be in Yes/No form. This will lead to what we call a decision tree. Now we have the number of questions needed to be asked for the purpose. We give illustration to describe the point.

Suppose we wish to codify any one person among the following: Lal Bahadur Shastri (Indian politician, male), P V Sindhu (Indian badminton player, female), Deepika Padukone (Indian actress, female), R K Narayan (Indian writer, male), Roger Federer (Swiss Tennis player, male), Hillary Clinton (American politician, female), Brad Pitt (American actor, male), J K Rowling (British author, female). In other words, we will mathematically quantify this information.

With this data we can build a structure by asking certain questions as depicted in the decision tree of figure 10.1 and by associating '0' with a Yes (Y = 0) answer and '1' with a No (N = 1) answer.

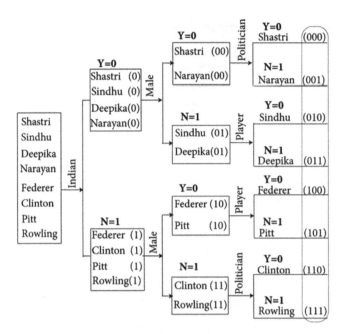

Figure 10.1. Decision tree.

To select one we start with the question, 'Are you an Indian?'

This divides the lot into two groups. One group of four Indians that we denote by Yes or 0 comprising Shastri, Sindhu, Deepika, Narayan and another group of four non-Indians that we denote by No or 1 comprising Federer, Clinton, Pitt, Rowling.

To subdivide we ask the question, 'Are you a male?'

This establishes a subdivision. We attach Yes or 0 for being a male and N or 1 for not being a male. So we have Shastri (Indian male = 00), Narayan (Indian male = 00), Sindhu, (Indian female = 01), Deepika (Indian female = 01), Federer (non-Indian male = 10), Pitt (non-Indian male = 10), Clinton (non-Indian female = 11), Rowling (non-Indian female = 11).

The last two questions : 'Are you a politician?' 'Are you a player?' causes further subdivision and codes are generated.

Shastri (Indian, male, politician = 000), Narayan (Indian, male, not a politician = 001), Sindhu (Indian, female, player = 010), Deepika (Indian, female, not a player = 011), Federer (non-Indian, male, player = 100), Pitt (non-Indian, male, not a player = 101), Clinton (non-Indian, female, politician = 110), Rowling (non-Indian, female, not a politician = 111).

Clearly the decision tree, as elaborated, is one way of quantifying information. This illustrates the number of bits of communication needed to convey or communicate to arrive at a particular answer out of a previously agreed set of alternatives.

10.3 Measure of information: Shannon's entropy

Consider a discrete random variable X corresponding to a physical process or event X. The possible outcomes of X are x_1 with probability p_1; x_2 with probability p_2; x_3 with probability p_3;.........x_M with probability p_M; so that

$$\sum_i p_i = 1.$$

As the measure of information is connected with uncertainty let us define a quantity

$H(p_1, p_2, p_3,....p_M) = $ average uncertainty associated with the event $X = x_i$

It thus represents the average uncertainty that is removed when we know the result of an experiment. To find a mathematical expression for this quantity we define an auxiliary function $f(M)$.

Consider a situation where all probabilities are the same

$$p_1 = p_2 = p_3 = ... = p_M$$

In other words since there are M events then

$$p_1 = p_2 = p_3 = ... = p_M = \frac{1}{M}$$

Let us define $f(M)$ as the average uncertainty associated with M equally likely events as

$f(M) = H(\frac{1}{M}, \frac{1}{M}, \frac{1}{M} .... \frac{1}{M}) = $ monotonic, continuously increasing non-negative.

We give two examples.

- In an unbiased coin tossing experiment we have only 2 events with probabilities

$$p_1 = \frac{1}{2}, p_2 = \frac{1}{2}$$

  So auxiliary function becomes

  $f(2) = H(\frac{1}{2}, \frac{1}{2}) = $ average uncertainty associated with a coin toss.

- In an unbiased die-casting experiment we have 6 events with probabilities

$$p_1 = \frac{1}{6}, p_2 = \frac{1}{6}, p_3 = \frac{1}{6}, p_4 = \frac{1}{6}, p_5 = \frac{1}{6}, p_6 = \frac{1}{6}.$$

So the auxiliary function becomes

$f(6) = H(\frac{1}{6}, \frac{1}{6}, \frac{1}{6}, \frac{1}{6}, \frac{1}{6}, \frac{1}{6}) = $ average uncertainty associated with die-throwing.

- $f(M)$ is a monotonically increasing function with increasing value of M, i.e. $f(6) > f(2)$ and in general if $M > M'$ then $f(M) > f(M')$.
- In the case of a single event there is no uncertainty and so $f(1) = 0$. This is another algebraic property satisfied by $f(M)$.

- Suppose we perform a simultaneous experiment of coin tossing plus die-throwing. Let X be tossing of a coin having two alternatives with equal probability and Y be casting a die having six alternatives with equal probability. Since we can get a Head with any of the die-throw outcomes (1,2,3,4,5,6) or tail with any of the die-throw outcomes (1,2,3,4,5,6) the number of alternatives shoots up. Number of events now becomes $M . N = 2.6 = 12$ alternatives with equal probability. And $f(12)$ is the average uncertainty associated with the simultaneous experiment of coin tossing and die-throwing.

Suppose the result of coin tossing is revealed. This will not remove the uncertainty in die-throwing and we will be left with the uncertainty $f(6)$. Thus we have $f(12) - f(2) = f(6) \Rightarrow f(12) = f(2) + f(6) \Rightarrow f(2.6) = f(2) + f(6)$.
In general

$$f(MN) = f(M) + f(N)$$

where the average uncertainty associated with the experiment X is $f(M)$, with the experiment Y is $f(N)$ and with the simultaneous experiment XY is $f(MN)$; since knowing the result of one event X (which has M alternatives) does not remove the uncertainty associated with the other event Y that is given by $f(N)$.

We shall explore the situation when we relax the condition of equal probability.

Grouping theorem

Consider a random variable X with M different outcomes in an experiment. The events are divided into two groups A and B. Of these groups A has r outcomes $x_1, x_2, \ldots\ldots x_r$ with respective probability being $p_1, p_2, \ldots\ldots p_r$ and group B has $M - r$ outcomes $x_{r+1}, x_{r+2}, \ldots\ldots\ldots x_M$ with respective probability being $p_{r+1}, p_{r+2}, \ldots\ldots p_M$.

In group A the outcome of the event $X = x_i$ is $\dfrac{p_i}{\sum\limits_{i=1}^{r} p_i}$ and in group B the outcome of the event $X = x_i$ is $\dfrac{p_i}{\sum\limits_{i=r+1}^{M} p_i}$.

Let us assume that Y is the result of the combined experiment in which we first choose the group and then the event. The probability of the event $Y = x_i$ in the combined experiment can be estimated using Bayes theorem in statistics.

Before the combined experiment is performed, the uncertainty associated with the outcome is $H(p_1, p_2, \ldots\ldots p_M)$. If the chosen group (either A or B) is revealed the uncertainty removed is $H(p_A, p_B) = H(\sum\limits_{i=1}^{r} p_i, \sum\limits_{i=r+1}^{M} p_i)$. Clearly the uncertainty remaining after choice of the group is revealed is given by the grouping theorem

$$H(p_1, p_2, \ldots\ldots p_M) - H\left(\sum_{i=1}^{r} p_i, \sum_{i=r+1}^{M} p_i \right)$$

$$= \sum_{i=1}^{r} p_i H\left(\frac{p_1}{\sum\limits_{i=1}^{r} p_i}, \frac{p_2}{\sum\limits_{i=1}^{r} p_i}, \ldots\ldots \frac{p_r}{\sum\limits_{i=1}^{r} p_i}\right) + \sum_{i=r+1}^{M} p_i H\left(\frac{p_{r+1}}{\sum\limits_{i=r+1}^{M} p_i}, \frac{p_{r+2}}{\sum\limits_{i=r+1}^{M} p_i}, \ldots\ldots \frac{p_M}{\sum\limits_{i=r+1}^{M} p_i}\right). \quad (10.1)$$

Example

To explain this let us consider two groups : group A in which there are 2 events with respective probability $p_1 = \frac{1}{2}$, $p_2 = \frac{1}{4}$; so that $P(A) = p_1 + p_2 = \frac{1}{2} + \frac{1}{4} = \frac{3}{4}$; group B in which there are 2 events with probability $p_3 = \frac{1}{8}$, $p_4 = \frac{1}{8}$ so that $P(B) = p_3 + p_4 = \frac{1}{8} + \frac{1}{8} = \frac{1}{4}$. Clearly then considering the total uncertainty for the whole event, according to the grouping theorem

LHS $= H(p_1, p_2, p_3, p_4) - H(p_1 + p_2, p_3 + p_4).$

$$= H\left(\frac{1}{2}, \frac{1}{4}, \frac{1}{8}, \frac{1}{8}\right) - H\left(\frac{3}{4}, \frac{1}{4}\right) = H\left(\frac{1}{2}, \frac{1}{4}, \frac{1}{8}, \frac{1}{8}\right) - H\left(\frac{1}{2} + \frac{1}{4}, \frac{1}{8} + \frac{1}{8}\right)$$

$$= \text{RHS} = \sum_{i=1}^{2} p_i H\left(\frac{p_1}{p_1 + p_2}, \frac{p_2}{p_1 + p_2}\right) + \sum_{i=3}^{4} p_i H\left(\frac{p_3}{p_3 + p_4}, \frac{p_4}{p_3 + p_4}\right)$$

$$= p_1 H\left(\frac{p_1}{p_1 + p_2}, \frac{p_2}{p_1 + p_2}\right) + p_2 H\left(\frac{p_1}{p_1 + p_2}, \frac{p_2}{p_1 + p_2}\right)$$

$$+ p_3 H\left(\frac{p_3}{p_3 + p_4}, \frac{p_4}{p_3 + p_4}\right) + p_4 H\left(\frac{p_3}{p_3 + p_4}, \frac{p_4}{p_3 + p_4}\right)$$

$$= \frac{1}{2} H\left(\frac{\frac{1}{2}}{\frac{1}{2} + \frac{1}{4}}, \frac{\frac{1}{4}}{\frac{1}{2} + \frac{1}{4}}\right) + \frac{1}{4} H\left(\frac{\frac{1}{2}}{\frac{1}{2} + \frac{1}{4}}, \frac{\frac{1}{4}}{\frac{1}{2} + \frac{1}{4}}\right) + \frac{1}{8} H\left(\frac{\frac{1}{8}}{\frac{1}{8} + \frac{1}{8}}, \frac{\frac{1}{8}}{\frac{1}{8} + \frac{1}{8}}\right) + \frac{1}{8} H\left(\frac{\frac{1}{8}}{\frac{1}{8} + \frac{1}{8}}, \frac{\frac{1}{8}}{\frac{1}{8} + \frac{1}{8}}\right)$$

$$= \frac{3}{4} H\left(\frac{2}{3}, \frac{1}{3}\right) + \frac{1}{4} H\left(\frac{1}{2}, \frac{1}{2}\right)$$

This is known as grouping theorem.

Information is a measure of uncertainty associated with a statement or message. When out of various possibilities associated with an event a particular event occurs, the amount of uncertainty that gets removed is a measure of the information.

It appears that information is defined in a negative sense in relation to the amount of uncertainty associated with the occurrence of an event.

We claim that a function which is constant times logarithm of M (i.e. $f(M) = C \log M, C > 0$) satisfies the following necessary algebraic properties.

- $f(M^2) = f(M. M) = f(M) + f(M) = 2f(M)$
- $f(M^k) = kf(M)$
- $f(M) = f[(M^{1/n})^n] = nf(M^{1/n}) \Rightarrow f(M^{\frac{1}{n}}) = \frac{1}{n} f(M)$
- $f(M^{p/n}) = pf(M)$

- So for any real number a, $f(M^a) = af(M)$.
- Again, since $f(1) = f(1.1) = f(1) + f(1) = 2f(1)$ it follows that $f(1) = 0$ and is consistent with the fact that $\log 1 = 0$ as $f(1) = C \log 1 = 0$. The significance of the result $f(1) = 0$ is that if an event is certain then there is no uncertainty.
- For $r =$ arbitrary positive integer and for any integral $M > 1$ and following relationship is satisfied for some integral k

$$M^k \leqslant 2^r \leqslant M^{k+1}$$

An example

Suppose $M = 4$, $r = 3$. So we are in search of a number k such that

$$4^k \leqslant 2^3 \leqslant 4^{k+1} \text{ i.e. } 4^k \leqslant 8 \leqslant 4^{k+1}.$$

Now for $k = 1$, $4 \leqslant 8 \leqslant 4^2$ holds.

Since $M^k \leqslant 2^r \leqslant M^{k+1}$, $f(M^k) \leqslant f(2^r) \leqslant f(M^{k+1})$. This can be recast as

$$kf(M) \leqslant rf(2) \leqslant (k+1)f(M).$$

$$\frac{k}{r} \leqslant \frac{f(2)}{f(M)} \leqslant \frac{k+1}{r} \tag{10.2}$$

Logarithmic function is also a monotonically increasing function. Consider

$$\log(M^k) \leqslant \log(2^r) \leqslant \log(M^{k+1})$$

$$\Rightarrow k \log M \leqslant r \log 2 \leqslant (k+1)\log M$$

$$\frac{k}{r} \leqslant \frac{\log 2}{\log M} \leqslant \frac{k+1}{r} \tag{10.3}$$

From equation (10.2) and equation (10.3) we find that $\frac{f(2)}{f(M)}$ and $\frac{\log 2}{\log M}$ lie between $\frac{k}{r}$ and $\frac{k+1}{r}$, the distance between them on the real line is less than $\frac{1}{r}$. As r is arbitrary we can assume it to have an indefinitely large value. In the limit they merge and so we can write

$$\frac{\log 2}{\log M} = \frac{f(2)}{f(M)}$$

$$\Rightarrow f(M) = C \log M \text{ where } C = \frac{f(2)}{f(M)} > 0.$$

The traditional choice in information theory is to set $C = 1$, which gives

$$f(M) = \log M.$$

The properties of logarithm were used in such a way that the base of logarithm did not matter and so we are free to make a choice as per our convenience. Since we deal with bits in computing it is traditional to take base of logarithm as two.

The uncertainty does not depend on the value that the random variable assumes but depends upon the uncertainty associated with an event. We wish to find an expression for the uncertainty function H.

Now the function $f(M)$ is the same function as H with the condition that the probabilities of possible events are all equal. Also, we showed that logarithm is a good function satisfying this property.

Consider s events, each with same probability $\frac{1}{s} = p_i$, but r of them belonging to group A and $s - r$ belonging to group B. From grouping theorem we can thus write

$$H\left(\frac{1}{s}, \frac{1}{s}, \frac{1}{s}, \ldots\ldots \frac{1}{s}\right) - H\left(\sum_{i=1}^{r} \frac{1}{s}, \sum_{i=r+1}^{s} \frac{1}{s}\right)$$

$$= \sum_{i=1}^{r} \frac{1}{s} H\left(\frac{\frac{1}{s}}{\sum_{i=1}^{r}\frac{1}{s}}, \frac{\frac{1}{s}}{\sum_{i=1}^{r}\frac{1}{s}}, \ldots\ldots, \frac{\frac{1}{s}}{\sum_{i=1}^{r}\frac{1}{s}}\right) + \sum_{i=r+1}^{s} \frac{1}{s} H\left(\frac{\frac{1}{s}}{\sum_{i=r+1}^{s}\frac{1}{s}}, \frac{\frac{1}{s}}{\sum_{i=r+1}^{s}\frac{1}{s}}, \ldots\ldots, \frac{\frac{1}{s}}{\sum_{i=r+1}^{s}\frac{1}{s}}\right)$$

$$H\left(\frac{1}{s}, \frac{1}{s}, \frac{1}{s}, \ldots\ldots \frac{1}{s}\right) - H\left(\frac{r}{s}, \frac{s-r}{s}\right) = \frac{r}{s} H\left(\frac{\frac{1}{s}}{\frac{r}{s}}, \frac{\frac{1}{s}}{\frac{r}{s}}, \ldots\ldots, \frac{\frac{1}{s}}{\frac{r}{s}}\right) + \frac{s-r}{s} H\left(\frac{\frac{1}{s}}{\frac{s-r}{s}}, \frac{\frac{1}{s}}{\frac{s-r}{s}}, \ldots\ldots, \frac{\frac{1}{s}}{\frac{s-r}{s}}\right)$$

Hence

$$H\left(\frac{1}{s}, \frac{1}{s}, \frac{1}{s}, \ldots\ldots \frac{1}{s}\right) - H\left(\frac{r}{s}, \frac{s-r}{s}\right) = \frac{r}{s} H\left(\frac{1}{r}, \frac{1}{r}, \ldots \frac{1}{r}\right) + \frac{s-r}{s} H\left(\frac{1}{s-r}, \frac{1}{s-r}, \ldots \frac{1}{s-r}\right)$$

Identifying

$$f(s) = H\left(\frac{1}{s}, \frac{1}{s}, \frac{1}{s}, \ldots, \frac{1}{s}\right), \quad f(r) = H\left(\frac{1}{r}, \frac{1}{r}, \ldots \frac{1}{r}\right), \quad f(s-r) = H\left(\frac{1}{s-r}, \frac{1}{s-r}, \ldots \frac{1}{s-r}\right)$$

we have

$$f(s) - H\left(\frac{r}{s}, \frac{s-r}{s}\right) = \frac{r}{s} f(r) + \frac{s-r}{s} f(s-r)$$

$$f(s) = H\left(\frac{r}{s}, \frac{s-r}{s}\right) + \frac{r}{s} f(r) + \frac{s-r}{s} f(s-r)$$

Using $f(M) = \log M$ and supposing that we have essentially two groups, $\frac{r}{s} = p$ and $1 - p = 1 - \frac{r}{s} = \frac{s-r}{s}$. Hence the uncertainty function or information function for two events with probability p, $1 - p$ is $H(p, 1 - p)$ and by grouping theorem we can write

$$f(s) = H(p, 1 - p) + pf(r) + (1 - p)f(s - r)$$

$$\log s = H(p, 1 - p) + p \log r + (1 - p)\log(s - r)$$

$$H(p, 1 - p) = \log s - p \log r - (1 - p)\log(s - r)$$

$$H(p, 1 - p) = -[- \log s + p \log r - p \log s + p \log s + (1 - p)\log(s - r)]$$

(adding and subtracting $p \log s$)

$$H(p, 1 - p) = -\left[-(1 - p)\log s + p \log \frac{r}{s} + (1 - p)\log(s - r)\right]$$

$$H(p, 1 - p) = -[(1 - p)\{\log(s - r) - \log s\} + p \log p]$$

$$H(p, 1 - p) = -\left[(1 - p)\log \frac{s-r}{s} + p \log p\right] = -(1 - p)\log\left(1 - \frac{r}{s}\right) - p \log p$$

$$H(p, 1 - p) = -(1 - p)\log(1 - p) - p \log p$$

$$H(\{p_1, p_2\}) = H = -p_1 \log p_1 - p_2 \log p_2$$

We can generalize the result to more than two events with probabilities p_i and write

$$H(\{p_i\}) = H = -\sum_i p_i \log p_i$$

This is the measure of the information uncertainty or the degree of uncertainty, also called Shanon entropy. This expression shows that the uncertainty associated with an event does not depend on the values that it assumes but depends on the probability of occurrence of the events.

For coin tossing (probability of Head is $\frac{1}{2}$, probability of Tail is $\frac{1}{2}$)

$$H\left(\frac{1}{2}, \frac{1}{2}\right) = -\sum_i p_i \log_2 p_i = -\frac{1}{2}\log_2 \frac{1}{2} - \left(1 - \frac{1}{2}\right)\log_2\left(1 - \frac{1}{2}\right) = -2 \cdot \frac{1}{2}\log_2\frac{1}{2}$$

$$=-\log_2 \frac{1}{2} = -\left(\log_2 1 - \log_2 2\right) = -\left(0 - \log_2 2\right) = 1$$

So one bit of uncertainty is associated with a coin toss where the unbiased coin shows either Head or Tail each with probability $\frac{1}{2}$. A single bit takes value 0 or 1. So if 0 indicates head and 1 indicates tail then there is 1 bit of uncertainty associated with it.

Plot of $H(p, 1 - p)$ against p for two events gives a curve as indicated in figure 10.2. Once p exceeds $\frac{1}{2}$ there is symmetry because the other event has higher probability.

If the coin is biased the uncertainty decreases because then we are more sure or certain about the face that the coin would show up.

We give an example.

Let X take the values $x_i (i = 1$ to $5)$ (random variable) with probability p_i as shown in table 10.1. This leads to the average uncertainty associated with this event to be $H = -\sum_i p_i \log p_i = 2.27$ bits.

Let us build the decision tree.

We can also regard the uncertainty as the minimum of number of questions having Yes–No format per event that can be asked to get the result in an effort to

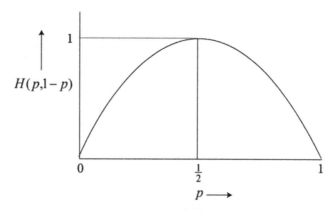

Figure 10.2. Variation of uncertainity with probability for two events.

Table 10.1. Estimate of uncertainty.

X :	x_1	x_2	x_3	x_4	x_5
p :	0.3	0.2	0.2	0.15	0.15

$$H = -\sum_i p_i \log p_i$$

$$H = -0.3\log(0.3) - 0.2\log(0.2) - 0.2\log(0.2)$$
$$- 0.15\log(0.15) - 0.15\log(0.15)$$

$$H = 2.27$$

remove the uncertainty. Let us find the average number of questions corresponding to the data of table 10.1 by building a decision tree as shown in figure 10.3. The number of queries being determined we can proceed to find the average number of questions by taking the product of actual number of questions and the corresponding probability.

So average number of questions is

$$(0.3)(2) + (0.2)(2) + (0.2)(2) + (0.15)(3) + (0.15)(3) = 2.3$$

While the decision tree gives the uncertainty to be 2.3 bits, the uncertainty function or Shannon entropy gives a value of 2.27 bits for the problem considered. It is clear that Shannon entropy gives a lower bound on the uncertainty associated with an event. The Shannon entropy actually provides the minimum bits of uncertainties that any code will have and is known as unique decipherable code.

Information is measured by a function, called Shannon entropy that gives the amount of uncertainty associated with an event. The uncertainty does not depend

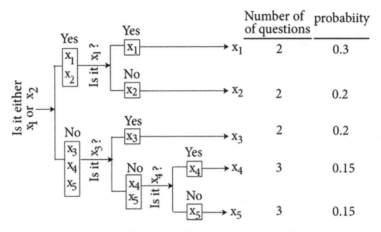

				Number of of questions	probabiity
			x_1	2	0.3
			x_2	2	0.2
			x_3	2	0.2
			x_4	3	0.15
			x_5	3	0.15

Figure 10.3. Decision tree to determine the number of queries.

upon the value which the random variable takes but depends on the uncertainty associated with an event.

$$H(\{p_i\}) = -\sum_i p_i \log_2 p_i \text{ (information entropy)}$$

In statistical physics, entropy refers to the amount of disorder or chaos that a system possesses. Boltzmann's definition of entropy refers to the thermodynamic probability which essentially represents the number of possible accessible configurations (microstates) corresponding to a given macro state

$$S = k \ln \Omega$$

where k = Boltzmann constant, Ω = thermodynamic probability, which is the number of microstates corresponding to a given macro state. This comes from the second law of thermodynamics. Macro state refers to the gross characteristic of an ensemble of systems. Corresponding to each macro state description, there are several microstates. In short, we have macroscopically identical systems distributed over various microstates.

Consider N particles in a given volume V. We mentally divide the volume into m identical cells. The microstate is defined by specifying in which cell the particles reside. Each shift in position of any particle gives a new microstate. The macrostate involves specification of number of particles which is a constant N. The thermodynamic probability Q represents the number of microstates corresponding to the given macro state.

Let n_1 particles reside in cell 1, n_2 particles in cell 2 and so on. The number of configurations (as obtained from the theory of permutations and combinations) is given by the thermodynamic probability

$$\Omega = \frac{N!}{n_1! n_2! \ldots \ldots n_m!}, \text{ where } \sum_i n_i = N.$$

Take log of both sides we get

$$\log \Omega = \log N! - \log(n_1! n_2! \ldots\ldots n_m!) = \log N! - \sum_i \log(n_i!)$$

Use Stirling approximation

$$\log N! = N \log N - N (\text{forlarge } N)$$

We thus have

$$\log \Omega = [N \log N - N] - \sum_i [n_i \log n_i - n_i]$$

$$= [N \log N - N] - \sum_i n_i \log n_i - \sum_i n_i$$

$$= [N \log N - N] - \sum_i n_i \log n_i - N = N \log N - \sum_i n_i \log n_i$$

Probability of finding particle in the i th cell is

$$p_i = \frac{n_i}{N} \quad \Rightarrow \quad n_i = N p_i$$

$$\log \Omega = N \log N - \sum_i (N p_i) \log(N p_i)$$

$$= N \log N - \sum_i (N p_i)\left[\log N + \log p_i \right]$$

$$= N \log N - N \log N \sum_i p_i - N \sum_i p_i \log p_i$$

As $\sum_i p_i = 1$ first two terms on RHS cancel out and we have

$$\log \Omega = -N \sum_i p_i \log p_i$$

Entropy is given by

$$S = k \log \Omega = k\left(-N \sum_i p_i \log p_i\right) = -kN \sum_i p_i \log p_i$$

Average entropy is

$$\frac{S}{N} = = -k \sum_i p_i \log p_i = k \sum_i p_i \log \frac{1}{p_i}$$

Let us consider the following distributions (microstates) corresponding to the macro state of 10^6 particles.

(a) All particles in any one single cell and none in any other cell. So $p_i = 1$ for some $i = j$ and $p_i = 0$ for $i \neq j$. So

$$S = -kp_j \log p_j - kp_{i \neq j} \log p_{i \neq j} = -kp_j \log 1 - k(0) \log p_{i \neq j} = 0$$

Clearly entropy is 0 in this case.

(b) Total number of ways particles can be distributed evenly in any two cells out of a total of $m = 10^6$ cells, given by

$$^{10^6}C_2 = \frac{10^6!}{2!(10^6 - 2)!} = \frac{10^6(10^6 - 1)!}{2} \cong \frac{10^{12}}{2} = 5 \times 10^{11}$$

Probability of a particle being in either cell is $\frac{1}{2}$ and so average entropy is

$$S = -kp_1 \log p_1 - kp_2 \log p_2 = -k\frac{1}{2} \log \frac{1}{2} - k\frac{1}{2} \log \frac{1}{2} = -2k\frac{1}{2} \log 2^{-1}$$

$$= k \log 2$$

In case (a) since the number of cells is 10^6 the probability of occupying a cell is 10^{-6}. Starting from all particles in a single cell (where entropy is zero) the probability that they migrate to two cells (when entropy is $k \log 2$) is

$$= \frac{\text{Total number of particles distributed evenly in any 2 cells}}{\text{Total number of cells for single cell occupancy or any 2 cell occupancy}}$$

$$= \frac{5 \times 10^{11}}{10^6 + 5 \times 10^{11}} = 1 - 10^{-5} \approx 1$$

Clearly the system tends to attain the state where entropy is maximum. In other words the system tends to equilibrate to a state of maximum entropy and in the process particles tend to occupy more states, thereby increasing chaos and disorder.

10.4 Statistical entropy and Shannon's information entropy

Statistical entropy is a measure of disorder in the system and it increases as disorder increases.

Shannon's information entropy is a measure of uncertainty associated with events which occur with different probabilities and it increases as uncertainty increases.

Clearly there is one-to-one correspondence both in terms of physical interpretation and in terms of their expressions we obtained in statistical physics ($S = k \log_e \Omega$) and information theory ($H = -\sum_i p_i \log_2 p_i$). We do not need the Boltzmann constant in the definition of Shannon entropy as we are not going to discuss statistical physics.

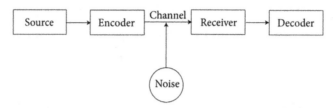

Figure 10.4. Schematic representation of communication system.

10.5 Communication system

We now discuss a typical communication system as shown in the schematic diagram in figure 10.4.

The source generates a message that is to be communicated. It is encoded in terms of binary digits through the encoder. It is received in the receiver after its passage through a channel (may be fibre optic channel or microwave channel). Obviously the receiver accepts it in a coded form. So it has to be decoded by the decoder and the original message is restored.

Now during the passage of the signal or information through the channel it might, in general pick up external noise with random disturbances or errors, which need to be filtered or eliminated, i.e. minimized.

We define the information capacity of a communication system as the rate of information usually measured in kilobits per second (kbps) that can be transported by a channel with least amount of error. This is lower than the raw information capacity which is the capacity of communication channel in the absence of noise. In practice the real information capacity is lower than the raw information capacity because there is presence of noise in the latter.

10.6 Shannon's noiseless coding theorem

Suppose in a communication the following letters occur with a frequency as indicated.

A = 40%, C = 30%, G =15%, T = 15%.

We want to codify the letters using a two-bit code (per letter) as follows

A = 00, C = 01, G = 10, T = 11.

This coding scheme has on an average two bits per letter since

0.4 × 2 bits +0.3 × 2 bits +0.15 × 2 bits +0.15 × 2 bits = 2 bits per letter.

Consider an alternative coding scheme

A = 0, C = 10, G = 110, T = 111.

In this coding scheme the average number of bits per letter is calculated as follows

0.4 × 1 bit + 0.3 × 2 bits + 0.15 × 3 bits + 0.15 × 3 bits = 1.9 bits per letter.

This 1.9 bits per letter is lower than the previous case of 2 bits per letter and holds an edge over the previous one because the second coding scheme saves 0.1 bit per letter.

We now discuss what we mean by optimal code and the maximum compression of bits given by Shannon's entropy where we do not have to worry about the specific coding. Shannon's entropy gives the minimum bits per letter that is called the maximum possible compression.

In the given examples the frequencies were A = 40%, C = 30%, G =15%, T = 15% and hence the corresponding probabilities are A = 0.4, C = 0.3, G = 0.15, T = 0.15 .

Let us calculate the Shanon entropy as follows

$$H = -\sum_i p_i \log_2 p_i$$

$$=-0.4 \log_2(0.4) - 0.3 \log_2(0.3) - 0.15 \log_2(0.15) - 0.15 \log_2(0.15) = 1.871$$

This is the maximum compression that is possible for any of the codes that we may write. This is called Shannon's noiseless coding theorem.

Shanon's noiseless coding theorem is applicable to all uniquely decipherable codes (prefix codes). It provides a limit of the average length of a code that can be carried with a high degree of faithfulness or fidelity over a noiseless channel.

10.7 Prefix code, binary tree

Prefix code (also referred to as prefix-free code or Huffman code) is a set of binary sequences in which no sequence is a prefix of any other sequence in it. Fixed length codes are prefix codes.

Example

Consider the sequence {01, 010, 10}. It is not a prefix code since 01 is a prefix in 010.

Consider the sequence {01, 100, 101}. It is a prefix code since 01 is not a prefix of the other two; 100 is not a prefix of the other two; 101 is not a prefix of the other two also.

We give a few more examples in table 10.2.

We can represent every prefix code in a binary tree (figure 10.5).

In a binary tree we start from the root. Every left branch is a 0 and every right branch is a 1. The tree ends in leaf or external codes. Any intermediate code is an internal code. External codes make up a prefix code.

A few examples are furnished in figure 10.6.

Table 10.2. Example of prefix code (see problem 11 for binary tree).

Set 1 : { 01 , 10, 11, 000, 001 }   Prefix code

Set 2 : { 01 , (00), 000, 10 }   Not a prefix code

Set 3 : { 00 , 01 , 10 , 11}   Prefix code
                                    (fixed length code)

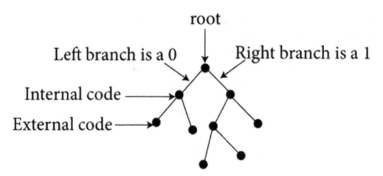

Figure 10.5. Binary tree.

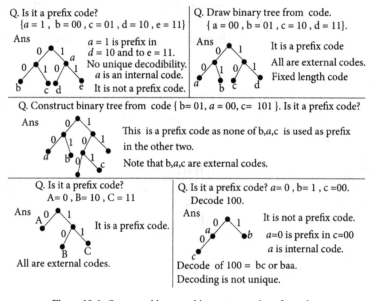

Q. Is it a prefix code?
{$a = 1$, $b = 00$, $c = 01$, $d = 10$, $e = 11$}

Ans

$a = 1$ is prefix in
$d = 10$ and to $e = 11$.
No unique decodibility.
a is an internal code.
It is not a prefix code.

Q. Draw binary tree from code.
{$a = 00$, $b = 01$, $c = 10$, $d = 11$}.

Ans

It is a prefix code
All are external codes.
Fixed length code

Q. Construct binary tree from code { $b = 01$, $a = 00$, $c = 101$ }. Is it a prefix code?

Ans

This is a prefix code as none of b,a,c is used as prefix in the other two.

Note that b,a,c are external codes.

Q. Is it a prefix code?
$A = 0$, $B = 10$, $C = 11$

Ans

It is a prefix code.

All are external codes.

Q. Is it a prefix code? $a = 0$, $b = 1$, $c = 00$.
Decode 100.

Ans

It is not a prefix code.
$a = 0$ is prefix in $c = 00$
a is internal code.

Decode of 100 = bc or baa.
Decoding is not unique.

Figure 10.6. Some problems on binary tree and prefix codes.

The seven lettered word QUANTUM can be coded using 21 bits using the code of figure 10.7 as against $7 \times 8 = 56$ bits (one byte per letter) that amounts to a significant compression of $\frac{21}{56} \times 100 = 37.5\%$.

As per noiseless coding theorem, for uniquely decipherable codes where the letter x_i occurs with probability p_i, having a length of code n_i and hence an average length of the letters $\sum_i n_i p_i$ maximum word compression would occur if

$$\sum_i n_i p_i \geqslant H$$

where the entropy function is $H = -\sum_i p_i \log_2 p_i$. So Shannon entropy gives the lower bound on the uncertainty associated with an event. So maximum length of a word cannot exceed the Shannon entropy in a classical system.

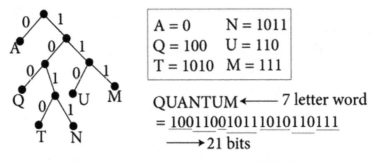

Figure 10.7. A binary tree to code the word QUANTUM.

We now make a direct extension of classical Shannon entropy, which characterizes uncertainties in a classical information system and move on to discuss a quantum entropy called Von Neumann entropy where classical ensembles will be replaced by quantum ones to get an idea of uncertainty of a quantum system.

10.8 Quantum information theory, Von Neumann entropy

Von Neumann entropy provides a measure of our ignorance or uncertainties of quantum systems. It generalizes the classical Shannon entropy to the case of quantum systems. An ensemble of quantum systems is described by a density matrix $\hat{\rho}$ in terms of which we have $\langle \hat{A} \rangle = Tr(\hat{A}\hat{\rho})$ and $Tr\hat{\rho} = 1$, $Tr\hat{\rho}^2 = 1$ for pure systems with $\hat{\rho}^2 = \hat{\rho}$. Here trace is a weighted sum of expectation values of $\hat{A}$ in all pure states of the ensemble.

Von Neumann entropy $S(\rho)$, provides a measure of the degree of mixedness or chaos of the ensemble. So for a pure system Von Neumann entropy $S(\rho)$ should be zero. This is analogous to the classical case in which Shannon entropy is zero if probability of an event is 1 or 0.

Let us define Von Neumann entropy $S(\rho)$ as

$$S(\rho) = -Tr(\hat{\rho} \log_2 \hat{\rho})$$

and $S(\rho) = 0$ for a well-ordered state..

We start with a pure system, like in the case of Shannon entropy. We mention some properties that the system must have.

- Entropy must be independent of the choice of basis in which quantum states are expressed. Again trace is basis independent. It is convenient to switch over to a basis in which $\hat{\rho}$ is diagonal. In such a basis $S(\rho)$ is given by

$$S(\rho) = -\sum_i \lambda_i \log_2 \lambda_i \geqslant 0$$

$\hat{\rho}$ is a positive operator with unit trace. The eigenvalues are $0 \leqslant \lambda_i \leqslant 1$ and $\sum_i \lambda_i = 1$ implying that entropy is a positive quantity.

For a pure system, $\lambda_i = 1$ for a specific i and $\lambda_i = 0$ otherwise. So

$$S(\rho) = -1 \log_2 1 - 0 \log_2 0 = 0$$

Clearly as only one eigenvalue of $\hat{\rho}$ is 1 and all others are zero $S(\rho) = 0$ for a pure state.

- If there are D non-zero eigenvalues of $\hat{\rho}$, since $0 \leqslant \lambda_i \leqslant 1$, the maximum value of $S(\hat{\rho})$ will occur when all the eigenvalues are equal, i.e. each eigenvalue is $\frac{1}{D}$. So maximum entropy is $\log_2 D$. Hence

$$S(\rho) \leqslant \log_2 D$$

Equality sign holds when all eigenvalues are equal.

- The less we know about how the state is prepared the higher will be the entropy. This is called mixing theorem of entropy and follows from the relation

$$S\left(\sum_i x_i \rho_i\right) \geqslant \sum_i x_i S(\rho_i)$$

for $x_i \geqslant 0$, $\sum_i x_i = 1$.

- Consider a system having two components AB. The entropy for the composite system can be shown to be

$$S(\rho_{AB}) \leqslant S(\rho_A) + S(\rho_B)$$

where the equality holds for $\hat{\rho}_{AB} = \hat{\rho}_A \otimes \hat{\rho}_B$ (i.e. for independent systems).

- For pure states the eigenvalues of density matrices are identical since for each density matrix one of the eigenvalues is 1 and all others are 0 (and this follows from the relation $Tr\hat{\rho}^2 = 1$). Thus $\hat{\rho}_A$ and $\hat{\rho}_B$ have the same eigenvalues and then the equality sign holds. So when we mix the systems A and B then $S(\rho_{AB})$ of the composite system cannot be greater than the sum of individual entropies. This is the standard mixing theorem.

- Consider a pure state, eigenvalues being 1,0. In a representation in which it is diagonal we have

$$\hat{\rho} = \begin{pmatrix} 1 & 0 \\ 0 & 0 \end{pmatrix}, \quad S(\rho) = 0$$

- Consider the state $|\psi> = \frac{1}{\sqrt{2}}(|0> + |1>)$. The pure state is

$$\hat{\rho} = \left|\psi><\psi\right| = \frac{1}{\sqrt{2}}(|0> + |1>)\frac{1}{\sqrt{2}}(<0|+<1|)$$

$$= \frac{1}{2}[|0><0| + |1><0| + |0><1| + |1><1|]$$

Using $|0> = \begin{pmatrix} 1 \\ 0 \end{pmatrix}$, $|1> = \begin{pmatrix} 0 \\ 1 \end{pmatrix}$ we get

$$\hat{\rho} = \frac{1}{2}\left[\binom{1}{0}(1 \ 0) \Big| + \binom{0}{1}(1 \ 0) + \binom{1}{0}(0 \ 1) + \binom{0}{1}(0 \ 1) \right]$$

$$= \frac{1}{2}\left[\begin{pmatrix} 1 & 0 \\ 0 & 0 \end{pmatrix} + \begin{pmatrix} 0 & 0 \\ 1 & 0 \end{pmatrix} + \begin{pmatrix} 0 & 1 \\ 0 & 0 \end{pmatrix} + \begin{pmatrix} 0 & 0 \\ 0 & 1 \end{pmatrix} \right] = \frac{1}{2}\begin{pmatrix} 1 & 1 \\ 1 & 1 \end{pmatrix}$$

Eigenvalues of this matrix are 1 and 0 and so $S(\rho) = 0$. Also $\hat{\rho}^2 = \hat{\rho}$, $Tr\hat{\rho}^2 = 1$.

- We now consider the maximally mixed state

$$\hat{\rho} = \begin{pmatrix} \frac{1}{2} & 0 \\ 0 & \frac{1}{2} \end{pmatrix} \quad (\hat{\rho}^2 \neq \hat{\rho}, \ Tr\hat{\rho}^2 < 1)$$

$$S(\rho) = -\frac{1}{2}\log_2\frac{1}{2} - \frac{1}{2}\log_2\frac{1}{2} = -\log_2\frac{1}{2} = \log_2 2 = 1$$

- Consider the 2-qubit state which is

$$|\psi> = \cos\theta|00> + \sin\theta|11>$$

It is a pure state the entropy of which is zero (i.e. $S^{AB} = 0$). The representation used here is $|A \ B>$ (A, B are the 2 qubits). We calculate $\hat{\rho}^A$ by taking partial trace over B of $\hat{\rho}_{AB}$.

$$\hat{\rho}^A = Tr_B(\hat{\rho}_{AB})$$

$$\hat{\rho}_{AB} = |\psi> <\psi|$$

$$= [\ \cos\theta|00> + \sin\theta|11>][\cos\theta <00|+ \sin\theta <11|]$$

$$= \cos^2\theta|00> <00| + \sin^2\theta|11> <11| + \cos\theta\sin\theta \ (\ |00> <11| + |11> <00| \ )$$

Tracing out B (the second element of the ket) we have

$$\hat{\rho}^A = Tr_B(\hat{\rho}_{AB}) = \cos^2\theta|0> <0| \ Tr|0> <0| + \sin^2\theta|1> <1| \ Tr|1> <1|$$

$$+ \cos\theta\sin\theta(|0> <1| \ Tr|0> <1| + |1> <0| \ Tr|1> <0|)$$

As trace of a ket followed by a bra is the product of bra with a ket we have

$$\hat{\rho}^A = \cos^2\theta|0> <0| < 0|0> +\sin^2\theta|1> <1|<1|1>$$
$$+ \cos\theta\sin\theta|0> <1| < 1|0> +|1> <0|<0|1>$$

Using $<0|0> = 1, <1|1> = 1, <0|1> = 0, <1|0> = 0$ we get

$$\hat{\rho}^A = Tr_B(\hat{\rho}_{AB}) = \cos^2\theta|0> <0| + \sin^2\theta|1> <1|$$

$$=\cos^2\theta\begin{pmatrix}1\\0\end{pmatrix}(1\ \ 0) + \sin^2\theta\begin{pmatrix}0\\1\end{pmatrix}(0\ \ 1) = \cos^2\theta\begin{pmatrix}1 & 0\\0 & 0\end{pmatrix} + \sin^2\theta\begin{pmatrix}0 & 1\\0 & 1\end{pmatrix}$$

$$\hat{\rho}^A = \begin{pmatrix}\cos^2\theta & 0\\0 & \sin^2\theta|\end{pmatrix}$$

It is a diagonal matrix, eigenvalues are $\cos^2\theta$ and $\sin^2\theta$. Also, trace is $\cos^2\theta + \sin^2\theta = 1$.

$$S^A = -\cos^2\theta\log_2(\cos^2\theta) - \sin^2\theta\log_2(\sin^2\theta)$$

$$S^A = -2\cos^2\theta\log_2(\cos\theta) - 2\sin^2\theta\log_2(\sin\theta)$$

Similarly, tracing out A we have

$$S^B = -2\cos^2\theta\log_2(\cos\theta) - 2\sin^2\theta\log_2(\sin\theta)$$

Hence

$$S^A + S^B = -4\cos^2\theta\log_2(\cos\theta) - 4\sin^2\theta\log_2(\sin\theta)$$

As $\cos\theta$ and $\sin\theta$ are less than 1, so logarithm turns out to be negative and hence

$$S^A + S^B > 0$$

But the pure state had $S^{AB} = 0$. Clearly the entropy of the mixed state AB is less than the sum of the entropies of the two component pure states A and B.

10.9 Further reading

[1] Ghosh D 2016 Online course on quantum information and computing (Bombay: Indian Institute of Technology)

[2] Nielsen M A and Chuang I L 2019 *Quantum Computation and Quantum Information* (Cambridge: Cambridge University Press)

[3] Nakahara M and Ohmi T 2017 *Quantum Computing* (Boca Raton, FL: CRC Press)

[4] Pathak A 2016 *Elements of Quantum Computing and Quantum Communication* (Boca Raton, FL: CRC Press)

10.10 Problems

1. *Shannon entropy and Von Neumann entropy refer to:*

(a) *Shannon entropy to classical system and Von Neumann entropy to quantum system.*

(b) *Shannon entropy to quantum system and Von Neumann entropy to classical system.*

(c) Both to classical system.
(d) Both to quantum system.

Ans. *(a)*

2. *The set { 00, 01, 10, 11, 110 } is which of the following?*
    (a) A prefix code(b) Not a prefix code, (c) a Prefix free code, (d) Huffman code.
    Ans. *(b).*

3. *Shannon information entropy is a measure of*
    (a) probability associated with an event, (b) uncertainty associated with event, (b) certainty associated with an event, (d) disorder associated with an event.
    Ans. *(b).*

4. *If a letter x_i occurs with a probability p_i in a code of length n_i and H is the entropy function, then which of the following is suggested for maximum word compression as per noiseless coding theorem?*
    (a) $\sum_i n_i p_i = H$ (b) $\sum_i n_i p_i > H$ (c) $\sum_i n_i p_i < H$ (d) $\sum_i n_i p_i \geqslant H$.
    Ans. *(d).*

5. *The number of bits n that is required to send m different messages is given by*
    (a)$n = \log_2 m$ (b)$2^n = m$ (c)$m = \log_2 n$ (d)$2^m = n$.
    Ans *(a), (b).*

6. *Stirling's approximation holds for*
    (a) small N, (b) large N, (c) any N, (d) N lying between10^4 and 10^6.
    Ans. *(b).*

7. *Which of the following is Shannon entropy?*
    (a)$k \log \Omega$ (b)$-Tr(\hat{\rho} \log \hat{\rho})$ (c)$Tr\hat{\rho}^2 \leqslant 1$ (d)$-\sum_i p_i \log p_i$.
    Ans. *(d).*

8. *Which of the following is Von Neumann entropy?*
    (a)$k \log \Omega$ (b)$-Tr(\hat{\rho} \log \hat{\rho})$ (c)$Tr\hat{\rho}^2 \leqslant 1$ (d)$-\sum_i p_i \log p_i$.
    Ans. *(b).*

9. *Which of the following is Boltzmann entropy?*
    (a)$k \log \Omega$ (b)$-Tr(\hat{\rho} \log \hat{\rho})$ (c)$Tr\hat{\rho}^2 \leqslant 1$ (d)$-\sum_i p_i \log p_i$.
    Ans. *(a).*

10. *Which is a wrong statement?*

(a) Probability of occurrence of an event is measured by Shannon entropy.
(b) Chaos in information is measured by Shannon entropy.
(c) Information is measured by Shannon entropy.
(d) Von Neumann entropy provides a measure of uncertainty predicted in Heisenberg's uncertainty principle.

Ans. *(c).*

11. *Construct the binary tree of set 1, set 2 and set 3 mentioned in* table 10.2.

Ans.
Binary tree shown in figure 10.8.

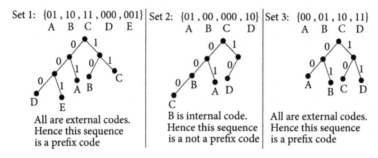

Set 1: {01 , 10 , 11 , 000 , 001}
A   B   C   D   E

All are external codes.
Hence this sequence
is a prefix code

Set 2: {01 , 00 , 000 , 10}
A   B   C   D

B is internal code.
Hence this sequence
is a not a prefix code

Set 3: {00 , 01 , 10 , 11}
A   B   C   D

All are external codes.
Hence this sequence
is a prefix code

Figure 10.8. Binary tree of problem 11 (refer to table 10.2).

Quantum Optics and Quantum Computation
An introduction
Dipankar Bhattacharyya and Jyotirmoy Guha

Chapter 11

EPR paradox and Bell inequalities

Classical physics describes nature at an ordinary level (the macroscopic scale) while quantum physics describes nature in the small (the microscopic or atomic and subatomic scale) where classical physics is inadequate.

11.1 EPR paradox

We address the question regarding how much reliable, dependable and robust quantum mechanics is in transmitting information over a quantum channel or over a public channel. To show this we need to establish the supremacy of quantum mechanics over its classical counterpart.

Einstein had reservations about the Copenhagen interpretation of quantum mechanics. This objection is famously known as Einstein–Podolsky–Rosen paradox or in short EPR paradox.

Consider the two-qubit Bell state $\frac{1}{\sqrt{2}}(|00> + |11>)$ which is maximally entangled and is one of the EPR pairs. $\frac{1}{\sqrt{2}}$ is the normalization factor. If we take reduced trace over either the first qubit or the second qubit we get a maximally mixed state $\rho = \frac{1}{2}$. The characteristic of entangled state is that if we make a local measurement on one of the qubits the other qubit automatically gets fixed and the information that could be there in the entangled state gets lost. So we get no information about how the system was prepared since measurement of the first qubit collapses the entangled second qubit and similarly measurement of the second qubit collapses the entangled first qubit also (section 4.7).

Einstein's argument or objection was as follows.

If qubit 1 and qubit 2 are separated by space-like distance then a measurement of qubit 1 should not affect the state of qubit 2, because the communicating agency or signal, then has to move with speed greater than the speed of light in free space and this is not a possibility as per the special theory of relativity. This is referred to as EPR paradox that challenges the Copenhagen interpretation (1935). As no information regarding measurement of the first qubit can reach the location of the second qubit it is inconceivable that the measurement of the first qubit determines the state of the second qubit. So the objection in a nutshell is: How can the second qubit collapse as a result of the first qubit measurement?

This correlation between two particles, irrespective of how widely they are separated (space-like for instance) is the so called 'spooky action at a distance'. At first glance it seems that an entangled state violates an axiom of special theory of relativity.

According to Einstein, a physical system should have an element of physical reality. If without disturbing the system in any way we can predict with certainty (i.e. with probability 1) the value of a physical quantity then there exists an element of physical reality corresponding to this physical quantity.

But quantum mechanics says that the value of a physical observable has no meaning independent of its measurement. In other words its value gets fixed the moment we have made a measurement.

Consider two systems. They interact over a time interval from say $t = 0$ to $t = T$, after which they separate and do not interact. The state of the combined system and their development or evolution is known from quantum mechanics by solving the Schrödinger equation, even for $t > T$. However, we cannot calculate the state in which either subsystem was left after the interaction without measurement, i.e. by collapse of wave function. Let us think of the following example.

Consider a particle disintegrating from rest into two equal masses. The process of disintegration is an internal process. Momentum will be conserved. Being at rest, in the centre of mass frame of reference its momentum is zero. By conservation of momentum the total momentum is equal to zero even after disintegration and so the particles would have equal and opposite momentum. So we may get simultaneous information (according to Einstein) regarding position and momentum of a particle. But according to Heisenberg's uncertainty principle we cannot fix the position and momentum simultaneously. Let us explain this elaborately.

Suppose we make measurement of momentum of particle 1 $(=\vec{p}_1)$ and position of particle 2 simultaneously (Heisenberg's uncertainty principle does not restrict such measurements). Since momentum is conserved this fixes the momentum of particle 2 $(\vec{p}_2 = -\vec{p}_1$ as total momentum is zero in the centre of mass frame of reference. Again since the centre of mass has to remain fixed we would be able to calculate the position of particle 1. This argument of Einstein points to the fact that we know information about position and momentum at the same instant. This was Einstein's primary objection to the Copenhagen's interpretation of quantum mechanics.

Einstein's interpretation rests on the definition of terms like 'realism' and 'locality'.

Realism

An object possesses property independent of measurement which only gets revealed when a measurement is made. In other words measurement is not the process that endowed the particle with an attribute or property. The process of measurement is simply a revealing process.

According to Einstein a physical theory must satisfy the requirement of local realism. In other words, a physical theory, in order that it is understandable, must have physical local reality.

Locality

A measurement on a subsystem (a combined system or of an entangled pair) at one place, cannot influence the property of another subsystem (other member) in another place.

According to Einstein there are some hidden variables in the problem which encode the property of a system.

Einstein insisted that an object must possess a property independent of the process of measurement, while in quantuim mechanics it is said that it is during measurement that the particle acquires a property. Einstin demands that a physical theory must satisfy the requirement of the dual property of locality and realism.

11.2 David Bohm's version of EPR paradox (1951)

Consider a composite system having total spin angular momentum $S = 0$. Suppose it decays into two spin-half particles. Since angular momentum is conserved, the total angular momentum of the two-particle system would still be $S = 0$. The pair of particles would be in a state after decay given by

$$\frac{1}{\sqrt{2}}(|\uparrow\downarrow> - |\downarrow\uparrow>)$$

where $| \uparrow > = $|up spin> and $| \downarrow > = $|down spin>. Clearly if S_z value of particle 1 is $+\frac{\hbar}{2}$ ($\uparrow$ i.e. up) that of the other is $-\frac{\hbar}{2}$ ($\downarrow$ i.e. down) and vice versa.

After decay, this two-qubit system gets separated by a large distance such that disturbance of one cannot have any influence on the state of the second one.

Suppose Alice has particle 1 and Bob has particle 2.

If Alice makes a measurement of spin along the z-direction and finds her particle to have $S_z = \frac{1}{2}\hbar$ (i.e. $\uparrow$ state), $\sigma_z = +1$, eigenstate $| \uparrow > = \begin{pmatrix} 1 \\ 0 \end{pmatrix}$ then if Bob measures the particle he is guaranteed to get $S_z = -\frac{1}{2}\hbar$(i.e. $\downarrow$ state), $\sigma_z = -1$, eigenstate $| \uparrow > = \begin{pmatrix} 0 \\ 1 \end{pmatrix}$. So Alice's measurement and Bob's measurement have a perfect anti-correlation. (Alice |up>, Bob |down>). Such perfect anti-correlation will not be there if, however, Alice and Bob measure their spin components in different directions.

Now let us suppose that Alice made a measurement along z direction and got a spin up $S_z = \frac{1}{2}\hbar$(i.e. $\uparrow$ state), $\sigma_z = +1$, eigenstate $| \uparrow > = \begin{pmatrix} 1 \\ 0 \end{pmatrix}$. But Bob decides to make measurement in the arbitrary direction $\hat{n}(\theta, \phi)$(making angle θ with z axis and $\phi=$ azimuthal angle) so that $\hat{n} = (\sin\theta\cos\phi, \sin\theta\sin\phi, \cos\theta)$. The result of measurement will depend on the angle θ that his axis of measurement makes with the z-axis. Let us find the probability of obtaining spin down state $S_z = -\frac{1}{2}\hbar$ (i.e. $\downarrow$ state), $\sigma_z = -1$, eigenstate $| \uparrow > = \begin{pmatrix} 0 \\ 1 \end{pmatrix}$ by Bob.

The state vector of a qubit corresponding to a point on Bloch sphere $\hat{n}$ (θ, ϕ) is

$$|n, +> = \begin{pmatrix} \cos\dfrac{\theta}{2} \\ e^{i\phi}\sin\dfrac{\theta}{2} \end{pmatrix} = e^{i\phi/2}\begin{pmatrix} \cos\dfrac{\theta}{2}e^{-i\phi/2} \\ \sin\dfrac{\theta}{2}e^{i\phi/2} \end{pmatrix} \text{(section 4.4)}$$

and corresponding opposite point $-\hat{n}$ $(\pi - \theta, \pi + \phi)$ on Bloch sphere is

$$|n, -> = \begin{pmatrix} \sin\dfrac{\theta}{2} \\ -e^{i\phi}\cos\dfrac{\theta}{2} \end{pmatrix} = e^{i\phi/2}\begin{pmatrix} \sin\dfrac{\theta}{2}e^{-i\phi/2} \\ -\cos\dfrac{\theta}{2}e^{i\phi/2} \end{pmatrix} \text{(section 4.4)}$$

If we look at $\hat{\sigma}_n$ (the component of Pauli matrix along an arbitrary direction $\hat{n}$) then the eigenstate of the Pauli matrix $\hat{\sigma}_n$ corresponding to the eigenvalue $+1$ and -1 are, respectively, $|n, +>$ and $|n, ->$. These eigenstates are orthonormalized. Let us construct a state using the eigenstates $|n, +>$ and $|n, ->$ as

$$|\chi> = \sin\frac{\theta}{2}|n, +> - \cos\frac{\theta}{2}|n, ->$$

and verify what it represents.

$$|\chi> = \sin\frac{\theta}{2}e^{i\phi/2}\begin{pmatrix} \cos\dfrac{\theta}{2}e^{-i\phi/2} \\ \sin\dfrac{\theta}{2}e^{i\phi/2} \end{pmatrix} - \cos\frac{\theta}{2}e^{i\phi/2}\begin{pmatrix} \sin\dfrac{\theta}{2}e^{-i\phi/2} \\ -\cos\dfrac{\theta}{2}e^{i\phi/2} \end{pmatrix}$$

$$= e^{i\phi/2}\begin{pmatrix} \sin\dfrac{\theta}{2}\cos\dfrac{\theta}{2}e^{-i\phi/2} \\ \sin^2\dfrac{\theta}{2}e^{i\phi/2} \end{pmatrix} - e^{i\phi/2}\begin{pmatrix} \sin\dfrac{\theta}{2}\cos\dfrac{\theta}{2}e^{-i\phi/2} \\ -\cos^2\dfrac{\theta}{2}e^{i\phi/2} \end{pmatrix}$$

$$= e^{i\phi/2}\begin{pmatrix} 0 \\ \sin^2\dfrac{\theta}{2}e^{i\phi/2} + \cos^2\dfrac{\theta}{2}e^{i\phi/2} \end{pmatrix} = e^{i\phi/2}\begin{pmatrix} 0 \\ e^{i\phi/2} \end{pmatrix}$$

$$|\chi> = e^{i\phi}\begin{pmatrix} 0 \\ 1 \end{pmatrix} \rightarrow \begin{pmatrix} 0 \\ 1 \end{pmatrix} = |\uparrow>$$

The overall global phase factor $e^{i\phi}$ can be overlooked. Obviously then we can write

$$|\uparrow> = \sin\frac{\theta}{2}|n, +> - \cos\frac{\theta}{2}|n, ->$$

So Bob will measure spin down state $S_z = -\frac{1}{2}\hbar$(i.e. $\downarrow$ state), $\sigma_z = -1$, eigenstate $|\uparrow> = \begin{pmatrix} 0 \\ 1 \end{pmatrix}$ with a probability $|\cos\frac{\theta}{2}|^2 = \cos^2\frac{\theta}{2}$. So Bob's measurement would depend on angle θ which the measurement axis makes with the z-axis. So the measurements are still anti-correlated. The anti-correlation is still there between Alice's measurement and Bob's measurement even though the two are measuring on two different systems. The anti- correlation is perfect if the axes are the same.

According to Einstein, a physical theory must satisfy the requirement of local realism. In other words a physical theory in order that it is understandable must have physical local reality.

The rival or conflicting points of view are as follows.

Quantum mechanics

A particle does not have properties independent of observation and quantum mechanics reveals the results with some probabilities through the process of measurement.

The Copenhagen interpretation is of the view that indeterminacy observed in nature is a fundamental law of nature and does not in any way reflect our inadequacy of scientific knowledge.

Hidden variables

The property of the system, though revealed during measurement was pre-ordained as it was encoded into the system through some hidden variables (hidden from our understanding process, perception or consciousness).

According to EPR, quantum mechanics fails to incorporate physical local reality and hence is an incomplete theory. Further interaction or the unavoidable disturbance that occurs between the measuring apparatus and the system being measured is local and its effect is confined to the place where the measurement is made and does not influence distant objects or processes.

11.3 Bell's (Gedanken) experiment: EPR and Bell's inequalities

Bell suggested a Gedanken experiment (thought experiment).

Consider one of the four Bell states for instance

$$|\psi^-> = \frac{1}{\sqrt{2}}(|01> - |10>)$$

Let us refer to particle number 1 by A because it is the qubit belonging to Alice and particle number 2 by B because it is the qubit belonging to Bob.

Let us take i th component of Pauli spin $\hat{\vec{\sigma}}$ acting on state $|\psi^->$ i.e.

$$(\hat{\sigma}_i^A + \hat{\sigma}_i^B)|\psi^->.$$

For $i = x$ we have

$$(\hat{\sigma}_x^A + \hat{\sigma}_x^B)|\psi^-> = (\hat{\sigma}_x^A + \hat{\sigma}_x^B)\frac{1}{\sqrt{2}}(|01> - |10>)$$

$$= \hat{\sigma}_x^A \frac{1}{\sqrt{2}}(|01> - |10>) + \hat{\sigma}_x^B \frac{1}{\sqrt{2}}(|01> - |10>)$$

Using $\hat{\sigma}_x|0> = |1>$, $\hat{\sigma}_x|1> = |0>$ we have ($\hat{\sigma}_x^A$ operates on first particle and $\hat{\sigma}_x^B$ operates on second particle)

$$(\sigma_x^A + \sigma_x^B)|\psi^-> = \frac{1}{\sqrt{2}}[(|11> -|00>) + (|00> -|11>)] = 0$$

For $i = y$ we have

$$(\hat{\sigma}_y^A + \hat{\sigma}_y^B)|\psi^-> = (\hat{\sigma}_y^A + \hat{\sigma}_y^B)\frac{1}{\sqrt{2}}(|01> -|10>)$$

$$= \hat{\sigma}_y^A \frac{1}{\sqrt{2}}(|01> -|10>) + \hat{\sigma}_y^B \frac{1}{\sqrt{2}}(|01> -|10>)$$

Using $\hat{\sigma}_y|0> = i|1>$, $\hat{\sigma}_y|1> = -i|0>$ we have ($\hat{\sigma}_y^A$ operates on first particle and $\hat{\sigma}_y^B$ operates on second particle)

$$(\hat{\sigma}_y^A + \hat{\sigma}_y^B)|\psi^-> = \frac{1}{\sqrt{2}}(i|11> +i|00>) + \frac{1}{\sqrt{2}}(-i|00> -i|11>) = 0.$$

For $i = z$ we have

$$(\hat{\sigma}_z^A + \hat{\sigma}_z^B)|\psi^-> = (\hat{\sigma}_z^A + \hat{\sigma}_z^B)\frac{1}{\sqrt{2}}(|01> -|10>)$$

$$= \hat{\sigma}_z^A \frac{1}{\sqrt{2}}(|01> -|10>) + \hat{\sigma}_z^B \frac{1}{\sqrt{2}}(|01> -|10>)$$

Using $\hat{\sigma}_z|0> = |0>$, $\hat{\sigma}_z|1> = -|1>$ we have ($\hat{\sigma}_z^A$ operates on first particle and $\hat{\sigma}_z^B$ operates on second particle)

$$(\hat{\sigma}_z^A + \hat{\sigma}_z^B)|\psi^-> = \frac{1}{\sqrt{2}}(|01> +|10>) + \frac{1}{\sqrt{2}}(-|01> -|10>) = 0.$$

Hence we end up with the result that

$$(\hat{\sigma}_i^A + \hat{\sigma}_i^B)|\psi^-> = 0 \quad \Rightarrow \quad \hat{\sigma}_i^B|\psi^-> = -\hat{\sigma}_i^A|\psi^->$$

It follows that $\hat{\sigma}_i$ operator flips the spin. So one can equivalently replace $\hat{\sigma}_i^B$ acting on particle 2 by $-\hat{\sigma}_i^A$ acting on particle 1.

The expectation value of the ith component of total spin $\hat{\sigma}_i^A + \hat{\sigma}_i^B$ in the Bell state is zero.

This result shows that if Alice measures spin along a direction say $\hat{a}$ and reveals it to Bob then Bob would know what result he would have got had he measured spin along $\hat{a}$, it would be negative of what Alice got.

Suppose Alice makes a measurement of spin along some direction $\vec{a} = \hat{a}$ and Bob measures along another direction $\vec{b} = \hat{b}$. Let us calculate the expectation value of $(\sigma^A . \hat{a})(\sigma^B . \hat{b})$ where the Pauli spin operators are denoted by $\hat{\sigma}^A = \sigma^A$ and $\hat{\sigma}^B = \sigma^B$. We consider

$$<\psi^-|(\sigma^A.\,\hat{a})(\sigma^B.\,\hat{b})|\psi^->$$

$$=-<\psi^-|(\sigma^A.\,\hat{a})(\sigma^A.\,\hat{b})|\psi^-> \text{ (since } \sigma_i^B|\psi^->=-\sigma_i^A|\psi^->)$$

$$=-\sum_{i,j}<\psi^-|\sigma_i^A\sigma_j^A|\psi^->a_ib_j \text{ (no reference to Bob henceforth)}$$

$$=-\sum_{i,j\neq i}<\psi^-|\sigma_i^A\sigma_j^A|\psi^->a_ib_j-\sum_{i,j=i}<\psi^-|\sigma_i^A\sigma_j^A|\psi^->a_ib_j = A + B$$

where A = six $i\neq j$ terms and B = three $i=j$ terms. The $i\neq j$ terms are
$A=-\sum_{i,j\neq i}<\psi^-|\sigma_i^A\sigma_j^A|\psi^->a_ib_j$. Let us evaluate

$$-\sum_{i,j\neq i}<\psi^-|\sigma_i^A\sigma_j^A|\psi^->$$

$$=-<\psi^-|\ \sigma_x^A\sigma_y^A + \sigma_y^A\sigma_x^A + \sigma_y^A\sigma_z^A + \sigma_z^A\sigma_y^A + \sigma_x^A\sigma_z^A + \sigma_z^A\sigma_x^A\ |\psi^->$$

$$=-<\psi^-|\ \sigma_x^A\sigma_y^A + \sigma_x^A\sigma_z^A + \sigma_y^A\sigma_z^A|\psi^->+<\psi^-|\sigma_y^A\sigma_x^A + \sigma_z^A\sigma_x^A + \sigma_z^A\sigma_y^A\ |\psi^->$$

Consider $-<\psi^-|\sigma_x^A\sigma_y^A + \sigma_x^A\sigma_z^A + \sigma_y^A\sigma_z^A|\psi->$

$$=-[(<01|-<10|)\sigma_x^A\sigma_y^A(|01>-|10>)$$
$$+(<01|-<10|)\sigma_x^A\sigma_z^A(|01>-|10>) + (<01|-<10|)\sigma_y^A\sigma_z^A(|01>-|10>)]$$

Using $\hat{\sigma}_x|0>=|1>$, $\hat{\sigma}_x|1>=|0>$, $\hat{\sigma}_y|0>=i|1>$, $\hat{\sigma}_y|1>=-i|0>$,

$$\hat{\sigma}_z|0>=|0>, \quad \hat{\sigma}_z|1>=-|1>$$

$$-<\psi^-|\sigma_x^A\sigma_y^A + \sigma_x^A\sigma_z^A + \sigma_y^A\sigma_z^A|\psi^->$$

$$=-[(<01|-<10|)\sigma_x^A(i|11>+i|00>) + (<01|-<10|)\sigma_x^A(|01>+|10>)$$
$$+(<01|-<10|)\sigma_y^A(|01>+|10>)]$$

$$=-[(<01|-<10|)\ i(|01>+i|10>)$$
$$+(<01|-<10|)(|11>+|00>) + (<01|-<10|)(i|11>-i|00>)]$$

$$= -i(<01|01>+<01|10>-<10|01>-<10|10>)$$
$$+(<01|11>+<01|00>-<10|11>-<10|00>)$$
$$+i(<01|11>-<01|00>-<10|11>+<10|00>)$$

Using ortho-normalization relations

$$-<\psi|\sigma_x^A\sigma_y^A + \sigma_x^A\sigma_z^A + \sigma_y^A\sigma_z^A|\psi>=-i(1-1)+0+0=0$$

Similarly $<\psi^-|\sigma_y^A\sigma_x^A + \sigma_z^A\sigma_x^A + \sigma_z^A\sigma_y^A\ |\psi^->=0$ also. Hence

$$A = -\sum_{i,j\neq i} <\psi^-|\sigma_i^A\sigma_j^A|\psi^-> a_ib_j = 0$$

$$<\psi^-|(\sigma^A.\,\hat{a})(\sigma^B.\,\hat{b})|\psi^->=A + B = B$$

$$=-\sum_{i,j=i} <\psi^-|\sigma_i^A\sigma_j^A|\psi^-> a_ib_j$$

$$=-\sum_{i} <\psi^-|(\sigma_i^A)^2|\psi^-> a_ib_i = -\sum_{i} <\psi^-|\psi^-> a_ib_i \text{ using } (\sigma_i^A)^2 = 1$$

$$<\psi^-|(\sigma^A.\,\hat{a})(\sigma^B.\,\hat{b})|\psi^->=-\sum_{i} a_ib_i \qquad \text{as } <\psi^-|\psi^->=1$$

$<\psi^-|(\sigma^A.\,\hat{a})(\sigma^B.\,\hat{b})|\psi^->=-\vec{a}.\,\vec{b}=-\cos\theta$ as $\vec{a} = \hat{a}$, $\vec{b} = \hat{b}$ are unit vectors.

This is because Bob is making a measurement along an axis which is making an angle θ with that of Alice's axis. We thus have shown through explicit calculation that if Alice is making a measurement along one direction $\hat{a}$ and Bob is making measurement along a different direction $\hat{b}$, then their results are correlated through a $\cos\theta$ factor (actually anti-correlated). The anti-correlation is perfect if Alice and Bob use the same axis ($\theta = 0°$, $\cos\theta = 1$).

Projection operators

Alice is making measurement along a particular direction and gets a particular result say +1. This means the eigenstate $|n, +>$ of Pauli matrix $\hat{\sigma}_n$ is being projected out. And if eigenvalue -1 is obtained then the eigenstate $|n, ->$ is projected out.

The projection operator for the state $|n, +>$ is $\hat{P}(n, +) = \frac{1}{2}(I + \hat{n}.\,\hat{\sigma})$ and the projection operator for the state $|n, ->$ is $\hat{P}(n, -) = \frac{1}{2}(I - \hat{n}.\,\hat{\sigma})$. Let us check the operation of these projection operators. Let us consider $\hat{P}(n, +) \mid\uparrow >=\frac{1}{2}(I + \hat{n}.\,\hat{\sigma})\mid\uparrow >$

$$\frac{1}{2}(I + \hat{n}.\,\hat{\sigma})\mid\uparrow > =\frac{1}{2}(I + \hat{n}.\,\hat{\sigma})|\sigma_z = +1 > =\frac{1}{2}(I + \hat{n}.\,\hat{\sigma})\begin{pmatrix}1\\0\end{pmatrix}$$

$$=\frac{1}{2}\begin{pmatrix}1\\0\end{pmatrix} + \frac{1}{2}(n_x\sigma_x + n_y\sigma_y + n_z\sigma_z)\begin{pmatrix}1\\0\end{pmatrix}$$

$$=\frac{1}{2}\begin{pmatrix}1\\0\end{pmatrix} + \frac{1}{2}n_x\begin{pmatrix}0 & 1\\1 & 0\end{pmatrix}\begin{pmatrix}1\\0\end{pmatrix} + \frac{1}{2}n_y\begin{pmatrix}0 & -i\\i & 0\end{pmatrix}\begin{pmatrix}1\\0\end{pmatrix} + \frac{1}{2}n_z\begin{pmatrix}1 & 0\\0 & -1\end{pmatrix}\begin{pmatrix}1\\0\end{pmatrix}$$

$$=\frac{1}{2}\begin{pmatrix}1\\0\end{pmatrix} + \frac{1}{2}n_x\begin{pmatrix}0\\1\end{pmatrix} + \frac{1}{2}n_y\begin{pmatrix}0\\i\end{pmatrix} + \frac{1}{2}n_z\begin{pmatrix}1\\0\end{pmatrix}$$

Using $n_x = \sin\theta\cos\phi$, $n_y = \sin\theta\sin\phi$, $n_z = \cos\theta$

$$\frac{1}{2}(I + \hat{n}.\,\hat{\sigma})\begin{pmatrix}1\\0\end{pmatrix} = \frac{1}{2}\begin{pmatrix}1\\0\end{pmatrix} + \frac{1}{2}\sin\theta\cos\phi\begin{pmatrix}0\\1\end{pmatrix} + \frac{1}{2}\sin\theta\sin\phi\begin{pmatrix}0\\i\end{pmatrix} + \frac{1}{2}\cos\theta\begin{pmatrix}1\\0\end{pmatrix}$$

$$=\frac{1}{2}\begin{pmatrix} 1 + \cos\theta \\ \sin\theta\cos\phi + i\sin\theta\sin\phi \end{pmatrix} = \frac{1}{2}\begin{pmatrix} 1 + \cos\theta \\ \sin\theta\ e^{i\phi} \end{pmatrix} = \frac{1}{2}\begin{pmatrix} 2\cos^2\dfrac{\theta}{2} \\ 2\sin\dfrac{\theta}{2}\cos\dfrac{\theta}{2}\ e^{i\phi} \end{pmatrix}$$

$$=\cos\frac{\theta}{2}e^{i\phi/2}\begin{pmatrix} \cos\dfrac{\theta}{2}e^{-i\phi/2} \\ \sin\dfrac{\theta}{2}\ e^{i\phi/2} \end{pmatrix} = \cos\frac{\theta}{2}e^{i\phi/2}|n, +>$$

Let us consider $\hat{P}(n, +)\ |\downarrow> = \frac{1}{2}(I + \hat{n}.\ \hat{\vec{\sigma}})|\uparrow>$

$$\frac{1}{2}(I + \hat{n}.\ \hat{\vec{\sigma}})|\uparrow> = \frac{1}{2}(I + \hat{n}.\ \hat{\vec{\sigma}})|\sigma_z = -1> = \frac{1}{2}(I + \hat{n}.\ \hat{\vec{\sigma}})\begin{pmatrix} 0 \\ 1 \end{pmatrix}$$

$$=\frac{1}{2}\begin{pmatrix} 0 \\ 1 \end{pmatrix} + \frac{1}{2}(n_x\sigma_x + n_y\sigma_y + n_z\sigma_z)\begin{pmatrix} 0 \\ 1 \end{pmatrix}$$

$$=\frac{1}{2}\begin{pmatrix} 0 \\ 1 \end{pmatrix} + \frac{1}{2}n_x\begin{pmatrix} 0 & 1 \\ 1 & 0 \end{pmatrix}\begin{pmatrix} 0 \\ 1 \end{pmatrix} + \frac{1}{2}n_y\begin{pmatrix} 0 & -i \\ i & 0 \end{pmatrix}\begin{pmatrix} 0 \\ 1 \end{pmatrix} + \frac{1}{2}n_z\begin{pmatrix} 1 & 0 \\ 0 & -1 \end{pmatrix}\begin{pmatrix} 0 \\ 1 \end{pmatrix}$$

$$=\frac{1}{2}\begin{pmatrix} 0 \\ 1 \end{pmatrix} + \frac{1}{2}n_x\begin{pmatrix} 1 \\ 0 \end{pmatrix} + \frac{1}{2}n_y\begin{pmatrix} -i \\ 0 \end{pmatrix} + \frac{1}{2}n_z\begin{pmatrix} 0 \\ -1 \end{pmatrix}$$

Using $n_x = \sin\theta\cos\phi$, $n_y = \sin\theta\sin\phi$, $n_z = \cos\theta$

$$\frac{1}{2}(I + \hat{n}.\ \hat{\vec{\sigma}})\begin{pmatrix} 0 \\ 1 \end{pmatrix} = \frac{1}{2}\begin{pmatrix} 0 \\ 1 \end{pmatrix} + \frac{1}{2}\sin\theta\cos\phi\begin{pmatrix} 1 \\ 0 \end{pmatrix} + \frac{1}{2}\sin\theta\sin\phi\begin{pmatrix} -i \\ 0 \end{pmatrix} + \frac{1}{2}\cos\theta\begin{pmatrix} 0 \\ -1 \end{pmatrix}$$

$$=\frac{1}{2}\begin{pmatrix} \sin\theta\cos\phi - i\sin\theta\sin\phi \\ 1 - \cos\theta \end{pmatrix} = \frac{1}{2}\begin{pmatrix} \sin\theta\ e^{-i\phi} \\ 1 - \cos\theta \end{pmatrix} = \frac{1}{2}\begin{pmatrix} 2\sin\dfrac{\theta}{2}\cos\dfrac{\theta}{2}\ e^{-i\phi} \\ 2\sin^2\dfrac{\theta}{2} \end{pmatrix}$$

$$=\sin\frac{\theta}{2}e^{-i\phi/2}\begin{pmatrix} \cos\dfrac{\theta}{2}e^{-i\phi/2} \\ \sin\dfrac{\theta}{2}\ e^{i\phi/2} \end{pmatrix} = \sin\frac{\theta}{2}e^{-i\phi/2}|n, +>$$

Similarly we have two other relations. We list down all of them in the following.

$$\hat{P}(\hat{n}, +)\begin{pmatrix} 1 \\ 0 \end{pmatrix} = \frac{1}{2}(I + \hat{n}.\ \hat{\vec{\sigma}})|\sigma_z = +1> = \cos\frac{\theta}{2}e^{i\phi/2}|n, +> \tag{11.1}$$

$$\hat{P}(\hat{n}, +)\begin{pmatrix} 1 \\ 0 \end{pmatrix} = \frac{1}{2}(I + \hat{n}.\ \hat{\vec{\sigma}})|\sigma_z = -1> = \sin\frac{\theta}{2}e^{-i\phi/2}|n, +> \tag{11.2}$$

$$\hat{P}(\hat{n},\ -)\begin{pmatrix}0\\1\end{pmatrix} = \frac{1}{2}(I - \hat{n}.\ \hat{\vec{\sigma}})|\sigma_z = -1> = \cos\frac{\theta}{2}e^{-i\phi/2}|n,\ ->\qquad (11.3)$$

$$\hat{P}(\hat{n},\ -)\begin{pmatrix}1\\0\end{pmatrix} = \frac{1}{2}(I - \hat{n}.\ \hat{\vec{\sigma}})|\sigma_z = +1> = -\sin\frac{\theta}{2}e^{i\phi/2}|n,\ ->\qquad (11.4)$$

We suppose that Alice makes measurement upon state $|\psi^->$ along axis $\hat{n}$ that makes angle θ_1 with $\hat{z}$ axis and Bob makes measurement upon state $|\psi^->$ along some direction $\hat{m}$ that makes angle θ_2 with $\hat{z}$ axis. Let us define

$\hat{P}^A(\hat{n},\ \pm) = \frac{1}{2}(1 \pm \hat{n}.\ \hat{\vec{\sigma}})$, $\hat{P}^B(\hat{m},\ \pm) = \frac{1}{2}(1 \pm \hat{m}.\ \hat{\vec{\sigma}})$ (superscript A for Alice and B for Bob)

Consider

$$\hat{P}^A(\hat{n},\ +)\hat{P}^B(\hat{m},\ +)|\psi^-> = \hat{P}^A(\hat{n},\ +)\hat{P}^B(\hat{m},\ +)\frac{1}{\sqrt{2}}(|01> -|10>)$$

$$= \frac{1}{\sqrt{2}}[\hat{P}^A(\hat{n},\ +)\hat{P}^B(\hat{m},\ +)|01> - \hat{P}^A(\hat{n},\ +)\hat{P}^B(\hat{m},\ +)|10>]$$

$$= \frac{1}{\sqrt{2}}[\hat{P}^A(\hat{n},\ +)|0> \hat{P}^B(\hat{m},\ +)|1> - \hat{P}^A(\hat{n},\ +)|1> \hat{P}^B(\hat{m},\ +)|0>]$$

Using equation (11.1) and equation (11.2) and noting that the projection operator $\hat{P}^A$ operates on the first qubit while the projection operator $\hat{P}^B$ operates on the second qubit we have $(|0> = \begin{pmatrix}1\\0\end{pmatrix}, |1> = \begin{pmatrix}0\\1\end{pmatrix})$

$$\hat{P}^A(\hat{n},\ +)\hat{P}^B(\hat{m},\ +)|\psi^->$$

$$= \frac{1}{\sqrt{2}}[\hat{P}^A(\hat{n},\ +)\begin{pmatrix}1\\0\end{pmatrix}\hat{P}^B(\hat{m},\ +)\begin{pmatrix}0\\1\end{pmatrix} - \hat{P}^A(\hat{n},\ +)\begin{pmatrix}0\\1\end{pmatrix}\hat{P}^B(\hat{m},\ +)\begin{pmatrix}1\\0\end{pmatrix}]$$

$$= \frac{1}{\sqrt{2}}[\cos\frac{\theta_1}{2}e^{i\phi/2}|n,\ +> \sin\frac{\theta_2}{2}e^{-i\phi/2}|m,\ +>$$

$$-\sin\frac{\theta_1}{2}e^{-i\phi/2}|n,\ +> \cos\frac{\theta_2}{2}e^{i\phi/2}|m,\ +>]$$

$$= \frac{1}{\sqrt{2}}[\cos\frac{\theta_1}{2}e^{i\phi/2} \sin\frac{\theta_2}{2}e^{-i\phi/2}|n+,\ m+>$$

$$-\sin\frac{\theta_1}{2}e^{-i\phi/2} \cos\frac{\theta_2}{2}e^{i\phi/2}|n+,\ m+>]$$

$$= \frac{1}{\sqrt{2}}(\sin\frac{\theta_2}{2}\cos\frac{\theta_1}{2} - \cos\frac{\theta_2}{2}\sin\frac{\theta_1}{2})|n+,\ m+>$$

$$= \frac{1}{\sqrt{2}}\sin(\frac{\theta_2}{2} - \frac{\theta_1}{2})|n+,\ m+>$$

$$\hat{P}^A(\hat{n}, +)\hat{P}^B(\hat{m}, +)|\psi^-> = \frac{1}{\sqrt{2}} \sin \frac{\theta}{2}|\hat{n} +, \hat{m} +> \tag{11.5}$$

where $\theta = \theta_2 - \theta_1$ is the angle between the two directions $\hat{n}$ and $\hat{m}$.

Similar calculations show that

$$\hat{P}^A(\hat{n}, +)\hat{P}^B(\hat{m}, -)|\psi^-> = \frac{1}{\sqrt{2}} \cos \frac{\theta}{2}|\hat{n} +, \hat{m} -> \tag{11.6}$$

Let us find the expectation value of different results by Alice and Bob when they choose different axes for measurement say along directions $\hat{n}$ and $\hat{m}$, respectively. Consider

$$<\psi^-|\hat{P}^A(\hat{n} +)\hat{P}^B(\hat{m} +)|\psi^-> = <\psi^-|[\hat{P}^A(\hat{n} +)]^2[\hat{P}^B(\hat{m} +)]^2|\psi^->$$

(using $\hat{P}^2 = \hat{P}$, i.e. square of projection operators are projection operators themselves)

$$<\psi^-|\hat{P}^A(\hat{n} +)\hat{P}^B(\hat{m} +)|\psi^-> = <\psi^-|\hat{P}^A(\hat{n} +)\hat{P}^A(\hat{n} +)\hat{P}^B(\hat{m} +)\hat{P}^B(\hat{m} +)|\psi^->$$

$$= <\psi^-|P^A(\hat{n} +)P^B(\hat{m} +)P^A(\hat{n} +)P^B(\hat{m} +)|\psi^->$$

Using equation (11.5) we get

$$<\psi^-|\hat{P}^A(\hat{n} +)\hat{P}^B(\hat{m} +)|\psi^-> = [\frac{1}{\sqrt{2}} <\hat{n} +, \hat{m} +| \sin \frac{\theta}{2}] [\frac{1}{\sqrt{2}} \sin \frac{\theta}{2}|\hat{n} +, \hat{m} +>]$$

$$= \frac{1}{2} \sin^2 \frac{\theta}{2} <\hat{n} +, \hat{m} +|\hat{n} +, \hat{m} +> = \frac{1}{2} \sin^2 \frac{\theta}{2} = \frac{1}{4}(1 - \cos \theta)$$

And

$$<\psi^-|\hat{P}^A(\hat{n} +)\hat{P}^B(\hat{m} -)|\psi^-> = <\psi^-|\hat{P}^A(\hat{n} +)\hat{P}^A(\hat{n} +)\hat{P}^B(\hat{m} -)\hat{P}^B(\hat{m} -)|\psi^->$$

$$= <\psi^-|P^A(\hat{n} +)P^B(\hat{m} -)P^A(\hat{n} +)P^B(\hat{m} -)|\psi^->$$

Using equation (11.6) we get

$$<\psi^-|\hat{P}^A(\hat{n} +)\hat{P}^B(\hat{m} -)|\psi^-> = [\frac{1}{\sqrt{2}} <\hat{n} +, \hat{m} -| \cos \frac{\theta}{2}] [\frac{1}{\sqrt{2}} \cos \frac{\theta}{2}|\hat{n} +, \hat{m} ->]$$

$$= \frac{1}{2} \cos^2 \frac{\theta}{2} <\hat{n} +, \hat{m} -|\hat{n} +, \hat{m} -> = \frac{1}{2} \cos^2 \frac{\theta}{2} = \frac{1}{4}(1 + \cos \theta)$$

Similarly we have two other relations. We list down all the expectation values in the following.

$$<\psi^-|P^A(\hat{n} +)P^B(\hat{m} +)|\psi^-> = <\psi^-|P^A(\hat{n} -)P^B(\hat{m} -)|\psi^-> = \frac{1}{2}\sin^2\frac{\theta}{2} = \frac{1}{4}(1 - \cos \theta)$$

$$<\psi^-|P^A(\hat{n} +)P^B(\hat{m} -)|\psi^-> = <\psi^-|P^A(\hat{n} -)P^B(\hat{m} +)|\psi^-> = \frac{1}{2}\cos^2\frac{\theta}{2} = \frac{1}{4}(1 + \cos \theta)$$

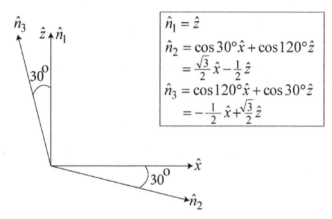

Figure 11.1. Specification of the direction $\hat{n}$

Let Alice's direction $\hat{n}$ be any one of the three directions shown in the diagram of figure 11.1.

When Alice makes measurement along any of these directions the spin state gets disturbed and so she cannot tell what the result would be if a different axis had been chosen. Suppose that Bob measures along $\hat{n}_1$ and communicates this result to Alice. Alice would thus know the result of her measurement along $\hat{n}_1$ as their measurements are anti- correlated. So instead Alice measures along $\hat{n}_2$ and obtains a result. The probability that Alice's measurements along $\hat{n}_1$ and $\hat{n}_2$ give the same result is equal to the probability that Alice and Bob get the same result when Alice measures along $\hat{n}_2$ and Bob measures along $-\hat{n}_1$ (due to their anti-correlation). Here the same result means either both + or both −. The probability is

$$P_{\text{same}}(++,\ --) = \frac{1}{2}\sin^2\frac{\theta}{2} + \frac{1}{2}\sin^2\frac{\theta}{2} = \sin^2\frac{\theta}{2} = \frac{1}{2}(1 - \cos\theta) = P_{\text{same}}(\hat{n}_1, \hat{n}_2)$$

For $\theta = 60°$ (= angle between $-\hat{n}_1$ and $\hat{n}_2$),

$$P_{\text{same}}(++,\ --) = \frac{1}{2}(1 - \cos 60°) = \frac{1}{2}.\,(1 - \frac{1}{2}) = \frac{1}{4} = P_{\text{same}}(\hat{n}_1, \hat{n}_2)$$

The probability of getting the same result for three pairs of axes is

$$= P_{\text{same}}(\hat{n}_1, \hat{n}_2) + P_{\text{same}}(\hat{n}_2, \hat{n}_3) + P_{\text{same}}(\hat{n}_3, \hat{n}_1) = \frac{1}{4} + \frac{1}{4} + \frac{1}{4} = \frac{3}{4} < 1$$

This result that the probability of getting the same result for three pairs of axes is less than 1 follows as per quantum mechanics.

If there are hidden variables in the problem, local realism does not give the anti-correlation between Bob's and Alice's results. So Bob's informing Alice about his result is immaterial and Alice can make measurement along the two axes because there is no collapse of the wave function now. Probabilities arise from a statistical distribution of the probabilities of different possible results.

Since there are three axes and two possible results, at least for one pair of axes, the result has to be the same (i.e. probability = 1). Hence

$$P_{\text{same}}(\hat{n}_1, \hat{n}_2) + P_{\text{same}}(\hat{n}_2, \hat{n}_3) + P_{\text{same}}(\hat{n}_3, \hat{n}_1) \geqslant 1$$

If quantum mechanics is correct we should have probability less than 1 and if hidden variable theory is correct we would have probability greater than or equal to 1. These are Bell's inequalities that are needed to be checked as they provide a confirmatory test regarding which one is right.

Discussion of EPR paradox is very central to quantum computing because in case Einstein's objections are upheld and if there are hidden variables in the problem the entire philosophy of doing computing through quantum mechanics would need alteration. It is needed to verify if Copenhagen interpretation of quantum mechanics is right or if there are quantities which are hidden from our perception or consciousness that actually encode information regarding physical properties of the system.

According to Copenhagen interpretation of quantum mechanics, when we do a measurement, any one of the eigenvalues of the operator corresponding to the physical alternatives will be realized with a probability. To be precise, the process of measurement gives a value. Einstein's idea of realism says that an object must have a physical property independent of the measurement process. Measurement process is simply a means of revealing the value of that physical quantity.

Secondly there is a question of locality also. If we make a measurement on one particle at one point in space and we make a second measurement on another particle at a far enough (space-like) distance then disturbance in the first measurement cannot influence the second measurement.

These dual properties of local realism are the cornerstone of the EPR paradox or the objection of Einstein, Podolsky and Rosen to the standard Copenhagen interpretation of quantum mechanics.

In the Gedanken experiment we put quantum mechanics and hidden variable theory to the test. We derived an inequality called Bell's inequality to check or provide a confirmatory test on which one is right.

Another inequality to test the above is CHSH inequality that we mention in the following.

11.4 Clauser, Horne, Shimony and Holt's inequality

An entangled state $|\psi^-\rangle = \frac{1}{\sqrt{2}}(|01\rangle - |10\rangle)$ is prepared and shared by Alice and Bob. The first particle is with Alice and the second particle is with Bob. Suppose Alice and Bob can measure any two properties of the particles they have. Alice measures properties denoted by Q, R while Bob measures properties denoted by S, T and each can assume values $+1$ or -1. To do this, assume that there are identical copies of entangled states available. So they can make experiments and take statistical data. Alice and Bob are separated by space-like distance.

Alice and Bob measure the properties at random but at the same time. The measurements are thus not causally related. Each measurement returns either $+1$ or $^-1$ for the properties Q, R, S, T.

After a series of measurements the results are tabulated. The results of measurements are denoted by Q, R, S, T(either $+1$ or -1). Data corresponding to measurement Q of Alice and S of Bob at the same time are filtered and collected. Likewise for the pairs Q, T; R, S and R, T.These are the various collections of pairs that are possible.

Let us consider the following expression

$$C = QS + RS + RT - QT = (Q + R)S + (R - Q)T$$

As Q and R can take values either $+1$ or -1 so one of these terms $Q + R$, $R - Q$ must be equal to 0 and the other ± 2. Thus one of $Q + R$, $R - Q$ is zero. Also S and T can take values either $+1$ or -1. We have shown the various possibilities in the following.

Q	R	C		C
1	1	2S	$\xrightarrow{S=\pm 1}$	± 2
1	-1	-2S	$\xrightarrow{S=\pm 1}$	∓ 2
-1	1	2T	$\xrightarrow{T=\pm 1}$	± 2
-1	-1	-2T	$\xrightarrow{T=\pm 1}$	∓ 2

The above leads to

$$C = QS + RS + RT - QT = \pm 2$$

Local hidden variable assumption has obviously crept in since we have imagined that values ± 1 can be assigned simultaneously to the four observables though it is impossible to measure both of Q, R or both of S, T.

This would hold if there are local hidden variables, i.e. an object has a property independent of its observation and the property is revealed on observation.

Again, since $|<C>| \leqslant <|C|>$ and since $<|C|> =<|\pm 2| > =2$ we can write

$$|<C>| \leqslant 2$$

So considering several sets of measurements the average gives

$$<QS + RS + RT - QT> =<QS> +<RS> +<RT> -<QT> \quad \leqslant 2$$

This is called the CHSH inequality obtained under the assumption of existence of hidden variable, i.e. under the assumption of local realism.

- Let us now proceed relying on quantum mechanics. Let us calculate this by using the definition of average in quantum mechanics. In quantum mechanics the four observables or properties (i.e. Q, R with Alice and S, T with Bob) are replaced by operators $\hat{Q}$, $\hat{R}$, $\hat{S}$, $\hat{T}$ defined by the following matrices which have eigenvalues ± 1.

$$\hat{Q} = \hat{Z} = \begin{pmatrix} 1 & 0 \\ 0 & -1 \end{pmatrix}, \quad \hat{R} = \hat{X} = \begin{pmatrix} 0 & 1 \\ 1 & 0 \end{pmatrix},$$

$$\hat{S} = \frac{1}{\sqrt{2}}(-\hat{Z} - \hat{X}) = \frac{1}{\sqrt{2}}\begin{pmatrix} -1 & -1 \\ -1 & 1 \end{pmatrix}, \quad \hat{T} = \frac{1}{\sqrt{2}}(\hat{Z} - \hat{X}) = \frac{1}{\sqrt{2}}\begin{pmatrix} 1 & -1 \\ -1 & -1 \end{pmatrix}$$

The following relations occur

$$\hat{Q}|0> = |0>, \quad \hat{Q}|1> = -|1>, \quad \hat{R}|0> = |1>, \quad \hat{R}|1> = |0>$$

$$\hat{S}|0> = \frac{1}{\sqrt{2}}\begin{pmatrix} -1 & -1 \\ -1 & 1 \end{pmatrix}\begin{pmatrix} 1 \\ 0 \end{pmatrix} = \frac{1}{\sqrt{2}}\begin{pmatrix} -1 \\ -1 \end{pmatrix} = -\frac{1}{\sqrt{2}}\begin{pmatrix} 1 \\ 1 \end{pmatrix}$$

$$= -\frac{1}{\sqrt{2}}[\begin{pmatrix} 0 \\ 1 \end{pmatrix} + \begin{pmatrix} 1 \\ 0 \end{pmatrix}] = -\frac{1}{\sqrt{2}}(|0> + |1>)$$

$$\hat{S}|1> = \frac{1}{\sqrt{2}}\begin{pmatrix} -1 & -1 \\ -1 & 1 \end{pmatrix}\begin{pmatrix} 0 \\ 1 \end{pmatrix} = \frac{1}{\sqrt{2}}\begin{pmatrix} -1 \\ 1 \end{pmatrix} = \frac{1}{\sqrt{2}}[\begin{pmatrix} 0 \\ 1 \end{pmatrix} - \begin{pmatrix} 1 \\ 0 \end{pmatrix}] = -\frac{1}{\sqrt{2}}[|0> - |1>]$$

$$\hat{T}|0> = \frac{1}{\sqrt{2}}\begin{pmatrix} 1 & -1 \\ -1 & -1 \end{pmatrix}\begin{pmatrix} 1 \\ 0 \end{pmatrix} = \frac{1}{\sqrt{2}}\begin{pmatrix} 1 \\ -1 \end{pmatrix} = \frac{1}{\sqrt{2}}[\begin{pmatrix} 1 \\ 0 \end{pmatrix} - \begin{pmatrix} 0 \\ 1 \end{pmatrix}] = \frac{1}{\sqrt{2}}(|0> - |1>)$$

$$\hat{T}|1> = \frac{1}{\sqrt{2}}\begin{pmatrix} 1 & -1 \\ -1 & -1 \end{pmatrix}\begin{pmatrix} 0 \\ 1 \end{pmatrix} = \frac{1}{\sqrt{2}}\begin{pmatrix} -1 \\ -1 \end{pmatrix} = -\frac{1}{\sqrt{2}}\begin{pmatrix} 1 \\ 1 \end{pmatrix}$$

$$= -\frac{1}{\sqrt{2}}[\begin{pmatrix} 1 \\ 0 \end{pmatrix} + \begin{pmatrix} 0 \\ 1 \end{pmatrix}] = -\frac{1}{\sqrt{2}}[|0> + |1>]$$

Average of QS in the entangled state $|\psi^->= \frac{1}{\sqrt{2}}(|01> - |10>)$ is

$$<\hat{Q}\hat{S}> = <\psi^-|\hat{Q}\hat{S}|\psi^-> = \frac{1}{\sqrt{2}}(<01| - <10|)\hat{Q}\hat{S}\frac{1}{\sqrt{2}}(|01> - |10>)$$

$$= \frac{1}{2}[<01|\hat{Q}\hat{S}|01> - <01|\hat{Q}\hat{S}|10> - <10|\hat{Q}\hat{S}|01> + <10|\hat{Q}\hat{S}|10>] \qquad (11.7)$$

$\hat{Q}$ acts on first qubit and $\hat{S}$ acts on second qubit. So we can write

$$<\hat{Q}\hat{S}> = \frac{1}{2}(<01|[\hat{Q}|0> \hat{S}|1>] - <01|[\hat{Q}|1> \hat{S}|0>]$$

$$- <10|[\hat{Q}|0> \hat{S}|1>] + <10|[\hat{Q}|1> \hat{S}|0>])$$

$$<\hat{Q}\hat{S}> = \frac{1}{2}(<01|\ [|0> (-\frac{1}{\sqrt{2}}(|0> -|1>))]$$

$$- <01|\ [(-|1>)(-\frac{1}{\sqrt{2}}(|0> +|1>))] - <10|\ [|0> (-\frac{1}{\sqrt{2}}(|0> -|1>))]$$

$$+ <10|\ [(-|1>)(-\frac{1}{\sqrt{2}}(|0> +|1>))])$$

$$<\hat{Q}\hat{S}> = \frac{1}{2}\ (- <01|\ [\frac{1}{\sqrt{2}}(|00> -|01>)] - <01|\ [\frac{1}{\sqrt{2}}(|10> +|11>)]$$

$$+ <10|\ [\frac{1}{\sqrt{2}}(|00> -|01>)] + <10|\ [(\frac{1}{\sqrt{2}}(|10> +|11>)]$$

$$<\hat{Q}\hat{S}> = \frac{1}{2\sqrt{2}}(-[<01|00> -<01|01>]-[<01|\ 10> +<01|11>]$$

$$+ [<10|\ 00> -<10|01>]+[<10|\ 10> +<10|11>])$$

$$<\hat{Q}\hat{S}> = \frac{1}{2\sqrt{2}}(1 + 1) = \frac{1}{\sqrt{2}}$$

Similarly, $<\hat{R}\hat{S}> = \frac{1}{\sqrt{2}}$, $<\hat{R}\hat{T}> = \frac{1}{\sqrt{2}}$, $<\hat{Q}\hat{T}> = -\frac{1}{\sqrt{2}}$

Thus

$<QS> +<RS> +<RT> -<QT> = \frac{1}{\sqrt{2}} + \frac{1}{\sqrt{2}} + \frac{1}{\sqrt{2}} - (-\frac{1}{\sqrt{2}}) = \frac{4}{\sqrt{2}} = 2\sqrt{2}$

This violates CHSH inequality.

To verify Bell's inequalities, experimental tests have been devised. Bell's inequalities have always been found to be violated. This provides support to quantum mechanics over local hidden variable theory.

11.5 Further reading

[1] Ghosh D 2016 Online course on quantum information and computing (Bombay: Indian Institute of Technology)
[2] Nielsen M A and Chuang I L 2019 *Quantum Computation and Quantum Information* (Cambridge: Cambridge University Press)
[3] Nakahara M and Ohmi T 2017 *Quantum Computing* (Boca Raton, FL: CRC Press)
[4] Pathak A 2016 *Elements of Quantum Computing and Quantum Communication* (Boca Raton, FL: CRC Press)

11.6 Problems

1. *Identify the incorrect statement.*
    (a) Classical physics describes systems in macroscopic scale.
    (b) Quantum mechanics describes systems in the microscopic scale.
    (c) Classical physics cannot describe microscopic systems.

(d) Quantum mechanics describes systems only in the microscopic scale.
   Ans. *(d).*

2. *An object possesses a property independent of the process of measurement. This is a statement made*
   (a) in quantum mechanics, (b) in classical mechanics,
   (c) by Einstein, (d) in Copenhagan interpretation.
   Ans. *(c).*

3. *Violation of Bell inequalities means which of the following are correct?*
   (a) Quantum mechanics, (b) Local hidden variable theory,
   (c) Bell states (entangled pairs), (d) Quantum computing.
   Ans. *(a).*

4. *Which of the following is true regarding the Gedanken experiment?*
   (a) EPR paradox is proposed and confirmed.
   (b) Quantum mechanics and hidden variable theory are put to test.
   (c) Quantum mechanics and special theory of relativity are put to test.
   (d) Einstein's objection to Copenhagan interpretation is upheld.
   Ans. *(b).*

5. *Physical local reality is an integral part of*
   (a) quantum mechanics, (b) EPR paradox, (c) relativity theory, (d) all theories.
   Ans. *(b).*

6. *Which is untrue about CHSH inequality?*
   (a) It is proposed by quantum mechanics, (b) It assumes local realism,
   (c) It is violated by quantum mechanics, (d) It assumes existence of hidden variables.
   Ans. *(a).*

7. *The indeterminacy of quantum mechanical theory is a fundamental law of nature and is not a reflection of our inadequacy of scientific knowledge. Identify the correct statement.*
   (a) This is EPR paradox.
   (b) Theory of local realism should be taken to solve for the indeterminacy of quantum mechanics.
   (c) Indeterminacy of quantum mechanical theory is due to hidden variables of the theory.
   (d) This is Copenhagan interpretation of quantum mechanics.
   Ans. *(d).*

Chapter 12

Cryptography—the art of coding

In this chapter we shall address an application of the principles of quantum computing, namely cryptography, which is the art of coding and decoding secret messages.

In Greek Kryptos means secret and Graphein means writing and they actually refer to what we call cryptography. Cryptography is the process of communicating a message in a secret manner. In this sophisticated and secure process the message appears to be unintelligible to any unauthorized party. While classical cryptographic protocols are conditionally secure, the quantum cryptography protocols are unconditionally secure, meaning thereby that it is an absolutely secure message transfer. The basic idea is to modify a message so as to make it unintelligible to anyone except the intended recipient.

The message is encoded, i.e. encrypted—and the disguised way of writing a code is referred to as cipher. The coded message is sent over a public channel (perhaps a fibre optic channel or a channel that may be open to outsiders) for being received and decrypted by the authorized receiver to covert the cipher text back to the original message that was communicated.

12.1 A bit of history of cryptography

Julius Caesar used to send cryptic messages to his soldiers and every letter in the message was advanced by three letters (e.g. letter A was actually D, B was actually E etc; so 'FIGHT' was actually 'ILJKW'). The recipient soldiers knew the key to this conversion, i.e. the process of mapping or transformation beforehand and decrypted the message. This kind of encoding is called Caesarian cipher. Obviously it is a simple technique of encoding and decoding.

During World War 2 the Germans used an electro-mechanical machine (Enigma) to send messages to troops. The Allied forces tried to break the code to unravel the message.

The art of code-breaking, i.e. decoding is called cryptanalysis. The subject of encryption or cryptography and cryptanalysis is called cryptology.

12.2 Essential elements of cryptography

Cryptography involves the following challenges which need to be addressed: they are the essential elements of cryptography.

Privacy

The art of keeping a message free of public scrutiny, i.e. maintaining secrecy is called privacy. Such secrecy might have a lifetime during which the message is not understandable by any unauthorized persons.

Data integrity

Message should not get corrupted during passage through a public channel. Stray noises should not affect the message or data. In the case of a conversation we do not want any eavesdropper to interfere or tap.

Non-repudiation

Suppose someone has sent a message and later there should be no scope of denial regarding that sending of message. One cannot decline his/her act of sending the message. Also, the receiver who received the message cannot deny or decline having received the message.

Authetication

When someone wishes to send a message or converse with someone or do a business transaction his/her identity should be verified or authenticated. For instance, a bank verifies its clients before offering permission to do a transaction. The bank uses various protective measures like password, one time password (OTP) etc. Authentication of sender and receiver is very important in the branch of cryptography.

Secure communication

Secure communication necessitates people at a distance sharing any amount of message or information without any interference or interception by a third party. To achieve this a set of procedures and protocols are established, collectively called cryptography. Elements like privacy, data integrity, non-repudiation and authentication are an inherent part of it.

Duration of secrecy

An ideal code would be one that is unbreakable, i.e. has infinite lifetime. But there is no such thing. Given enough time every code can be broken.

Also, it is not always necessary that the content of a message needs to be an eternal secret. Duration of secrecy depends upon security needs of the content or message. For instance, an instruction regarding a particular financial transaction should not be made public till the transaction is over.

12.3 One-time pad

This uses a random sequence of binary digits as a code that is used only once (one-time code) and so no conclusion can be reached on the code by any repetitive pattern. It is a code which is known to be robust and is called a one-time pad or Vernam cipher. Its principle is as follows.

Suppose two friends (a sender and a receiver) have identical copies of a book. One friend can pick up a word from the book on page 19, row 05, word number from left being 07 to form a code 190507 and send it over. The other friend (for whom it was intended) can break the code. Using such code once does not run the risk of an eavesdropper identifying the book or document. But if the friends use the same code (or code pattern) repeatedly, an eavesdropper can identify the book that the friends are using and the message they are sharing.

A one-time pad is a code which a sender and receiver agree to use once only. It consists of a random sequence of binary digits to be used as codes for messages, to be used only once.

Suppose P refers to the plain text (message or data) that is being intended to be sent by sender and it is represented by a string of binary digits (table 12.1). The sender and the receiver possess a key denoted by K which is basically a sequence of random binary digits. One then can define a cipher text C as ($\oplus$ is addition modulo two operation)

$$C = P \oplus K \text{ and bitwise we can write } C_i = P_i \oplus K_i$$

and send this as message.

As $C_i = P_i \oplus K_i \Rightarrow P_i = C_i \oplus K_i$ it follows that at the receiver's end the receiver can recover the original message by performing an addition modulo two operation as

$$C \oplus K = P \text{ or } C_i \oplus K_i = P_i$$

This is elaborated in table 12.1. Use of key K_i once is secured and is a one-time pad.

Table 12.1. One time coding: use of key K once is secure.

	XOR (addition modulo 2)			
	$\begin{aligned} 0 \oplus 0 &= 0 \\ 0 \oplus 1 &= 1 \\ 1 \oplus 0 &= 1 \\ 1 \oplus 1 &= 0 \end{aligned}$ (the rule)			
First set				
P	1101	1010	1101	1001 (text to be sent)
K	1111	0110	1010	1100 (key)
C	0010	1100	0111	0101 (Message sent = cipher text = P $\oplus$ K)
C $\oplus$ K	1101	1010	1101	1001 (decoded text = P)
Second set (same key K used)				
P'	1110	1100	1001	1101 (text to be sent)
K	1111	0110	1010	1100 (key)
C'	0001	1010	0011	0001 (Message sent = cipher text = P'$\oplus$ K)
C'$\oplus$ K	1110	1100	1001	1101 (decoded text = P')

Table 12.2. Use of key multiple times can lead to code breaking.

(P)	1101	1010	1101	1001	(text to be sent)	first set
(P')	1110	1100	1001	1101	(text to be sent)	second set
P $\oplus$ P'	0011	0110	0100	0100		
C	0010	1100	0111	0101	(Message sent = cipher text =P $\oplus$ K)	
C'	0001	1010	0011	0001	(Message sent = cipher text =P'$\oplus$K)	
C $\oplus$ C'	0011	0110	0100	0100	(same as P $\oplus$ P')	

However, if one uses the key K a second time say for another plain text P' and gets a cipher text $C' = P' \oplus K$ then the code of the key can be broken since then one can gather sufficient additional information like $C + C' = P + P'$ as shown in table 12.2.

12.4 RSA cryptosystem

A code that is widely used both in military communication and financial transactions was invented by Rivest, Shamir and Adleman in 1977 at MIT, and it is called RSA cryptosystem. Here a public key encryption is used for encoding and is sent over the internet. It helps one to establish the identity of the sender and the receiver and the data remains absolutely confidential. RSA uses a trap door function.

A trap door is a one-way door that opens in one direction only, it is difficult to open in another direction. It operates like a valve or rectifier which is a unidirectional device or a trap door device.

In the language of computer science an easy problem is one that can be computed or solved in polynomial time.

f: An easy problem to calculate (i.e. can be calculated in polynomial time)

f^{-1}: A hard problem (i.e. cannot be calculated in polynomial time)

For instance multiplication of two large prime numbers can be done in polynomial time and so is an easy problem, whereas splitting a composite number into two large prime numbers is a hard problem. In other words the inverse problem of integer factorization is difficult. The Euclid algorithm (where repeated division is done) takes $\sqrt{N}$ steps for calculating factors of N.

In RSA, the trap door function is multiplication of two large prime numbers which is an easy problem and so to say a one-way function because the inverse process of factorization is difficult. The RSA cryptosystem depends heavily on the belief that factorization of a large number into its prime factors in polynomial time is extremely difficult. Classical cryptography based on the RSA algorithm is reasonably robust.

The reason for ascribing such a name trap door is because it represents a situation in which it is easy to fall through but not easy to climb back from the trapped state and hence the name trap door.

Keys and key distribution

To perform sending of message and receipt of message by intended person one has to operate with a key (K) that should be known only to the sender of the message and the intended recipient. So the key is a secret since it is not known by the public and so noone, other than the sender/receiver, can interfere with the message exchanged. Also, for ensuring secrecy, it is needed that the same key should not be used multiple times but only once. So for perfect cryptosystem one has to use a one-time pad (OTP) and distribute the key efficiently called quantum key distribution (QKD). In fact using quantum effects, key distributions can be done with perfect secrecy. In a perfect cryptosystem one has thus to ensure quantum computing that uses quantum key distribution plus OTP, i.e.

$$QC = QKD + OTP.$$

In the road to RSA encryption we require knowledge of some theorems that are discussed now.

12.5 Fermat's little theorem

Consider a prime number p and also consider any number a such that a is not divisible by p,.i.e. GCD(a, p) = 1 then

$$a^{p-1} = 1 \ (\text{mod} \ p).$$

✓ Example

Let us take $a = 2, p = 5$, five is a prime and two is not divisible by five. So Fermat' little theorem is applicable.

$$2^{5-1} = 2^4 = 16 = 1 (\text{mod} \ 5)$$

$$5 \ ) \ 16 \ (\ 3$$
$$\underline{15}$$
$$1 \longleftarrow \text{Remainder}$$

Proof
Consider

$$(x_1 + x_2)^2 = x_1^2 + x_2^2 + 2x_1x_2 = x_1^2 + x_2^2 + (\text{term divisible by 2})$$
$$= (x_1^2 + x_2^2) \ \text{mod} \ 2$$

$$(x_1 + x_2)^3 = x_1^3 + 3x_1^2x_2 + 3x_1x_2^2 + x_2^3 = x_1^3 + x_2^3 + (\text{terms divisible by 3})$$
$$= (x_1^3 + x_2^3) \ \text{mod} \ 3$$

Generalizing we can write

$$(x_1 + x_2)^p = =(x_1^p + x_2^p) \ \text{mod} \ p$$

Extending this result up to x_a

$$(x_1 + x_2 + ... + x_a)^p = (x_1^p + x_2^p + ... + x_a^p) \bmod p$$

Put

$$x_1 = x_2 = ... = x_a = 1$$
$$(1 + 1 + ... \ a \ \text{times})^p = (1^p + 1^p + ...a \ \text{times}) \bmod p$$
$$= (1 + 1 + ...a \ \text{times}) \bmod p$$

$$a^p = a \bmod p$$

Dividing by a we have

$$a^{p-1} = 1 \bmod p$$

which is Fermat's little theorem.

Table 12.3 shows some application of Fermat's little theorem.

12.6 Euler theorem

Definition of Euler Totient function

If n is a positive integer then $\phi(n)$ is the number of positive integers less than n that are co-prime to n. Two numbers are co-prime if they have no common factors, i.e. their GCD value is 1.

Example:

$\phi(10) = 4$ since the four numbers 1,3,7,9, are co-prime to 10.

$\phi(6) = 2$ since the two numbers 1,5 are co-prime to 6.

Table 12.3. Some application of Fermat's little theorem.

Find remainder (least residue) in the following divisions using Fermat Little theorem. ($a^{p-1} = 1 \bmod p$)	
(1) $2^{16} by 17$	(2) $2^{50} by 17$

$a=2, p=17$, 2 not divisible by 17
 By Fermat Little theorem
 	
 $2^{17-1} = 1 \bmod 17$ Check
 $2^{16} = 1 \bmod 17$ $17)2^{16} = 65536 (3855$
     $\underline{65535}$
      1
	$a=2, p=17$, 2 not divisible by 17
	 $2^{17-1} = 1 \bmod 17$ (by Fermat
	 $2^{16} = 1 \bmod 17$ Little theorem)
	 $2^{50} = 2^{16.3+2} = (2^{16})^3 \ 2^2$
	    $= 1^3 2^2 = 4 \bmod 17$

(3) $4^{532} by 11$	(4) $7^{2001} by 5$

$a=4, p=11$, 4 not divisible by 11
 $4^{11-1} = 1 \bmod 11$ (by Fermat
 $4^{10} = 1 \bmod 11$ Little theorem)
 $4^{532} = 4^{10.53+2} = (4^{10})^{53} 4^2$
    $= 1^{53} 4^2 \bmod 11$
    $= 16 \bmod 11$
    $= 5 \bmod 11$
	$a=7, p=5$, 7 not divisible by 5
	 $7^{5-1} = 1 \bmod 5$ (by Fermat
	 $7^4 = 1 \bmod 5$ Little theorem)
	 $7^{2001} = 7^{4.500+1} = (7^4)^{500} 7^1$
	    $= 1^{500} 7^1 \bmod 5$
	    $= 7 \bmod 5$
	    $= 2 \bmod 5$

$\phi(18) = 6$ since the six numbers 1,5,7,11,13,17 are co-prime to 18.

$\phi(7) = 6 = 7 - 1$ since the six numbers 1,2,3,4,5,6, are co-prime to 7. We note that seven is a prime number and that $\phi(7) = 7 - 1 = 6$. So we can write the following in general

If p is a prime number then $\phi(p) = p - 1$.

Another example

$$\phi(13) = 13 - 1 = 12.$$

All the 12 numbers below 13 i.e. 1,2,3,4,5,6,7,8,9,10,11,12 are co-prime to 13, i.e. all these 12 numbers have GCD value with 13 to be 1.

A few properties of Euler Totient function

$\phi(mn) = \phi(m)\phi(n)$ for $GCD(m, n) = 1$(i.e. m, n are co-prime)

$\phi(p^m) = p^m - p^{m-1}$ for $p=$ prime number, $m \geqslant 1$ its power. There are p^{m-1} numbers that are divisible by p.

✓ **Example**

$$\phi(6) = \phi(3.2) = \phi(3)\phi(2) = (3 - 1)(2 - 1) = 2.1 = 2$$

In the set $\{1,2,3,4,5,6\}$ there are two co-prime numbers to 6 viz. 1,5.

✓ **Example**

$$\phi(3^2) = 3^2 - 3^{2-1} = 9 - 3 = 6.$$

In the set $\{1,2,3,4,5,6,7,8,9\}$, $3^{2-1} = 3$ numbers viz. 3,6,9 are divisible by 3. And the $\phi(9) = 6$ numbers viz. 1,2,4,5,7,8 are co-prime to 9.

✓ **Example**

$$\phi(2^3) = 2^3 - 2^{3-1} = 8 - 4 = 4.$$

Let us study the set $\{1,2,3,4,5,6,7,8\}$. Now $2^{3-1} = 4$ numbers viz. 2,4,6,8 are divisible by 2. And the $\phi(8) = 4$ numbers viz. 1,3,5,7 are co-prime to 8 i.e. $GCD(1.8) = 1$, $GCD(3,8) = 1$, $GCD(5,8) = 1$, $GCD(7,8) = 1$.

✓ **Example**

$$\phi(18) = \phi(9.2) = \phi(3^2.2) = \phi(3^2)\phi(2) = (3^2 - 3^{2-1})(2 - 1) = 6.1 = 6$$

Table 12.4 shows a few more examples.

Statement of Euler theorem:

If n is a positive integer and a, n are co-prime i.e. $GCD(a, n) = 1$, then

$$a^{\phi(n)} = 1 \bmod n$$

where $\phi(n)$ is Euler Totient function.

Explanation

Take $n = 10$, $a = 3$ and $GCD(3, 10) = 1$(10 and 3 are co-prime)

Then by Euler theorem

$$3^{\phi(n)} = 1 \bmod 10$$

Table 12.4. Some problems on Euler Totient function.

Find $\phi(20)$

$\phi(20) = \phi(4.5) = \phi(2^2.5) = (2^2 - 2^{2-1})(5-1) = (4-2).4 = 8$

(The numbers are $1,3,7,9,11,13,17,19$)

Find $\phi(42)$

$\phi(42) = \phi(3.2.7) = \phi(3)\phi(2)\phi(7) = (3-1)(2-1)(7-1) = 2.1.6 = 12$

Find $\phi(240)$	Find $\phi(49)$
$\phi(240) = \phi(16.5.3) = \phi(2^4.5.3)$	$\phi(49) = \phi(7^2)$
$\qquad = \phi(2^4)\phi(5)\phi(3)$	$\qquad = 7^2 - 7^{2-1} = 49 - 7$
$\qquad = (2^4 - 2^{4-1})(5-1)(3-1)$	$\qquad = 42$
$\qquad = (16-8).4.2$	
$\qquad = 64$	$\phi(41) = 41 - 1 = 40$

$\phi(32) = \phi(2^5) = 2^5 - 2^{5-1} = 32 - 16 = 16$

$\phi(35) = \phi(5.7) = \phi(5)\phi(7) = (5-1)(7-1) = 4.6 = 24$

$\phi(600) = \phi(8.3.25) = \phi(2^3.3.5^2) = \phi(2^3)\phi(3)\phi(5^2)$

$\qquad = (2^3 - 2^{3-1})(3-1)(5^2 - 5^{2-1})$

$\qquad = (8-4).2.(25-5) = 4.2.20 = 160$

$$\phi(n) = \phi(10) = \phi(5.2) = \phi(5)\phi(2) = (5-1)(2-1) = 4$$

(since $\phi(p) = p - 1$, for $p=$ prime number)

Physically we see from the set $\{1,2,3,4,5,6,7,8,9,10\}$ that the four numbers $1,3,7,9$ are co-prime to 10.

So as per Euler theorem we can write $3^4 = 1 \bmod 10$.

Let us check if it holds.

Let us find

$$3^4 = 81 \bmod 10$$
$$= 1 \bmod 10$$

```
10 ) 81 ( 8
     80
   ――――――
      1 ←――Remainder
```

So we have verified Euler's theorem for $n = 10$, $a = 3$.

We now prove Euler's theorem.

First we provide proof of Euler's theorem for $n = 10$, $a = 3$ and then the general proof.

Proof of Euler's theorem for $n = 10$, $a = 3$

Consider $n = 10$, $a = 3$ and GCD(3, 10) = 1

$\phi(10) = 4$ and from the set $\{1,2,3,4,5,6,7,8,9,10\}$ the four numbers $1,3,7,9$ are co-prime to 10.

Consider the set of $\phi(10) = 4$ numbers that are co-prime to $n = 10$.

$$S = \{1, 3, 7, 9\}$$

Multiply S by $a = 3$ to get

$$S' = 3S = \{3, 9, 21, 27\}$$

We note that the numbers 3, 9, 21, 27 are also co-prime to $n = 10$
Take mod n, i.e. mod 10 to get

$$S'' = S' \bmod 10$$
$$= \{ 3 \bmod 10, 9 \bmod 10, 21 \bmod 10, 27 \bmod 10 \} \tag{12.1}$$

$$= \{3, 9, 1, 7\} \tag{12.2}$$

We note that the numbers 3, 9, 1, 7 are co-prime to $n = 10$.
On rearranging

$$S'' = \{ 1, 3, 7, 9 \} = S \tag{12.3}$$

we get back the same set S.

We identify from a study of equation (12.1), (12.2), (12.3), i.e. from $S'' = S$ the following

$$3 \bmod 10 = 3$$
$$9 \bmod 10 = 9$$
$$21 \bmod 10 = 1$$
$$27 \bmod 10 = 7$$

Taking product

$$(3 \bmod 10)(9 \bmod 10)(21 \bmod 10)(27 \bmod 10) = 3.9.1.7$$
$$(3.1 \bmod 10)(3.3 \bmod 10)(3.7 \bmod 10)(3.9 \bmod 10) = 3.9.1.7$$
$$3^4(1 \bmod 10)(3 \bmod 10)(7 \bmod 10)(9 \bmod 10) = 3.9.1.7$$
$$3^4(1.3.7.9)\bmod 10 = 3.9.1.7 = 1.3.7.9$$
$$3^4 \bmod 10 = 1, \text{ cancelling out}(1.3.7.9)$$
$$3^4 = 1 \bmod 10$$

which is Euler's theorem.

General proof of Euler's theorem
Let n be a positive integer and a, n are co-prime i.e. GCD$(a, n) = 1$.
$\phi(n)$ counts the positive integers less than n that are co-prime to n.
Consider the set of $\phi(n)$ numbers that are co-prime to n.
$S = \{x_1, x_2, x_3......x_p\}$ where we denote $p = \phi(n)$ for convenience of writing.
Multiply S by a to get

$$S' = aS = \{ax_1, ax_2, ax_3.......ax_p\}$$

We note that the numbers $ax_1, ax_2, ax_3.......ax_p$ are also co-prime to n

Take mod n to get

$$S'' = S' \bmod n = \{\ ax_1 \bmod n,\ ax_2 \bmod n,\ ax_3 \bmod n.......ax_p \bmod n\ \}$$

We note that the numbers $ax_1 \bmod n$, $ax_2 \bmod n$, $ax_3 \bmod n.......ax_p \bmod n$ are co-prime to n.

On rearranging we get back the same set S, i.e.

$$S'' = S$$

We identify from a study of $S'' = S$ the following

$$ax_1 \bmod n = x_1$$
$$ax_2 \bmod n = x_2$$
$$...$$
$$ax_p \bmod n = x_p$$

Taking product

$$(ax_1 \bmod\ n)(ax_1 \bmod\ n)........(ax_p \bmod\ 10) = x_1 x_1....x_p$$

$$a^p(x_1 x_2.....x_p) \bmod n = x_1 x_2.....x_p$$

$a^{\phi(n)} \bmod n = 1$ where we put back $p = \phi(n)$ and $(x_1 x_2.....x_p)$ gets cancelled.

$$a^{\phi(n)} = 1\ \bmod n.$$

This is Euler's theorem.

Table 12.5 illustrates the theorem further.

Proof of Fermat's Little theorem from Euler's theorem.

Fermat's little theorem is a special case of Euler's theorem where n is a prime number.

For prime number n we rewrite Euler's theorem

$$a^{\phi(n)} = 1\ \bmod n \xrightarrow{n=p=\text{prime number}} a^{\phi(p)} = 1\ \bmod n$$

Using $\phi(p) = p - 1$ for prime number

$$a^{p-1} = 1\ \bmod n$$

This is Fermat's little theorem (derived from Euler's theorem.)

12.7 Chinese remainder theorem

Statement

Suppose we have a set of integers: n_1, n_2, $n_3.............n_k$ such that no pair of them has any common factor (i.e. they are pairwise co-prime). Consider another set of integers b_1, b_2, $b_3.............$, b_k. The system of equations $x = b_i\ (\bmod\ n_i)$ for $1 \leqslant i \leqslant k$ will possess a unique solution w.r.t mod N given by

$$x = \sum_{i=1}^{k} b_i N_i x_i \quad (\bmod\ N)$$

Table 12.5. Euler theorem (an example).

✓Example

Write down Euler's theorem for $n = 100, a = 77$ and $GCD(77,100) = 1$

Ans $\phi(100) = \phi(25.4) = \phi(5^2 2^2) = (5^2 - 5^{2-1})(2^2 - 2^{2-1})$

$= (25-5)(4-2) = 20.2 = 40$

Then by Euler 's theorem we have $77^{40} = 1 \bmod 100$

✓Prove by Euler theorem $5^6 = 1 \bmod 18$

Ans Consider $n = 18, a = 5$ and $GCD(5,18) = 1$

$\phi(18) = 6$ and from the set $\{1,2,3,...,18\}$ the 6 numbers 1,5,7,11,13,17 are co-prime to 18.

Consider the set of $\phi(18) = 6$ numbers that are co-prime to 18.

$S = \{1,5,7,11,13,17\}$

Multiply S by $a = 5$ to get

$S' = 5S = \{5,25,35,55,65,85\}$

We note that the numbers 5,25,35,55,65,85 are also co-prime to 18.

Take mod 18 to get

$S'' = S \bmod 18$

$= \{5 \bmod 18, 25 \bmod 18, 35 \bmod 18, 55 \bmod 18, 65 \bmod 18, 85 \bmod 18\}$

$= \{5, 7, 17, 1, 11, 13\}$

We note that the numbers 5,7,17,1,11,13 are co-prime to 18.

On rearranging

$S'' = \{1,5,7,11,13,17\} = S$ (we get back the same set S)

We identify $5 \bmod 18 = 5$

$25 \bmod 18 = 7$

$35 \bmod 18 = 17$

$55 \bmod 18 = 1$

$65 \bmod 18 = 11$

$85 \bmod 18 = 13$

Taking product

$(5 \bmod 18)(25 \bmod 18)(35 \bmod 18)(55 \bmod 18)(65 \bmod 18)(85 \bmod 18) = 5.7.17.1.11.13$

$(5.1 \bmod 18)(5.5 \bmod 18)(5.7 \bmod 18)(5.11 \bmod 18)(5.13 \bmod 18)(5.17 \bmod 18) = 5.7.17.1.11.13$

$5^6 (1 \bmod 18)(5 \bmod 18)(7 \bmod 18)(11 \bmod 18)(13 \bmod 18)(17 \bmod 18) = 5.7.17.1.11.13$

$5^6 (1.5.7.11.13.17) \bmod 18 = 5.7.17.1.11.13 = 1.5.7.11.13.17$

$5^6 \bmod 18 = 1$ (cancelling (1.5.7.11.13.17))

$5^6 = 1 \bmod 18$

---

where $N = \prod_{i=1}^{k} n_i = n_1. \, n_2........n_k$, $N_i = \frac{N}{n_i}$, $x_i = N_i^{-1} \bmod n_i$.

Explanation

Consider

$$x = b_1 \bmod n_1$$
$$x = b_2 \bmod n_2.$$
$$x = b_3 \bmod n_3$$

where n_1, n_2, n_3 are pairwise co-prime. To find x.

Let us put the data in a table for convenience.

We define $N = n_1 n_2 n_3$, $N_i = \dfrac{N}{n_i}$.

So $N_1 = \dfrac{N}{n_1} = n_2 n_3$, $N_2 = \dfrac{N}{n_2} = n_1 n_3$, $N_3 = \dfrac{N}{n_3} = n_1 n_2$

Mod w.r.t	Remainders b_i	$N_i = \dfrac{N}{n_i}$	Inverse of $N_i x_i$	$b_i N_i x_i$
n_1	b_1	$N_1 = n_2 n_3$	$x_1 = N_1^{-1} \bmod n_1$	$b_1 N_1 x_1$
n_2	b_2	$N_2 = n_1 n_3$	$x_2 = N_2^{-1} \bmod n_2$	$b_2 N_2 x_2$
n_3	b_3	$N_3 = n_1 n_2$	$x_3 = N_3^{-1} \bmod n_3$	$b_3 N_3 x_3$

$$x = \sum_{i=1}^{3} b_i N_i x_i \ \bmod N$$

Tables 12.6, 12.7 and 12.8 give a few illustrations.

12.8 RSA algorithm

Suppose Bob considers two arbitrarily large prime numbers p, q (say 128 bit numbers), $p \neq q$. The product of these is $N = pq$ and we note that it is very hard to factorize N. In fact by the definition of trap door function it is not possible to factorize within polynomial time. So p, q are only known to Bob.

Euler's Totient function is

$$\phi(N) = \phi((pq)) = (p - 1)(q - 1)$$

As p, q are not known by the public one cannot evaluate $\phi(N) = (p - 1)(q - 1)$ though N is a number that is publicly known.

We shall use low bit number for illustration purpose.

◆ Consider the prime number $p = 7$, $\phi(7) = 7 - 1 = 6$ and another prime number $q = 5$, $\phi(5) = 5 - 1 = 4$.

Now $N = pq = (7)(5) = 35$.

Between 1 and 34 there are 4 multiples of 7 (viz. 7,14,21,28,) and 6 multiples of 5 (viz. 5,10,15,20,25,30) i.e. a total of 10 numbers which are not co-prime to 35. Hence

$$\phi(35) = 34 - 10 = 24 = (6)(4) = (7 - 1)(5 - 1)$$

Suppose Bob chooses a number e (such that $1 < e < N$), co-prime with $\phi(N)$ i.e. $GCD(e, \phi(N)) = 1$. The set (N, e) is called the public key or public code which means that anyone has access to these two numbers.

◆ In our example suppose Bob chooses $e = 7$ and the Bob's public code is $(N, e) = (35, 7)$.

Table 12.6. Illustration of Chinese remainder theorem.

Q. Consider $x = 3 \bmod 5$, $x = 1 \bmod 7$, $x = 6 \bmod 8$. Find x.

Ans. $n_1 = 5, n_2 = 7, n_3 = 8$ pairwise co-prime. $N = n_1 n_2 n_3 = 5.7.8 = 280$

Mod w.r.t	Remainders	$N_i = \dfrac{N}{n_i}$	$x_i = $ Inverse of N_i	$b_i N_i x_i$
$n_1 = 5$	$b_1 = 3$	$N_1 = n_2 n_3$ $= 7.8 = 56$	$x_1 = N_1^{-1} \bmod n_1$ $= 56^{-1} \bmod 5 = 1$ Write $56x \bmod 5 = 1$ (use $56 \bmod 5 = 1$) Check $56 \text{①} \bmod 5 = 1$	$b_1 N_1 x_1$ $= 3.56.1$ $= 168$
$n_2 = 7$	$b_2 = 1$	$N_2 = n_3 n_1$ $= 8.5 = 40$	$x_2 = N_2^{-1} \bmod n_2$ $= 40^{-1} \bmod 7 = 3$ Write $40x \bmod 7 = 1$ (use $40 \bmod 7 = 5$) $5x \bmod 7 = 1$ $5.\text{③} \bmod 7 = 1$ Check $40.\text{③} \bmod 7$ $= 120 \bmod 7 = 1$	$b_2 N_2 x_2$ $= 1.40.3$ $= 120$
$n_3 = 8$	$b_3 = 6$	$N_3 = n_1 n_2$ $= 5.7 = 35$	$x_3 = N_3^{-1} \bmod n_3$ $= 35^{-1} \bmod 8 = 3$ Write $35x \bmod 8 = 1$ (use $35 \bmod 8 = 3$) $3x \bmod 8 = 1$ $3.\text{③} \bmod 8 = 1$ Check $35.\text{③} \bmod 8$ $= 105 \bmod 8 = 1$	$b_3 N_3 x_3$ $= 6.35.3$ $= 630$

$$x = \sum_{i=1}^{3} b_i N_i x_i \bmod N = (168 + 120 + 630) \bmod 280 = 918 \bmod 280$$

$$x = \boxed{78} \bmod 280$$
Ans

$$280 \overline{)918} \big(3$$
$$\underline{840}$$
$$78$$

Check: $78 = 3 \bmod 5$
$78 = 1 \bmod 7$
$78 = 6 \bmod 8$

$5 \overline{)78}(15$ $7 \overline{)78}(11$ $8 \overline{)78}(9$
$\underline{75}$ $\underline{77}$ $\underline{72}$
$\ \ 3$ $\ \ 1$ $\ \ 6$

If Alice wishes to send a message m called plain text (say $m < N$) to Bob, she has to encode m using the encoder e as (encryption process)

$c = m^e \bmod N$, c is called cipher text corresponding to m.

Table 12.7. Another illustration of Chinese remainder theorem.

Q. Consider $x=5 \mod 7$, $x=3 \mod 10$, $x=7 \mod 9$. Find x.

Ans. $n_1=7, n_2=10, n_3=9$ pairwise co-prime. $N=n_1 n_2 n_3 = 7.10.9=630$

Mod w.r.t	Remainders	$N_i = \dfrac{N}{n_i}$	$x_i =$ Inverse of N_i	$b_i N_i x_i$
$n_1=7$	$b_1=5$	$N_1=n_2 n_3$ $=10.9=90$	$x_1=N_1^{-1} \mod n_1$ $=90^{-1} \mod 7 =6$ Write $90x \mod 7=1$ (use $90 \mod 7=6$) $6x \mod 7 =1$ $6.\textcircled{6} \mod 7 =1$ Check $90\textcircled{6} \mod 7$ $=540 \mod 7=1$	$b_1 N_1 x_1$ $=5.90.6$ $=2700$
$n_2=10$	$b_2=3$	$N_2=n_3 n_1$ $=9.7=63$	$x_2=N_2^{-1} \mod n_2$ $=63^{-1} \mod 10=7$ Write $63x \mod 10=1$ (use $63 \mod 10=3$) $3x \mod 10 =1$ $3.\textcircled{7} \mod 10 =1$ Check $63\textcircled{7} \mod 10$ $=441 \mod 10=1$	$b_2 N_2 x_2$ $=3.63.7$ $=1323$
$n_3=9$	$b_3=7$	$N_3=n_1 n_2$ $=7.10=70$	$x_3=N_3^{-1} \mod n_3$ $=70^{-1} \mod 9 =4$ Write $70x \mod 9=1$ (use $70 \mod 9=7$) $7x \mod 9 =1$ $7.\textcircled{4} \mod 9 =1$ Check $70\textcircled{4} \mod 9$ $=280 \mod 9=1$	$b_3 N_3 x_3$ $=7.70.4$ $=1960$

$$x = \sum_{i=1}^{3} b_i N_i x_i \mod N = (2700+1323+1960) \mod 630 = 5983 \mod 630$$

$$x = \textcircled{313} \mod 630$$
Ans

$$630 \overline{\smash{)}5983} \big(9$$
$$\underline{5670}$$
$$313$$

Check: $313 = 5 \mod 7$
$313 = 3 \mod 10$
$313 = 7 \mod 9$

$7\overline{\smash{)}313}(44 \quad \underline{308} \quad 5$

$10\overline{\smash{)}313}(31 \quad \underline{310} \quad 3$

$9\overline{\smash{)}313}(34 \quad \underline{306} \quad 7$

Table 12.8. A third illustration of Chinese remainder theorem.

Q. Consider $x=5$ mod 7, $x=3$ mod 11, $x=10$ mod 13. Find x.

Ans. $n_1=7, n_2=11, n_3=13$ pairwise co-prime. $N=n_1 n_2 n_3 = 1001$

Mod w.r.t	Remainders	$N_i = \dfrac{N}{n_i}$	x_i = Inverse of N_i	$b_i N_i x_i$
$n_1 = 7$	$b_1 = 5$	$N_1 = n_2 n_3$ $=143$	$x_1 = N_1^{-1} \bmod n_1$ $=143^{-1} \bmod 7 = 5$	$b_1 N_1 x_1$ $=3575$
$n_2 = 11$	$b_2 = 3$	$N_2 = n_3 n_1$ $=91$	$x_2 = N_2^{-1} \bmod n_2$ $= 91^{-1} \bmod 11 = 4$	$b_2 N_2 x_2$ $=1092$
$n_3 = 13$	$b_3 = 10$	$N_3 = n_1 n_2$ $=77$	$x_3 = N_3^{-1} \bmod n_3$ $= 77^{-1} \bmod 13$ $= 12$ ✱	$b_3 N_3 x_3$ $=9240$

$$x = \sum_{i=1}^{3} b_i N_i x_i \ \bmod N = 13907 \bmod 1001 = 894 \leftarrow \qquad 1001 \overline{)13907} (13$$
$$\underline{13013}$$
$$894$$

Ans (894) i.e. $894 = 5$ mod 7
$894 = 3$ mod 11
$894 = 7$ mod 13

✱ 77^{-1} mod 13

Write $77x$ mod $13=1$

(use 77 mod $13=12$)

$12x$ mod $13 =1$

We inspect that :

$13.(1)=13 \quad 12.(1)=12$

Clearly $12.1 = 12$ *falls short* of $13.1 = 13$ by 1. Since w.r.t mod 13, the number -1 and $13 - 1 = 12$ are equivalent let us check with 12 as follows.

$(-d \bmod n = (n - d) \bmod n)$

$77.\boxed{12}$ mod 13
$= 924$ mod $13 = 1$

$13 \overline{)924} (71$
$\underline{923}$
1

As $77.12 = 1 \pmod{13}$, $77^{-1} = 12$

◆ Suppose the message to be sent is $m = 3$ (called the plain text or original message) As $e = 7$, $N = 35$ the encoded or encrypted format of Alice is $c = 3^7$ mod $35 = 2187$ mod $35 = 17$ mod 35. This is the cipher text or encrypted text).

Bob decodes the message using his decoder key generated as (decryption process)

$$ed = 1 \bmod \phi(N)$$

We emphasize that only Bob knows d which is his decoder key or decryption key. The set (N, d) is called the private key or private code.

◆ As in our example $e = 7$, $N = 35$, $\phi(N) = \phi(35) = 24$ we thus have

$$ed = 1 \mod \phi(N)$$

$\Rightarrow 7d = 1 \mod 24 \Rightarrow d = 7^{-1} \mod 24 = 7$
(writing $7x \mod 24 = 1$, on inspection we find that for $x = 7$
$7(7) \mod 24 = 49 \mod 24 = 1$. So $d = 7$)

To decode, Bob performs the following operation

$$c^d = m^{ed} \mod N = m \mod N \quad (\text{as } ed = 1 \mod \phi(N))$$

$$m = c^d \mod N$$

So the original message or plain text has been decrypted.

◆ As $c = 17 \mod 35$, $d = 7$ we have $m = c^d \mod N = 17^7 \mod 35 = 3$ which was the original message sent by Alice.(Details of solution has been done in table 12.9)

Using Fermat's little theorem and Chinese remainder theorem we can establish the theoretical basis or proof of the above cryptosystem. So the cryptosystem involves encoding a plain or normal text, then sending it. The receiver receives the message in coded form, decrypts it to the original form.

Table 12.10 summarizes essential features of the RSA algorithm.

Table 12.11 explains the RSA algorithm.

Tables 12.12 and 12.13 contain problems on the RSA algorithm.

The classical encryption process relies on the fact that factorization of large composite numbers in polynomial time is extremely difficult or virtually impossible. But once quantum computers become a reality, factoring a composite number using Shor's algorithm will change the perception.

Table 12.9. Evaluation of $17^7 \mod 35$.

$17^1 \mod 35 = 17$

$17^2 \mod 35 = 289 \mod 35 = 9 \mod 35$

```
35 ) 289 ( 8
     280
    ─────
      9 ←
   Remainder
```

```
35 ) 81 ( 2
     70
    ─────
     11 ←
   Remainder
```

$17^4 \mod 35 = (17^2)^2 \mod 35 = 9^2 \mod 35$

$\qquad = 81 \mod 35 = 11 \mod 35$

$17^7 \mod 35 = 17^{2+4+1} \mod 35$

$\qquad = (17^2 \mod 35)(17^4 \mod 35)(17^1 \mod 35)$

$\qquad = 9 . 11 . 17 \mod 35 = 1683 \mod 35 = 3 \mod 35$

```
35 ) 1683 ( 48
     1680
    ─────
       3 ←
   Remainder
```

Table 12.10. Essential features of RSA algorithm.

RSA algorithm

1. Select 2 prime numbers p, q and p ≠ q.

2. Evaluate N = pq = Block size

3. Evaluate Euler's Totient function ϕ (N)= (p−1) (q−1)

4. Choose a number 1< e < N co-prime with ϕ(N) , GCD (e, ϕ(N)) = 1
   e = encryption key or encoding key

5. d = decryption key or decoder key ,d = e^{-1} mod ϕ (N)
   or ed = 1 mod ϕ(N)

6. Public key or public code (N,e) . Private key or private code (N,d)

7. c = cipher text, c = m^e mod N (encryption)

   m = plain text (original message),

   m = c^d mod N (decryption)

12.9 Quantum cryptography

Let us use quantum ideas for the subject of coding.

Wiesner's idea (1960) of quantum money

We can use polarized light to store information and for this we can use four bases. Horizontal (X-axis, ↔), Vertical (Y-axis, ↕), 45° diagonal (with +ve X-axis), 135° diagonal with +ve X-axis.

Suppose in each currency note we embed a sequence of light traps randomly. Light traps are tiny devices which capture and store a single photon. Each trap can have any one of the four polarization states. For instance, for 25 light traps there would be 4^{25} different configurations or possibilities.

So a particular currency note of certain denomination (say Rs.500) bearing a particular number (say C 19007) is entered in the database with the shown polarization states of the 25 light traps (figure 12.1). As there are too many options (4^{25}) currency notes are necessarily not identically coded (all have different numbers, different polarization states and duplication is not needed). And so identification of fake or duplicated currency note can be easily done by comparing with the database and identifying any mismatch. The significant point is that quantum mechanics can be used for encryption, decryption, verification of codes.

If quantum mechanics is employed in cryptography things become secure because a message or data cannot be copied as per no-cloning theorem. It is not possible to make a copy of a random quantum state.

Also, each single qubit is a linear combination or superposition of two states. So any attempt to intercept and read the message would disturb the system (as per uncertainty principle) and the interceptor would get a message that bears no sense.

The interceptor thus, upon unauthorized interception or measurement of a message cannot reconstruct the original message because of the irreversible nature of quantum measurement.

Table 12.11. Explanation of RSA algorithm.

RSA algorithm – a problem

Consider 2 prime numbers p=13, q=11

N=pq=13.11=143

$\phi(N) = \phi(143) = \phi(13.11) = (13-1)(11-1) = 120$

Select e=13. GCD(13,ϕ(N))=GCD (13,120)=1

$d = e^{-1} \bmod \phi(N) = 13^{-1} \bmod 120$

Write

$13x \bmod 120 = 1$

$13\widehat{(37)} \bmod 120$

$= 481 \bmod 120 = 1$

$x=37$ i.e. d=37

120.2=240	13.18=234
120.3=360	13.27=351
120.4=480	13.37=481

Public key=(N,e)=(143,13) | Private key=(N,d)=(143,37)

m given. Find c. | *c given. Find m.*

Cipher text = c= m^e mod N | Plain text = m= c^d mod 143

Take m=13 | Take c=52

$c = 13^{13} \bmod 143 = 52$ | $m = 52^{37} \bmod 143 = 13$

$13^1 = 13 \bmod 143$	$52^1 = 52 \bmod 143$
$13^2 = 169 \bmod 143 = 26 \bmod 143$	$52^2 = 2704 \bmod 143 = 130 \bmod 143$
$13^4 = (13^2)^2 = 26^2 \bmod 143$	$52^4 = (52^2)^2 = 130^2 \bmod 143$
$= 676 \bmod 143 = 104 \bmod 143$	$= 16900 \bmod 143 = 26 \bmod 143$
$13^8 = (13^4)^2 = 104^2 \bmod 143$	$52^8 = (52^4)^2 = 26^2 \bmod 143$
$= 10816 \bmod 143 = 91 \bmod 143$	$= 676 \bmod 143 = 104 \bmod 143$
$13^{13} = 13^{8+4+1} \bmod 143$	$52^{16} = (52^8)^2 = 104^2 \bmod 143$
$= 13^8 13^4 13^1 \bmod 143$	$= 10816 \bmod 143 = 91 \bmod 143$
$= 91.104.13 \bmod 143$	$52^{32} = (52^{16})^2 = 91^2 \bmod 143$
$= 123032 \bmod 143$	$= 8281 \bmod 143 = 130 \bmod 143$
$= 52 \bmod 143$	$52^{37} = 52^{32+4+1} \bmod 143$
	$= 52^{32} 52^4 52^1 \bmod 143$
	$= 130.26.52 \bmod 143$
	$= 175760 \bmod 143$
	$= 13 \bmod 143$

Let us discuss some protocols for sending secure communication. If Alice and Bob are certain that there is no eavesdropper they can communicate through a one-time pad (the Vernam Cipher) involving a string of bits, identical copies of which are possessed by Alice and Bob. This pad is for one-time usage.

12.10 Protocol of quantum cryptography

A protocol is a set of rules governing the exchange of messages over a channel. A security protocol is a special protocol designed to ensure that security properties are met during communications. There are security protocols like BB84, B92 etc. The

Table 12.12. Illustration of RSA algorithm.

<u>A problem on RSA algorithm</u>

Consider two prime numbers p=7, q=17 . Take public key e=5.
If plain text value is 6 then what will be cipher text value according to
RSA algorithm ? Also calculate plain text value from cipher text.

<u>Ans</u>
Consider 2 prime numbers p=7, q=17

$N=pq=7.17=119$

$\phi(N)=\phi(119)=\phi(7.17)=(7-1)(17-1)=96$

Select e=5. GCD(5, $\phi(N)$)=GCD (5,96)=1

$d=e^{-1} \bmod \phi(N)= 5^{-1} \bmod 96$

$d=77 \leftarrow$

Write 5x mod 96=1
We find that $96.1=96$; $5.19=95$.
$5.19=95$ *falls short* of $96.1=96$ by 1.
W.r.t mod 96, -19 and $96-19=77$ are
equivalent.
      Hence $5.(77) \bmod 96$
        $= 385 \bmod 96 = 1$

$$96)\overline{385}(4$$
$$\underline{384}$$
$$1$$

As w.r.t mod 96 we have
$5.77 = 1 \Rightarrow 5^{-1} = (77)$

Given m. Find c.

Public key=(N,e)=(119, 5)

Cipher text = $c= m^e \bmod N$
    Take m=6

$c=6^5 \bmod 119 = 41$

$41^1 = 41 \bmod 119$

$41^2 = 1681 \bmod 143 = 15 \bmod 119$

$41^4 = (41^2)^2 = 15^2 \bmod 143$
    $= 225 \bmod 119 = 106 \bmod 119$

$41^8 = (41^4)^2 = 106^2 \bmod 119$
    $= 11236 \bmod 119 = 50 \bmod 119$

$41^{16} = (41^8)^2 = 50^2 \bmod 119$
    $= 2500 \bmod 119 = 1 \bmod 119$

$41^{32} = (41^{16})^2 = 1 \bmod 119$

$41^{64} = (41^{32})^2 = 1 \bmod 119$

Given c. Find m..

Private key=(N,d)=(119, 77)

Plain text = $m= c^d \bmod N$
    Take c=41

$m=41^{77} \bmod 119 = 6$

$41^{77} = 5 2^{64+8+4+1} \bmod 119$

$= 41^{64} 41^8 41^4 41^1 \bmod 119$

$= 1.50.106.41 \bmod 119$

$= 217300 \bmod 119$

$= 6 \bmod 119$

purpose of these protocols is to generate a secret key that is to be shared between sender (say Alice) and receiver (say Bob) with a possible eavesdropper (say Eve) trying to intercept the message shared. The key itself does not carry any message but is used to encrypt a message and hence secrecy of the key is an important factor just like the secrecy of the message.

The quantum key distribution in quantum cryptography uses three quantum mechanical characteristics, namely quantum superposition, quantum entanglement and the uncertainty principle.

Table 12.13. Illustration of on RSA algorithm.

<u>Another problem on RSA algorithm</u>

Consider p=11, q=3 two prime numbers. Take public key e=7.
If cipher text value is 30 then what will be plain text value according to
RSA algorithm ? Also calculate plain text value from cipher text.
  <u>Ans</u>
Consider 2 prime numbers p=11, q=3

$N=pq=11.3=33$

$\phi(N)=\phi(33)=\phi(11.3)=(11-1)(3-1)=20$

Select e=7. GCD(7,ϕ(N))=GCD (7,20)=1

$d=e^{-1} \mod \phi(N)=7^{-1} \mod 20$ | Write 7x mod 20=1

$d=3$ ← | 7.③ mod 20=1

   Public key=(N,e)=(33, 7)
   Private key=(N,d)=(33, 3)

$6^1 \mod 33 = 6 \mod 33$

$6^2 \mod 33 = 36 \mod 33$
            $= 3 \mod 33$

Given c find m.

Plain text = $m = c^d \mod 33$
         Take c=30 (cipher)

$6^4 \mod 33 = (6^2)^2 \mod 33$
            $= 3^2 \mod 33$
            $= 9 \mod 33$

       $m = 30^3 \mod 33$
          $= 27000 \mod 33 = \boxed{6}$

$6^7 \mod 33 = 6^{4+2+1} \mod 33$

Given m find c.
            $= 6^4 6^2 6^1 \mod 33$

  Cipher text = $c = m^7 \mod N$
            $= 9.3.6 \mod 33$

       Take m=6 (plain text)
            $= 162 \mod 33$

         $c = 6^7 \mod 33$
            $= 30 \mod 33$

         $= \boxed{30}$ ←

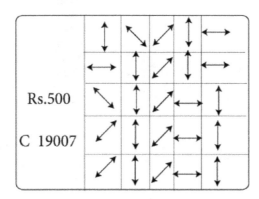

Figure 12.1. Currency notes using light traps.

12.10.1 BB84 protocol

This protocol was suggested by Bennette and Brassard in 1984 and hence the
nomenclature. It is the first quantum cryptography protocol.

Quantum key distribution can be done by using polarization states of photon and using a channel such as an optical fibre to send them. If Eve (an unauthorized party) intercepts and measures the quantum state sent by Alice, the state would collapse into a state determined by Eve's measuring device. So Eve cannot measure and copy the state because quantum states cannot be cloned. The security of key sharing rests on this fact that the presence of Eve can be detected and the protocol aborted if it is suspected that some interceptor (Eve, say) is trying to intercept the message.

Calcite crystals are standard polarizers that can be used to generate polarized photon states. Let us mention what happens when such photons, polarized along a specific direction fall on a second polarizer (called analyzer, as these are also calcite crystals used to study polarization states of photons) (figure 12.2). If axis of analyzer is parallel to the direction of polarized photons there will be full (100%) transmission of photons. If axis of analyzer is perpendicular to the direction of polarized photons there will be zero (0%) transmission of photons. If the axis of the analyzer makes angle say ϕ to the direction of polarized photons, the intensity I of the transmitted photon beam will be reduced and is given by Malus law $I \ \alpha \ \cos^2 \phi$. In other words the probability of passage of photons through the analyzer is $\cos^2 \phi$. (For a single photon the probability of passing through is $\cos^2 \phi$ since the photon cannot split up into parts.)

A set of four polarizers are used in two base states. Alice has a set of two polarizers and Bob has a set of two polarizers. It has been agreed upon by them that the sender (Alice) has to encode the message in one of the two bases while the

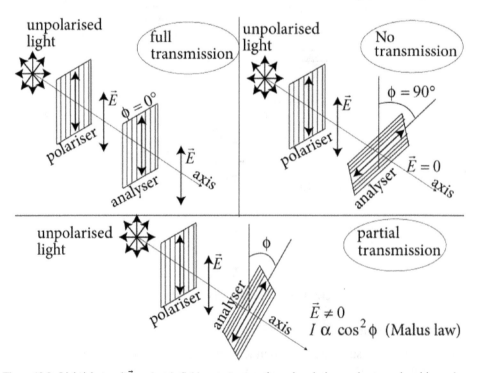

Figure 12.2. Light/photon (($\vec{E}$ = electric field vector)passes through polarizer and gets analyzed in analyzer.

recipient (Bob) has to decode the message using those two bases. This is indicated in the diagram of figure 12.3.

Alice makes a random use of bases as follows.

Suppose Alice tosses a coin.

If Head occurs she uses horizontal/vertical basis ($\leftrightarrow/\updownarrow$ or $\oplus$), i.e. the computational bases for encoding. She encodes bit 0 as $|0>$ (i.e. horizontal basis representing single photon polarized along X-axis) and bit 1 as $|1>$ (i.e. vertical basis representing single photon polarized along Y-axis).

If Tail occurs she uses diagonal basis (denoted by $\otimes$) and encodes bit 0 as $|+>$ (i.e. diagonal basis representing single photon polarized along 45° to X-axis) and $|->$ (i.e. diagonal basis representing single photon polarized along 135° to X-axis).

In terms of computational basis the diagonal bases are

$$|+>=\frac{|0>+|1>}{\sqrt{2}}, \quad |->=\frac{|0>-|1>}{\sqrt{2}}$$

and in terms of diagonal basis the computational bases are

$$|0>=\frac{|+>+|->}{\sqrt{2}}, \quad |1>=\frac{|+>-|->}{\sqrt{2}},$$

Upon receiving the message Bob has to decode it. Bob does not know about the basis used by Alice. Actually, the basis used by Alice should not be publicly known as then it might be used by an eavesdropper (say Eve) who can interrupt the message on its way. This is why Bob too has to toss a coin to select the base states for decoding.

Suppose Alice is using the horizontal/vertical base states and Eve comes to know of it. So Eve will align her polarizer accordingly and when a bit emerges she gets information and allows the bit to pass to Bob. Clearly both Bob as well as the unauthorized Eve receive the message sent by Alice and surely the secrecy has been violated. In a public channel cryptosystem there is always the risk of some eavesdropping which is undesirable and should be eliminated.

Bob also makes a random use of bases by tossing a coin. If Head occurs he uses horizontal/vertical basis ($\leftrightarrow/\updownarrow$), i.e. the computational bases for decoding. He encodes bit 0 as $|0>$ and bit 1 as $|1>$. If Tail occurs he uses diagonal basis and encodes bit 0 as $|+>$ and bit 1 as $|->$.

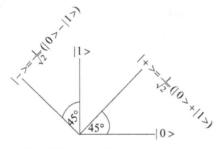

Bases used in BB84 protocol
(4 bases are: Horizontal base, vertical base, diagonal bases)

Figure 12.3. BB84 protocol of quantum cryptography.

Suppose Alice wishes to send the sequence

Qubit sent	1	0	0	1	1	0	0	0
Basis	⊕	⊗	⊗	⊕	⊗	⊕	⊕	⊗

Suppose Bob gets this distribution based upon his tosses

Qubit sent	1	0	0	1	1	0	0	0
Basis	⊕	⊗	⊗	⊕	⊗	⊕	⊕	⊗
Bob	⊕	⊕	⊗	⊗	⊕	⊗	⊕	⊗

Matching or agreement occurs at first bit (whch is 1), third bit (which is 0), seventh bit (which is 0) and the eigth bit (which is 0). In these bits bases were the same for Alice and Bob.

Qubit sent	1	0	0	1	1	0	0	0
Basis	⊕	⊗	⊗	⊕	⊗	⊕	⊕	⊗
Bob	⊕	⊕	⊗	⊗	⊕	⊗	⊕	⊗
Correct bits received	1		0				0	0

When the basis does not match or agree, Alice could have sent anything but Bob has an equal probability of measuring either 0 or 1. If Alice used ↔/↕ basis to encode say 0 (which in the diagonal basis is $|0> = \frac{|+> + |->}{\sqrt{2}}$) when Bob makes a measurement he has equal probability of measuring $|+>$ (corresponding to bit 0) and $|->$ (corresponding to bit 1). Clearly even if there is disagreement in the choice of bases Bob would get 50% correct results (provided the data is sent randomly and in large amount).

We suppose that Bob has the following distribution:

Qubit sent	1	0	0	1	1	0	0	0
Basis	⊕	⊗	⊗	⊕	⊗	⊕	⊕	⊗
Bob	⊕	⊕	⊗	⊗	⊕	⊗	⊕	⊗
Bob's distribution	✓ 1	✗ 1	✗ 0	✗ 0	1	0	✓ 0	✓ 0

agree
probabilistically

The tick or right sign indicates the correct results that Bob gets due to coincidence of bases. The cross sign indicates results that Bob gets due to disagreement in the choice of bases. The boxed results indicate results that Bob gets which are correct because of the random nature of the process (i.e. due to statistical or probabilistic factor) inspite of the fact that choice of bases were different. (Something like this: In a multiple choice question (MCQ) set, if we mark all answers to be A we would get at least some correct and indeed 25% correct if a large number of questions are set having answers randomly distributed between choices A,B,C and D.)

Assuming that a large number of bits are sent by Alice (and there is no eavesdropper like Eve say) we expect that Alice and Bob both randomly use computational basis and diagonal basis 50% of the time (= 50% of the tosses). And when the bases do not agree Bob still gets 50% of the 50% disagreements to be correct and this is 25%. Obviously thus Bob correctly receives 50%+25%=75% correct results that are sent by Alice.

Now Alice uses a public classical channel (telephone say) to announce the sequence of the basis used by her (assuming no eavesdropper), e.g. Alice declares the sequence as ⊕⊗⊗⊕⊗⊕⊕⊗. Bob now identifies and removes the results that correspond to disagreement in bases (i.e. removes 50% of the results including the results that matched the sent bits through chance, i.e. through a statistical agreement). This is done as Bob does not have an idea regarding which bits would agree by chance. So now Bob is certain to possess an identical copy of 50% of the bits sent by Alice (and he is assured that there is no eavesdropper).

Qubit sent	1	0	0	1	1	0	0	0
Basis	⊕	⊗	⊗	⊕	⊗	⊕	⊕	⊗
Bob identifies the common bases	✓⊕	⊕	✓⊗	⊗	⊕	⊗	✓⊕	✓⊗

If there were an eavesdropper, say Eve, there is every possibility that she would tap the data sent by Alice, make an identical copy of the same, and send it to Bob. But this is prohibited by the no-cloning theorem. In classical communication the presence of Eve cannot be detected with certainty. In quantum mechanics, the process of measurement itself disturbs the system as per Heisenberg's uncertainty principle. If Eve measures one observable the canonically conjugate observable, due to its non commuting nature would become random (i.e. indeterminate). This effort of Eve would alert us to her unauthorised entry and attempt to intervene.

Let us now consider the effect of an eavesdropper Eve who intrudes in an unauthorised manner trying to deal with and manipulate the data sent by Alice. Eve does not know what basis is being used by Alice or by Bob. So Eve too tries to rely on randomly chosen basis.

Eve therefore tosses a coin and creates a sequence of her own as shown (which is a combination of the computational basis and the diaogonal basis).

Qubit sent by Alice	1	0	0	1	1	0	0	0
Basis	⊕	⊗	⊗	⊕	⊗	⊕	⊕	⊗
Eve	⊕	⊕	⊗	⊗	⊗	⊕	⊗	⊕
	✓		✓				✓	✓
Bob	⊕	⊕	⊗	⊗	⊕	⊗	⊕	⊗

As a large number of bits are sent and due to the random nature of things, there will be 50% agreement due to coincidence of the basis chosen by Alice and Eve and a further 25% agreement due to statistical nature of things (probabilistic factor) even if the bases chosen are non coincident. This is shown in the following.

Qubit sent by Alice	1	0	0	1	1	0	0	0
Basis	⊕	⊗	⊗	⊕	⊗	⊕	⊕	⊗
	✓		✓		✓	✓		
Eve	⊕	⊕	⊗	⊗	⊗	⊕	⊗	⊕
	1	[0]	0	0	1	0	[0	0]

probabilistic agreement

Bob	✓		✓				✓	✓
	⊕	⊕	⊗	⊗	⊕	⊗	⊕	⊗

Now Bob would receive data depending upon agreement of the bases used by all the three. If bases used by Alice, Eve and Bob agree then Bob gets the correct result as indicated by the double arrow. In other words in such a situation whatever Alice sends will be received by Bob correctly.

Qubit sent by Alice	1	0	0	1	1	0	0	0
Basis	⊕	⊗	⊗	⊕	⊗	⊕	⊕	⊗
	✓		✓		✓	✓		
Eve	⊕	⊕	⊗	⊗	⊗	⊕	⊗	⊕
	1	[0]	0	0	1	0	[0	0]

probabilistic agreement

Bob	⫽		⫽				✓	✓
	⊕	⊕	⊗	⊗	⊕	⊗	⊕	⊗

We note that out of eight cases of choice of bases by Alice, Eve and Bob, two times there happen to be agreement (in the considered situation). In other words in $\frac{2}{8} = \frac{1}{4}$

th of the cases (i.e. in 25% of cases) Bob's basis matches with those of Alice's and Eve's. This means the probability of coincidence of bases now drops from the earlier 50% (leaving out the probabilistic agreement that made it 75%) to 25%, i.e. a drop by $(50 - 25)\% = 25\%$.

Alice chooses her bits randomly which means bit 0 is encoded as $|0>$ with a probability $\frac{1}{2}$ and as $|+>$ with equal probability. The density matrix of Alice's bit 0 is $\frac{1}{2}[|0><0| + |+><+|]]$. And bit 1 is encoded as $|1>$ with a probability $\frac{1}{2}$ and as $|->$ with equal probability. The density matrix of Alice's bit 1 is $\frac{1}{2}[|1><1| + |-><-|]]$. This is how the BB84 protocol works.

Alice chooses $4n + \delta$ number of bits, encodes them using random bases and sends to Bob. And Bob measures them using random bases. Alice declares over a public channel the sequence of bases she used to transmit data (and not the bits).

At least in $2n$ cases their bases would agree. The extra δ bits are taken to compensate for the possibility of actual agreement being less than 50%. Bob rejects the bits that correspond to disagreements in bases. And in the absence of Eve these $2n$ bits sent by Alice and received by Bob agree. With the remaining $2n$ bits they would perform some checks to determine if there is an eavesdropper.

Alice now picks a random sequence of n bits (called the check bits) out of the $2n$ bits and declares the location and content (i.e. the bits sent and the position of bits) over a public channel. Ideally, Bob and Alice should get the same bits on comparison. Bob still has n bits left to be identified. Now Bob makes a comparison of what Alice sent and what he received. Ideally they should get the same bits (as bases are identified to be the same). On comparison, if Bob finds that the agreement is not within acceptable error then he is sure that there is an eavesdropper interrupting or actively operating in the public channel. If the disturbance is intolerable then the process should be kept in abeyance (protocol is aborted) and resumed at an opportune time as most likely Eve would not be interrupting all the time.

If there is agreement between the bits within tolerable error then the process can be continued and the remaining n bits can be sent using a one-time pad for secure communication. This n bit string (one-time pad) can be used as the shared key.

With the shared string Alice and Bob does information reconciliation and a privacy amplification after which they go ahead with the protocol.

Information reconciliation is actually a process of error correction applied on the keys with Alice and Bob to ensure that they have identical keys. This is done by a process of cascading whereby they divide the entire key strings into blocks and carry out parity checks (counting the number of 1s in a string of bits) on each block, adds them (through addition modulo two operation). If the results of parity checks done by Bob and Alice do not agree, the blocks for which the parity mismatch occurs are identified. Such blocks are subdivided and the process repeated till location of error is pinpointed. This process might reveal an insignificant amount of information to the eavesdropper and does not harm Bob and Alice. Bob can discard the blocks after being sure that they contain errors without much effect. It is clear that the entire process reveals some information to Eve but that is insignificant, since only a few bits may be known to Eve correctly.

Privacy amplification is now done. Alice picks up a random subset of strings shared by her and Bob and they both compute parity check. The parity they calculate in various blocks now become their secret key (instead of the earlier string sent by Alice to Bob). This is the principle of privacy amplification. This new string so generated is less in size and completely confidential.

12.10.2 B92 protocol

B92 protocol is a simpler version of BB84 protocol and was devised by Bennet.

In this protocol Alice does not use coin tosses but encodes her basis in a pre-determined manner using a non-orthogonal basis $|0>$ and $|+>$. She encodes the bit $a = 0$ in computational basis $|0>$(that corresponds to a horizontally polarized photon) but bit $a = 1$ in a basis not orthogonal to it (i.e. not in the vertical basis $|1>$) but say in the diagonal basis $|+>= \frac{|0>+|1>}{\sqrt{2}}$ (that corresponds to a photon polarized along angle $\theta = 45°$, say). The diagram in figure 12.4 depicts the encoding of Alice and Bob. Suppose a represents bits, a' represents result of toss.

Bob uses coin toss approach and if the result of toss is Head (i.e. if $a' = 0$) he would use computational basis $(|0 >, |1>)$ to measure the incoming photon while if the result of coin toss is a Tail (i.e. if $a' = 1$) Bob would use diagonal basis $(|+>, | ->)$.

Consider the following situations:

Case 1. $a' = 0$.

Bob uses computational basis $(|0 >, |1>)$. If Bob measures 1 (i.e. his state $|1>$) Alice could not have sent bit 0 because then Alice's $|0>$ is orthogonal to Bob's $|1>$. So Alice cannot send 0 but must have sent 1, i.e. from state $|+>$ using the 45° polarized photon ($a = 1$). If Bob measures 0 (i.e. his state $|0>$) then Alice might have sent either 0 or 1 because both $|0>$ and $|+>$ of Alice have components along Bob's $|0>$. So no definite conclusion can be made.

In other words, if Bob used $a' = 0$ and measures 1 then Alice must have sent 1 ($a = 1$).

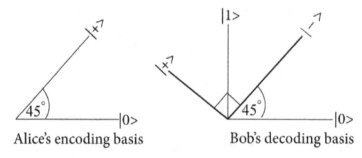

Alice's encoding basis   Bob's decoding basis

Figure 12.4. Bases used in B92 protocol. Bob's $|0>$ is parallel to Alice's $|0>$. Bob's $|->$ is parallel to Alice's $|+>$.

Case 2. $a' = 1$.

Bob uses diagonal basis ($|+>$, $|->$). If Bob measures 1 (i.e. his state $|+>$) Alice could not have sent bit 1 because Alice's $|+>$ is orthogonal to Bob's $|+>$. So Alice cannot send 1 but must have sent 0, i.e. from state $|0>$ using the horizontally polarized photon ($a = 0$). If Bob measures 0 (i.e. his state $|->$) then Alice might have sent either 0 or 1 because both $|0>$ and $|+>$ of Alice have components along Bob's $|->$. So no definite conclusion can be made.

In other words, if Bob used $a' = 1$ and measures 1 then Alice must have sent 0 ($a = 0$).

A definite conclusion can be reached whenever Bob measures 1 (for $a' = 0$, $a = 1$ and $a' = 1$, $a = 0$). Thus the final key is the bases which Alice and Bob used (and not the bits that Alice generated and Bob measured).

12.10.3 Ekert protocol using EPR pairs (E91)

In BB84 and B92 protocols, the key bits appear to have been originated by Alice. In EPR protocol the key is generated from a fundamentally random process involving the properties of entanglement.

In this protocol, instead of Alice preparing the states and sending them to Bob, a third party, say Charlie prepares an EPR pair and sends one qubit to Alice and its correlated entangled partner to Bob. He chooses an EPR singlet of spin states

$$|\psi> = \frac{1}{\sqrt{2}}(\alpha(1)\beta(2) - \alpha(2)\beta(1))$$

Here $\alpha(1)$, $\alpha(2)$ correspond to spin up and $\beta(1)\beta(2)$ correspond to spin down states of spin $\frac{1}{2}$ particle. ($|\alpha> = |\uparrow>$, $\beta = |\downarrow>$)

A Stern–Gerlach apparatus with a particular direction of magnetic induction can be used as analyzer to measure the spin states of the particle. Alice and Bob uses three coplanar axes to measure the spin of the particles that are coming towards them. The axes for Alice and Bob are (a_1, a_2, a_3) and (b_1, b_2, b_3), respectively, to measure the spin of the particles. Assuming the particles to travel in the z direction, the plane of the axes will be the X–Y plane. In figure 12.5 the axes a_1, a_2, a_3 make

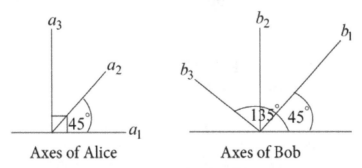

Axes of Alice   Axes of Bob

Figure 12.5. Bases used in Eckert 91 protocol.

$0°$, $45°$, $90°$, respectively, with the x-axis and the axes b_1, b_2, b_3 make $45°$, $90°$, $135°$, respectively, with the x-axis.

There are six base pairs that can be thought of

$$(a_2, \ b_1); \ (a_3, \ b_2); \ (a_1, \ b_1); \ (a_1, \ b_3); \ (a_3, \ b_1); \ (a_3, \ b_3).$$

Out of these there are two base pairs of compatible bases, namely (a_2, b_1) ; (a_3, b_2) because if Alice generates a string Bob will measure the complementary string. The probability that Alice and Bob choose compatible bases to measure the spin of their particles is $\frac{2}{6} = \frac{1}{3}$. In these cases if Alice detects her particle to be in the spin up state then Bob will find his particle to be in the spin down state and vice versa. For instance, if Alice used a_2 while Bob used b_1 or if Alice used a_3 while Bob used b_2, suppose Alice generates the string 1110001101 then Bob would find the string to be 0001110010 where the bit 0 has been used to represent spin up state and the bit 1 to represent spin down state.

After measurement, Alice and Bob announce the axes they used for measurement and keep the corresponding bits as forming their shared key.

The remaining four base pairs (a_1, b_1); (a_1, b_3); (a_3, b_1); (a_3, b_3) are not used for key generation as they correspond to incompatible axes. They can be used to detect the presence of Eve.

12.11 Further reading

[1] Ghosh D 2016 Online course on quantum information and computing (Bombay: Indian Institute of Technology)
[2] Nielsen M A and Chuang I L 2019 *Quantum Computation and Quantum Information* (Cambridge: Cambridge University Press)
[3] Nakahara M and Ohmi T 2017 *Quantum Computing* (Boca Raton, FL: CRC Press)
[4] Pathak A 2016 *Elements of Quantum Computing and Quantum Communication* (Boca Raton, FL: CRC Press)

12.12 Problems

1. *Which of the following algorithm can be used to break the RSA crypto system?*

   Grover's algorithm, Shor's algorithm, Simon's algorithm, DeutschJozsa algorithm.

   Ans.

   Shor's algorithm.

2. *Justify that the following options are Euler's Totient function?*

   $$\phi(21) = 12, \ \phi(35) = 24, \ \phi(13) = 12$$

3. *Consider a basket full of apples the number of which you have to determine. It is known that if the apples are taken out 3 or 5 or 7 at a time there would, respectively, be 2 or 3 or 2 left. Solve using the Chinese remainder theorem.*

Ans.

x = number of apples. Construct

$x = 3m + 2 \Rightarrow x \bmod 3 = 2 \Rightarrow x = 2 \bmod 3 (\text{compare } x = b_1 \bmod n_1)$

$x = 5m + 3 \Rightarrow x \bmod 5 = 3 \Rightarrow x = 3 \bmod 5 (\text{compare } x = b_2 \bmod n_2)$

$x = 7m + 2 \Rightarrow x \bmod 7 = 2 \Rightarrow x = 2 \bmod 7 (\text{compare } x = b_3 \bmod n_3)$

$n_1 = 3, \ n_2 = 5, \ n_3 = 7$ are pairwise co-primes.

We define $N = n_1 n_2 n_3 = 3.5.7 = 105, \ N_i = \dfrac{N}{n_i}$.

Mod w.r.t	Remainders b_i	$N_i = \dfrac{N}{n_i}$	Inverse of $N_i x_i$	$b_i N_i x_i$
$n_1 = 3$	$b_1 = 2$	$N_1 = n_2 n_3 = 35$	$x_1 = N_1^{-1} \bmod n_1 = 35^{-1} \bmod 3 = 2*$	$b_1 N_1 x_1 = 140$
$n_2 = 5$	$b_2 = 3$	$N_2 = n_3 n_2 = 21$	$x_2 = N_2^{-1} \bmod n_2 = 21^{-1} \bmod 5 = 1$	$b_2 N_2 x_2 = 63$
$n_3 = 7$	$b_3 = 2$	$N_3 = n_1 n_2 = 15$	$x_3 = N_3^{-1} \bmod n_3 = 15^{-1} \bmod 7 = 1$	$b_3 N_3 x_3 = 30$

$$x = \sum_{i=1}^{3} b_i N_i x_i \ \bmod N = (140 + 63 + 30) \bmod 105 = 23 \bmod 105$$

Possible answers are $105n + 23 \, (n = \text{integer})$

$= 23, 128, 233...$

Check for 23

$3) \overline{23} (7$ $5) \overline{23} (4$ $7) \overline{23} (3$
   $\underline{21}$ $\underline{20}$ $\underline{21}$
    2 3 2

$* \quad 35^{-1} \bmod 3$

Write $35x \bmod 3 = 1$

Use $35 \bmod 3 = 2$

$2x \bmod 3 = 1$

$2.(2) \bmod 3 = 1$

Check:

$35.(2) \bmod 3$
$= 70 \bmod 3$
$= 1$

$3) \overline{70} (23$
   $\underline{69}$
    1

As in mod 3 we have $35.2 = 1 \Rightarrow 35^{-1} = 2$

4. *The number 851 is known to be the product of 37 and 23. In RSA algorithm find the acceptable encryption exponent out of the following choices: 5,7,9,11.*

   Ans.

   $N = 851 = (37)(23); \ \phi(N) = \phi(851) = (37 - 1)(23 - 1) = (36)(22)$
   $ed = 1 \bmod \phi(N)$ and $\text{GCD}(e, \ \phi(N)) = 1$ i.e. $\text{GCD}(e, \ 36 \times 22) = 1$.
   As 5, 7 do not have common factor with $36 \times 22 \ e = 5, \ e = 7$ are possible choices.

5. *In RSA algorithm $N = 91, \ e = 5$ is the public key. If $m = 10$ find the public message.*

   Ans.

   $N = 91, \ e = 5, m = 10,$
   cipher text $c = m^e \bmod N = 10^5 \bmod 91$

$= (10^2)^2 .10 \bmod 91 = (10^2 \bmod 91)^2 (10 \bmod 91) = (9^2 \bmod 91)(10 \bmod 91)$

$$=810 \bmod 91 = 82$$

6. *Identify the decryption key d if you are to apply RSA algorithm with N = 247 and e = 7 and N is privately known to have the factors 19 and 13.*

   Ans.

   $N = 247, e = 7, ed = 1 \bmod \phi(N)$

$$\phi(N) = \phi(247) = \phi(19.13) = \phi(19)\phi(13) = (19 - 1)(13 - 1) = 18.12 = 216$$

$$7d = 1 \bmod 216, \quad d = 7^{-1} \bmod 216,$$

   Write $7x \bmod 216$. As $7.31 \bmod 216 = 1 \Rightarrow 7^{-1} = 31$. So $d = 31$

7. *In BB84 protocol Alice uses the bases DHHHDDHDH to encode the bit string 100100011 where H = horizontal/vertical basis and D= diagonal basis. Assuming no interceptor in the channel find the shared code if Bob measures 110110001 in his basis DDHDHDDHH.*

Alice basis	D	H	H	H	D	D	H	D	H
Bob basis	D	D	H	D	H	D	D	H	H
Basis matches	✓		✓			✓			✓
Bob receives code	1	1	0	1	1	0	0	0	1
Shared code	1		0			0			1

8. *In BB84 protocol Alice communicates to Bob and an unauthorized eavesdropper Eve is active who uses the same bases. Alice and Bob retain only the bases that they check and find to be identical. They pick randomly m check bits and verify whether Bob received the bits sent by Alice. Find the probability that the eavesdropper can get undetected.*

   Ans.

   Presence of Eve is detected if (a) Eve uses a different basis than that of Alice and Bob and (b) Eve gets a different result from the result obtained by Bob.

   $P(\text{Eve using different basis}) = \dfrac{1}{2}$, $P(\text{Eve getting different result}) = \dfrac{1}{2}$

   (for $\quad\quad\quad\quad\quad\quad\quad\quad\quad\quad\quad P(\text{Eve getting detected}) = \frac{1}{2} \cdot \frac{1}{2} = \frac{1}{4}$,

   $P(\text{Eve not getting detected}) = 1 - \frac{1}{4} = \frac{3}{4}$ (for 1 bit)

   For m bits, $P(\text{Eve not getting detected}) = (\frac{3}{4})^m$ (independent events, hence multiplied).

9. *In BB84 protocol suppose the basis in which Alice prepares bits and the basis in which Bob measures them are same (say H/V basis). An eavesdropper Eve is active in the communication process and uses the bases*

$$|E+> = \frac{\sqrt{3}}{2}|0> + \frac{1}{2}|1> \text{ and } |E-> = -\frac{1}{2}|0> + \frac{\sqrt{3}}{2}|1>$$

and measures the bits she intercepts from Alice and sends to Bob. Find the probability that there would be any disagreement between Alice and Bob regarding the message shared.

Ans.

Let us find the following:

$$P(\text{Eve receives/sends 0 using E + basis}) = |<0|E+>|^2 = \frac{3}{4}$$

$$P(\text{Eve receives/sends 0 using E - basis}) = |<0|E->|^2 = \frac{1}{4}$$

$$P(\text{Eve receives/sends 1 using E + basis}) = |<1|E+>|^2 = \frac{1}{4}$$

$$P(\text{Eve receives/sends 1 using E - basis}) = |<1|E->|^2 = \frac{3}{4}$$

$$P(\text{Eve chooses any one basis}) = \frac{1}{2}$$

Disagreements can occur if Alice sends 0 and Bob measures 1 or if Alice sends 1 and Bob measures 0. In either case Eve can receive a 0 or a 1 and also can use either of the bases $|E+>$, $|E->$.

Let us use the notation:

$P(A\ E\ B)$ = Prob. that Alice sends bit A, Eve measures bit E in some basis | E ±>, and Bob measures bit B ≠ A)

As $P(A\ E\ B) = |<A|E>|^2|<B|E>|^2$ and since $|<0|E+>|^2|<1|E+>|^2$ $= \frac{3}{4} \cdot \frac{1}{4} = \frac{3}{16}$ and $|<0|E->|^2|<1|E->|^2 = \frac{1}{4} \cdot \frac{3}{4} = \frac{3}{16}$.

Hence the probability of disagreement would be

$P(\text{Eve uses any one basis})[P(001) + P(011) + P(100) + P(110)]$

$$= \frac{1}{2}(\frac{3}{16} + \frac{3}{16} + \frac{3}{16} + \frac{3}{16}) = \frac{3}{8}.$$

10. *In BB84 protocol suppose the basis in which Alice prepares bits and the basis in which Bob measures them are the same (say $|\pm>$ basis). An eavesdropper Eve is active in the communication process and uses the bases*

$$|E+>=\frac{\sqrt{3}}{2}|0> +\frac{1}{2}|1> \text{ and } |E->=-\frac{1}{2}|0> +\frac{\sqrt{3}}{2}|1>$$

and measures the bits she intercepts from Alice and sends to Bob. Find the probability that there would be any disagreement between Alice and Bob regarding the message shared.

Ans.

Let us find the following:

$$P(\text{Eve receives/sends} + \text{using E} + \text{basis}) = |<+|E+>|^2$$
$$P(\text{Eve receives/sends} + \text{using E} - \text{basis}) = |<+|E->|^2$$
$$P(\text{Eve receives/sends} - \text{using E} + \text{basis}) = |<-|E+>|^2$$
$$P(\text{Eve receives/sends} - \text{using E} - \text{basis}) = |<-|E->|^2$$

$$P(\text{Eve chooses any one basis}) = \frac{1}{2}$$

The basis of Alice/Bob is $|+>=\frac{|0>+|1>}{\sqrt{2}}, |->=\frac{|0>-|1>}{\sqrt{2}}$

So $|0> =\frac{|+>+|->}{\sqrt{2}}, |1> =\frac{|+>-|->}{\sqrt{2}}$

Eve's basis in terms of basis of Alice/Bob

$$|E+>=\frac{\sqrt{3}}{2}|0> +\frac{1}{2}|1> = \frac{\sqrt{3}}{2}\frac{|+>+|->}{\sqrt{2}} + \frac{1}{2}\frac{|+>-|->}{\sqrt{2}}$$

$$= \frac{\sqrt{3}+1}{2\sqrt{2}}|+>+\frac{\sqrt{3}-1}{2\sqrt{2}}|->$$

$$|E->=-\frac{1}{2}|0> +\frac{\sqrt{3}}{2}|1> =-\frac{1}{2}\frac{|+>+|->}{\sqrt{2}} + \frac{\sqrt{3}}{2}\frac{|+>-|->}{\sqrt{2}}$$

$$= \frac{\sqrt{3}-1}{2\sqrt{2}}|+>+\frac{-\sqrt{3}-1}{2\sqrt{2}}|->$$

Hence

$$|<+|E+>|^2 = \left|\frac{\sqrt{3}+1}{2\sqrt{2}}\right|^2, |<-|E+>|^2 = \left|\frac{\sqrt{3}-1}{2\sqrt{2}}\right|^2$$

$$|<+|E->|^2 = \left|\frac{\sqrt{3}-1}{2\sqrt{2}}\right|^2, |<-|E->|^2 = \left|\frac{-\sqrt{3}-1}{2\sqrt{2}}\right|^2$$

As $P(A \ E \ B) = |<A|E>|^2|<B|E>|^2$ and since

$$|<+|E+>|^2|<-|E+>|^2 = \left|\frac{\sqrt{3}+1}{2\sqrt{2}}\right|^2\left|\frac{\sqrt{3}-1}{2\sqrt{2}}\right|^2 = \frac{1}{16} \text{ and}$$

$$|<+|E->|^2|<-|E->|^2 = \left|\frac{\sqrt{3}-1}{2\sqrt{2}}\right|^2 \left|\frac{-\sqrt{3}-1}{2\sqrt{2}}\right|^2 = \frac{1}{16}$$

Hence the probability of disagreement would be

$$P(\text{Eve uses any one basis})[P(--+) + P(-++) + P(+--) + P(++-)]$$

$$=\frac{1}{2}(\frac{1}{16} + \frac{1}{16} + \frac{1}{16} + \frac{1}{16}) = \frac{1}{8}.$$

11. *In B92 protocol Alice sends a string of qubits*

$$|0>, |+>, |0>, |0>, |0>, |+>, |+>|0>, |0>, |0>, |+>, |+>$$

and Bob measures them to be

$$|->, |+>, |0>, |0>, |1>, |+>, |->|+>, |0>, |->, |1>, |0>$$

Identify which qubits are to be kept for error checking and key generation.
Ans.
Only those qubits where Bob's outcome is either $|->$ or $|1>$ are to be kept for error checking and key generation.
Qubits obtained by Bob are

qubit No.	1	2	3	4	5	6	7	8	9	10	11	12
Qubit	$|->$	$|+>$	$|0>$	$|0>$	$|1>$	$|+>$	$|->$	$|+>$	$|0>$	$|->$	$|1>$	$|0>$
	✔				✔		✔			✔	✔	

Qubits for error checking and key generation are at qubit numbers 1,5,7,10,11.

12. *In B92 after error checking Bob's states are denoted by the run of qubits*

$$|->, |1>, |1>, |->, |->, |1>, |->|1>, |->, |->.$$

Identify the key that is being distributed.
Ans.
Bob's outcome $|->$ corresponds to Alice's input state $|0>$ with bit value 0 and $|1>$ corresponds to Alice's input state $|+>$ with bit value 1.
So the key is 0110010100.

IOP Publishing

Quantum Optics and Quantum Computation
An introduction
Dipankar Bhattacharyya and Jyotirmoy Guha

Chapter 13

Experimental aspects of quantum computing

Let us begin by stating that classical processors have to meet several sophisticated requirements. And it is evident that non-classical processing has to satisfy a much more stringent hardware requirement. The requirements for experimentally implementing quantum computing are collectively called DiVincenzo's criteria or constraints as follows. There are five primary criteria and two additional criteria needed for a viable computer.

1. There must be a scalable physical system with well-defined and perfectly characterized qubits. This means that when several qubits are put together the system should remain well behaved. A system may involve several kinds of qubits. Use of several types of qubits simultaneously is a good way of achieving a viable quantum computer.

2. The starting situation or starting environment is of paramount importance in every system and this is expected to be a simple state (ground state). Initialization of a quantum system to a given initial quantum state (ground state) is needed, say $|000....0>$. From here one can transform the system to a desired state.

3. A quantum computer is fragile against external disturbance called de-coherence. De-coherence is a very hard obstacle and should be overcome to get a quantum computer. Interaction of the system with the environment causes quantum state degradation and affects time available for quantum computation. Gate operations need to be executed before decay of state (i.e. before de-coherence time which may be ~µs). One has to ensure that the gate operation time is much shorter than the decoherence time, for instance a typical gate operation time would be ~ps. Then the system can perform $\frac{\text{decoherence time}}{\text{gate operation time}} \sim \frac{\mu s}{ps} = \frac{10^{-6}}{10^{-12}} = 10^6$ gate operations before the quantum state decays. (Actually 10^5 gates are needed to factorise 21 into 3 and 7 by using Shor's algorithm.) Clearly the main aim is to prolong or lengthen the de-

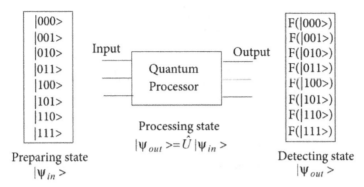

Figure 13.1. Plan of ideal quantum computer.

coherence, the methods of which are for instance closed loop control method, open loop control method etc.

4. A universal set of quantum gates should be implementable on the quantum computer (one qubit and two qubit gates).

5. Qubit specific measurement capability should be ensured. The state after execution of a quantum algorithm must be measured to extract the outcome of the computation. In the case of de-coherence, gate operation error and so on, the operation or computation has to be repeated to achieve reliability.

It is needed to send and store quantum information to construct quantum data processing work. This requires two more criteria to be ensured for the network needed. They are as follows.

6. The ability to interconvert stationary qubits and flying bits. This is needed because a working quantum computer may involve several kinds of qubits and distributed quantum computing might be necessary.

7. The ability of faithfully transmitting flying qubits between specified locations. This is necessary for quantum communication, quantum key distributuion and distributed quantum computing.

Physical implementations

There are a large number of possible physical systems that can be considered as potential candidates for a viable quantum computer. Some of them are as follows.

Linear optics, nuclear magnetic resonance, ion trap quantum computing, quantum dots etc. Figure 13.1 shows a plan of a quantum computer.

13.1 Basic principle of nuclear magnetic resonance quantum computing

A single molecule is a perfect example of an ideal physical system that can be used as a quantum computer where nuclear spins of individual atoms represent qubits. The quantum behavior of the spins can be exploited to perform quantum computation.

Take for instance the choloroform molecule (figure 13.2) in which carbon and hydrogen nuclei represent two qubits. Applying radio frequency pulse to the

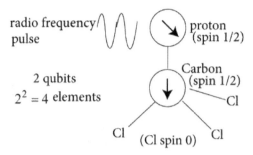

Figure 13.2. Chloroform molecule.

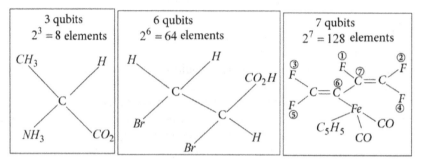

Figure 13.3. NMR quantum computing.

hydrogen nucleus, the corresponding qubit can be made to rotate from $|0>|0>=|00>$ state to a superposition state $\frac{|0>+|1>}{\sqrt{2}}|0>=\frac{|00>+|10>}{\sqrt{2}}$. Interactions through chemical bonds allow multiple-qubit logic to be performed.

If spin states of single molecules could be controlled and measured then the nuclear spins would be ideal qubits and the single molecules could be treated as ideal quantum computers. Nuclear magnetic resonance (NMR) makes this possible using large ensembles of molecules at room temperature, but at the expense of signal loss due to inefficient preparation procedure because it is difficult to ensure that a particular hydrogen/proton having a particular orientation w.r.t the carbon nucleus will interact with the applied field (i.e. r.f. pulse). With choloroform, two qubits have been achieved and with other molecules a higher number of qubits was also possible, figure 13.3.

One of the major drawbacks in NMR quantum computing lies in falling off of the available state polarizations with increase in the number of qubits and solutions to NMR quantum computing is not scalable to useful number of spins. Quantum noise is a great problem as it overwhelms the quantum signal for moderate number of qubits at room temperature. NMR is a slow process and resetting the qubit takes a lot of time, and frequency at which one can do computing is quite slow.

NMR is the interaction of nuclear spins put in a magnetic field with electro-magnetic radiation in the radio frequency region. It is a frequently used spectro-scopic technique. Nuclear spin along the field is aligned and denoted as $|\uparrow>=|0>$

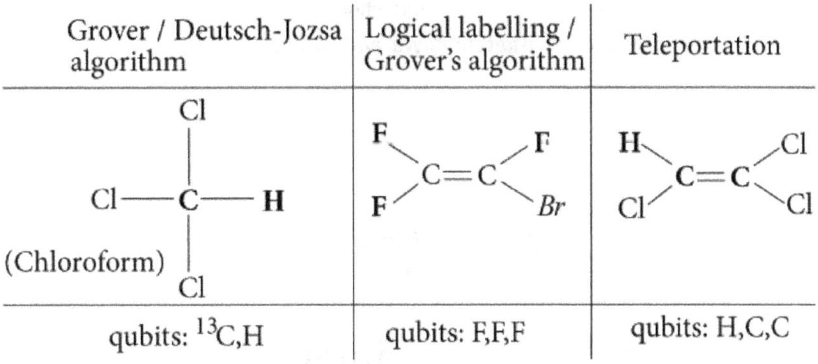

Grover / Deutsch-Jozsa algorithm	Logical labelling / Grover's algorithm	Teleportation
Cl | Cl———C———H (Chloroform) | Cl	F⟍ ⟋F C=C F⟋ ⟍Br	H⟍ ⟋Cl C=C Cl⟋ ⟍Cl
qubits: ^{13}C,H	qubits: F,F,F	qubits: H,C,C

Figure 13.4. Some molecules used in NMR quantum computer and the operation they perform.

and those against are anti-aligned and denoted as $| \downarrow > = |1>$. Under the externally applied static magnetic field a spin will precess about the field direction. Such precessional movement will generate an oscillating field that can be measured and give information regarding state of the spin. Upon application of magnetic field oscillating with radio frequency the spin will gradually move from one state (up/down) to the other (down/up).

The nucleus of carbon-12 has no spin and chloroform used contains carbon-13. Carbon is oriented so that it definitely points up, parallel to the magnetic field. The spin of the hydrogen/proton can be parallel or anti-parallel to a vertically applied magnetic field. The radio frequency pulse can rotate the carbon's spin downward into the horizontal plane.

The parallel spin has lower energy than the anti-parallel spin, by an amount that depends on the strength of the externally applied magnetic field. In the normal state opposing spins are present in equal numbers in a liquid state NMR. The applied field favours creation of parallel spins and so imbalance between the two states generates and the excess is measured.

A basic limitation of chloroform computer is its small number of qubits as this depends on the number of atoms in the molecule employed. Also, in NMR single spins are too weak to measure (tendency to align with field is weak) and one must consider ensembles.

Some molecules used in an NMR quantum computer and the operation they perform are shown in figure 13.4.

13.2 Further reading

[1] Nielsen M A and Chuang I L 2019 *Quantum Computation and Quantum Information* (Cambridge: Cambridge University Press)

IOP Publishing

Quantum Optics and Quantum Computation
An introduction
Dipankar Bhattacharyya and Jyotirmoy Guha

Chapter 14

Light–matter interactions

In this chapter, we will discuss light–matter interaction. The main essence of this topic is how the electromagnetic (e-m) radiation field interacts with an atom or molecule whose transition frequency is resonant with the e-m field frequency. In general, there are three approaches by which we can describe this phenomenon, classical approach, semi-classical and quantum mechanical approach. In classical formulation, the atom is treated as Hertzian dipoles and light as classical waves by using Maxwell's equations; in the semi-classical description, an atom is considered quantum mechanically (quantum two-level system for simplicity) and field is still a classical wave. Light and atom are both treated quantum mechanically in the full quantum mechanical formulation. Here we are going to discuss the semi-classical treatment and in chapters 16 and 17, we will discuss full quantum mechanical theory. It is always useful to start with a simple atomic system and a two-level atomic system is the simplest form of the atom. Here we assume the atom has two energy levels and the e. m. field is in resonance with the transition frequency of the atom. Other levels (in the case of multi-level systems) are far detuned with respect to the field. The main advantage of taking a sample of the two-level atom is to derive a complete analytical expression for the atomic medium polarization or susceptibility (χ). In the case of a multilevel atomic system, it is not possible to calculate the analytic expression of the medium susceptibility in most cases.

The two-level approximation is based on the resonance phenomenon. According to the classical picture during light–atom interaction the atomic dipole is induced in the atom that oscillates and re-radiates energy which is the same as that of the incident frequency. There would be a resonant oscillation of the atomic dipole if the light frequency matches the natural frequency of the atom and the light will be strong. But at off-resonance, i.e. for a large mismatch of the light frequency and the natural frequency of atom, interaction is small and the magnitude of the driven oscillation will be small. In other words, the light–atom interaction is stronger for

doi:10.1088/978-0-7503-2715-2ch14

the case of resonant transition as compared to off-resonant transitions. Hence off-resonant transitions can be ignored as an approximation.

The spin $\frac{1}{2}$ magnetic dipole is a true two-level system in nuclear magnetic resonance having simple decay mechanisms. This is a true two-level system with relatively simple decay mechanisms. However, its response can differ significantly depending on level decay rates.

14.1 Interaction Hamiltonian

The interaction of an atom (for simplicity we assume it is a one-electron system) with an external e.m. field can be represented by the following Hamiltonian

$$\mathbf{H} = \frac{1}{2m}[\vec{p} - e\vec{A}(\vec{r}, t)]^2 + e\phi(\vec{r}, t) + V(r) \tag{14.1}$$

Here $\vec{A}(\vec{r}, t)$ and $\phi(\vec{r}, t)$ are the vector and scalar potentials of the external field, c is the speed of the light in free space, $\hat{p}$ is the momentum of the electron and is defined quantum-mechanically as $\hat{p} = -i\hbar\vec{\nabla}$, $\hbar$ is the reduced Planck's constant (Planck's constant divided by 2π), $\hbar = 1.054 \times 10^{-34}$ joule-seconds. e is the charge of the electron and $e > 0$, $V(r)$ is the central potential experienced by an electron due to the static nucleus. The motion of the electron can be described by the time-dependent Schrödinger equation ($H\psi(\vec{r}, t) = i\hbar\frac{\partial\psi(\vec{r}, t)}{\partial t}$). Here the Hamiltonian takes the form of equation (14.1).

$$i\hbar\frac{\partial\psi(\vec{r}, t)}{\partial t} = \left[\frac{1}{2m}[\vec{p} - e\vec{A}(\vec{r}, t)]^2 + V(r) + e\phi(\vec{r}, t)\right]\psi(\vec{r}, t) \tag{14.2}$$

The functions $\vec{A}(\vec{r}, t)$ and $\phi(\vec{r}, t)$ are the gauge-dependent potentials and they can be represented by the following transformations

$$\vec{A}(\vec{r}, t) \rightarrow \vec{A}' = \vec{A} + \vec{\nabla}\zeta(\vec{r}, t) \quad \text{and}$$

$$\phi(\vec{r}, t) \rightarrow \phi' = \phi - \frac{\partial\varsigma(\vec{r}, t)}{\partial t}, \tag{14.3}$$

here $\varsigma(\vec{r}, t)$ is an arbitrary scalar function. The gauge-independent quantities are the electric ($\vec{E}$) and magnetic $\vec{B}$ fields. They can be written in terms of $\vec{A}(\vec{r}, t)$ and $\phi(\vec{r}, t)$ as $\vec{E} = -\vec{\nabla}\phi(\vec{r}, t) - \frac{\partial\vec{A}(\vec{r}, t)}{\partial t}$ and $\vec{B} = \vec{\nabla} \times \vec{A}(\vec{r}, t)$.

By using the transformations equation (14.3) in equation (14.2) the time-dependent Schrödinger equation becomes,

$$i\hbar\frac{\partial\psi(\vec{r}, t)}{\partial t} = \left[\frac{1}{2m}[\vec{p} - e\vec{A}(\vec{r}, t) - e\vec{\nabla}\zeta(\vec{r}, t)]^2 + V(r) + e\phi(\vec{r}, t)\right.$$
$$\left. - e\frac{\partial\varsigma(\vec{r}, t)}{\partial t}\right]\psi(\vec{r}, t) \tag{14.4}$$

To solve equation (14.4) we apply the unitary transformation in the wave function $(\psi(\vec{r},\, t))$ with a unitary operator $\hat{R}$ such that

$$\phi(\vec{r},\, t) = \hat{R}\psi(\vec{r},\, t) \tag{14.5}$$

Suppose the unitary operator is $\hat{R} = \exp(\frac{ie}{\hbar}\zeta(\vec{r},\, t))$. By using this new wave function in the time-dependent Schrödinger equation we get,

$$i\hbar\frac{\partial\phi(\vec{r},\, t)}{\partial t} = \exp\left(\frac{ie}{\hbar}\zeta(\vec{r},\, t)\right)\left[i\hbar\frac{\partial\psi(\vec{r},\, t)}{\partial t} - e\frac{\partial\zeta(\vec{r},\, t)}{\partial t}\psi(\vec{r},\, t)\right]$$

$$=\exp\left(\frac{ie}{\hbar}\zeta(\vec{r},\, t)\right)\left[H\psi(\vec{r},\, t) - e\frac{\partial\zeta(\vec{r},\, t)}{\partial t}\psi(\vec{r},\, t)\right]$$

$$=\{\frac{1}{2m}[\vec{p} - e\vec{A}(\vec{r},\, t) - e\vec{\nabla}\zeta(\vec{r},\, t)]^2 + V(r) + e\phi(\vec{r},\, t)\}\phi(\vec{r},\, t) - e\frac{\partial\zeta(\vec{r},\, t)}{\partial t}\phi(\vec{r},\, t)$$

$$i\hbar\frac{\partial\phi(\vec{r},\, t)}{\partial t} = H'\phi(\vec{r},\, t) \tag{14.6}$$

Here the new Hamiltonian $H' = \{\frac{1}{2m}[\vec{p} - e\vec{A}(\vec{r},\, t) - e\vec{\nabla}\zeta(\vec{r},\, t)]^2 + V(r) + e\phi(\vec{r},\, t) - e\frac{\partial\zeta(\vec{r},\, t)}{\partial t}\}$.

By choosing a suitable gauge we can simplify the problem and in this case, we are working with a radiation gauge in which $\vec{\nabla}.\,\vec{A}(\vec{r},\, t) = 0$ and $\phi(\vec{r},\, t) = 0$. The vector potential $(\vec{A}(\vec{r},\, t))$ satisfies the wave equation $\nabla^2\vec{A} - \frac{1}{c^2}\frac{\partial^2\vec{A}}{\partial t^2} = 0$, if no sources of radiation are present near the atom. The solution of the wave equation is in the following form: $\vec{A}(\vec{r},\, t) = \vec{A}(\vec{R},\, t)e^{i(\vec{k}.\vec{r}-\omega t)} + c.\,c$, here $\vec{\kappa} = \frac{2\pi}{\lambda}$, wave vector of the radiation field and c.c. is indicating the complex conjugate. $\vec{R}$ is the position of the atomic centre of mass. If the wavelength (λ) of the radiation field is much larger than the atomic dimension $(|\,\vec{r}\,|)$, then we can easily ignore the spatial variation of the field. Normally, atomic dimension is in the order of few Angstroms and λ the radiation field is of the order of a few hundred nanometers (400–700 nm), so $\lambda \gg |\,\vec{r}\,|$. So we can assume that the entire atom is submerged in the radiation field. This approximation is known as *dipole approximation* and it is often used in quantum optics. Under this approximation $(\vec{k}.\,\vec{r} \ll 1)$ we can write the vector potential as $\vec{A}(\vec{r},\, t) = \vec{A}(\vec{R},\, t)$, i.e. it is spatially uniform.

If we take the gauge function $\zeta(\vec{r},\, t) = -\vec{A}(\vec{R},\, t).\,\vec{r}$, we get the following equations

$$\vec{\nabla}\zeta(\vec{r},\, t) + \vec{A}(\vec{R},\, t) = 0$$

$$e\frac{\partial\zeta(\vec{r},\, t)}{\partial t} = -e\vec{r}.\,\frac{\partial\vec{A}(\vec{R},\, t)}{\partial t}$$

$$=e\vec{r}.\ \vec{E}(\vec{R},\ t),\ \because\ \frac{\partial\vec{A}(\vec{R},\ t)}{\partial t} = -\vec{E}(\vec{R},\ t) \tag{14.7}$$

The first equation of equation (14.7) says that the vector potential vanishes and the second equation (14.7) tells us about the appearance of a new interaction potential due to the interaction of electric field and dipole moment of the atom ($\vec{\mu} = e\vec{r}$). Using the outcomes of equation (14.7) in the expression of **H′**, we will get the simplified expression of total Hamiltonian.

$$\mathbf{H'} = -\frac{\hbar^2}{2m}\nabla^2 + V(r) - e\vec{r}.\ \vec{E}(\vec{R},\ t) = H_0 + H_I \tag{14.8}$$

In equation (14.8) H_0 is the unperturbed Hamiltonian which describes the isolated (static) system and H_I is the electromagnetic interaction Hamiltonian. Expressions of the H_0 and H_I are given below,

$$H_0 = -\frac{\hbar^2}{2m}\nabla^2 + V(r)$$

$$H_I = -e\vec{r}.\ \vec{E}(\vec{R},\ t) = -\vec{\mu}.\ \vec{E}(\vec{R},\ t) \tag{14.9}$$

The unitary transformation applied in equation (14.5) is the simplest form of a gauge-invariant theory for the electromagnetic field. It is also known as U(1) gauge theory since the transformation of the state ($\phi(\vec{r},\ t)$) corresponds to multiplication with a Unitary 1×1 matrix (= complex number of modulus 1). A special unitary SU (2) gauge theory is applied by Steven Weinberg and Abdus Salam for the unification of weak and electromagnetic interaction.

14.2 Rabi oscillations

In figure 14.1 we have shown a two-level atomic system, the lower level can be specified as |a> and excited level as |b> with respective energies $E_a = \hbar\omega_a$ and $E_b = \hbar\omega_b$. The transition frequency of this two-level system is $\omega_0 = \frac{(E_a - E_b)}{\hbar}$, an electromagnetic wave $E(t) = E_0 \cos \nu t$ interacts with this two-level system. The frequency of the field is very close to the transition frequency of the two-level atom (i.e. $\nu \approx \omega_0$). Under *dipole approximation*, the atom interaction Hamiltonian of this can be written from equation (14.9).

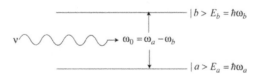

Figure 14.1. Electromagnetic wave of frequency ω interacting with a two-level atom having a resonant frequency ω_0.

$$H_I = -\mu_{ab}E_0\cos\nu t = -\frac{\mu_{ab}E_0}{2}(e^{i\nu t} + e^{-i\nu t}) \tag{14.10}$$

where $\mu_{ab} = \mu_{ba}^* = e\langle a|r|b\rangle$ is the matrix element of the electric dipole moment. If H_0 is the unperturbed Hamiltonian of the atom, the total Hamiltonian (H) of the system can be written as,

$$H = H_0 + H_I = \begin{pmatrix} H_{aa} & H_{ab} \\ H_{ba} & H_{bb} \end{pmatrix} \tag{14.11}$$

Here, $H_0 = \begin{pmatrix} \hbar\omega_a & 0 \\ 0 & \hbar\omega_b \end{pmatrix}$, and $H_I = \begin{pmatrix} 0 & -\frac{\mu_{ab}E_0}{2}(e^{i\nu t} + e^{-i\nu t}) \\ -\frac{\mu_{ab}E_0}{2}(e^{-i\nu t} + e^{i\nu t}) & 0 \end{pmatrix}$.

The complete wave function of the time-dependent Schrödinger equation is $\psi(\vec{r}, t) = \sum_i C_i(t)e^{-i\omega_i t}U_i(r)$ where 'i' runs for all energy states of the system. The time-dependent wave function of the two-level atom can be reduced to,

$$\psi(\vec{r}, t) = C_a(t)e^{-i\omega_a t}U_a(r) + C_b(t)e^{-i\omega_b t}U_b(r) \tag{14.12}$$

where C_a and C_b are the wave function amplitude coefficients for the ground and excited states of the two-level atom. Later we will see that these coefficients are proportional to the population of the ground and excited states, i.e. $|C_a|^2 = N_a$ (no. of the particles in the ground state) and $|C_b|^2 = N_b$ (no. of the particles in the excited state). Now our aim is to solve the time-dependent Schrödinger equation $(H\psi(\vec{r}, t) = i\hbar\frac{\partial\psi(\vec{r}, t)}{\partial t})$ for a two-level atom interacting with a resonant electromagnetic field. We have seen in equation (14.11) the total Hamiltonian consists of two parts unperturbed and perturbed Hamiltonian. The solution of the unperturbed Hamiltonian (H_0) is $H_0U_i(r) = E_iU_i(r) = \hbar\omega_iU_i(r)$, $\{i = a, b\}$. By putting the wave function given in equation (14.11) in the time dependent Schrödinger equation with the total Hamiltonian (equation (14.12)).

$$(H_0 + H_I)(C_a\ e^{-i\omega_a t}U_a + C_b\ e^{-i\omega_b t}U_b) = i\hbar((-i\omega_aC_a + \dot{C_a})e^{-i\omega_a t}U_a$$
$$+ (-i\omega_bC_b + \dot{C_b})e^{-i\omega_b t}U_b) \tag{14.13}$$

By using the solution of the unperturbed Hamiltonian in equation (14.13) we get

$$(C_a\ H_I\ U_ae^{-i\omega_a t} + C_b\ H_I\ U_be^{-i\omega_b t}) = i\hbar(\dot{C_a}\ U_ae^{-i\omega_a t} + \dot{C_b}\ U_be^{-i\omega_b t}) \tag{14.14}$$

Multiplying by $U_a{}^*$ both sides of the equation (14.14) and integrating overall space we find that:

$$\dot{C_a} = -\frac{i}{\hbar}(C_a\ \int U_a^*\ H_{aa}U_ad^3r + C_be^{-i\omega_0 t}\ \int U_a^*\ H_{ab}U_bd^3r) \tag{14.15}$$

Similarly, by multiplying by U_b^* both sides of the equation (14.14) and integrating we will get,

$$\dot{C}_b = -\frac{i}{\hbar}(C_a e^{i\omega_0 t} \int U_{ab}^* \ H_{ba}U_a d^3r + C_b \int U_b^* \ H_{bb}U_b d^3r) \qquad (14.16)$$

Here we used orthogonal relation of the eigenfunctions, which is $\int U_i^*(r)U_j(r)d^3r = \delta_{ij}$, δ_{ij} is the Kronecker delta function. Due to parity conservation, the first term of the left side of the equations (14.15) and (14.16) is zero, i.e. $\mu_{aa} = \mu_{bb} = 0$. By putting the value of the H_{ab} and H_{ba} from equation (14.11) in (14.15) and (14.16) and performing a little algebra we get the following,

$$\dot{C}_a = i(\frac{\mu_{ab}E_0}{2\hbar})[e^{i(\nu-\omega_0)t} + e^{i(\nu+\omega_0)t}]C_b \qquad (14.17a)$$

$$\dot{C}_b = i(\frac{\mu_{ab}E_0}{2\hbar})[e^{-i(\nu-\omega_0)t} + e^{-i(\nu+\omega_0)t}]C_a \qquad (14.17b)$$

Here we take $\mu_{ab} = \mu_{ba}$, because $\mu_{ab} = \mu_{ba}^*$ the dipole moment is a measurable quantity so it must be real. The above equations are the rate equations for C-coefficients, we will solve them in weak and strong field cases. The exact solution is possible for strong field cases and in weak field approximation the method is applicable. Now we can introduce a parameter known as *Rabi flopping frequency* we define as $\Omega_R = (\frac{\mu_{ab}E_0}{\hbar})$, dimension of the Ω_R is frequency. Professor I I Rabi in 1939 first studied this effect in a similar two-level system, namely spin $\frac{1}{2}$ magnetic dipole in nuclear magnetic resonance (NMR). Rabi frequency has nothing to do with the oscillation frequency of the atom or the frequency of the applied electric field, later we will see this is a frequency in which an atom can oscillate between two energy levels under the action of e.m. field.

14.3 Weak field case

In this case, a weak field is interacting with the atom and since the field is weak so the interaction energy (H_I) is small. We can consider the effect of this weak interaction energy as a perturbation on the atom. In the interaction of visible light with atoms, mostly all atoms are present at the ground level and few atoms are in the excited state. When a weak field interacting with the atoms cannot transfer a large number of atoms in the excited state, most of the atoms remain at the ground level. So here we can consider all the atoms are at ground level.

$$C_a(0) = 1, \text{ and } C_b(0) = 0 \qquad (14.18)$$

We assume that $C_a(t)$ remains the same all the time as it was in $t = 0$, since the population change caused by the field is small. Though the value of $C_b(t)$ will be changed, so we can write.

$$\dot{C}_a(t) = 0 \qquad (14.19)$$

$$\dot{C}_b(t) = i\left(\frac{\mu_{ab}E_0}{2\hbar}\right)[e^{-i(\nu-\omega_0)t} + e^{-i(\nu+\omega_0)t}]$$

After integrating the above equation, we get the value of $C_b(t)$.

$$C_b(t) = i\left(\frac{\mu_{ab}E_0}{2\hbar}\right)\left[\frac{e^{-i(\nu-\omega_0)t}}{-i(\nu-\omega_0)} + \frac{e^{-i(\nu+\omega_0)t}}{-i(\nu+\omega_0)}\right] \tag{14.20}$$

In the on-resonance case, absorption occurs, i.e. $\nu \cong \omega_0$, hence $(\nu - \omega_0)$ is very small, whereas $(\nu + \omega_0) \approx 2\omega_0$, so in equation (14.20) the first term is present and the second term can be neglected. This is the same as neglecting $e^{i\nu t}$ term of $\cos\nu t$ in the electric field.

In the interaction energy (equation (14.10)) we can neglect the term($e^{i\nu t}$) since we are working in optical frequencies ($\nu \sim 10^{15}$ radians/sec) and field frequency is a near resonance to transition frequency of the atom. So the sum frequency ($\omega_0 + \nu$) term is safely omitted in comparison to the difference frequency ($\nu - \omega_0$) term. Neglecting the energy nonconserving terms and retaining the energy-conserving terms is known as *rotating wave approximation* (RWA). After dropping the second term in equation (14.20) that leads to,

$$\mid C_b(t) \mid^2 = \left(\frac{\mu_{ab}E_0}{2\hbar}\right)^2 \frac{\sin^2(\nu-\omega_0)t/2}{[(\nu-\omega_0)t/2]^2} \tag{14.21}$$

The above equation represents the probability of absorption in a two-level atom, it will give the maxima at resonance when $\nu = \omega_0$. In figure 14.2, we have plotted the absorption probability with detuning ($\nu - \omega_0$). Central maxima occur at zero detuning and other small maxima are present at different time t.

14.4 Strong field case: Rabi oscillations

In the case of a strong field the electric field amplitude E_0 is high so the interaction energy is not weak. Here we cannot treat the interaction as a perturbation, we will follow Rabi's method of the exact solution by using RWA dropping the second term containing ($\nu + \omega_0$) in equation (14.17a) and (14.17b). Now differentiating equation (14.17b) ($\dot{C}_b(t)$) of and then using the value of $\dot{C}_a$ we get,

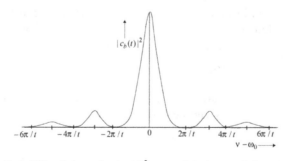

Figure 14.2. Probability of absorption $\mid c_b(t)\mid^2$ versus detuning graph for a two-level atom.

$$\ddot{C}_b + i(\nu - \omega_0) \ \dot{C}_b + \frac{\Omega_R^2}{4}C_b = 0 \tag{14.22}$$

Equation (14.18) has solutions of the form $C_b(t) = Ae^{i\xi t}$. Substituting this solution in equation (14.22) we find the possible values of ξ

$$\xi_\pm = \frac{1}{2}\left\{ \delta \pm (\delta^2 + \Omega_R^2)^{\frac{1}{2}} \right\} \tag{14.23}$$

In equation (14.23), $\delta = (\nu - \omega_0)$ is the detuning parameter if $\delta = 0$. Then the field is exactly on-resonance with the transition frequency of the atom (ω_0). When $\delta > 0$ field frequency is known as blue detuned compared to the transition frequency ω_0 and if $\delta < 0$ field frequency is red detuned. Now with the values of ξ the general solution of the equation (14.22) can be written as,

$$C_b(t) = A_+e^{i\xi t} + A_-e^{-i\xi t} \tag{14.24}$$

To find the value of A_+ and A_- we can apply the initial condition $C_a(0) = 1$ and $C_b(0) = 0$ all populations are the in-ground state when no field is applied. With this initial condition, we can find the values of $A_\pm$,

$$A_\pm = \pm\frac{\Omega_R}{2}[(\delta^2 + \Omega_R^2)^{-\frac{1}{2}}] \tag{14.25}$$

Final solution of the equation (14.17a and 14.17b) is

$$C_b(t) = i\frac{\Omega_R}{(\delta^2 + \Omega_R^2)^{\frac{1}{2}}}e^{\frac{i\delta t}{2}}\sin(\frac{(\delta^2 + \Omega_R^2)^{\frac{1}{2}}}{2}t) \tag{14.26a}$$

$$C_a(t) = e^{\frac{i\delta t}{2}}[\cos\left(\frac{(\delta^2 + \Omega_R^2)^{\frac{1}{2}}}{2}t\right) - i\frac{\delta}{(\delta^2 + \Omega_R^2)^{\frac{1}{2}}}\sin\left(\frac{(\delta^2 + \Omega_R^2)^{\frac{1}{2}}}{2}t\right)] \tag{14.26b}$$

The difference between the eigenfrequencies of equation (14.23) is $\xi_+ - \xi_- = (\delta^2 + \Omega_R^2)^{\frac{1}{2}} = \Omega_G$ defined as the *generalized Rabi flopping frequency*. At on-resonance (i.e. $\delta = 0$) the generalized Rabi frequency (Ω_G) is equal to the Rabi frequency (Ω_R). The product of the eigenfrequencies is proportional to the square of the on-resonant Rabi frequency $\xi_+\xi_- = -(\frac{\Omega_R}{2})^2$.

The probabilities of the atom in the excited state ($|b>$) and ground state ($|a>$) can be found from equations (14.26a and 14.26b),

$$| C_b(t) |^2 = \frac{\Omega_R^2}{\Omega_G^2}sin^2(\frac{\Omega_G t}{2}) \tag{14.27a}$$

$$| C_a(t) |^2 = cos^2\left(\frac{\Omega_G t}{2}\right) + \frac{\delta^2}{\Omega_G^2}sin^2(\frac{\Omega_G t}{2}) \tag{14.27b}$$

For the on-resonance ($\delta = 0$) case, we have from equation (14.27a, 14.27b)

$$| C_b(t) |^2 = \sin^2\left(\frac{\Omega_R t}{2}\right) = (1 - \cos\Omega_R t)/2 \qquad (14.28a)$$

$$| C_a(t) |^2 = \cos^2\left(\frac{\Omega_R t}{2}\right) = (1 + \sin\Omega_R t)/2 \qquad (14.28b)$$

The dynamics of these probabilities are plotted in figure 14.3, it is seen from the figure that at $t = (\frac{\pi}{\Omega_R})$ atom is in the excited state ($|b>$). When $t = (\frac{2\pi}{\Omega_R})$ atom returns to the ground state ($|a>$) so as a result, the atom oscillates between the ground and excited states sinusoidally with an oscillation frequency ($\frac{\Omega_R}{2\pi}$). This oscillatory behaviour of the atom is known as *Rabi oscillation or Rabi flopping*, this phenomenon is observed in the strong e-m field case. A $\pi-$ pulse is a pulse of $\frac{\pi}{\Omega_R}$ the duration that takes the atom from the lower state to the upper state and then the system does not revert back if there is no spontaneous emission. This pulse changes the phase of the system by π. If a continuous-wave laser is present, the system will flip back and forth with Rabi-frequency.

From equation (14.27a,b) we can see at off-resonance condition (i.e. $\delta \neq 0$) the Rabi oscillation increases and the atom oscillates with a frequency ($\Omega_G = (\delta^2 + \Omega_R^2)^{\frac{1}{2}}$) though the amplitude of the oscillation decreases.

If field frequency is far-resonance ($\delta > 0$) then from equation (14.27a) we get,

$$| C_b(t) |^2 = \frac{\Omega_R^2}{\delta^2} \sin^2\left(\frac{\delta t}{2}\right) \qquad (14.29)$$

14.5 Damping phenomena

In the discussion of atom–field interaction, we do not yet consider the decay or damping process. Though in the real situation, decay or damping of atoms from excited to ground state always occurs. There are mainly two reasons due to which atoms can decay from an excited state to a ground state, the first reason is spontaneous decay and the second is due to collisions among the atoms and other dephasing phenomena. We are aware of the spontaneous decay which is related to

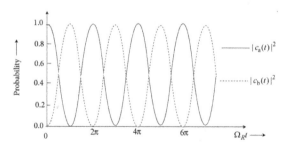

Figure 14.3. Occupancy probability of the level $|a>$ and $|b>$ with time at on-resonance ($\delta = 0$) condition.

the Einstein A coefficient if an atom spends T_1 time in the excited state then relax to the ground state. This damping time of population decay is known as the *longitudinal relaxation* rate. The time which is connected to the dephasing damping process can be represented as T_2 the *transverse relaxation* rate.

In figure 14.4, we have added such relaxation rates to the excited and ground level, γ_a is the decay rate of the ground level ($|a>$) and γ_b is the decay rate of the excited level ($|b>$). Here these relaxation rates are spontaneous decay rates of the level $|a>$ and $|b>$ and it is the inverse of T_1. We are dealing with semi-classical theory so we can add the decay rates (γ_a, γ_b) phenomenologically in equation (14.17a and 14. b) after dropping the ($\nu + \omega_0$) term.

$$\dot{C}_a(t) = -\frac{1}{2}\gamma_a C_a(t) + i\left(\frac{\Omega_R}{2}\right)e^{i(\nu-\omega_0)t}C_b(t) \tag{14.30}$$

$$\dot{C}_b(t) = -\frac{1}{2}\gamma_b C_b(t) + i\left(\frac{\Omega_R}{2}\right)e^{-i(\nu-\omega_0)t}C_a(t) \tag{14.31}$$

The $\frac{1}{2}$ terms present in the decay rates of both equations (14.30) and (14.31) so that the probability $|C_b|^2$ decays exponentially as exp($-\omega_b t$) when no electric field is present ($E_0 = 0$). If we assume the level $|a>$ is the true ground level of the two-level system then $\gamma_a = -\gamma_b$, so that the total occupation probability of the system is conserved ($|C_a|^2 + |C_b|^2 = 1$). By using the initial conditions $C_a(0) = 1$ and $C_b(0) = 0$, we will get from equation (14.30).

$$C_a(t) = C_a(0)\exp\left(-\frac{\gamma_a t}{2}\right) = \exp\left(-\frac{\gamma_a t}{2}\right) \tag{14.32}$$

Putting the value of $C_a(t)$ from equation (14.32) into equation (14.31) and solving the equation by integrating factor method, we have

$$\frac{d}{dt}\left(C_b(t)\exp\left(-\frac{\gamma_b t}{2}\right)\right) = i\frac{\Omega_R}{2}\exp[-i(\nu-\omega_0)t]\exp[-(\frac{\gamma_a+\gamma_b}{2})t] \tag{14.33}$$

The perturbative solution of the equation (14.33) is

$$C_b^1(t) = \exp\left[-\left(\frac{\gamma_b}{2}\right)t\right]\left(i\frac{\Omega_R}{2}\right)\left[\frac{\exp(-i(\nu-\omega_0)t+\left(\frac{\gamma_b-\gamma_a}{2}\right)t]-1}{-i(\nu-\omega_0)+\left(\frac{\gamma_b-\gamma_a}{2}\right)}\right] \tag{14.34}$$

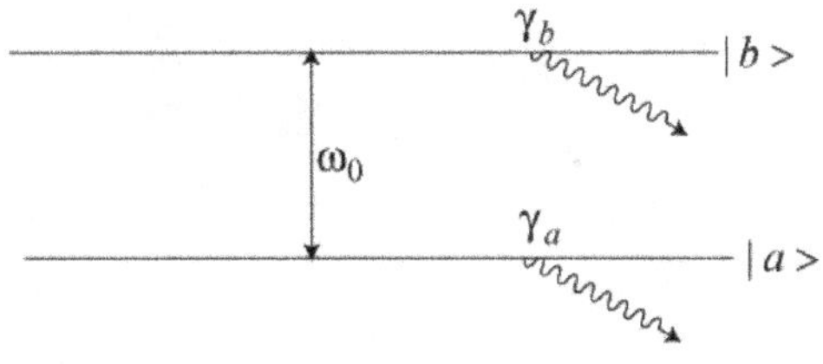

Figure 14.4. Electromagnetic wave of frequency ν interacting with a two level atom with decay parameters γ_a and γ_b.

Here $C_b^1(t)$ is the first order perturbative solution where we have written up to the first order of Ω_R. The probability of the atom in an excited state ($|b>$) is mainly $|C_b|^2$, given by

$$|C_b(t)|^2 \simeq |C_b^1(t)|^2 = \left(\frac{\Omega_R}{2}\right)^2 \left[\frac{\exp(-\gamma_a)t - 2\exp\left\{-\frac{(\gamma_a + \gamma_b)t}{2}\right\}\cos\delta t + \exp(-\gamma_b)t}{\delta^2 + \left\{\frac{(\gamma_b - \gamma_a)}{2}\right\}^2}\right] \quad (14.35)$$

In equation (14.34), δ is the detuning parameter as defined earlier. It is easily shown that this formula reduces to the undamped case (equation (14.29)) if we put $\gamma_a = \gamma_b = 0$ in equation (14.35).

For simplicity, if we take $\gamma_a = \gamma_b = \gamma$ then equation (14.35) reduces to

$$|C_b(t)|^2 = \left(\frac{\Omega_R}{2}\right)^2 \exp(-\gamma t) \left(\frac{\sin\left(\frac{\delta t}{2}\right)}{\frac{\delta}{2}}\right)^2 \quad (14.36)$$

The plot of $|C_b(t)|^2$ versus Ω_R is shown in figure 14.5 from equation (14.36) for different values of the damping constant (γ). If $\gamma = 0$ it gives the Rabi-oscillation without damping case (same as in figure 14.2), for higher values of γ, Rabi-oscillation is damped due to $\exp(-\gamma t)$. At a very large damping rate ($\gamma \gg \Omega_R$) no oscillation is observed.

14.6 The density matrix

As we discussed in the atom–field interaction semi-classical section, the wavefunction for the two-level system in the time-dependent Schrödinger equation can be written as,

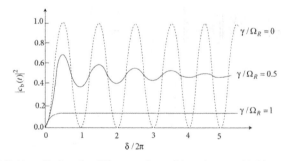

Figure 14.5. Damped Rabi oscillations for different values of damping rate (γ) (a) $\gamma = 0$ (b) $\gamma = 0.5$ (c) $\gamma = 1$ (Ω_R = constant).

$$\psi(\vec{r}, t) = C_a\ (t)U_a(r) + C_b\ (t)U_b(r) \tag{14.37}$$

where $C_i(t)$ $(i = a, b)$ is the wavefunction amplitude in i-th level. The equivalent state vector of the system is,

$$|\ \psi(t) > = C_a\ (t)\ |\ a > + C_b\ (t)\ |\ b> \tag{14.38}$$

Now we introduce the density matrix operator as,

$$\rho = |\ \psi(t) > <\psi(t)| \tag{14.39}$$

which can be written as an $n{\times}n$ matrix (n is the number of wave functions that describe the system); for a two-level system it is a $2{\times}2$ matrix. The density matrix elements are defined by:

$$\rho_{ij} = \ <U_i\ |\ \rho\ |\ U_j > = <U_i\ |\ \psi(t) > <\psi(t)\ |\ U_j > = C_iC_j^* \tag{14.40}$$

The density matrix elements can be related to observable quantities, $|C_i|^2$ is the probability of finding atoms in the state $|i>$, which lies between 0 to 1 if the system is closed. The off-diagonal matrix elements $C_iC_j^*$ represents coherence since they depend on the phase difference between C_i and C_j. In the matrix notation, the density matrix operator for a two-level system can be written as,

$$\rho = \begin{pmatrix} C_aC_a^* & C_aC_b^* \\ C_bC_a^* & C_bC_b^* \end{pmatrix} = \begin{pmatrix} |\ C_a\ |^2 & C_aC_b^* \\ C_bC_a^* & |\ C_b\ |^2 \end{pmatrix} = \begin{pmatrix} \rho_{aa} & \rho_{ab} \\ \rho_{ba} & \rho_{bb} \end{pmatrix} \tag{14.41}$$

The off-diagonal terms (ρ_{ab} and ρ_{ba}) are present if the atoms are coherent superposition state and in the case of statistical mixtures, these terms are zero. In a statistical mixture, atoms are either in the upper state ($|b>$) or in the lower state ($|a>$) so $|C_a| = 1$ and $|C_b| = 0$ or the other way round. Conversely, if the atom is in a superposition state both $|C_a|$ and $|C_b|$ exist in the atomic wavefunctions. So, the off-diagonal terms are present, equation (14.41) is the density matrix operator for the superposition state where all four elements are non-zero.

In terms of density matrix operator (equation (14.39)), the expectation value of an operator $\hat{\Theta}$ can be written in the following way,

$$\langle\hat{\Theta}\rangle = \ <\sum_i C_iU_i\ |\ \hat{\Theta}\ |\ \sum_j C_jU_j > = \sum_{i,j} C_i^*C_j < U_i\ |\ \hat{\Theta}\ |\ U_j>$$

$$= \sum_{i,j}\rho_{ji}\hat{\Theta}_{ij} = \sum_j (\rho\hat{\Theta})_{jj} = Tr(\rho\hat{\Theta}) \tag{14.42}$$

The $Tr(\rho\hat{\Theta})$ is the trace of the product of ρ and operator $\hat{\Theta}$. In the case of a two-level system, equation (2.42) can be written as $\langle\hat{\Theta}\rangle = (\rho_{aa}\hat{\Theta}_{aa} + \rho_{ab}\hat{\Theta}_{ab} + \rho_{bb}\hat{\Theta}_{bb} + \rho_{ba}\hat{\Theta}_{ba})$. The density operator ρ defines an $(n \times n)$ matrix in which the diagonal elements are real and the off-diagonal elements are complex we have total n^2 elements in the ρ matrix. We can remove one of these equations by using the fact that $Tr(\rho\hat{\Theta}) = \sum_{i=1}^{n}\rho_{ii} = 1$, for a two-level system $\rho_{aa} + \rho_{bb} = 1$. This ensures the conservation of total probability which means the system is closed. By using this conservation of probability we can omit one equation (i.e. $\rho_{aa} = 1 - \rho_{bb}$). Now we shall deal with (n^2-1) linear equations which describe our entire system. Hence we shall have, for a two-level system, three linear equations, a three-level system (TLS), eight linear equations and so on. We can write down the properties of a density matrix,

$$\rho^2 = \rho \text{ Projector} \qquad (14.43a)$$

$$\rho^+ = \rho \text{ i. e. the density matrix is a Hermitian operator.} \qquad (14.43b)$$

$$Tr(\rho) = 1 \text{ Normalization condition} \qquad (14.43c)$$

$$Tr(\rho^2)\begin{cases} =1 \text{ for pure state} \\ <1 \text{ for mixed state} \end{cases} \qquad (14.43d)$$

$$\rho \geqslant 0 \text{ Positivity of density operator} \qquad (14.43e)$$

Equation (14.43a) follows from the definition of the density matrix and normalization condition (equation (14.43c)) and it implies that density matrix describes a pure state. It is very straightforward to prove equation (14.43b) by using equation (14.39), in connection with matrix elements using $\rho_{ij}^* = \rho_{ji}$. The normalization condition (equation (14.43c)) of the density matrix has already been proved. If Tr $(\rho^2) = 1$ then the state of a physical system is called the pure state and if $Tr(\rho^2) < 1$ then the state is mixed. The last property (equation (14.43e)) of the density matrix means the eigenvalues of the density matrix are greater than or equal to zero. We can show easily that $<\phi|\rho|\phi> = <\phi|\psi><\psi|\phi> = |<\phi|\psi>|^2 \geqslant 0$. This is an important property since probabilities are always greater than or equal to zero.

14.7 Pure and mixed states

In the case of a pure state density matrix express a single atom or molecule. It is a pure ensemble that is not a superposition of two different ensembles though, in reality, we deal with a system that is a mixed ensemble of many systems. Suppose, we have N different identical systems following the same equation of motion but they all have different phases. The density matrix of a pure state is the same as in equation (14.39), i.e. $\rho = |\psi><\psi|$. We can describe the 'mixed state' by

$$\rho = \sum_{j=1}^{n}P_j|\Psi_j> <\Psi_j| \qquad (14.44)$$

Here P_j is the probability of the atoms staying in the state $|\psi_j>$. It means the system is a mixture of several pure states. If for a particular state $P_j = 1$ and for all other states it is equal to zero then the state is known as pure state and density matrix operator is defined as in equation (14.39). Though the mixed state is a statistical feature of an ensemble, the expectation value of an operator can still be calculated in the same manner as we did in equation (14.42).

$$\langle \hat{\Theta} \rangle = \sum_j P_j < \psi_j \mid \hat{\Theta} \mid \psi_j > = \sum_j P_j \sum_l <\psi_j \mid \hat{\Theta} \mid l > <l \mid \psi_j>$$

$$= \sum_l <l \mid \sum_j P_j \mid \psi_j > <\psi_j \mid \hat{\Theta} \mid l > = \sum_l (\rho\hat{\Theta})_{ll} = Tr(\rho\hat{\Theta}) \tag{14.45}$$

So we can see from equation (14.42) and equation (14.45) that for the mixed state as well as in the case of the pure state the expectation value of an operator gives the same result.

14.8 Equation of motion of the density operator

Equation of motion of the density matrix operator can be easily derived from the Schrödinger equation. First, we take the time derivative of equation (14.44) and then use the Schrödinger equation (e.g. $\mid \dot{\psi} > = -\frac{i}{\hbar} H \mid \psi>$)

$$\dot{\rho} = \sum_j P_j \{\mid \dot{\psi}_j > <\psi_j \mid + \mid \psi_j > <\dot{\psi}_j\mid\} = -\frac{i}{\hbar} \sum_j P_j \{H \mid \psi_j > <\psi_j \mid - \mid \psi_j > <\psi_j \mid H\}$$

$$= -\frac{i}{\hbar}[H, \rho] \tag{14.46}$$

Here $[H, \rho] = H\rho - \rho H$, is the commutator of **H** and ρ. Equation (14.46) is known as *Liouville* or *Von Neumann* equation of motion for the density matrix. This equation is more widespread than the Schrödinger equation. In Schrödinger's equation, a particular state vector is used but here density operator is used. So this equation provides information both about statistical and quantum mechanical aspects. The *ij*-th matrix element of equation (14.46) is given by

$$\dot{\rho}_{ij} = -\frac{i}{\hbar} < i \mid H\rho - \rho H \mid j > = -\frac{i}{\hbar}\sum_l \{<i\mid H \mid l> <l \mid \rho \mid j > -<i \mid \rho \mid l> <l \mid H \mid j > \}$$

$$= -\frac{i}{\hbar} \sum_l \{H_{il}\rho_{lj} - \rho_{il}H_{lj}\} \tag{14.47}$$

Equation (14.47) is useful to calculate the density matrix elements of the multi-level systems. In this formulation, we have not considered the decay process. So the density matrix elements containing the population term (e.g. ρ_{ii}, ρ_{jj}) and polarization term (ρ_{ij}) don't have to contain any loss parameter. In the next section, we add the decay parameters in the *Liouville* or *Von Neumann* equation to get a more general equation of motion.

14.9 Inclusion of decay phenomena

In equation (14.46) no decay parameters are added though in the real situation atoms can decay from excited states due to spontaneous emission, because of collisions or other phenomena. The finite lifetime of the atomic states can be incorporated by adding the decay terms phenomenologically in the density matrix equation of motion. By adding a relaxation matrix (Γ) in equation (14.46) we can absorb the decay process of the atomic system. Γ can be written as the following equation

$$<i \mid \Gamma \mid j> = \gamma_{ii}\delta_{ij} \qquad (14.48)$$

By adding the relaxation matrix (e.g. equation (14.48)) in the equation (14.46), the Liouville equation modifies as

$$\dot{\rho} = -\frac{i}{\hbar}[H, \rho] + \{\Gamma, \rho\} \qquad (14.49)$$

where $\{\Gamma, \rho\} = (\Gamma\rho + \rho\Gamma)$. After inclusion of the decay process phenomenologically in the Liouville equation we get equation (14.49) and this equation is known as the *master equation*. With the decay term, the density matrix equation for *ij*-th element becomes.

$$\dot{\rho}_{ij} = -\frac{i}{\hbar} \sum_l \{H_{il}\rho_{lj} - \rho_{il}H_{lj}\} + \sum_l \{\Gamma_{il}\rho_{lj} + \rho_{il}\Gamma_{lj}\} \qquad (14.50)$$

Now we can calculate the density matrix equation of motion of the two-level atom interacting with a stationary e-m field from equation (14.50). The Hamiltonian of the system is the same as given in equation (14.10).

$$\dot{\rho}_{bb} = -\gamma_{bb}\rho_{bb} + \frac{i}{2\hbar} (\mu_{ab}E_0\rho_{ab}e^{-i\nu t} - c.\ c.\ ) \qquad (14.51)$$

$$\dot{\rho}_{aa} = -\gamma_{aa}\rho_{aa} - \frac{i}{2\hbar} (\mu_{ab}E_0\rho_{ab}e^{-i\nu t} - c.\ c.\ ) \qquad (14.52)$$

$$\dot{\rho}_{ba} = -(\gamma_{ba} + i\omega_0)\rho_{ba} - \frac{i}{2\hbar} \ \mu_{ab}E_0e^{-i\nu t}(\rho_{bb} - \rho_{aa}) \qquad (14.53)$$

γ_{aa} and γ_{bb} are the population relaxation rates of the level $|a>$ and $|b>$ as shown in the equation (14.48), $\gamma_{ba} = \frac{(\gamma_{aa} + \gamma_{bb})}{2}$, is the coherence relaxation rate. In equations (14.51)–(14.53) we do not consider all the interactions that can affect the coherent optical dynamics, one such process being atom–atom elastic collision in the gas. In a gas, when an atom collides with another atom their energy levels undergo random Stark shifts. Though their energy levels do not change their state, the decay rate of

γ_{ba} becomes higher due to this elastic collision. The decay rates γ_{aa} and γ_{bb} are not changed much. The expression of the modified decay rate of γ_{aa} can be calculated in a simple manner from equation (14.53) by adding the random Stark shift $\delta\omega(t)$ to the transition frequency(ω_0) of the two systems and ignoring the atom–field interaction term (i.e. $\mu_{ab}E_0 = 0$) for simplicity. The density matrix equation of motion for the density matrix element ρ_{ba} becomes.

$$\dot{\rho}_{ba} = -[\gamma_{ba} + i(\omega_0 + \delta\omega(t))]\rho_{ba} \tag{14.54}$$

Formal integration of equation (14.54), gives us

$$\rho_{ba}(t) = exp\big(-(\gamma_{ba} + i\omega_0)t - i\int_0^t \delta\omega(t')dt'\big]\rho_{ba}(0) \tag{14.55}$$

In equation (14.55) all terms except $\delta\omega(t)$ are fixed. We take an ensemble average of this equation over the random variations in $\delta\omega(t)$. This ensemble average only affects the last term in the exponential of equation (14.55).

$$<\exp\left[-i\int_0^t \delta\omega(t')dt'\right]>$$
$$=\Big\langle 1 - i\int_0^t dt_1\delta\omega(t_1) - \frac{1}{2}\int_0^t dt_1\int dt_2\delta\omega'(t_1)\delta\omega(t_2) + \frac{(-i)^{2n}}{(2n)!} \tag{14.56}$$
$$\int_0^t dt_1\ldots\ldots\int_0^t dt_{2n}\delta\omega(t_1)\ldots\ldots\ldots\delta\omega(t_{2n}) + \ldots\ldots\ldots\Big\rangle$$

The sign of $\delta\omega(t)$ is positive as well as negative so ensemble average of $< \delta\omega(t)) > = 0$. The variation of $\delta\omega(t)$ is more rapid compared to the other changes which happen to be in the time scale of $1/\gamma_{ba}$. Using a Markoff approximation for the pair of correlation functions we can write

$$<\delta\omega(t')\delta\omega(t) > =2\gamma_{deph}\delta(t - t'), \tag{14.57}$$

here γ_{deph} is a constant. We can assume that $\delta\omega(t)$ is described by a Gaussian random process, so according to the fluctuations-dissipation theorem we get

$$<\exp\left[-i\int_0^t \delta\omega(t')dt'\right] > =\exp(-\gamma_{deph}t) \tag{14.58}$$

Using equation (14.58) in equation (14.55), we get the expression of $\rho_{ba}(t)$ as,

$$\rho_{ba}(t) = \exp\big(-(i\omega_0 + \gamma_{ba} + \gamma_{deph})t\big]\rho_{ba}(0) \tag{14.59}$$

In equation (14.59) total relaxation rate for dephasing can be written as $\Gamma_{Ph} = \left(\frac{(\gamma_{aa} + \gamma_{bb})}{2} + \gamma_{deph}\right)$. Density matrix equation of motion of the coherence term (ρ_{ba}) becomes,

$$\dot{\rho}_{ba} = -(\Gamma_{Ph} + i\omega_0)\rho_{ab} - \frac{i}{\hbar} \ \mu_{ab}E_0 e^{-i\nu t}(\rho_{bb} - \rho_{aa}) \tag{14.60}$$

In this manner in equation (14.60) we incorporated the collision phenomenon and the collision/pressure broadening linewidth Γ_{Ph}.

14.10 Vector model of density matrix equations of motion

The density matrix equations of motion (equations (14.51)–(14.53)) for a two-level system can be written in the form of Bloch equations under certain conditions, in common with equations for gyroscopic precession. The Bloch equations originate from the studies of NMR. In the Bloch equations, it is assumed that diagonal and off-diagonal decay rates are the same. The diagonal decay rates are related to the longitudinal decay time (T_1) and the off-diagonal decay with transverse relaxation time (T_2) defined as,

$$\gamma_{aa} = \gamma_{bb} = \frac{1}{T_1} \tag{14.61}$$

$$\gamma_{ba} = \frac{1}{T_2} \tag{14.62}$$

Generally $T_2 \leqslant T_1$, so the coherence decay occurs much faster than the population decay. In equation (14.10) we have already presented the expression of the atom–field interaction Hamiltonian under rotating wave approximation as,

$$H_I = -\frac{\mu_{ab}E_0}{2}e^{-i\nu t} \tag{14.63}$$

Multiplying by $e^{i\nu t}$ both sides of the equation (14.49) we get

$$\dot{\rho}_{ba}e^{i\nu t} = -(\gamma_{ba} + i\omega_0)\rho_{ba}e^{i\nu t} - \frac{i}{2\hbar}\mu_{ab}E_0(\rho_{bb} - \rho_{aa}) \tag{14.64}$$

If we consider the limiting case where diagonal and off-diagonal decay rates are the same $\gamma_{aa} = \gamma_{bb} = \gamma_{ba} = \Gamma = \frac{1}{T}$, equation (14.60) can be written as

$$\frac{d}{dt}(\rho_{ba}e^{i\nu t}) = -[\Gamma + i(\omega_0 - \nu)]\rho_{ba}e^{-i\nu t} - \frac{i}{2\hbar}\mu_{ab}E_0(\rho_{bb} - \rho_{aa}) \tag{14.65}$$

Similarly, equations (14.47) and (14.48) can be written as,

$$\frac{d}{dt}(\rho_{bb}) = -\Gamma\rho_{bb} + \frac{i}{2\hbar}\mu_{ab}E_0(\rho_{ab}e^{-i\nu t} - \rho_{ba}e^{i\nu t}) \tag{14.66}$$

$$\frac{d}{dt}(\rho_{aa}) = -\Gamma\rho_{aa} - \frac{i}{2\hbar}\mu_{ab}E_0(\rho_{ab}e^{-i\nu t} - \rho_{ba}e^{i\nu t}) \tag{14.67}$$

From these three equations (14.61)–(14.63), we can now introduce three new variables in the form

$$B_1 = (\rho_{ba}e^{i\nu t} + \rho_{ab}e^{-i\nu t}) \tag{14.68}$$

$$B_2 = -i(\rho_{ba}e^{i\nu t} - \rho_{ab}e^{-i\nu t}) \tag{14.69}$$

$$B_3 = (\rho_{bb} - \rho_{aa}) \tag{14.70}$$

From equations (14.68 and 14.69) we can write $(B_1 + iB_2) = 2\rho_{ba}e^{i\nu t}$, putting this value in equation (14.65)

$$\frac{d}{dt}(B_1 + iB_2) = -[\Gamma + i(\omega_0 - \nu)](B_1 + iB_2) - \frac{i}{\hbar}\mu_{ab}E_0 B_3 \tag{14.71}$$

By comparing the coefficients of the real and imaginary part of the equation (14.71) we get

$$\frac{d}{dt}B_1 = -\Gamma B_1 + (\omega_0 - \nu)B_2 = -\frac{B_1}{T} + \delta B_2 \tag{14.72a}$$

$$\frac{d}{dt}B_2 = -\Gamma B_2 - (\omega_0 - \nu)B_1 - \frac{\mu_{ab}E_0}{\hbar}B_3 = -\frac{B_2}{T} - \delta B_1 - \Omega_R B_3 \tag{14.72b}$$

$$\frac{d}{dt}B_3 = -\Gamma B_3 + \frac{\mu_{ab}E_0}{\hbar}B_2 = -\frac{B_3}{T} + \Omega_R B_2 \tag{14.72c}$$

In the equations (14.72a)–(14.72c) we can replace $(\omega_0 - \nu)$ by δ detuning parameter and $\frac{\mu_{ab}E_0}{\hbar}$ by Ω_R, the Rabi frequency as defined in the earlier section.

These variables (B_1, B_2 and B_3) are mutually independent and vary slowly in the optical period so we can think of them as components of a vector $\vec{B}$. If we assume $\hat{n}_1$, $\hat{n}_2$, n_3 are three mutually perpendicular unit vectors, then $\vec{B} = (\hat{n}_1 B_1 + \hat{n}_2 B_2 + \hat{n}_3 B_3)$ called the *Bloch vector*. Equations (14.72a)–(14.72c) are the optical Bloch equations that can be expressed as

$$\frac{d}{dt}\vec{B} = -\frac{\vec{B}}{T} + \vec{\Omega} \times \vec{B} \tag{14.73}$$

Here $\vec{\Omega}_B = \Omega_R \hat{n}_1 - \delta \hat{n}_3$ is the precession frequency of the vector $\vec{B}$ having magnitude $\Omega_B = [\delta^2 + \Omega_R^2]^{\frac{1}{2}}$. Equation (14.73) is the equation of motion of a gyroscope. The elements of the density matrix (in the limiting case above) can be identified as the components of a Bloch vector $\vec{B}$ which executes simple precession about an effective field Ω shown in figure 14.6. Due to the process of decay, the length of $\vec{B}$ shrinks with time as $\exp[-t/T]$. At $\delta = 0$ (on-resonance condition) precession frequency (Ω_B) becomes Rabi frequency (Ω_R) and the $\vec{B}$ vector rotates around the direction of $\hat{n}_1$. If $\delta \neq 0$, the off-resonance case Bloch vector will precess around in the direction of $\hat{n}_1$ and $\hat{n}_3$ plane with frequency $\Omega_B = [\delta^2 + \Omega_R^2]^{\frac{1}{2}}$. In this case, the complete transition from the ground state to the excited state or vice versa is not possible.

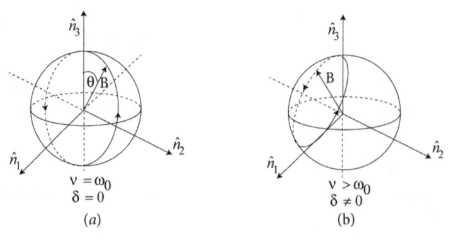

Figure 14.6. Precession of a Bloch vector. (a) $\delta = 0$ on-resonance condition, the Bloch vector is precessed in the direction of $\hat{n}_1$ with precession frequency Ω_R. (b) At off-resonance condition ($\delta \neq 0$) precession is around the direction of and $\hat{n}_1$ and $\hat{n}_3$ plane with precession frequency Ω_B.

14.11 Power broadening and saturation of the spectrum

By using rotating wave approximation from equations (14.65–14.68) we can write the set of optical Bloch equations for a two-level atomic system as,

$$\frac{d}{dt}(\rho_{bb}) = -\Gamma \rho_{bb} + \frac{i\,\Omega}{2}(\tilde{\rho}_{ab} - \tilde{\rho}_{ba}) \tag{14.74a}$$

$$\frac{d}{dt}(\rho_{aa}) = +\Gamma \rho_{bb} - \frac{i\,\Omega}{2}(\tilde{\rho}_{ab} - \tilde{\rho}_{ba}) \tag{14.74b}$$

$$\frac{d}{dt}(\tilde{\rho}_{ba}) = -[\Gamma + i\delta]\tilde{\rho}_{ba} - \frac{i\,\Omega}{2}(\rho_{bb} - \rho_{aa}) \tag{14.74c}$$

Here we used a new variable $\rho_{ba} = \tilde{\rho}_{ba}e^{-i\Omega t}$ to remove all the time dependence. This type of frame is known as a co-rotating frame and the corresponding basis is familiar as the rotating basis, which we shall denote by $\tilde{\rho}_{ij}$'s on transformed matrix elements. δ and Ω are the detuning parameter and Rabi frequency was defined earlier. We assumed our two-level system to be a closed system. So the population of the excited state ($|b>$) decay to the ground state ($|a>$) with decay rate Γ. The population is conserved here, i.e. $\rho_{aa} + \rho_{bb} = 1$. This is the reason we call it a closed system. In the steady-state limit where $\frac{d\rho}{dt} = 0$ the value of $\tilde{\rho}_{ba}$ can be written from equation (14.74c) as

$$\tilde{\rho}_{ba} = -\frac{i\,\Omega}{2}\frac{(\rho_{bb} - \rho_{aa})}{(\Gamma + i\delta)} \tag{14.75}$$

The population difference can be written as $(\rho_{bb} - \rho_{aa}) = \Delta N = \frac{1}{s}$, where s is the saturation parameter and is given by

$$s = \frac{\Omega^2}{2\,|(\Gamma + i\delta)|} = \frac{\Omega^2}{2(\Gamma^2 + \delta^2)} = \frac{s_0}{[1 + \left(\dfrac{\delta}{\Gamma}\right)^2]} \tag{14.76}$$

's_0' can be defined as an on-resonance saturation parameter, $s_0 = \Omega^2/2\,\Gamma^2$. On-resonance saturation parameter s_0 can be written in another way viz. $s_0 = I/I_S$. Here 'I' is the intensity of the e-m field and I_S is the saturation intensity of an atomic or molecular transition. When the saturation parameter is low ($s \ll 1$), the population is mostly in the ground state $\Delta N = 1$. For a high value of 's,' the population is equally distributed among the ground and excited state ($\rho_{bb} = \rho_{aa} = 1/2$). The population in the excited state decays with a rate γ_{sc} and in steady-state, the excitation rate and decay rate are the same. So the total scattering rate γ_{sc} of the light from the e.m. radiation is given by

$$\gamma_{sc} = \Gamma\tilde{\rho}_{bb} = \frac{2s_0\Gamma}{(1 + s_0 + \left(\dfrac{\delta}{\Gamma}\right)^2)} \tag{14.77}$$

At very high intensities $s_0 \gg 1$, γ_{sc} saturates to Γ. By taking s_0 outside, the above equation can be written as,

$$\gamma_{sc} = \left(\frac{s_0}{1 + s_0}\right)\left(\frac{2\Gamma}{1 + \left(\dfrac{\delta}{\Gamma_P}\right)^2}\right) \tag{14.78}$$

where $\Gamma_p = \Gamma\sqrt{1 + s_0}$ is called the power broadened linewidth of the transition. Due to saturation, the linewidth of the transition is broadened from its natural linewidth Γ to its power-broadened value Γ_p. This phenomenon is called the power broadening of the spectral profile. Figure 14.7 shows the variation of γ_{sc} with e.m. field detuning (δ) for different values of saturation parameter s_0. It is clear from figure 14.7 that for large s_0, the absorption continues to increase in the wings of the absorption spectra with the increase of field intensity. Although at the higher s_0 half of the population gets transferred from ground state to excited state, the absorption at the centre of the curve is saturated but wings are not saturated.

When the resonant laser beam travels through a sample of atoms due to scattering with the atoms there is a loss in intensity of the laser beam. The amount of scattered power per unit volume is $n\hbar\omega_0\gamma_{sc}$, where n is the number of atoms per unit volume and ω_0 is the resonance frequency of the atom. So, we can write

$$\frac{dI}{dz} = -n\hbar\omega_0\gamma_{sc} = -n\hbar\omega_0\frac{2s_0\Gamma}{1 + s_0 + \left(\dfrac{\delta}{\Gamma}\right)^2} = -n\hbar\omega_0\frac{2\Gamma\left(\dfrac{I}{I_s}\right)}{1 + s_0 + \left(\dfrac{\delta}{\Gamma}\right)^2} \tag{14.79}$$

$$= -n\sigma I = -\alpha I$$

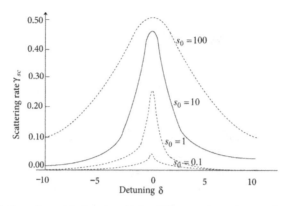

Figure 14.7. The variation of γ_{sc} with detuning (δ) for different values of saturation parameter (s_0). With increasing the value of s_0 the spectrum becomes more and more power broadened.

The e-m field propagates along the z-direction through the atomic sample. In equation (14.79) σ is the scattering cross-section and α is the absorption coefficient. Saturation intensity 'I_s' can be defined as $I_s = \frac{\pi hc\Gamma}{\lambda_0^2}$ where λ_0 is the resonance wavelength. By using the expression of I_s we can write the expression of scattering cross-section as,

$$\sigma \cong \frac{2\lambda_0^2}{\pi} \frac{1}{1 + s_0 + \left(\frac{\delta}{\Gamma}\right)^2} \tag{14.80}$$

At the low intensity of the em-field or laser field, i.e. $s_0 \ll 1$, $\sigma \cong \frac{2\lambda_0^2}{\pi} \frac{1}{1 + \left(\frac{\delta}{\Gamma}\right)^2}$ and at exact resonance condition ($\delta = 0$) the expression of scattering cross-section is reduced to $\frac{2\lambda_0^2}{\pi}$. In the strong field intensity regime, $I \gg I_s$ the absorption coefficient (α) goes to zero though it does not mean absorption vanishes. In this intensity limit, we have $\alpha I \to 2n\hbar\omega_0\Gamma$ using this expression of the absorption coefficient in equation (14.79) we get

$$\frac{dI}{dz} = -\alpha I = -2n\hbar\omega_0\Gamma \tag{14.81}$$

The solution of the equation (14.81) is $I(z) = I(0) - 2(n\hbar\omega_0\Gamma)z$. The field intensity is reduced with an increase in the propagation length. If the field intensity is less than the saturation intensity, i.e. $I \ll I_s$ then 'I' decreases exponentially with propagation length $I(z) = I(0)e^{-\alpha z}$. Here $I(0)$ is the intensity at $z = 0$.

14.12 Spectral line broadening mechanism

The measured spectral line generated due to spontaneous emission from a group of atoms is perfectly monochromatic. For measurements by using a high-resolution laser spectrometer, or a Fourier transform infrared (FTIR) spectrometer, a spectral distribution $I(\nu)$ of the absorbed or emitted intensity around the centre frequency ν_0 correspondings to an atomic or molecular transition are observed. This intensity distribution function $I(\nu)$ around the resonant frequency is ν_0 called the line shape. The shape of any spectroscopic transition depends on one or more perturbing mechanisms in different experimental conditions. The broadness of any spectral profile may give rise to the linewidth parameter containing valuable information about the collision dynamics of the associated atoms or molecules and also about the interaction forces and relevant molecular parameters. A minimum width inherent in any spectral transition is said to be the natural linewidth of the transition. Other significant contributions to the line broadening mechanism come from the thermal motion of the constituent spikes and their associative collision phenomena. There are two distinct types of line broadenings: (1) homogeneous and (2) inhomogeneous. In the homogeneous broadening, all the single atoms or molecules in an ensemble respond in the same manner during interaction with the radiation field. The probability of transition of all atoms or molecules at a particular frequency of radiation is the same. So all of them generate the same spectrum. Since we cannot distinguish the atoms or molecules from one another, they are indistinguishable. On the other hand in inhomogeneous broadening, due to different velocities of atoms, each atom sees a shifted radiation frequency. So individual atoms act heterogeneously during interaction with the field and different velocity groups of an atom contribute to different parts of the spectrum.

In the case of solids, atoms are stuck at a particular position and cannot move as freely as a gaseous molecule so Doppler or pressure broadenings are not applicable here. A different kind of inhomogeneous broadening is observed here due to the presence of local defects. The spectrum is broadened due to the interaction of atoms with the surrounding environment, known as environmental broadening.

The lineshape of the spectrum generated due to homogeneous broadening is Lorentzian and for inhomogeneous cases, it is a Gaussian profile. Now the principal causes of the line broadening along with the line shape functions are discussed briefly in the following sections.

14.13 Natural broadening

According to the Heisenberg energy–time uncertainty relation, we can write $\Delta E \, \Delta t \geqslant \hbar$. So the energy of a state would be exactly defined only when Δt is equal to infinity. In reality, Δt is finite so a state is represented by energy spread over ΔE. By putting $\Delta t = \tau$, we call τ the radiative lifetime of the excited state. The corresponding line broadening is called 'natural broadening', the amount of which in the angular-frequency unit is $\Delta\omega = \frac{\Delta E}{\hbar} \geqslant \frac{1}{\tau}$.

Now we can see the lineshape of the natural broadening. The radiation emitted due to the spontaneous emission process from the excited state per unit volume per unit time can be written as

$$I(t) = A_{ba} N_b \hbar \omega_0 \exp(-t/\tau) \tag{14.82}$$

Here A_{ba} is the Einstein A-coefficient, N_b is the number of atoms in the excited level (b) and τ is the radiative lifetime. $I(t)$ is also known as the intensity of spontaneous emission. The electric field associated with the spontaneous emission can be found from the expression $I(t)$ since intensity is proportional to the electric field square.

$$E(t) = E_0\exp(-t/2\tau)\exp(i\omega_0 t) \qquad (14.83)$$

Here E_0 is the amplitude of the electric field at $t = 0$. Equation (14.83) shows that the oscillating electric field loses its amplitude exponentially with time. The first exponential term represents the damping factor. To find the electric field in the frequency domain we need to take the Fourier transform of equation (14.83). The electric field in the frequency domain is

$$E(\omega) = E_0 \int_0^\infty \exp(i\omega_0 t)\exp(-t/2\tau)\exp(-i\omega t)dt = E_0\frac{1}{i(\omega - \omega_0) + \dfrac{1}{2\tau}} \qquad (14.84)$$

The intensity spectrum is proportional to $|E(\omega)|^2$. So we get

$$I(\omega) \propto \frac{1}{(\omega - \omega_0)^2 + (\dfrac{1}{2\tau})^2} \qquad (14.85)$$

This lineshape function is normalized to unity, i.e. $\int I(\omega)d\omega = 1$, by applying this condition we can get the expression of $I(\omega)$ as,

$$I(\omega) = \frac{1}{2\pi\tau}\frac{1}{(\omega - \omega_0)^2 + (\dfrac{1}{2\tau})^2} \qquad (14.86)$$

Equation (14.86) is a Lorentzian line shape so the nature of the *natural broadening* or *radiative broadening* is Lorentzian. The full-width at half maxima (FWHM) of this broadening is $\Delta\omega_F = \frac{1}{\tau}$. In the case of rubidium D$_2$ transitions, the spontaneous decay rate of the excited level is approximately $\tau \approx 28$ ns so the natural linewidth $(\frac{1}{2\pi\tau})$ is in the order of 6 MHz. The line shape of the spectrum given in equation (14.86) has been plotted in figure 14.8.

14.14 Collision or pressure broadening

Atoms or molecules in the gas regularly collide with each other or with the wall of the container. Due to this atom–atom collision, the energy level of an atom gets perturbed due to the collision with another atom. This interrupts the light emission process and may reduce the effective lifetime of the excited state. If the mean collision duration time (τ_c) is less than the radiative time (τ) then we need to change the τ by τ_c in the expression of $\Delta\omega_F$. It will introduce another line broadening known as *collision broadening*. The value of τ_c can be calculated from the kinetic theory of gases,

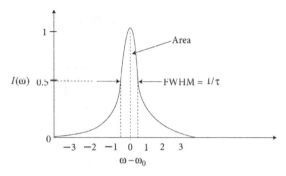

Figure 14.8. The plot of the Lorentzian lineshape function ($I(\omega)$) with frequency. The peak of the function is at the transition frequency ω_0.

$$\tau_c \approx \frac{1}{\sigma_c P} \sqrt{\frac{\pi m K T}{8}} \tag{14.87}$$

Here P is the pressure, σ_c is the collisional cross-section, m is the mass of the atom. The collisional linewidth $\Delta\omega = \frac{1}{\sigma_c}$ is proportional to the pressure (P). This is the reason for collisional broadening being also called *pressure broadening*. The typical value of τ_c at standard temperature and pressure (STP) is 10^{-10} s, which is much smaller than the radiative lifetime (10^{-6} s). To reduce the pressure broadening in the experiment of atomic spectroscopy, people reduced the vapour pressure of the atomic sample in the atomic vapour cell. Rb, Cs, Na, etc alkali atoms vapour cells are available commercially with micro-Torr pressure inside the cell. Inert buffer gases like argon, neon, helium, nitrogen, etc may be added in the atomic vapour cell containing Rb, Cs, and Na atoms to reduce the pressure broadening.

14.15 Inhomogeneous broadening or Doppler broadening

Let us consider a gaseous system, weakly influenced by collisions, irradiated by electromagnetic (e.m.) field of frequency ν and velocity 'c'. Suppose an atom of this system is moving with velocity 'v' ($v \ll c$). If the atom is moving away from the field then the frequency of the radiation appears to be red-shifted from the transition frequency of the atom due to the Doppler shift.

$$\nu \cong \left(1 - \frac{v}{c} + \left(\frac{v}{c}\right)^2 - \dots \right)\nu_0 \tag{14.88}$$

here ν_0 is the transition frequency of the atom. In the same fashion if the atom is moving towards the field then the shift of the radiation frequency is blue,

$$\nu \cong \left(1 + \frac{v}{c} + \left(\frac{v}{c}\right)^2 + \dots \right)\nu_0 \tag{14.89}$$

Since $v \ll c$, equations (14.88) and (14.89) are reduced to $\nu = \left(1 - \frac{v}{c}\right)\nu_0$ (moving away from the field) and $\nu = \left(1 + \frac{v}{c}\right)\nu_0$ (moving towards the field). If we assume that atoms are moving along z-direction towards the field, then from equation (14.89) the frequency shift seen by the atom would be $\nu = \left(1 + \frac{v_z}{c}\right)\nu_0$. Here v_z is the velocity of the atom in the z-direction. When the frequency of the e.m. field is ν it will be resonant with the atoms whose velocity is

$$v_z = \left(\frac{\nu - \nu_0}{\nu_0}\right)c \tag{14.90}$$

If we assume that atoms have Maxwell–Boltzmann's (M–B) velocity distribution then the probability distribution function is

$$P(v_z) = \sqrt{\left(\frac{M}{2\pi KT}\right)} \; e^{-\left(\frac{Mv_z^2}{2KT}\right)}dv_z = \frac{1}{u\sqrt{\pi}}e^{-\left(\frac{v_z^2}{u^2}\right)}dv_z \tag{14.91}$$

where K is the Boltzmann's constant and 'M' is the mass of the atom and $u = \sqrt{2KT/M}$ is the most probable speed of the atom. In figure 14.9 absorption spectrum of an ensemble of moving atoms is shown. The width of the absorption spectrum (i.e. Doppler width) is much broader than the homogeneous linewidth of the spectrum (see figure 14.9). Using equation (14.90) we can deduce the probability function $g_D(\nu)$ in frequency(ν) space

$$g_D(\nu) = \frac{c}{u\nu_0\sqrt{\pi}} \; e^{-\left\{\frac{c^2}{u^2}\frac{(\nu-\nu_0)^2}{\nu_0^2}\right\}} \tag{14.92}$$

This is a Gaussian line shape function and it represents the distribution of atoms having various velocity components in the z-direction (both positive and negative z). The maximum value of $g_D(\nu)$ occurs at $\nu = \nu_0$ and this function falls to half of its maximum value at $(\nu - \nu_0) = \delta_{1/2}$, where

$$ln2 = \left(\frac{c}{u\nu_0}\delta_{1/2}\right)^2 \tag{14.93}$$

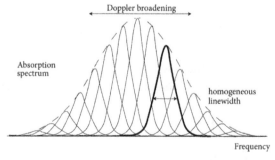

Figure 14.9. Absorption spectrum for an ensemble of moving atoms is much broader than the absorption spectrum of a single velocity class, which has a natural linewidth set by the spontaneous decay rate.

The Doppler broadened line width, i.e. full-width at half maxima (FWHM) is $\Delta\nu_D = 2\delta_{1/2}$ is given by

$$\Delta\nu_D = \frac{2\nu_0}{c}\ u\ \sqrt{ln2} \cong 1.7\ (\frac{\nu_0}{c}\ u) \tag{14.94}$$

It is clear from equation (14.94) that an atom of lower mass will have larger Doppler broadening width and this broadening can be reduced to some extent, by lowering the temperature T. This broadening can be removed by producing the sample in the form of an atomic or molecular beam where all atoms are moving in a particular direction with a restricted range of velocities. So we see that atomic motion gives rise to a broadening of the transition.

The atomic spectroscopy is often limited by the Doppler broadening. In most cases, the natural linewidth of a transition is much lower than its Doppler width. As an example in the case of ^{87}Rb D_2-lines ($5^2S_{1/2} \rightarrow 5^2P_{3/2}$) at room temperature ($T = 300$ K), the value of transition wavelength is $\lambda_0 = 780$ nm and so $\nu_0 = \frac{3 \times 10^8}{780 \times 10^{-9}} = 3.85 \times 10^{14}$ Hz, M $= 87 \times 1.67 \times 10^{-27}$ kg. By using equation (14.94) we can calculate the Doppler width (FWHM) ^{87}Rb atoms at room temperature to be ≈ 513 MHz, whereas the natural width of the ^{87}Rb atom is of the order of 6 MHz. The most interesting and effective way to remove the Doppler broadening is Lamb-dip spectroscopy (see chapter 15) in which the group of atoms moving with zero velocity in the direction of the propagation field only participates in the transition. But nowadays, by using the method of laser cooling and trapping almost a Doppler-free line shape can be produced.

14.16 Further reading

[1] Meystre P and Sergent III M 2009 *Elements of Quantum Optics* 3rd edn (Berlin: Springer)
[2] Fox M 2006 *Quantum Optics: An Introduction* (Oxford: Oxford University Press)
[3] Stenholm S 2005 *Foundation of Laser Spectroscopy* (New York: Dover)
[4] Rand S C 2010 *Non-linear and Quantum Optics* (Oxford: Oxford University Press)
[5] Loudon R 2001 *Quantum Theory of Light* 3rd edn (Oxford: Oxford University Press)
[6] Scully M O and Zubairy S 1997 *Quantum Optics* (Cambridge: Cambridge University Press)
[7] Ghosh P N 2018 *Laser Physics and Spectroscopy* (Boca Raton, FL: CRC Press)

14.17 Problems

1. Establish that at the on-resonance condition (i.e. at $\delta = 0$) under rotating wave approximation (RWA), equations (14.17a and 14.17b) are exactly solvable. Find the solutions.
2. Write down the density matrix for gas of two-level atoms at temperature T.

3. A strong on-resonant electromagnetic field interacts with a two-level atom. Calculate the time duration of a pulse for which the populations of the two levels are the same.

4. Consider Rabi oscillations when the light is not exactly resonant with the transition frequency. Equations (14.17a) and (14.17b), in the rotating wave approximation, becomes

$$\dot{C}_a = i(\frac{\Omega_R}{2})e^{i\delta t}C_b \text{ and } \dot{C}_b = i(\frac{\Omega_R}{2})e^{-i\delta t}C_a \text{ here } \delta = (\nu - \omega_0).$$

Answer the following questions.

i) Derive the following equation

$$\ddot{C}_b + i\delta\dot{C}_b + \frac{\Omega_R^2}{4}C_b = 0$$

ii) By assuming the trial solution $C_b = Ae^{-ift}$, s how is the general solution of the above differential equation:

$C_b(t) = A_+e^{-if_+t} + A_-e^{-if_-t}$? Here A_+, A_- are constants and

$$f_\pm = \frac{(\delta \pm \Omega_R)}{2} \text{ and } \Omega_G = (\delta^2 + \Omega_R)^{\frac{1}{2}}.$$

iii) Consider an initial condition where all the atoms are in the ground state at $t = 0$, (i.e. $C_a(0) = 1$, and $C_b(0) = 0$) prove that the transition probability is given by $| C_b(t) |^2 = \frac{\Omega_R^2}{2\Omega_G^2}(1 - cos\Omega_G t)$.

5. The C-coefficients for the interaction of a two-level atom with a strong field can be written after adding the decay constants phenomenologically as

$$\dot{C}_a = i\left(\frac{\Omega_R}{2}\right)e^{i(\nu-\omega_0)t}C_b - -\frac{1}{2}\gamma_a C_a(t)$$

$$\dot{C}_b = i\left(\frac{\Omega_R}{2}\right)e^{-i(\nu-\omega_0)t}C_a - -\frac{1}{2}\gamma_b C_b(t)$$

Use the Rabi method of the exact solution and show that the eigen frequencies are

$$f_\pm = -\frac{1}{2}\left[(\nu - \omega_0) - \frac{i}{2}(\gamma_a + \gamma_b)\right] \pm \frac{1}{2}[\left\{(\nu - \omega_0) - \frac{i}{2}(\gamma_a - \gamma_b)\right\}^2 + \Omega_R^2]^{1/2}$$

Calculate the oscillation time period in the resonant case.

6. Show that the Doppler width of Na D_1 transition ($\lambda = 589.0$ nm) at 500 K is approximately 0.02 Å.

7. If in an unperturbed two-level system occupation probability in the ground state is 0.8, write down the density matrix for this system.

8. Show that for a pure state

$$\rho^+ = \rho,$$

$$Tr(\rho) = 1,$$

$$\rho^2 = \rho,$$

$$Tr(\rho^2) = 1,$$

Explain whether the last two equations are true for the mixed state also? Justify.

9. Derive the rate equations for the Bloch vector components without using RWA. i.e. $H_I = -\frac{\mu_{ab}E_0}{2}(e^{-i\nu t} + c.\, c)$.

10. Consider a two-level system interacting with a radiation field detuned (Δ) from the resonance frequency. Suppose the field is switched off ($E_0 = 0$). Show that the time evolution of the vector components R_1 and R_2 can be represented by a rotation matrix that shows sinusoidal oscillation with exponential decay.

$$\begin{pmatrix} R_1(t) \\ R_2(t) \end{pmatrix} = \exp\left(-\frac{t}{T_2}\right)\begin{pmatrix} \cos\Delta t & -\sin\Delta t \\ \sin\Delta t & \cos\Delta t \end{pmatrix}\begin{pmatrix} R_1(0) \\ R_2(0) \end{pmatrix}$$

Also, show that the population difference decays exponentially.

$$R_3(t) = R_3(0)\exp\left(-\frac{t}{T_2}\right)$$

IOP Publishing

Quantum Optics and Quantum Computation
An introduction
Dipankar Bhattacharyya and Jyotirmoy Guha

Chapter 15

Laser spectroscopy and atomic coherence

Precision spectroscopy of atoms and molecules became a reality with the advent of the laser in the 1960s. Using the laser, which is a very narrow linewidth, tunable and coherent light source, it was possible to resolve atomic and molecular spectra with very high resolution. This led to a better understanding of the structure of atoms and molecules. Precision laser spectroscopy has established itself as a well sought after field now and a wide range of different spectroscopic techniques have been developed. Some exciting developments of the past two decades like *laser cooling and trapping of atoms* which culminated in 1995 with the realisation of *Bose–Einstein condensation in alkali gases* could be possible because of epoch breaking advances in laser spectroscopy. For instance, a tool for locking the lasers to particular atomic lines is accomplished by the technique of Doppler-free saturated absorption spectroscopy that is used in all the labs dealing with laser cooling and trapping of atoms.

In this chapter, we will first discuss the interaction of the electromagnetic field with a moving two-level atom. In the last chapter, we have discussed Doppler broadening, which is generated due to the thermal motion of atoms. The linewidth of the Doppler broadened profile is a few hundred MHz while the natural linewidths of the most alkali atoms are of the order of a few MHz. So the Doppler broadening is a common limit in precision laser spectroscopy. This spectral resolution limitation can be overcome by using saturated absorption spectroscopy and we will discuss this phenomenon here.

In the last section, we will discuss in detail some coherent phenomena like *coherent population trapping* (EIT), *electromagnetically induced transparency* (EIA), *electromagnetically induced absorption* (EIA) etc. These coherent phenomena originated due to atomic coherence. We also describe how one can reduce the group velocity of light passing through a coherent medium.

doi:10.1088/978-0-7503-2715-2ch15

15.1 Moving two-level atoms in a travelling wave field

Here we consider two-level atoms, the atom is moving along the z-direction with the velocity v. A travelling wave in the form of $E(z, t) = E_0 \cos(kz - \Omega t)$, interacting with the two-level atoms is shown in figure 15.1. Since the atom is in motion so the time derivative $(\frac{d}{dt})$ should be replaced by $\left(\frac{d}{dt}\right) \rightarrow \left(\frac{\partial}{\partial t}\right) + v\left(\frac{\partial}{\partial z}\right)$. By using the *Master equation* (equation (14.46)) we can write the density matrix equation of motion of this system.

$$\left(\frac{\partial}{\partial t} + v\frac{\partial}{\partial z}\right)\rho_{aa} = -\gamma_a\rho_{aa} + \Lambda_a + (\frac{iE_0\mu}{\hbar})\cos(kz - \Omega t)(\rho_{ab} - \rho_{ba}) \qquad (15.1a)$$

$$\left(\frac{\partial}{\partial t} + v\frac{\partial}{\partial z}\right)\rho_{bb} = -\gamma_b\rho_{bb} + \Lambda_b - (\frac{iE_0\mu}{\hbar})\cos(kz - \Omega t)(\rho_{ab} - \rho_{ba}) \qquad (15.1b)$$

$$\left(\frac{\partial}{\partial t} + v\frac{\partial}{\partial z} + i\omega\right)\rho_{ba} = -\gamma_{ba}\rho_{ba} - (\frac{iE_0\mu}{\hbar})\cos(kz - \Omega t)(\rho_{bb} - \rho_{aa}) \qquad (15.1c)$$

Here 'ω' is the transition frequency of the two-level atoms, Λ_a and Λ_b are incoherent pumping rates per unit time to the level $|a>$ and $|b>$, and $\mu_{ab} = \mu_{ba} = \mu$ is the dipole moment. A transient solution of the above equations (15.1a–15.1c) is possible by the time-integration of the above questions. Here we are not following this technique, we are interested in the steady-state solution since in our case continuous wave (CW) electromagnetic field is interacting with the atom.

If the atom is a free atom, i.e. when no e-m field is present, the off-diagonal matrix element oscillates with a frequency of ω. To remove this we perform a transformation,

$$\rho_{ba} = \tilde{\rho}_{ba}e^{i(kz-\Omega t)} \qquad (15.2)$$

By using this transformation, the derivative part of equation (15.1c) gives us

$$\left(\frac{\partial}{\partial t} + v\frac{\partial}{\partial z}\right)\rho_{ba} = -i(\Omega - kv)\rho_{ba} \qquad (15.3)$$

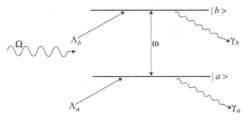

Figure 15.1. Schematic diagram of a two-level system with energy level $|a>$ and $|b>$ transition frequency ω interacting with a travelling wave field. Λ_a and Λ_b are incoherent pumping rate in the level $|a>$ and $|b>$.

By using the *rotating wave approximation* (RWA) we can write

$$e^{i(kz-\Omega t)}\cos(kz - \Omega t) = \frac{1}{2} \tag{15.4}$$

From equation (15.1c), we can write

$$(\gamma_{ba} - i(\Omega - kv) + i\omega)\tilde{\rho}_{ba} = -(\frac{iE_0\mu}{2\hbar})(\rho_{bb} - \rho_{aa})$$

$$\tilde{\rho}_{ba} = -(\frac{iE_0\mu}{2\hbar})\frac{(\rho_{bb} - \rho_{aa})}{(\gamma_{ba} + i(\delta + kv))} \tag{15.5}$$

The steady-state solution of the population can be derived from equation (15.1a–15.1b) by putting $\left(\frac{\partial \rho_{ii}}{\partial t}\right) = 0$ $(i = a, b)$.

$$\gamma_a\rho_{aa} = \Lambda_a + \frac{1}{2\gamma_{ba}}\left(\frac{E_0\mu}{\hbar}\right)^2(\rho_{bb} - \rho_{aa})L(\delta + kv) \tag{15.6a}$$

$$\gamma_b\rho_{bb} = \Lambda_b - \frac{1}{2\gamma_{ba}}\left(\frac{E_0\mu}{\hbar}\right)^2(\rho_{bb} - \rho_{aa})L(\delta + kv) \tag{15.6b}$$

Here $L(\delta + kv)$ is dimensionless Lorentzian lineshape defined as $L(\delta + kv) = \frac{\gamma_{ba}^2}{(\gamma_{ba}^2 + (\delta + kv)^2)}$, and $\delta = (\omega - \Omega)$ is the detuning parameter as defined earlier. In this Lorentzian function, resonance has occurred at $\delta = -kv$. If the atom is moving with a velocity 'v' in the same direction as the travelling wave then due to the Doppler shift it is now absorbed to give radiation at a higher frequency $\Omega = (\omega + kv)$. Now, by using equation (15.5) and equation (15.6a–15.6b), we can derive the steady-state population difference as,

$$(\rho_{bb} - \rho_{aa}) = \frac{\bar{N}}{[1 + 2I\eta\ L(\delta + kv)]} \tag{15.7}$$

Here $\bar{N} = \left(\frac{\Lambda_a}{\gamma_a} - \frac{\Lambda_b}{\gamma_b}\right)$ is the difference in occupation probabilities or unperturbed population, i.e. the difference between levels 'a' and 'b' when no field is present, $I = \frac{1}{2\gamma_a\gamma_b}\left(\frac{E_0\mu}{\hbar}\right)^2$ and $\eta = (\frac{\gamma_a + \gamma_b}{\gamma_{ba}})$. '$I$' is the dimensionless field intensity parameter and 'η' is the saturation decay parameter. If we assume that the population distribution of the atom is the Maxwell–Boltzmann (M–B) then we can write

$$\bar{N}(v) = \frac{\Lambda_0}{\sqrt{\pi}u}\exp(-\frac{v^2}{u^2}) \tag{15.8}$$

Λ_0 is the constant, u is $1/e$ width of the velocity distribution defined as $u = \sqrt{\frac{2k_BT}{M}}$, k_B is the Boltzmann constant, T is the temperature and M is the atomic mass. This is also known as Gaussian distribution. By putting the value of the $L(\delta + kv)$ in equation (15.7) we get

$$
\begin{aligned}
(\rho_{bb} - \rho_{aa}) &= \frac{\bar{N}(v)\left(\gamma_{ba}^2 + (\delta + kv)^2\right)}{\left[\gamma_{ba}^2 + (\delta + kv)^2 + 2I\eta\gamma_{ba}^2\right]} \\
&= \bar{N}(v)\left[1 - \frac{2I\eta\gamma_{ba}^2}{(\delta + kv)^2 + \gamma_{ba}^2(1 + 2I\eta)}\right]
\end{aligned}
\tag{15.9}
$$

It is clear from equation (15.9) that the deviation from the unperturbed population $(\bar{N}(v))$ is Lorentzian with the power broadened line-width $\Gamma_P = \gamma_{ba}\sqrt{(1 + 2I\eta)}$. As the intensity of the radiation field is increasing, the width (Γ_P) is increasing due to the intensity parameter term 'I' present in it. This phenomenon is known as power or saturation broadening.

The behaviour of the population difference (equation (15.9)) is shown in figure 15.2. It is seen from the figure that a hole appears in the population difference. In other words, we can say due to atom–field interaction a hole is burned the population difference curve. This hole is known as the *Bennet hole*. The physical reason for this hole is the transition of a particular group of the atom from the lower level ($|a>$) to the upper level ($|b>$) whose velocity satisfied the condition ($v = -\delta/k$).

The population of the lower ($|a>$) and the upper level ($|b>$) can be written as,

$$
\rho_{aa}(v) = \frac{\Lambda_a}{\gamma_a} + \bar{N}(v)\frac{I\gamma_b\gamma_{ba}}{(\delta + kv)^2 + \Gamma_P^2}
\tag{15.10}
$$

$$
\rho_{bb}(v) = \frac{\Lambda_b}{\gamma_b} - \bar{N}(v)\frac{I\gamma_a\gamma_{ba}}{(\delta + kv)^2 + \Gamma_P^2}
\tag{15.11}
$$

Equations (15.10–15.11) are plotted in figure 15.3 that shows how the population changes in each level of the atom due to the interaction with the field. A group of atoms with proper velocity accomplish the resonance and are removed from the lower level ($|a>$), which creates a hole in the population distribution curve. The same group of atoms move to the upper level and generate a peak in the population distribution curve, as shown in figure 15.3.

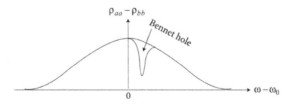

Figure 15.2. The spectral hole-burning phenomena in a Doppler broadened two-level atom interacting with a single travelling wave.

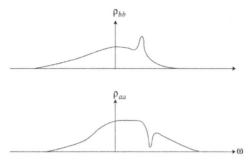

Figure 15.3. The individual population distribution of the levels $|a>$ and $|\underline{b}>$, a hole appears in the lower level and a peak is created in the upper level.

By using equations (15.5) and (15.7) the imaginary part of the coherence term ($\tilde{\rho}_{ab}$) becomes,

$$Im(\tilde{\rho}_{ba}) = \frac{1}{2i}(\tilde{\rho}_{ba} - \tilde{\rho}_{ab}) = -(\frac{E_0\mu}{2\hbar})\frac{\gamma_{ba}}{(\gamma_{ba}^2 + (\delta + kv)^2)}\frac{\bar{N}}{[1 + 2I\eta \; L(\delta + kv)]}$$

$$= -(\frac{E_0\mu}{2\hbar})\frac{\bar{N}\gamma_{ba}}{(\gamma_{ba}^2 + (\delta + kv)^2)\left[1 + \dfrac{2I\eta\gamma_{ba}^2}{(\delta + kv)^2 + \gamma_{ba}^2(1 + 2I\eta)}\right]} \qquad (15.12)$$

$$= -(\frac{E_0\mu}{2\hbar})\frac{\bar{N}\gamma_{ba}}{((\delta + kv)^2 + \Gamma_P^2)}$$

Equation (15.12) is a power broadened Lorentzian lineshape of the two-level atom, the linewidth above the Lorentzian function is Γ_P as defined earlier. This response is coming due to all atoms moving with a particular velocity. To calculate the response from all atoms in the medium we need to take into account all possible velocities of the atoms.

Hence the total macroscopic polarization in the medium can be derived by integrating equation (15.5) overall possible velocities.

$$P(z, t) = N\mu_{ab}\int_{-\infty}^{\infty}(\tilde{\rho}_{ba}e^{i(kz-\Omega t)} + \tilde{\rho}_{ab}e^{-i(kz-\Omega t)})dv$$

$$= N\mu_{ab}\int_{-\infty}^{\infty}[(\tilde{\rho}_{ba} + \tilde{\rho}_{ab})\cos(kz - \Omega t) + i(\tilde{\rho}_{ba} - \tilde{\rho}_{ab})\cos(kz - \Omega t)]dv \qquad (15.13)$$

Here N is the number of atoms per unit volume. Polarization of the medium can also be written in terms of the complex susceptibility (χ) of the medium.

$$P(z, t) = N\mu_{ab}\frac{E_0\varepsilon_0}{2}\{\chi(\omega)e^{i(kz-\Omega t)} + \chi^*(\omega)e^{-i(kz-\Omega t)}\} \qquad (15.14)$$

From equations (15.13) and (15.14), we can write

$$\chi(\omega) = \frac{2N\mu}{E_0\varepsilon_0} \int\limits_{-\infty}^{\infty} \rho_{ba}(v)dv = -i\frac{N\mu^2}{\varepsilon_0\hbar} \int\limits_{-\infty}^{\infty} \frac{[\gamma_{ba} - i(\delta + kv)]}{(\delta + kv)^2 + \gamma_{ba}^2(1 + 2I\eta)}N(v)dv \quad (15.15)$$

$N(v)$ has the form as defined in equation (15.8). The susceptibility of the medium is a complex quantity, the real (χ') and imaginary χ'' parts of the linear susceptibility (quantities related by the Kramers–Kronig relations) can be written as,

$$\chi = \chi' + i\chi'' \quad (15.16)$$

The real part of the susceptibility is a measure of the dispersion in the medium and the imaginary part of the susceptibility is related to the losses and gain in the medium. From equation (15.15) we can write the imaginary or absorptive part as,

$$\chi'' = \frac{N\mu^2\Lambda_0}{\varepsilon_0\hbar\sqrt{\pi}\,u} \int\limits_{-\infty}^{\infty} \left[\frac{\gamma_{ba}}{(\delta + kv)^2 + \gamma_{ba}^2(1 + 2I\eta)}\right]\exp(-\frac{v^2}{u^2})dv \quad (15.17)$$

The integral in equation (15.17) is a convolution of Lorentzian and a Gaussian function, known as *Voigt profile*. Analytical solution of this integral is possible when the linewidth (ku) of the Gaussian distribution function is much larger than the Lorentzian linewidth (γ_{ab}). It has a simple asymptotic form,

$$\chi'' = \frac{N\mu^2\Lambda_0 e^{-(\frac{\Delta}{ku})^2}}{\varepsilon_0\hbar\sqrt{\pi}\,ku\sqrt{(1 + 2I\eta)}} \int\limits_{-\infty}^{\infty} \left[\frac{\Gamma_P}{x^2 + \Gamma_P^2}\right]dx = \sqrt{\pi}\frac{N\mu^2\Lambda_0 e^{-(\frac{\delta}{ku})^2}}{\varepsilon_0\hbar\,ku\sqrt{(1 + 2I\eta)}} \quad (15.18)$$

This limit is known as the Doppler limit, i.e. $ku > > \Gamma_P \equiv \gamma_{ba}\sqrt{(1 + 2I\eta)}$ and also permits asymptotic evaluation of the dispersion part of the susceptibility.

$$\begin{aligned}
\chi' &= -\frac{N\mu^2\Lambda_0}{\varepsilon_0\hbar\sqrt{\pi}\,ku} \int\limits_{-\infty}^{\infty} \left[\frac{(\delta + kv)e^{-(\frac{v}{u})^2}}{(\delta + kv)^2 + \Gamma_P^2}\right]dv = -\frac{N\mu^2\Lambda_0}{\varepsilon_0\hbar\sqrt{\pi}\,ku} \int\limits_{-\infty}^{\infty} \frac{e^{-(\frac{x-\delta}{ku})^2}x}{x^2 + \Gamma_P^2}dx \\
&= -\frac{N\mu^2\Lambda_0}{\varepsilon_0\hbar\sqrt{\pi}\,ku}e^{-(\frac{\delta}{ku})^2} \int\limits_{-\infty}^{\infty} \frac{e^{(\frac{2\delta x}{k^2u^2})}xe^{-(\frac{x}{ku})^2}}{x^2 + \Gamma_P^2}dx \\
&= -\frac{2N\delta\mu^2\Lambda_0}{\varepsilon_0\hbar\sqrt{\pi}\,k^3u^3}e^{-(\frac{\delta}{ku})^2} \int\limits_{-\infty}^{\infty} \frac{x^2}{x^2 + \Gamma_P^2}e^{-(\frac{x}{ku})^2}dx \\
&= -\frac{2N\delta\mu^2\Lambda_0}{\varepsilon_0\hbar\,k^2u^2}e^{-(\frac{\delta}{ku})^2}
\end{aligned} \quad (15.19)$$

From the above discussion, we can notice that at the low intensity $I \to 0$, both real and imaginary parts of the susceptibility (χ' and χ'') are constant. So the polarization is to follow the linear relationship with the electric field intensity, $P = \varepsilon_0\chi E$. This is observed in conventional linear spectroscopy.

In the case of high intensities $I \gg 1$, equation (15.18) shows $\chi'' \propto \frac{1}{\sqrt{I}}$ or $\chi'' \propto \frac{1}{E}$ indicating that the absorption becomes nonlinear in the Doppler broadened system. This behaviour is quite different from the absorption saturation in the homogeneously broadened system. In the homogeneously broadened system $\chi'' \propto \frac{1}{I}$ for $I \gg 1$. The dispersive part (χ') at high intensity, however, does not saturate. This is because all the atoms are contributing to the dispersion, even in the low-intensity limit ($I \ll 1$). Power broadening only affects the absorptive part (equation (15.18)), where only resonantly interacting atoms are participating.

15.2 Moving atoms in a standing wave

If the electromagnetic wave is confined between two mirrors, the radiation travels back and forth between the two mirrors. Under this condition, the electromagnetic wave inside the cavity can be described as a standing wave field. This wave can be present as,

$$E(z, t) = E_0 \cos \Omega t \sin kz \qquad (15.20)$$

For a moving two-level atom interacting with the standing wave field, the density matrix equations are similar to equations (14.57a)–(14.57c). The interaction suffers change only,

$$\left(\frac{\partial}{\partial t} + v \frac{\partial}{\partial z} \right) \rho_{aa} = -\gamma_a \rho_{aa} + \Lambda_a + \left(\frac{iE_0\mu}{\hbar} \right) \cos \Omega t \sin kz (\rho_{ab} - \rho_{ba}) \qquad (15.21a)$$

$$\left(\frac{\partial}{\partial t} + v \frac{\partial}{\partial z} \right) \rho_{bb} = -\gamma_b \rho_{bb} + \Lambda_b - \left(\frac{iE_0\mu}{\hbar} \right) \cos \Omega t \sin kz (\rho_{ab} - \rho_{ba}) \qquad (15.21b)$$

$$\left(\frac{\partial}{\partial t} + v \frac{\partial}{\partial z} + i\omega \right) \rho_{ba} = -\gamma_{ba} \rho_{ba} - \left(\frac{iE_0\mu}{\hbar} \right) \cos \Omega t \sin kz (\rho_{bb} - \rho_{aa}) \qquad (15.21c)$$

Now if we introduce the transformation $\rho_{ba} = \tilde{\rho}_{ba} e^{-i\Omega t}$, in the same manner as in equation (14.58), in the interaction picture and by applying the RWA we get,

$$\left(\frac{\partial}{\partial t} + v \frac{\partial}{\partial z} \right) \rho_{aa} = -\gamma_a \rho_{aa} + \Lambda_a + \left(\frac{iE_0\mu}{2\hbar} \right) \sin kz (\tilde{\rho}_{ab} - \tilde{\rho}_{ba}) \qquad (15.22a)$$

$$\left(\frac{\partial}{\partial t} + v \frac{\partial}{\partial z} \right) \rho_{bb} = -\gamma_b \rho_{bb} + \Lambda_b - \left(\frac{iE_0\mu}{2\hbar} \right) \sin kz (\tilde{\rho}_{ab} - \tilde{\rho}_{ba}) \qquad (15.22b)$$

$$\left(\frac{\partial}{\partial t} + v \frac{\partial}{\partial z} \right) \tilde{\rho}_{ba} = -(\gamma_{ba} + i\delta) \tilde{\rho}_{ba} - \left(\frac{iE_0\mu}{2\hbar} \right) \sin kz (\rho_{bb} - \rho_{aa}) \qquad (15.22c)$$

Here $\delta = (\omega - \Omega)$ is the detuning parameter. The standing wave consists of two oppositely propagating travelling waves and so we can express it as,

$$E(z, t) = \frac{E_0}{2}[\sin(kz + \Omega t) + \sin(kz - \Omega t)] \tag{15.23}$$

From physical considerations, we consider the lowest order effect to be the sum of the effects caused by each wave separately. We identify the off-diagonal density matrix elements to be,

$$\tilde{\rho}_{ba} = \rho_+ e^{ikz} + \rho_- e^{-ikz} \tag{15.24}$$

Here the effect of the two counter-propagating beams on the off-diagonal matrix elements taken separately is through the terms ρ_+ and ρ_-. Here we also assume there is no spatial dependence on population terms (ρ_{aa} and ρ_{bb}). This means that we are taking their volume average value. By using equation (15.24) in equation (15.22c) and applying an approximation similar to RWA $e^{\pm ikz}\sin kz = \pm\frac{i}{2}$, we will get,

$$\left(\frac{d}{dt}\rho_+ + ikv\rho_+\right)e^{ikz} + \left(\frac{d}{dt}\rho_- - ikv\rho_-\right)e^{ikz} = -(\gamma_{ba} + i\delta)(\rho_+ e^{ikz} + \rho_- e^{-ikz})$$
$$-\left(\frac{iE_0\mu}{2\hbar}\right)\sin kz(\rho_{bb} - \rho_{aa}) \tag{15.25}$$

Comparing the coefficients of e^{ikz} and e^{-ikz} of the equation (15.25) we can derive expression of $\frac{d}{dt}\rho_+$ and $\frac{d}{dt}\rho_-$.

$$\frac{d}{dt}\rho_+ = -[\gamma_{ba} + i(\delta + kv)]\rho_+ - \left(\frac{E_0\mu}{4\hbar}\right)(\rho_{bb} - \rho_{aa}) \tag{15.26a}$$

$$\frac{d}{dt}\rho_- = -[\gamma_{ba} + i(\delta - kv)]\rho_- + \left(\frac{E_0\mu}{4\hbar}\right)(\rho_{bb} - \rho_{aa}) \tag{15.26b}$$

$$\frac{d}{dt}\rho_{aa} = -\gamma_a\rho_{aa} + \Lambda_a - \left(\frac{E_0\mu}{4\hbar}\right)(\rho_+ + \rho_+^* - \rho_- - \rho_-^*) \tag{15.26c}$$

$$\frac{d}{dt}\rho_{bb} = -\gamma_b\rho_{bb} + \Lambda_b + \left(\frac{E_0\mu}{4\hbar}\right)(\rho_+ + \rho_+^* - \rho_- - \rho_-^*) \tag{15.26d}$$

Equations (15.26c) and (15.26d) are obtained from equations (15.22a) and (15.22b). Since the off-diagonal density matrix elements are mainly affected by phase perturbation, as a result, they have a short relaxation time. So we can assume that the off-diagonal terms, ρ_+ and ρ_- can reach their equilibrium value much faster than the population terms ρ_{aa} and ρ_{bb}. Under the steady-state condition, we can put time derivates of equations (15.26a)–(15.26d) equal to zero.

$$\rho_\pm = \mp\frac{E_0\mu}{4\hbar}\frac{(\rho_{bb} - \rho_{aa})}{[\gamma_{ba} + i(\delta \pm kv)]} \tag{15.27}$$

$$\left(\rho_+ + \rho_+^* - \rho_- - \rho_-^*\right) = 2Re(\rho_+ - \rho_-)$$

$$= \frac{E_0\mu}{2\hbar}(\rho_{bb} - \rho_{aa})\left[\frac{-\gamma_{ba}}{\gamma_{ba}^2 + (\delta + kv)^2} - \frac{\gamma_{ba}}{\gamma_{ba}^2 + (\delta - kv)^2}\right] \quad (15.28)$$

$$= \frac{E_0\mu}{2\hbar\gamma_{ba}}(\rho_{bb} - \rho_{aa})[L(\delta + kv) + L(\delta - kv)]$$

here $L(\delta \pm kv) = \frac{\gamma_{ba}^2}{\gamma_{ba}^2 + (\delta \pm kv)^2}$, dimensionless Lorentzian lineshape. Population terms ρ_{aa} and ρ_{bb} becomes,

$$\rho_{aa} = \frac{\Lambda_a}{\gamma_a} - \left(\frac{1}{8\gamma_{ba}\gamma_a}\right)\left(\frac{E_0\mu}{\hbar}\right)^2 (\rho_{bb} - \rho_{aa})[L(\delta + kv) + L(\delta - kv)] \quad (15.29)$$

$$\rho_{bb} = \frac{\Lambda_b}{\gamma_b} + \left(\frac{1}{8\gamma_{ba}\gamma_b}\right)\left(\frac{E_0\mu}{\hbar}\right)^2 (\rho_{bb} - \rho_{aa})[L(\delta + kv) + L(\delta - kv)] \quad (15.30)$$

$$(\rho_{bb} - \rho_{aa}) = \frac{\bar{N}(v)}{1 + \frac{I}{2}\eta[L(\delta + kv) - L(\delta - kv)]} \quad (15.31)$$

Here $\bar{N}$, I and η have the same expression and same meaning as discussed in the earlier section. The occupation probability $\bar{N}$ is replaced by the Gaussian distribution function $\bar{N}(v)$ as described in equation (15.8). If we compare the expression of population difference of a moving atom in a single travelling wave field (equation (15.7)) with two counter-propagating fields (equation (15.31)) we can see here that two holes are burnt in the Gaussian distribution at velocities $v = \pm\delta/k$. These holes are described by Lorentzian function with a width of γ_{ba}.

The off-diagonal matrix elements presented in equation (15.27) can be written as

$$\rho_\pm = \mp\left(\frac{E_0\mu\bar{N}(v)}{4\hbar}\right)\frac{[\gamma_{ba} - i(\delta \pm kv)]}{[1 + \frac{I}{2}\eta][L(\delta + kv) - L(\delta - kv)][\gamma_{ba}^2 + (\delta \pm kv)^2]} \quad (15.32)$$

Figure 15.4(a) shows that the population difference curve of a two-level atomic system interacting with a standing wave has two *Bennet holes*. Two oppositely propagating travelling waves have pumped the atoms from the ground state ($|a>$) to the excited state ($|b>$) whose velocities are $v = +(\delta/k)$ and $=-(\delta/k)$. The height of each hole is controlled by the field intensity parameter 'I' and saturation parameter η.

It is clear from figure 15.4 that the separation between the two holes is equal to $(2\delta/k)$, so if we increase the detuning (δ), holes move away from each other. This process will continue until they reach the tail of the population distribution curve. The large detunings ($\delta \gg \gamma_{ba}$) confirm our assumption that the two oppositely moving travelling waves interact with the atoms independently. When detuning is small ($\delta \ll \gamma_{ba}$) two holes approaching each other can overlap at resonance. In this

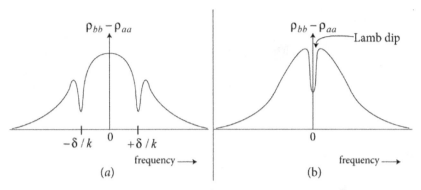

Figure 15.4. (a) Two Lorentzian shaped holes are created by each counter-propagating wave field in the population difference curve. (b) The Lamb dip is created at $\delta = 0$ (on resonance condition), two holes merge at resonance.

case, both waves interact with the same velocity group of an atom (i.e. $v = 0$). So our simple assumption of non-interacting waves does not hold good when $\delta = 0$, though this simple assumption can describe the physical process correctly.

In the same manner, as we did in the earlier section (equations (15.14) and (15.15)) we can define the susceptibility of the medium as,

$$\chi_\pm = \pm \frac{4N\mu}{E_0\varepsilon_0} \int_{-\infty}^{\infty} \rho_\pm(v)dv = -\frac{N\mu^2\Lambda_0}{\varepsilon_0\hbar\sqrt{\pi}\,u}$$

$$\int_{-\infty}^{\infty} \frac{[i\gamma_{ba} + (\delta \pm kv)]}{[1 + \frac{1}{2}I\eta][L(\delta + kv) - L(\delta - kv)][\gamma_{ba}^2 + (\delta \pm kv)^2]} \exp(-\frac{v^2}{u^2})dv \qquad (15.33)$$

The analytical expression of equation (15.33) can be found by using Doppler limit $ku \gg \gamma_{ba}$, as we did in the last section. We will do this in the two limiting cases of large detuning and small detuning or the on-resonance case.

Case I: Large detuning $\delta \gg \gamma_{ba}$.

In this case, the one Lorentzian in the denominator becomes unity (i.e. $\delta \cong ku$, $L(\delta - kv) \cong 1$) and we are left with.

$$\chi_\pm = \frac{N\mu^2\Lambda_0}{\varepsilon_0\hbar k\sqrt{\pi}\,u} \int_{-\infty}^{\infty} \frac{[i\gamma_{ba} + (\delta \pm x)]}{[(\delta \pm x)^2 + \gamma_{ba}^2(1 + \frac{1}{2}I\eta)]} \exp(-\frac{x^2}{k^2u^2})dx \qquad (15.34)$$

The imaginary and real part of the susceptibility can be derived in the same way as in equations (15.16)–(15.19).

$$\chi_\pm'' = \frac{N\mu^2\Lambda_0}{\varepsilon_0\hbar k\sqrt{\pi}\ u} \int\limits_{-\infty}^{\infty} \frac{\gamma_{ba}}{[(\delta \pm x)^2 + \gamma_{ba}^2(1 + \frac{1}{2}I\eta)]} \exp(-\frac{x^2}{k^2u^2})dx$$

$$= \frac{N\mu^2\Lambda_0\sqrt{\pi}\ \exp e^{-(\frac{\delta}{ku})^2}}{\varepsilon_0\hbar k\ u\sqrt{(1 + \frac{1}{2}I\eta)}}$$

(15.35)

The real part of the susceptibility ($\chi_\pm'$),

$$\chi_\pm' = \frac{N\mu^2\Lambda_0}{\varepsilon_0\hbar k\sqrt{\pi}\ u} \int\limits_{-\infty}^{\infty} \frac{(\delta \pm x)}{[(\delta \pm x)^2 + \gamma_{ba}^2(1 + \frac{1}{2}I\eta)]} \exp(-\frac{x^2}{k^2u^2})dx$$

$$= \frac{2\delta N\mu^2\Lambda_0 e^{-(\frac{\delta}{ku})^2}}{\varepsilon_0\hbar\ k^2u^2}$$

(15.36)

The large detuning limit, i.e. the expression of equations (15.35) and (15.36) are the same as the expression of a single travelling wave as we obtained in the last section. So here the two travelling waves interact with the atom in a completely independent way similar to the interaction with a single travelling wave though the saturation parameter may be different.

Case II: Small detuning condition $\delta \ll \gamma_{ba}$

At very small detuning or the perfectly on-resonance condition ($\delta \approx 0$) both counter-propagating travelling waves interact with the same group of velocity ($v = 0$) atoms. So our simple approximation of equation (15.24) is no longer valid but we can calculate the expression of susceptibility under the condition when detuning is very close to zero. If $\delta \ll \gamma_{ba}$, we can see from equation (15.33) that the two Lorentzians ($L(\delta + kv)$ and $L(\delta - kv)$) overlap upon each other and we get,

$$\chi_\pm = \frac{N\mu^2\Lambda_0}{\varepsilon_0\hbar k\sqrt{\pi}\ u} \int\limits_{-\infty}^{\infty} \frac{[i\gamma_{ba} \pm x]}{[x^2 + \gamma_{ba}^2(1 + \frac{1}{2}I\eta)]} \exp(-\frac{x^2}{k^2u^2})dx$$

(15.37)

The imaginary part of the susceptibility ($\chi_\pm''$) can be written as,

$$\chi_\pm'' = \frac{N\mu^2\Lambda_0\sqrt{\pi}}{\varepsilon_0\hbar k\ u\sqrt{(1 + \frac{1}{2}I\eta)}}$$

(15.38)

And the real part of the susceptibility vanishes with detuning so that the medium is dispersionless at resonance.

$$\chi_\pm' = 0$$

(15.39)

15.3 Lamb dip

In the last section, we discussed how two Bennett holes are created in the population distribution curve due to the interaction of two counter-propagating travelling waves. The position of the holes can be shifted by changing the detuning (δ). If the detuning is very low then these holes are coming close to each other. At the on-resonance condition (i.e. $\delta = 0$) two holes overlap on each other and a strong dip appears in the middle of the population distribution curve (figure 15.4(b)). This dip is known as *Lamb dip*, the technique by which Lamb dip is observed is known as *Saturation Absorption Spectroscopy* (SAS). The physical process behind this can be explained by using figure 15.4(a). In figure 15.4(a) we can see the two holes are burnt by two counter-propagating beams. One hole due to interaction of one beam with the atoms having a velocity $v_+ = +\delta/k$. At the same time, another hole is created due to the counter-propagating beam with the velocity of the atom around $v_- = -\delta/k$. If the velocity of v_+ and v_- is different then the hole generated by one beam would not affect the hole created due to another beam. When the laser frequency is very near to the resonance frequency, i.e. $\delta \approx 0$ then we have $v_+ = v_- = 0$, which means both beams interact with the same velocity group of atoms. One beam will burn a hole for the atoms with velocities near zero and the same atoms would involve the counter-propagating beam at the same frequency, as if two holes are overlapping each other at resonance (figure 15.4(b)). The width of this Lamb dip is close to the natural width of the transition if the power broadening and instrumental broadening etc are negligible.

The Doppler broadening limits the high-resolution absorption spectroscopy. This broadening is generated due to the thermal motion of the atomic or molecular sample. The atom moving with velocity v in the direction of the laser beam will be blue shifted ($\Omega' = (\omega + kv)$) and the oppositely moving atom will be red-shifted ($\Omega'' = (\omega - kv)$) from the resonance frequency ω. Due to this broadening, it is impossible to resolve the hyperfine splitting of the atomic transitions. Doppler broadening is much larger than the natural linewidth of the hyperfine lines. In the SAS experimental technique, two counter-propagating laser beams from the same source propagate through the atomic medium. Since these beams are taken from the same source they have the same frequency. The basic experimental arrangement of the SAS is shown in figure 15.5. Both laser beams should overlap inside the atomic vapour cell. An atomic vapour cell is a cell in which atomic species like rubidium (Rb), caesium (Cs), sodium (Na), etc are sealed under an ultra-high vacuum. The laser beam whose absorption is measured by using a photodetector (figure 15.5) is known as the probe beam. The second beam which is coming from the opposite direction and overlapped with the probe beam in the centre of the cell is called a pump beam. The power of the pump beam is chosen much stronger than the probe beam so that it can significantly change the population of the ground level. This change in the population is measured by using a probe beam. In the measured probe absorption spectrum for a two-level atomic system, a saturated absorption dip or Lamb dip appears at the centre of the Doppler broadened profile.

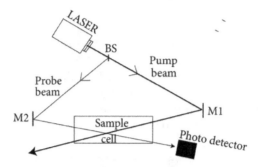

Figure 15.5. Experimental arrangement of the saturation absorption spectroscopy. BS: beam splitter, M1, M2: Mirror, sample cell contains atomic vapour (Rb, Cs etc).

15.4 Crossover resonances

The analysis of the last section is true only for a two-level system. But in general, we encounter the interaction of electromagnetic radiation with a multilevel system and the simplest multilevel system is a three-level system. A three-level system has a common lower or upper level. We shall first consider a three-level system with a common ground level ($|1>$) and two closely spaced upper levels ($|2>$ and $|3>$) commonly known as the V-type system. The allowed transitions are $|1> \rightarrow |2>$ and $|1> \rightarrow |3>$ and the transition frequencies are ω_1 and ω_2, respectively. We consider here closely spaced upper levels, i.e., the upper-level separation is within the Doppler width. If $\Delta\nu_D$ is the Doppler width (FWHM) then we can write mathematically $|\omega_2 - \omega_1| < \Delta\nu_D$. Now if this V-type system interacts with counter-propagating pump–probe wave fields (as shown in figure 15.6(a)) having a frequency ω, two holes are burnt for each of these allowed transitions ($|1> \rightarrow |2>$ and $|1> \rightarrow |3>$) due to the interaction of the atoms with the pump and probe field (figure 15.5(b)). These holes are burnt at velocities $v = \pm\frac{(\omega - \omega_1)}{k}$ and $v = \pm\frac{(\omega - \omega_2)}{k}$ ('light' and 'dark' holes as presented in figure 15.6(c)), respectively. So if we observe the velocity distribution curve of the atoms on the ground level ($|1>$) ρ_{11} we will see four holes are present there.

When the standing wave frequency scans the whole spectral-line contour, two holes appear at the resonance frequencies of the two allowed transitions (ω_1 and ω_2). This is due to the overlapping of holes in the line centre for each of the transitions. The 'light' holes generated due to field on ω_1 overlap with each other and similarly 'dark' holes due to the field on ω_2 overlap each other (see figure15.6(d)). Besides this, an extra hole appears at a frequency exactly equal to the mid of the two transition frequencies, i.e. $\omega_{12} = \frac{(\omega_1 + \omega_2)}{2}$, this is due to the overlapping of 'light' with 'dark' holes. In another way, we can say that the atoms moving with velocity $v \approx \pm\frac{(\omega_2 - \omega_1)}{k}$ can absorb the pump and probe beams simultaneously whose frequencies are ω_{12}. This dip is known as the *crossover resonance* that appears when a multilevel system interacts with a monochromatic standing wave field.

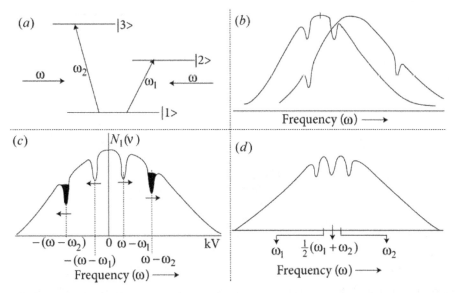

Figure 15.6. Creation of crossover resonances in a V-type system interacting with standing wave fields with a frequency ω. (a) Level scheme, (b) the Doppler contour shape for two allowed transitions, (c) the velocity distribution curve of the ground level $|1>$, (d) the shape of the saturated absorption spectrum.

15.5 Atomic coherence phenomena

In the previous sections, we restricted our discussion to two-level atomic systems but in a real situation, we would have to analyze multilevel systems. So now we shall discuss the interaction of two electromagnetic fields with the multilevel atomic systems. The simplest multilevel system is a three-level system (TLS). TLSs are of three types: V-type, lambda (Λ), and cascade or ladder (Ξ)-type systems, as shown in figure 15.7. In a TLS, two levels are connected by a strong field which can modify the absorption and dispersion properties of the atomic system. This field is called the *control field* or *pump field*. The effect of the control field in the atomic system is observed by a relatively weak field known as the *probe field* or *signal field*. Unlike saturation spectroscopy, here two laser sources are used instead of one. So here we do not observe any crossover resonances.

When laser fields interact with three- or multilevel atomic systems, many interesting phenomena can be observed like *Coherent Population Trapping* (CPT), *Electromagnetically Induced Transparency* (EIT), *Electromagnetically Induced Absorption* (EIA), *Lasing Without Inversion* (LWI) etc. Out of all these phenomena generated due to atomic coherences, in this section we will mostly concentrate on EIT and CPT phenomena and their applications, later we briefly discuss EIA. We consider a Λ-type TLS having one common upper or excited level $|3>$ and two lower or ground levels $|1>$ and $|2>$, as shown in figure 15.7(b), with energies, $\hbar\omega_1$, $\hbar\omega_2$, and $\hbar\omega_3$, respectively. $|1> \rightarrow |3>$ and $|2> \rightarrow |3>$ are two dipole allowed transitions. The transition frequencies are ω_{13} and ω_{23}. The strong coherent pump or control field of frequency ω_c with electric field amplitude E_c couples the $|2> \rightarrow |3>$ transition.

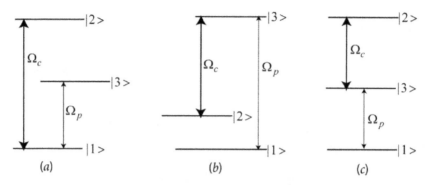

Figure 15.7. Three possible configurations for the three-level atomic system with two laser fields having Rabi-frequencies Ω_c (pump) and Ω_p (probe). (a) V-type; (b) Lambda (Λ)-type and (c) cascade-type (Ξ) system.

A weak probe field of frequency ω_p with electric field amplitude E_p couples the $|1> \rightarrow |3>$ transition whose dispersion and absorption signals are of interest. The control and probe Rabi frequencies are defined as $\Omega_c = \frac{E_c \mu_{23}}{\hbar}$ and $\Omega_p = \frac{E_p \mu_{13}}{\hbar}$ where μ_{23} and μ_{13} are the electric dipole moments of the two allowed transitions. The Hamiltonian of the atom interacting with two fields can be described as, $H = H_0 + H_I$.

The unperturbed and perturbed Hamiltonians are

$$H_0 = \hbar\omega_1 |1><1| + \hbar\omega_2 |2><2| + \hbar\omega_3 |3><3| \tag{15.40}$$

$$H_I = -\frac{\hbar\Omega_c}{2}\{|2><3| e^{i\omega_c t} + c.c.\} - \frac{\hbar\Omega_p}{2}\{|1><3| e^{i\omega_p t} + c.c.\} \tag{15.41}$$

The density matrix equation of motion can be derived from the master equation (equation (14.46)). In this system, the operator Γ is defined in the matrix form as,

$$\begin{pmatrix} 0 & 0 & 0 \\ 0 & \Gamma_{32} & 0 \\ 0 & 0 & \Gamma_{31} \end{pmatrix} \tag{15.42}$$

In the case of Λ-type system atoms can decay from level $|3>$ to level $|1>$ and $|2>$ with a spontaneous decay rate Γ_{31} and Γ_{32} but atoms cannot decay from the ground states $|1>$ and $|2>$ so decay rate is zero. The optical Bloch equations (OBEs) are expressed in the co-rotating frame we shall use '~' notation (Tilde) on the off-diagonal matrix elements (ρ_{ij}). The equations are

$$\frac{\partial\rho_{11}}{\partial t} = \Gamma_{31}\rho_{33} + i\frac{\Omega_p}{2}(\tilde{\rho}_{31} - \tilde{\rho}_{13}) \tag{15.43}$$

$$\frac{\partial\rho_{22}}{\partial t} = \Gamma_{32}\rho_{33} + i\frac{\Omega_c}{2}(\tilde{\rho}_{32} - \tilde{\rho}_{23}) \tag{15.44}$$

$$\frac{\partial\rho_{33}}{\partial t} = -(\Gamma_{31} + \Gamma_{32})\rho_{33} - i\frac{\Omega_p}{2}(\tilde{\rho}_{31} - \tilde{\rho}_{13}) - i\frac{\Omega_c}{2}(\tilde{\rho}_{32} - \tilde{\rho}_{23}) \tag{15.45}$$

$$\frac{\partial \tilde{\rho}_{21}}{\partial t} = i(\Delta_p - \Delta_c)\tilde{\rho}_{21} - i\frac{\Omega_p}{2}\tilde{\rho}_{23} + i\frac{\Omega_c}{2}\tilde{\rho}_{31} \qquad (15.46)$$

$$\frac{\partial \tilde{\rho}_{31}}{\partial t} = i\left(\Delta_p + i\frac{\Gamma_{31} + \Gamma_{32}}{2}\right)\tilde{\rho}_{31} + i\frac{\Omega_p}{2}(\rho_{11} - \rho_{33}) + i\frac{\Omega_c}{2}\tilde{\rho}_{21} \qquad (15.47)$$

$$\frac{\partial \tilde{\rho}_{32}}{\partial t} = i\left(\Delta_c + i\frac{\Gamma_{31} + \Gamma_{32}}{2}\right)\tilde{\rho}_{32} + i\frac{\Omega_c}{2}(\rho_{22} - \rho_{33}) + i\frac{\Omega_p}{2}\tilde{\rho}_{12} \qquad (15.48)$$

Another three equations are $\tilde{\rho}_{ji} = \tilde{\rho}_{ij}^*$ to give a total of sixteen equations. The off-diagonal matrix elements $\tilde{\rho}_{ji}(j \neq i)$ represent the coherence terms with the correlation between the states $|j\rangle$ and $|i\rangle$. ρ_{ii} ($i = 1, 2, 3$) is the population of the state $|i\rangle$, as the population of the system is conserved, hence $\sum_i \rho_{ii} = 1$. In the above equations (15.46)–(15.48), Δ_c and Δ_p are the probe and pump detuning defined as $\Delta_c = (\omega_c - \omega_{23})$ and $\Delta_p = (\omega_p - \omega_{13})$. We assume that initially, all the atoms are in the lower ground state $|1\rangle$, i.e.

$$\rho_{11} = 1, \quad \rho_{22} = 0 \text{ and } \quad \rho_{33} = 0 \qquad (15.49)$$

By using these approximations and under steady-state conditions (i.e. time derivatives vanish), equations (15.46)–(15.48) become

$$i(\Delta_p - \Delta_c)\tilde{\rho}_{21} - i\frac{\Omega_p}{2}\tilde{\rho}_{23} + i\frac{\Omega_c}{2}\tilde{\rho}_{31} = 0 \qquad (15.50a)$$

$$i\left(\Delta_p + i\frac{\Gamma_{31} + \Gamma_{32}}{2}\right)\tilde{\rho}_{31} + i\frac{\Omega_p}{2} + i\frac{\Omega_c}{2}\tilde{\rho}_{21} = 0 \qquad (15.50b)$$

$$i\left(\Delta_c + i\frac{\Gamma_{31} + \Gamma_{32}}{2}\right)\tilde{\rho}_{32} + i\frac{\Omega_p}{2}\tilde{\rho}_{12} = 0 \qquad (15.50c)$$

From these three equations, one can solve for the probe coherence term $\tilde{\rho}_{31}$ under weak probe regime ($\Omega_c \gg \Omega_p$), we can keep the terms with the first order of Ω_p and any order of Ω_c and the analytical solution for $\tilde{\rho}_{31}$ is:

$$\tilde{\rho}_{31} = -\frac{2\Omega_p(\Delta_p - \Delta_c)}{4(\Delta_p - \Delta_c)\left(\Delta_p + i\,\dfrac{\Gamma_{31} + \Gamma_{32}}{2}\right) - \Omega_c^2} \qquad (15.51)$$

Hence, the expression of the matrix element ρ_{31} which is related to probe absorption and dispersion signal is a function of the Ω_c, Ω_p, spontaneous decay rates of the different energy levels, and the detuning of the pump and probe fields.

The polarization of a medium is expressed as the expectation value of the dipole operator and assuming the number of atoms per unit volume N, it is given by

$$P = N\langle \mu \rangle = N\mathrm{Tr}(\rho\mu).$$

$$P = N\left(\mu_{31}\tilde{\rho}_{13}e^{i\omega_p t} + \mu_{32}\tilde{\rho}_{23}e^{i\omega_c t} + \mu_{13}\tilde{\rho}_{31}e^{-i\omega_p t} + \mu_{23}\tilde{\rho}_{32}e^{-i\omega_c t}\right) \tag{15.52}$$

Again for the linear susceptibility χ, the polarization can also be written in the following manner:

$$P = \epsilon_0\chi E = \frac{\epsilon_0\chi(\omega_p)E_p}{2}(e^{i\omega_p t} + e^{-i\omega_p t}) + \frac{\epsilon_0\chi(\omega_c)E_c}{2}(e^{i\omega_c t} + e^{-i\omega_c t}) \tag{15.53}$$

By using the above two equations (15.52) and (15.53) we can get the relation $\frac{\epsilon_0\chi(\omega_p)E_p}{2} = N\mu_{13}\tilde{\rho}_{31}$, and the susceptibility of the medium becomes

$$\chi = \frac{2N\mu_{13}}{\epsilon_0 E_p}\tilde{\rho}_{31} \tag{15.54}$$

By putting the analytical expression of $\tilde{\rho}_{31}$ from equation (15.51) to equation (15.54) we get the expression of analytical linear susceptibility.

$$\chi = -\frac{2N\mu_{13}}{\epsilon_0 E_p}\frac{2\Omega_p(\Delta_p - \Delta_c)}{4(\Delta_p - \Delta_c)\left(\Delta_p + i\frac{\Gamma_{31} + \Gamma_{32}}{2}\right) - \Omega_c^2} \tag{15.55}$$

Susceptibility is a complex quantity that can be written as $\chi = (\chi' + i\chi'')$, χ' is the real and χ'' is the imaginary part of χ.

$$\chi' = -\frac{2N\mu_{13}}{\epsilon_0 E_p}\frac{8\Omega_p\Delta_p(\Delta_p - \Delta_c)^2 - 2\Omega_p\Omega_c^2(\Delta_p - \Delta_c)}{16\left\{\Delta_p^2 + \left(\frac{\Gamma_{31} + \Gamma_{32}}{2}\right)^2\right\}(\Delta_p - \Delta_c)^2 - 8\Delta_p(\Delta_p - \Delta_c)\Omega_c^2 + \Omega_c^4} \tag{15.56}$$

$$\chi'' = \frac{2N\mu_{13}}{\epsilon_0 E_p}\frac{8\left(\frac{\Gamma_{31} + \Gamma_{32}}{2}\right)\Omega_p(\Delta_p - \Delta_c)^2}{16\left\{\Delta_p^2 + \left(\frac{\Gamma_{31} + \Gamma_{32}}{2}\right)^2\right\}(\Delta_p - \Delta_c)^2 - 8\Delta_p(\Delta_p - \Delta_c)\Omega_c^2 + \Omega_c^4} \tag{15.57}$$

The imaginary part of linear susceptibility is proportional to the absorption spectra of the field, but the real part of it is proportional to the dispersion spectra in the medium. The refractive index of the medium can be defined as $n_r = (1 + \frac{\chi'}{2})$ and the probe absorption coefficient $\alpha = \frac{\omega_p}{c}\chi''$. The transmission coefficient of the probe field is $T = e^{(-\chi'' kL)}$, where L is the length of the medium and $k = \frac{\omega_p}{c}$ is the wavenumber. If the control field is on-resonance with the transition $|2\rangle \rightarrow |3\rangle$ then we put $\Delta_c = 0$. In figure 15.8 the dotted curve represents probe absorption spectra and the solid curve represents probe dispersion of the medium. At zero probe detuning, i.e. $\Delta_p = 0$ (probe field is on-resonance with the transition $|1\rangle \rightarrow |3\rangle$) the

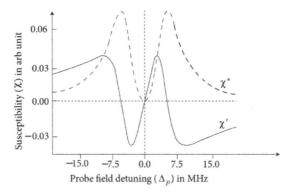

Figure 15.8. Plot of probe field susceptibility (χ) as a function of probe detuning (Δ_p) in presence of control field and control field is on-resonance, i.e. $\Delta_c = 0$ for the stationary atoms.

probe absorption of the probe field reduces drastically. Both χ' and χ'' are zero at $\Delta_p = 0$ so the medium becomes transparent in presence of a strong control field. This phenomenon is induced due to the presence of the electromagnetic field known as *Electromagnetically Induced Transparency* (EIT). The dip observed in the χ'' curve at the vicinity of $\Delta_p = 0$ is known as the *EIT window*. This EIT window is extremely narrow and has many potential applications in different areas of quantum optics. Much fundamental experimental research has been performed on EIT; we will discuss some of this in the section of application of EIT. Many review articles are available on this interesting topic, interested readers may see the further reading section of this chapter.

Considering the control field to be absent ($\Omega_c = 0$ and $\Delta_c = 0$) the real and imaginary parts of the linear susceptibility equations (15.56) and (15.57) are

$$\chi' = -\frac{2N\mu_{13}}{\epsilon_0 E_p} \frac{\Omega_p \Delta_p}{2\left\{\Delta_p^2 + \left(\frac{\Gamma_{31} + \Gamma_{32}}{2}\right)^2\right\}} \tag{15.58}$$

$$\chi'' = \frac{2N\mu_{13}}{\epsilon_0 E_p} \frac{\left(\frac{\Gamma_{31} + \Gamma_{32}}{2}\right)\Omega_p}{2\left\{\Delta_p^2 + \left(\frac{\Gamma_{31} + \Gamma_{32}}{2}\right)^2\right\}} \tag{15.59}$$

The dotted line curve and the solid line curve represent the probe absorption and probe dispersion spectra, respectively, in absence of the control field for stationary atoms. In this situation, the line shape of the absorption curve is typical Lorentzian and under this condition, the EIT window vanishes as expected. The slope of the χ' in figures 15.8 and 15.9 are different, in the EIT case the slope is increasing with detuning. A very steep normal dispersion profile is found in EIT, whereas when the control field is not present the slope is decreasing with probe detuning so anomalous dispersion is observed.

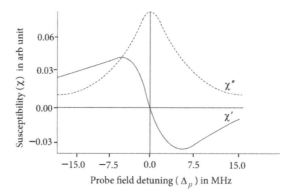

Figure 15.9. Plot of susceptibility (χ) as a function of detuning (Δ_p) of the probe of the probe field in the absence of the control field for stationary atoms. The dotted curve represents the imaginary part of susceptibility (χ'') and the solid curve represents the real part of susceptibility (χ'')

15.6 EIT Hamiltonian of the system

EIT cannot be explained classically as it is a quantum mechanical effect. However, people try to explain it by the interference that occurs between the probability amplitudes of the control and probe transitions. The probe absorption spectra reduce due to the destructive interference between the transition pathways when the probe and control fields are on-resonant with the energy levels. This type of phenomenon can also be noticed in other systems such as quantum beats. There are two possible transition pathways by which the atoms can move from the ground state $|1\rangle$ to the excited state $|3\rangle$. These two different pathways are $|1\rangle \rightarrow |3\rangle$ and $|1\rangle \rightarrow |3\rangle \rightarrow |2\rangle \rightarrow |3\rangle$ (figure 15.7(b)). These two pathways interfere destructively under suitable conditions. The EIT phenomenon is prominent in the Λ-type system, whereas it is diminished in Ξ-type or V-type systems due to the high coherence decay rate of the dipole forbidden transition.

The concept of atom–field interaction is well understood in the semi-classical approach, therefore, we shall employ the semi-classical treatment to find out the absorption and dispersion spectra of the system. First of all, we shall build up the total Hamiltonian of the system and by using this Hamiltonian we shall look for the concept of dressed states to interpret EIT. As we discussed earlier, the total Hamiltonian of the system can be written as $H = (H_0 + H_I)$, here H_0 and H_I are called unperturbed Hamiltonian or bare-atom Hamiltonian and interaction Hamiltonian of the system, respectively. Bare-atom Hamiltonian can be written as

$$H_0 = \left(\sum_i |i\rangle\langle i| \right) H_0 \left(\sum_i |i\rangle\langle i| \right) = \begin{pmatrix} \hbar\omega_1 & 0 & 0 \\ 0 & \hbar\omega_2 & 0 \\ 0 & 0 & \hbar\omega_3 \end{pmatrix} \quad (15.60)$$

We can form a complete orthogonal basis consisting of the three states $|i\rangle$, each with an eigenvalue $\hbar\omega_i (i = 1, 2, 3)$ for the Λ-type TLS. The interaction Hamiltonian can be written as,

$$H_I = -\mu E = -\left(\sum_i |i\rangle\langle i|\right)\mu\left(\sum_i |i\rangle\langle i|\right)E = -E\begin{pmatrix} \mu_{11} & \mu_{12} & \mu_{13} \\ \mu_{21} & \mu_{22} & \mu_{23} \\ \mu_{31} & \mu_{32} & \mu_{33} \end{pmatrix} \quad (15.61)$$

Here μ is the dipole moment, again the transition $|1\rangle \rightarrow |2\rangle$ is dipole forbidden so we choose $\mu_{12} = \mu_{21} = 0$ and we can remove all the diagonal matrix elements ($\mu_{ii} = 0$, $i = 1, 2, 3$) because the atoms with spherically symmetric wave functions have no permanent dipole moments. Using all these assumptions, we rewrite the interaction Hamiltonian as:

$$H_I = -E\begin{pmatrix} 0 & 0 & \mu_{13} \\ 0 & 0 & \mu_{23} \\ \mu_{31} & \mu_{32} & 0 \end{pmatrix} \quad (15.62)$$

To find the total Hamiltonian (EIT Hamiltonian) of the system due to the field vector $\mathbf{E}$ and the interaction Hamiltonian H_I, we use the rotating wave approximation (RWA); as we explained earlier, it neglects any rapidly oscillating term of the Hamiltonian. We will use the time evolution operator to transform H_I into the interaction picture having the form:

$$U(t) = e^{iH_U t/\hbar} = \begin{pmatrix} e^{i\omega_1 t} & 0 & 0 \\ 0 & e^{i\omega_2 t} & 0 \\ 0 & 0 & e^{i\omega_3 t} \end{pmatrix} \quad (15.63)$$

Using the above equation (15.63) involving transformation to H_I, we get

$$UH_I U^\dagger$$

$$= -E\begin{pmatrix} e^{i\omega_1 t} & 0 & 0 \\ 0 & e^{i\omega_2 t} & 0 \\ 0 & 0 & e^{i\omega_3 t} \end{pmatrix}\begin{pmatrix} 0 & 0 & \mu_{13} \\ 0 & 0 & \mu_{23} \\ \mu_{31} & \mu_{32} & 0 \end{pmatrix}\begin{pmatrix} e^{-i\omega_1 t} & 0 & 0 \\ 0 & e^{-i\omega_2 t} & 0 \\ 0 & 0 & e^{-i\omega_3 t} \end{pmatrix} \quad (15.64)$$

$$= -E\begin{pmatrix} 0 & 0 & \mu_{13}e^{i(\omega_1-\omega_3)t} \\ 0 & 0 & \mu_{23}e^{i(\omega_2-\omega_3)t} \\ \mu_{31}e^{i(\omega_3-\omega_1)t} & \mu_{32}e^{i(\omega_3-\omega_2)t} & 0 \end{pmatrix}$$

Two electric fields interact with the Λ-type TLS control, (E_c) and probe (E_p), and the electric field $\mathbf{E}$ can be written as,

$$E = E_p\cos(\omega_p t) + E_c\cos(\omega_c t) = \frac{E_p}{2}(e^{i\omega_p t} + e^{-i\omega_p t}) + \frac{E_c}{2}(e^{i\omega_c t} + e^{-i\omega_c t}) \quad (15.65)$$

Using this expression into the transformed interaction Hamiltonian and applying the RWA, i.e. omitting the rapidly varying terms compared to slow varying terms.

We can write the matrix elements as, $(UH_IU^\dagger)_{13} = -\frac{1}{2}E_p\mu_{13}e^{i(\omega_1-\omega_3+\omega_p)t}$, $(UH_IU^\dagger)_{23} = -\frac{1}{2}E_c\mu_{23}e^{i(\omega_2-\omega_3+\omega_c)t}$, $(UH_IU^\dagger)_{31} = -\frac{1}{2}E_p\mu_{31}e^{i(\omega_3-\omega_1-\omega_p)t}$ and $(UH_IU^\dagger)_{32} = -\frac{1}{2}E_c\mu_{32}e^{i(\omega_3-\omega_2-\omega_c)t}$. Now, we can come back to the interaction Hamiltonian again using the $U(t)$. In the Schrödinger picture, the interaction Hamiltonian becomes,

$$H_I = U^\dagger(UH_IU^\dagger)U = -\frac{\hbar}{2}\begin{pmatrix} 0 & 0 & \Omega_p e^{i\omega_p t} \\ 0 & 0 & \Omega_c e^{i\omega_c t} \\ \Omega_p e^{-i\omega_p t} & \Omega_c e^{-i\omega_c t} & 0 \end{pmatrix}. \tag{15.66}$$

Here, $\Omega_p = \frac{E_p\mu_{31}}{\hbar}$ and $\Omega_c = \frac{E_c\mu_{23}}{\hbar}$ are, respectively, the probe and control Rabi frequencies. Combining the interaction Hamiltonian with bare-atom Hamiltonian we can get the complete EIT Hamiltonian.

$$H_{EIT} = H_0 + H_I = \frac{\hbar}{2}\begin{pmatrix} 2\omega_1 & 0 & -\Omega_p e^{i\omega_p t} \\ 0 & 2\omega_2 & -\Omega_c e^{i\omega_c t} \\ -\Omega_p e^{-i\omega_p t} & -\Omega_c e^{-i\omega_c t} & 2\omega_3 \end{pmatrix} \tag{15.67}$$

Now we want to remove all the time dependence from the EIT Hamiltonian and for that, we change it in the co-rotating frame. In the co-rotating frame corresponding basis is known as the rotating basis, which we shall denote with ~'s (Tilde) on transformed matrix elements. The new basis $|\tilde{m}'\rangle$ with the new EIT Hamiltonian $\tilde{H}_{EIT}$ in the corotating frame, is connected to the old basis $|m'\rangle$ by the transformation relation $|\tilde{m}'\rangle = \tilde{U}(t)|m'\rangle$. The unitary matrix $\tilde{U}(t)$ is given below:

$$\tilde{U}(t) = \begin{pmatrix} e^{-i\omega_p t} & 0 & 0 \\ 0 & e^{-i\omega_c t} & 0 \\ 0 & 0 & 1 \end{pmatrix} \tag{15.68}$$

This transformation must be unitary and the Schrödinger equation is satisfied by the co-rotating Hamiltonian,

$$\tilde{H}_{EIT}|\tilde{m}'\rangle = i\hbar\frac{\partial}{\partial t}|\tilde{m}'\rangle$$

$$= \left(i\hbar\frac{\partial \tilde{U}}{\partial t}\tilde{U}^\dagger + \tilde{U}H_{EIT}\tilde{U}^\dagger\right)\tilde{U}|m'\rangle \tag{15.69}$$

The transformed EIT Hamiltonian ($\tilde{H}_{EIT}$) in the co-rotating frame can simplify to the following form,

$$\tilde{H}_{\text{EIT}} = \left(i\hbar \frac{\partial \tilde{U}}{\partial t} \tilde{U}^\dagger + \tilde{U} H_{\text{EIT}} \tilde{U}^\dagger \right)$$

$$= i\hbar \begin{pmatrix} -i\omega_p e^{-i\omega_p t} & 0 & 0 \\ 0 & -i\omega_c e^{-i\omega_c t} & 0 \\ 0 & 0 & 0 \end{pmatrix} \begin{pmatrix} e^{i\omega_p t} & 0 & 0 \\ 0 & e^{i\omega_c t} & 0 \\ 0 & 0 & 1 \end{pmatrix}$$

$$+ \frac{\hbar}{2} \begin{pmatrix} e^{-i\omega_p t} & 0 & 0 \\ 0 & e^{-i\omega_c t} & 0 \\ 0 & 0 & 1 \end{pmatrix} \begin{pmatrix} 2\omega_1 & 0 & -\Omega_p e^{i\omega_p t} \\ 0 & 2\omega_2 & -\Omega_c e^{i\omega_c t} \\ -\Omega_p e^{-i\omega_p t} & -\Omega_c e^{-i\omega_c t} & 2\omega_3 \end{pmatrix} \begin{pmatrix} e^{i\omega_p t} & 0 & 0 \\ 0 & e^{i\omega_c t} & 0 \\ 0 & 0 & 1 \end{pmatrix} \qquad (15.70)$$

$$\tilde{H}_{\text{EIT}} = \frac{\hbar}{2} \begin{pmatrix} 2(\omega_1 + \omega_p) & 0 & -\Omega_p \\ 0 & 2(\omega_2 + \omega_c) & -\Omega_c \\ -\Omega_p & -\Omega_c & 2\omega_3 \end{pmatrix}$$

This new or effective EIT Hamiltonian is free from the time parts. We can express it in terms of control and probe detuning if we add a multiple of the identity matrix (I) with the $\tilde{H}_{\text{EIT}}$, keeping the physical results unchanged. We add $-\hbar(\omega_1 + \omega_p)I$ with the $\tilde{H}_{\text{EIT}}$ to write:

$$\tilde{H}_{\text{EIT}} = -\frac{\hbar}{2} \begin{pmatrix} 0 & 0 & \Omega_p \\ 0 & 2(\Delta_p - \Delta_c) & \Omega_c \\ \Omega_p & \Omega_c & 2\Delta_p \end{pmatrix} \qquad (15.71)$$

where $\Delta_p = \omega_p - (\omega_3 - \omega_1) = (\omega_p - \omega_{13})$ and $\Delta_c = \omega_c - (\omega_3 - \omega_2) = (\omega_c - \omega_{23})$ are the probe and control field detuning, respectively.

15.7 Dressed states picture

It is very easy to derive the effective Hamiltonian for a two-level system interacting with one on-resonant e-m field.

$$\tilde{H}_{\text{Eff}} = -\frac{\hbar}{2} \begin{pmatrix} -\delta & -\Omega_R \\ \Omega_R & \delta \end{pmatrix} \qquad (15.72)$$

Here δ is the detuning of the applied field and Ω_R is the Rabi frequency as defined in chapter 14. Now we can use this equation (15.72) to explain the dressed state. The above Hamiltonian($\tilde{H}_{\text{Eff}}$) is presented in the bare state bases of the two levels say $\{|a>, |b>\}$. But, sometimes it is convenient to express it on an (atom + field) basis where the field dresses the bare atomic levels. This basis is called the dressed state basis. Now we will diagonalize the Hamiltonian matrix of equation (15.72) and after diagonalization, we get eigenvalues as,

$$\Omega_G = \pm(\delta^2 + \Omega_R^2)^{\frac{1}{2}} \qquad (15.73)$$

This is called a generalized Rabi frequency. The transformation matrix becomes.

$$R = \begin{pmatrix} \cos\theta & \sin\theta \\ -\sin\theta & \cos\theta \end{pmatrix} \tag{15.74}$$

Here, $\cos\theta = \dfrac{\Omega_R}{\sqrt{(\Omega_G - \delta)^2 + \Omega_R^2}}$ and $\sin\theta = \dfrac{(\Omega_G - \delta)}{\sqrt{(\Omega_G - \delta)^2 + \Omega_R^2}}$

Corresponding eigenvectors are calculated by an effective rotation of the uncoupled basis states $|a>$ and $|b>$ as,

$$\begin{pmatrix} |+> \\ |-> \end{pmatrix} = R.\,\bar{U} = \begin{pmatrix} \cos\theta \mid a > +\sin\theta \mid b> \\ -\sin\theta \mid a > +\cos\theta \mid b> \end{pmatrix} \tag{15.75}$$

The states $(|+>$ and $|->)$ are the dressed states of the two-level system. The energy corresponding to these frequencies has a shift from the energy of the bare state and is known as the AC-Stark shift. In the case of a TLS, the transformation matrix (R) would be a three-dimensional rotational matrix.

Now we will discuss the dressed states generated in the EIT process for which we can take the effective Hamiltonian for EIT ($\tilde{H}_{EIT}$). Since now we are dealing with a TLS, to express the eigenstates we consider the two mixing angles θ and ϕ. They depend simply upon the Rabi frequencies as well as on the one-photon detuning ($\Delta_p = \Delta$) and the Raman or two-photon detuning ($\Delta_R = (\Delta_p - \Delta_c)$). The Raman resonance condition is $\Delta_R = 0$, under this condition the mixing angles can be written as

$$\tan\theta = \frac{\Omega_p}{\Omega_c} \tag{15.76a}$$

$$\tan 2\phi = \frac{\sqrt{\mid \Omega_p \mid^2 + \mid \Omega_c \mid^2}}{\Delta_R} \tag{15.76b}$$

Solving the $\tilde{H}_{EIT}$ matrix (equation (15.71)) at Raman resonance condition, we have obtained the eigenvalues of the Hamiltonian as,

$$\lambda_0 = 0 \tag{15.77a}$$

$$\lambda_\pm = \frac{\hbar}{2}[\Delta \pm \sqrt{\Delta^2 + \mid \Omega_p \mid^2 + \mid \Omega_c \mid^2}] \tag{15.77b}$$

The corresponding eigenstates are written in terms of the bare states as,

$$\mid 0 > = \cos\theta \mid 1 > -\sin\theta \mid 2> \tag{15.78a}$$

$$\mid +> = \sin\theta\sin\phi \mid 1 > +\cos\phi \mid 3 > +\cos\theta\sin\phi \mid 2> \tag{15.78b}$$

$$\mid -> = \sin\theta\sin\phi \mid 1 > -\sin\phi \mid 3 > +\cos\theta\cos\phi \mid 2> \tag{15.78c}$$

The states $|\pm>$ contain all the states bare state components (i.e. $|1>$, $|2>$ and $|3>$) while the bare state $|3>$ has no contribution in the state $|0>$. The state $|0>$ is therefore called the dark state because there is no possibility of excitation to the excited state $|3>$ from it.

In the case of EIT considering a weak probe beam, i.e. $\Omega_p < <\Omega_c$ and then $\tan\theta < <1$ and $\sin\theta \rightarrow 0$, $\cos\theta \rightarrow 1$. If the probe field is on-resonance $\Delta_p = 0$, then from equation (15.76a) $\tan2\phi = \infty \Rightarrow \phi = \frac{\pi}{4}$, equation (15.78a)–(15.78c) becomes,

$$|0> = |1> \tag{15.79a}$$

$$|+> = \frac{1}{\sqrt{2}}(|2>+|3>) \tag{15.79b}$$

$$|-> = \frac{1}{\sqrt{2}}(|2>-|3>) \tag{15.79c}$$

The above states (equations (15.79a–15.79c)) are the dressed states created due to the EIT phenomenon with the condition of a strong pump beam and weak probe beam.

15.8 Coherent population trapping

If two CW fields to a Λ-type three-level atomic system are proposed in a coherent superposition of states, then the superposition of states is stable with absorption from the coherent radiation field. Therefore, this phenomenon is known as Coherent Population Trapping (CPT). The EIT Hamiltonian, for a Λ-type configuration consists of two terms unperturbed and the interaction Hamiltonian defined in equation (15.40) and equation (15.41). It is evident that the eigenstates of the unperturbed Hamiltonian (H_0), $|1\rangle$, $|2\rangle$, and $|3\rangle$ are not eigenstates of the H but the eigenstates of H are a linear combination of the states $|1\rangle$, $|2\rangle$, and $|3\rangle$. If we assume that the two fields are on-resonance (i.e. $\omega_p = (\omega_3 - \omega_1) = \omega_{13}$ and $\omega_c = (\omega_3 - \omega_2) = \omega_{23}$ with their associated transitions), then the three eigenstates of H are,

$$|C_1\rangle = \frac{1}{\sqrt{2}}\left(+|3\rangle + \frac{\Omega_p}{\sqrt{\Omega_p^2 + \Omega_c^2}}|1\rangle + \frac{\Omega_c}{\sqrt{\Omega_p^2 + \Omega_c^2}}|2\rangle\right) \tag{15.80}$$

$$|C_2\rangle = \frac{1}{\sqrt{2}}\left(|3\rangle - \frac{\Omega_p}{\sqrt{\Omega_p^2 + \Omega_c^2}}|1\rangle - \frac{\Omega_c}{\sqrt{\Omega_p^2 + \Omega_c^2}}|2\rangle\right) \tag{15.81}$$

$$|NC\rangle = \frac{\Omega_c}{\sqrt{\Omega_p^2 + \Omega_c^2}}|1\rangle - \frac{\Omega_p}{\sqrt{\Omega_p^2 + \Omega_c^2}}|2\rangle \tag{15.82}$$

The eigenstate $|NC\rangle$ has no component of the excited state $|3\rangle$, so there is no coupling between these two states although the eigenstates $|C_1\rangle$ and $|C_2\rangle$ are coupled with state $| \, 3 \rangle$. In this manner, any population that falls in an eigenstate $| \, NC \rangle$ cannot be transferred to the state $| \, 3 \rangle$. Therefore, these two eigenstates $| \, C_1 \rangle$ and $| \, C_2 \rangle$ are known as coupling states and state $| \, NC \rangle$ is called the non-coupling state. Over a while, depending on the spontaneous emission from the state $| \, 3 \rangle$, all the population of the atomic system gets coherently trapped in the state $| \, NC \rangle$. Hence this phenomenon is known as CPT and this non-coupling state $| \, NC \rangle$ is also called a *dark state*.

CPT has many useful fields of application. The excited states generally have less lifetime than the ground state. CPT was largely enhanced in metrology and recently CPT was used for realizing all-optical miniaturized *atomic clocks. Velocity selective coherent population trapping* (VSCPT) is another important application of CPT.

EIT phenomenon is similar to CPT. The basic difference between CPT and EIT lies in the relative strength of the probe and control beams. For CPT, the two coupling fields are of approximately equal Rabi frequency ($\Omega_p \approx \Omega_c$) and for EIT one of the fields (control) is much stronger than the other (probe) beam ($\Omega_c \gg \Omega_p$).

Therefore, the theoretical treatment for the formation of EIT can be obtained from equations (15.80)–(15.82) by applying the condition, i. e. $\Omega_c \gg \Omega_p$. Under this condition, the non-coupling state (dark state) becomes: $| \, NC \rangle = \dfrac{\Omega_c}{\sqrt{\Omega_p^2 + \Omega_c^2}} | \, 1 \rangle \approx | \, 1 \rangle$

(see equation (15.79)). In this situation, mainly the ground state is converted into the dark state and the decoupled dark state changes the stationary state of both the bare-atom Hamiltonian and EIT Hamiltonian. So all the population of the system becomes trapped in the ground state $| \, 1 \rangle$. The excitation of the atoms from the ground state $| \, 1 \rangle$ is not possible by the weak probe field anymore and the absorption of the probe field drastically diminishes. We have shown it mathematically in the earlier section.

The modification to the probe absorption spectra in an EIT medium caused by EIT is accompanied by modification of the probe dispersion profile. The group velocity of the medium is changed due to the modification of the dispersion of the medium and the reduction of the group velocity of a probe pulse is caused by EIT. The group velocity of $c/165$ in lead (Pb) vapour and $c/3000$ in caesium (Cs) vapour were reported. An investigation on one single-photon pulse to 'coherently control or manipulate the quantum state' of another pulse was proposed for the EIT medium. The light pulse stored as atomic excitations, i.e. dark state polaritons, was experimentally demonstrated where the magnetic moment of the polariton is seen to be around 5.1×10^{-24} J T^{-1}.

Another very important phenomenon is the storage of light within the dispersive medium. Researchers used EIT to slow down a pulse of light in sodium D_1 transition at a temperature of 0.9 μK.

15.9 Electromagnetically induced absorption (EIA)

In the presence of the control beam, if the absorption of a probe beam is increased, then this type of phenomenon is called electromagnetically induced absorption (EIA). In the case of EIA generally, two control beams are used to generate the

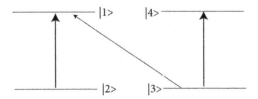

Figure 15.10. Schematic diagram of N-type configuration. Here, the levels |3>→|1> connect by a weak field (probe beam), |2> →|1> and |3>→4> connect by two strong fields (control beam).

absorption in the probe beam. In the EIT process the probe beam transparency is increased but here absorption is increased. This phenomenon needs an N-type energy level scheme as shown in figure 15.10. Here two control beams are acting, one control beam is between $|2> \rightarrow |1>$ transition and the second control beam couples $|3> \rightarrow |4>$ transition. The weak probe beam is connected to the $|3> \rightarrow |1>$ transition.

There are two different mechanisms for the formation of EIA, one is the transfer of coherence (TOC) and the other one is the transfer of population (TOP). The control beam and the probe beam must have different polarizations for the formation of EIA due to the transfer of coherence (EIA-TOC) but for EIA due to the transfer of population (EIA-TOP), the polarization of the control and the probe beam should be the same. When the control and the probe laser beams have different polarization, TOC can occur and also has a significant amount of control beam induced population transfer in the excited states. The most important criterion on the excited and ground hyperfine states are, $F_e = F_g + 1$ and $F_g > 0$ (F_e is the total angular momentum quantum number of the excited state, whereas F_g is that of the ground state). For N-type configuration, EIA-TOC requires a minimum of two degenerate ground levels and at least two degenerate excited states, as shown in figure 15.10.

The coherence has been created between the degenerate excited states. A transfer of coherence is caused by the spontaneous emission from the excited states to the ground states that lead to the formation of EIA. When the control Rabi frequency is increased, the excited state population is also increased, so the coherence of the excited state increases. Therefore, this phenomenon leads to an increased TOC to the ground state and we get the EIA. If the Rabi frequency of one of the control beams is zero, the probe absorption spectra would show EIT for the Λ-type or V-type system. So, we can conclude that if TOC is reduced significantly, there is the corresponding lowering of the predicated absorption, until zero TOC transparency (EIT) is predicted instead of absorption (EIA).

15.10 Further reading

[1] Scully M O and Zubairy S 1997 *Quantum Optics* (Cambridge: Cambridge University Press)
[2] Stenholm S 2005 *Foundation of Laser Spectroscopy* (New York: Dover)
[3] Foot C J 2005 *Atomic Physics* (Oxford: Oxford University Press)

[4] Letokhov V S and Chebotaev V P 1977 *Laser Spectroscopy* (Berlin: Springer)

[5] Demtroder W 2008 *Laser Spectroscopy* (Berlin: Springer)

[6] Ghosh P N *Laser Physics and Spectroscopy* (Boca Raton, FL: CRC Press)

[7] Harris S E 1997 Electromagnetically induced transparency *Phys. Today* **50** 36 (and references therein)

[8] Fleischhauer M, Imamoglu A and Marangos J P 2005 Electromagnetically induced transparency: optics in coherent media *Rev. Mod. Phys.* **77** 633

[9] Lezama A, Barreiro S and Akulshin A M 1999 Electromagnetically induced absorption *Phys. Rev. A* **62** 1033

[10] Arimondo E 1996 Coherent population trapping in laser spectroscopy *Prog. Opt.* **35** 247

[11] Vestergaard Hau L, Harris S E, Dutton Z and Behroozi C H 1999 Light speed reduction to 17 metres per second in an ultracold atomic gas *Nature* **397** 594–8

[12] Schlosberg H R and Javan A 1966 Saturation behaviour of a Doppler-broadened transition involving levels with closely spaced structure *Phys. Rev.* **150** 267

15.11 Problems

1. By using the density matrix equation for ij-th element of the density matrix

$$\dot{\rho}_{ij} = -\frac{i}{\hbar} \sum_l (H_{il}\rho_{lj} - \rho_{il}H_{lj})$$

derive the following equations of the density matrix of a two-level stationary atom interacting with a standing wave field $E(z, t) = E_0 \cos \Omega t \sin kz$ after adding the decay terms phenomenologically.

$$\frac{\partial \rho_{aa}}{\partial z} = -\gamma_a \rho_{aa} + \left(\frac{iE_0\mu}{\hbar}\right)\cos(kz - \Omega t)\sin kz(\rho_{ab} - \rho_{ba})$$

$$\frac{\partial \rho_{bb}}{\partial z} = -\gamma_b \rho_{bb} - \left(\frac{iE_0\mu}{\hbar}\right)\cos(kz - \Omega t)\sin kz(\rho_{ab} - \rho_{ba})$$

$$\frac{\partial \rho_{ba}}{\partial z} = (i\omega + \gamma_{ba})\rho_{ba} - \left(\frac{iE_0\mu}{\hbar}\right)\cos(kz - \Omega t)\sin kz(\rho_{bb} - \rho_{aa})$$

2. In problem 15.1, rewrite the density matrix equations under rotating wave approximation (RWA), then show that under the steady-state condition population difference is

$$(\rho_{bb} - \rho_{aa}) = \frac{\bar{N}}{[1 + 2I\eta \; L(\delta)\sin^2 kz]}$$

The constants $\bar{N}$, I and η are defined in the text, $L(\delta) = \dfrac{\gamma_{ba}^2}{(\gamma_{ba}^2 + (\delta)^2)}$, a dimensionless Lorentzian function. Give a plot of $(\rho_{bb} - \rho_{aa})$ versus z.

3. Show that the off-diagonal density matrix element ρ_{ba} of the problem 15.1 under steady-state condition ($\frac{\partial \rho_{ba}}{\partial z} = 0$) has the following form with the assumption $\rho_{ba} = \tilde{\rho}_{ba} e^{-i\Omega t}$.

$$\tilde{\rho}_{ba} = -\frac{E_0 \mu}{2\hbar} \frac{\bar{N}(\gamma_{ab} + i\delta)}{(\delta^2 + \gamma_{ba}[1 + 2I\eta \quad L(\delta)sin^2 \, kz]^{1/2})}$$

4. A two-level atom moving with a velocity v interacts with an electromagnetic field that is detuned from the resonance frequency by an amount Δ. Find the condition when the ratio of the unsaturated population difference ($\rho_{bb} - \rho_{aa}$) and the saturated or unperturbed population difference $\bar{N}$ is $\frac{1}{2}$ for the resonant case $\Delta = -kv$.

5. Consider a cascade (Ξ) type TLS as shown in figure 15.7(c). Two resonant e-m fields are interacting between level |1> → |3> and |2> → |3>. Write down the nine-density matrix equation of motion for this system.

6. Show that the eigenstates of the Hamiltonian of equation (15.74) are the following

$$| C_1 \rangle = \frac{1}{\sqrt{2}} \left(+| 3 \rangle + \frac{\Omega_p}{\sqrt{\Omega_p^2 + \Omega_c^2}} | 1 \rangle + \frac{\Omega_c}{\sqrt{\Omega_p^2 + \Omega_c^2}} | 2 \rangle \right)$$

$$| C_2 \rangle = \frac{1}{\sqrt{2}} \left(| 3 \rangle - \frac{\Omega_p}{\sqrt{\Omega_p^2 + \Omega_c^2}} | 1 \rangle - \frac{\Omega_c}{\sqrt{\Omega_p^2 + \Omega_c^2}} | 2 \rangle \right)$$

$$| NC \rangle = \frac{\Omega_c}{\sqrt{\Omega_p^2 + \Omega_c^2}} | 1 \rangle - \frac{\Omega_p}{\sqrt{\Omega_p^2 + \Omega_c^2}} | 2 \rangle$$

Find the corresponding eigenvalues.

Chapter 16

Quantum theory of radiation

Until now we have followed semi-classical treatment to treat atom–field interaction where the electromagnetic (e-m) field is treated classically and the atom is assumed to have quantized energy states. This description is usually adequate to give a reasonable explanation for a large number of problems. But there are a few cases where a full quantum mechanical description is essential to explain the phenomena properly. Spontaneous emission, squeezing of light, the Lamb shift, the anomalous gyromagnetic moment of the electron, resonance fluorescence, etc are examples of such processes where we arrive at a wrong conclusion by adopting the semi-classical approach. In this chapter, we shall aim to quantize the radiation field in free space.

16.1 Maxwell's equations

Electric and magnetic field responses in a medium can be combined by using Maxwell's equation in classical electrodynamics. Maxwell's equation in the free space can be written as,

$$\nabla \times \boldsymbol{E} = -\frac{\partial \boldsymbol{B}}{\partial t} \tag{16.1}$$

$$\nabla \times \boldsymbol{B} = \mu_0 \varepsilon_0 \frac{\partial \boldsymbol{E}}{\partial t} \tag{16.2}$$

$$\nabla \cdot \boldsymbol{B} = 0 \tag{16.3}$$

$$\nabla \cdot \boldsymbol{E} = 0 \tag{16.4}$$

where **E** and **B** are the electric and magnetic fields. μ_0 and ε_0 are the free-space permittivity and permeability, respectively. In free space velocity of light c can be written as $c = \frac{1}{\sqrt{\mu_0 \varepsilon_0}} = 2.998 \times 10^8$ m s^{-1}. If A is the vector potential then $\boldsymbol{B} = \nabla \times A$, by putting the value of **B** in equation (16.1), we get

doi:10.1088/978-0-7503-2715-2ch16

$$\nabla \times \left(\mathbf{E} + \frac{\partial A}{\partial t} \right) = 0 \tag{16.5}$$

We may write, E as, $\mathbf{E} = -\frac{\partial A}{\partial t} - \nabla\phi$, here ϕ is known as scalar potential. Using this value of, E in equation (16.4) we get.

$$\nabla^2\phi + \frac{\partial}{\partial t}(\nabla . \mathbf{A}) = \mathbf{0} \tag{16.6}$$

The electric and magnetic fields are not altered if we make the gauge transformations, $\mathbf{A}' = \mathbf{A} + \nabla\zeta$ and $\phi' = \phi - \frac{\partial\zeta}{\partial t}$, here '$\zeta$' is arbitrary scalar function. So the selection of gauge is arbitrary we may choose ζ such that $\nabla.A = 0$ known as Coulomb gauge. The scalar potential ϕ satisfies the equation $\nabla^2\phi = 0$, one of the solutions to this equation is $\phi = 0$. Then we will get $\mathbf{E} = -\frac{\partial A}{\partial t}$, by using equation (16.2) we can write

$$\nabla \times (\nabla \times A) = -\mu_0\varepsilon_0\frac{\partial^2 A}{\partial t^2} \tag{16.7}$$

By using the vector identity $\nabla \times (\nabla \times A) = \nabla(\nabla . \mathbf{A}) - \nabla^2 A$, we have

$$\nabla^2\mathbf{A} = \mu_0\varepsilon_0\frac{\partial^2 A}{\partial t^2} = \frac{1}{c^2}\frac{\partial^2 A}{\partial t^2} \tag{16.8}$$

This is the wave equation for the electromagnetic field in free space. The solution of the wave equation (equation (16.8)) is $A(r) = A(0) \exp(ik.r)$, $\mathbf{A}(0)$ is the value of $A(r)$ at $r = 0$. By using the condition $\nabla.A = 0$ we can write $k.A(0) = 0$, the wave amplitude $\mathbf{A}(0)$ is perpendicular to the propagation vector k. This specifies the transverse nature of the e-m field, there can be only two independent polarizations $\mathbf{A}(0)$ for each k.

16.2 The electromagnetic field in a cavity

Here we consider the electromagnetic radiation field in a three-dimensional cubical cavity having length 'L'. The volume of the cavity is $V = L^3$. The walls of the cavity are perfectly conducting, where the field vanishes, at $x,y,z = 0$ and $x,y,z = L$. The electric field inside the cavity will form a standing wave, as shown in figure 16.1, standing wave only along y-direction is shown. So we can apply the periodic boundary condition. Then we will have

$$\mathbf{A}(x = 0, y, z) = \mathbf{A}(x = L, y, z), \ \mathbf{A}(x, y = 0, z) = \mathbf{A}(x, y = L, z),$$
$$\mathbf{A}(x, y, z = 0) = \mathbf{A}(x, y, z = L) \tag{16.9}$$

This leads to

$$e^{ik_xL} = e^{ik_yL} = e^{ik_zL} = 1 \tag{16.10}$$

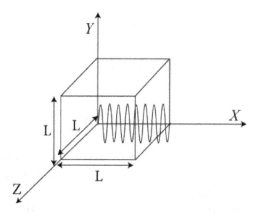

Figure 16.1. Electromagnetic radiation confined in a cubical cavity having a length L.

Hence

$$k_x = \frac{2\pi\alpha_x}{L}, \; k_y = \frac{2\pi\alpha_y}{L} \text{ and } k_z = \frac{2\pi\alpha_z}{L} \quad \alpha_x, \alpha_y, \alpha_z = 0, \pm 1, \pm 2, \pm 3, \ldots \quad (16.11)$$

Therefore, the complete solution of the vector potential can be written as,

$$\vec{A}(r, t) = \sum_{\gamma=-\infty}^{\infty} [\mathbf{q}_\gamma(t)\mathbf{A}_\gamma(\vec{r}) + \mathbf{q}_\gamma^*(t)\mathbf{A}_\gamma^*(\vec{r})] \quad (16.12)$$

Here, $\mathbf{A}_\gamma(\vec{r}) = \hat{e}_\gamma e^{i\vec{k}_\gamma \cdot \vec{r}}$ and $q_\gamma(t) = \left| q_\gamma \right| e^{-i\omega_\gamma t}$, γ implies the different modes of the electric field, i.e. a set of integer values including zero for α_x, α_y, α_z. $\hat{e}_\gamma$ is the direction of polarization and $\left| q_\gamma \right|$ is the magnitude of the vector potential corresponding to the set of γ values. $\mathbf{q}_\gamma^*(t)$ and $\mathbf{A}_\gamma^*(\vec{r})$ are, respectively, the complex conjugates of $\mathbf{q}_\gamma(t)$, $\mathbf{A}_\gamma(\vec{r})$ and so $A(\vec{r}, t)$ is real. We also assume that $k_{-\gamma} = -k_\gamma$ and $\omega_{-\gamma} = \omega_\gamma$. This implies $-\gamma$ terms in the equation (16.12) are the travelling waves moving the opposite to the direction of the waves with $+\gamma$. The orthogonal nature of the vector $\mathbf{A}_\gamma(\vec{r})$ can be expressed as

$$\iiint \mathbf{A}_\gamma \cdot \mathbf{A}_\mu^* d\tau = \iiint \mathbf{A}_\gamma \cdot \mathbf{A}_{-\mu} d\tau = V\delta_{\gamma,\mu} \quad (16.13)$$

Here we integrated over the whole volume of the cavity. We can write the electric and magnetic field in terms of vector potential as,

$$\mathbf{E} = -\frac{\partial A}{\partial t} = \sum_\gamma \mathbf{E}_\gamma \quad (16.14)$$

$$\mathbf{H} = \frac{1}{\mu_0}\nabla \times A = \sum_\gamma \mathbf{H}_\gamma \quad (16.15)$$

where

$$\mathbf{E}_\gamma = i\omega_\gamma\Big[q_\gamma(t)\mathbf{A}_\gamma(\vec{r}) - q_\gamma^*(t)\mathbf{A}_\gamma^*(\vec{r})\Big] \tag{16.16}$$

$$\mathbf{H}_\gamma = \frac{i}{\mu_0}k_\gamma \times \Big[q_\gamma(t)\mathbf{A}_\gamma(\vec{r}) - q_\gamma^*(t)\mathbf{A}_\gamma^*(\vec{r})\Big] \tag{16.17}$$

The total energy of the radiation field inside the cavity is given by,

$$\mathbf{H} = \frac{1}{2}\iiint(\epsilon_0\mathbf{E}.\,\mathbf{E} + \mu_0\mathbf{H}.\,\mathbf{H})d\tau \tag{16.18}$$

Now the first part of the equation (16.18)

$$\frac{1}{2}\iiint\epsilon_0\mathbf{E}.\,\mathbf{E}d\tau = -\frac{1}{2}\epsilon_0\iiint\sum_{\gamma,\mu}\omega_\gamma\omega_\mu\,[q_\gamma q_\mu\iiint\mathbf{A}_\gamma.\mathbf{A}_\mu d\tau$$

$$- q_\gamma q_\mu^*\iiint\mathbf{A}_\gamma.\,\mathbf{A}_\mu^* d\tau - q_\gamma^* q_\mu\iiint\mathbf{A}_\lambda^*.\mathbf{A}_\mu d\tau + q_\gamma^*\;q_\mu^*\iiint\mathbf{A}_\lambda^*.\,\mathbf{A}_\mu^*\;\;d\tau]$$

$$= -\frac{1}{2}\epsilon_0 V\sum_{\gamma,\mu}\omega_\gamma\omega_\mu[q_\gamma q_\mu\delta_{\gamma,-\mu} - q_\gamma q_\mu^*\;\delta_{\gamma,\mu} - q_\gamma^* q_\mu\delta_{\gamma,\mu} + q_\gamma^* q_\mu^*\delta_{\gamma,-\mu}] \tag{16.19}$$

$$\frac{1}{2}\iiint\epsilon_0\mathbf{E}.\,\mathbf{E} = -\frac{1}{2}\epsilon_0 V\sum_\gamma\omega_\gamma^2[q_\gamma q_{-\gamma} + q_\gamma^* q_{-\gamma}^* - 2q_\gamma q_\gamma^*]$$

Similarly, we can evaluate the second part of the equation (16.18),

$$\frac{1}{2}\iiint\mu_0\mathbf{H}.\,\mathbf{H}d\tau = -\frac{1}{2\mu_0^2}\sum_{\gamma\mu}\iiint[k_\gamma \times \{q_\gamma(t)\mathbf{A}_\gamma(\vec{r}) - q_\gamma^*(t)\mathbf{A}_\gamma^*(\vec{r})\}].\,[k_\mu\times$$

$$\{q_\mu(t)\mathbf{A}_\mu(\vec{r}) - q_\mu^*(t)\mathbf{A}_\mu^*(\vec{r})\}]d\tau \tag{16.20}$$

$$= \frac{1}{2}V\epsilon_0\sum_\gamma\omega_\gamma^2[q_\gamma q_{-\gamma} + q_\gamma^* q_{-\gamma}^* + 2q_\gamma q_\gamma^*]$$

here we have used the vector identity $(A \times B).(C \times D) = (A.C)(B.D) - (B.C)(A.D)$ and $k_\gamma^2 = \frac{v_\gamma^2}{c^2} = \mu_0\epsilon_0\omega_\gamma^2$.

Thus, the total classical Hamiltonian for the electromagnetic field in a cubical cavity with volume V can be written as

$$\mathbf{H} = -\frac{1}{2}\epsilon_0 V\sum_\gamma\omega_\gamma^2[q_\gamma q_{-\gamma} + q_\gamma^* q_{-\gamma}^* - 2q_\gamma q_\gamma^*] + \frac{1}{2}V\epsilon_0\sum_\gamma\omega_\gamma^2[q_\gamma q_{-\gamma} + q_\gamma^* q_{-\gamma}^* + 2q_\gamma q_\gamma^*]$$

$$= 2\epsilon_0 V\sum_\gamma\omega_\gamma^2 q_\gamma q_\gamma^* \tag{16.21}$$

We can see that the total energy of the classical field is the sum of energies of the individual modes in the cavity. Now we introduce two dimensionless variables,

$$Q_\gamma = (\epsilon_0 V)^{\frac{1}{2}}\left[q_\gamma(t) + q_\gamma^*(t)\right] \tag{16.22}$$

$$P_\gamma = -i\left(\epsilon_0 V\omega_\gamma^2\right)^{\frac{1}{2}}\left[q_\gamma(t) - q_\gamma^*(t)\right] \tag{16.23}$$

Then we can write the value of $q_\gamma(t)$ and $q_\gamma^*(t)$ as,

$$Q_\gamma(t) = \left(4\epsilon_0 V\omega_\gamma^2\right)^{-\frac{1}{2}}\left[\omega_\gamma Q_\gamma + iP_\gamma\right] \tag{16.24}$$

$$Q_\gamma^*(t) = \left(4\epsilon_0 V\omega_\gamma^2\right)^{-\frac{1}{2}}\left[\omega_\gamma Q_\gamma - iP_\gamma\right] \tag{16.25}$$

From equations (16.24) and (16.25) we can get $(P_\gamma^2 + \nu_\gamma^2 Q_\gamma^2) = 4\epsilon_0 V\omega_\gamma^2 q_\gamma(t)q_\gamma^*(t)$. Substituting this into equation (16.21) we get the expression of total Hamiltonian.

$$\mathbf{H} = \frac{1}{2}(P_\gamma^2 + \omega_\gamma^2 Q_\gamma^2) = \sum_\gamma \mathbf{H}_\gamma \tag{16.26}$$

Here $\mathbf{H}_\gamma = \frac{1}{2}(P_\gamma^2 + \omega_\gamma^2 Q_\gamma^2)$, this Hamiltonian is similar to the Hamiltonian of the linear harmonic oscillator of mass unity and frequency ω_γ. It leads to the conclusion that the electromagnetic field consists of an infinite number of linear harmonic oscillators with each mode and a particular direction of polarization. These P_γ and Q_γ are the conjugate variables. One thing we need to remember is that though they are like momentum and position coordinates it is just an analogy. There are not real momentum or position coordinates in the usual sense. Now we can consider P_γ and Q_γ as two Hermitian operators (i.e. $\hat{P}_\gamma$ and $\hat{Q}_\gamma$) that satisfy the following communication relations.

$$\left[\hat{P}_\gamma, \hat{Q}_\gamma\right] = \hat{P}_\gamma\hat{Q}_\gamma - \hat{Q}_\gamma\hat{P}_\gamma = i\hbar$$

$$\left[\hat{P}_\gamma, \hat{Q}_\zeta\right] = 0, \text{ for } \gamma \neq \xi$$

$$\left[\hat{P}_\gamma, \hat{Q}_\gamma\right] = \left[\hat{P}_\zeta, \hat{Q}_\zeta\right] = 0$$

$$\left[\hat{P}_\gamma, \hat{Q}_\zeta\right] = i\hbar\delta_{\gamma\zeta}$$

Now we introduce dimensionless operators,

$$\hat{a}_\gamma(t) = \frac{1}{\sqrt{2\hbar\omega_\gamma}}\left[\omega_\gamma\hat{Q}_\gamma(t) + i\hat{P}_\gamma(t)\right] \tag{16.28}$$

$$\hat{a}_\gamma^+(t) = \frac{1}{\sqrt{2\hbar\omega_\gamma}}\left[\omega_\gamma\hat{Q}_\gamma(t) - i\hat{P}_\gamma(t)\right] \tag{16.29}$$

This proceeds to $\left[\hat{a}_\gamma, \hat{a}_\gamma^+\right] = 1$. It is clear from equations (16.28) and (16.29) that $\hat{a}_\gamma$ and $\hat{a}_\gamma^+$ are proportional to q_γ and q_γ^*. So their time dependence is the same as $\hat{a}_\gamma$ which has the time dependence in the form of $e^{-i\omega_\gamma t}$, in a similar way $\hat{a}_\gamma^+$ is $e^{i\omega_\gamma t}$.

$$\hat{a}_\gamma(t) = \hat{a}_\gamma(0)e^{-i\omega_\gamma t} \tag{16.30}$$

$$\hat{a}_\gamma^+(t) = \hat{a}_\gamma^+(0)e^{i\omega_\gamma t} \tag{16.31}$$

From equations (16.28) and (16.29) we can derive

$$\hat{P}_\gamma(t) = i\left(\frac{\hbar\omega_\gamma}{2}\right)^{\frac{1}{2}} \left[\hat{a}_\gamma^+(t) - \hat{a}_\gamma(t)\right] \tag{16.32}$$

$$\hat{Q}_\gamma(t) = \left(\frac{\hbar}{2\omega_\gamma}\right)^{\frac{1}{2}} \left[\hat{a}_\gamma^+(t) + \hat{a}_\gamma(t)\right] \tag{16.33}$$

By putting the value of $\hat{P}_\gamma(t)$ and $\hat{Q}_\gamma(t)$ in equation (16.26) we get the expression of the Hamiltonian as,

$$\begin{aligned}
H &= \sum_\gamma \frac{1}{2}\hbar\omega_\gamma\left[\hat{a}_\gamma^+(t)\hat{a}_\gamma(t) + \hat{a}_\gamma(t)\hat{a}_\gamma^+(t)\right] \\
&= \sum_\gamma \frac{1}{2}\hbar\omega_\gamma\left[\hat{a}_\gamma^+\hat{a}_\gamma + \hat{a}_\gamma\hat{a}_\gamma^+\right] = \sum_\gamma\left[\hat{a}_\gamma^+\hat{a}_\gamma + \frac{1}{2}\right]\hbar\omega_\gamma
\end{aligned} \tag{16.34}$$

Here we have used the communication relation of $\hat{a}_\gamma$ and $\hat{a}_\gamma^+$. If can define $\hat{a}_\gamma^+\hat{a}_\gamma = n_\gamma$ known as number operator and then the Hamiltonian is

$$H = \sum_\gamma\left(n_\gamma + \frac{1}{2}\right)\hbar\omega_\gamma, \quad n_\gamma = 0, 1, 2, 3\ldots\ldots \tag{16.35}$$

From the quantum mechanical point of view, we can visualize the radiation field as consisting of an infinite number of simple harmonic oscillators. The energy increment or decrement of each oscillator can occur via integral multiples of $\hbar\omega_\gamma$. If we consider $\hbar\omega_\gamma$, as the energy of a *photon*, it follows that each oscillator can have energy corresponding to n_γ photons, or in other words, the γ–th mode is occupied by n_γ photons. It is clear from the equation (16.35) when $n_\gamma = 0$, a state in which none of the modes is occupied for all values γ. Each mode has an energy equal to $\frac{\hbar\omega_\gamma}{2}$ known as *zero-point energy*. An electromagnetic wave consists of an infinite number of modes and so the zero-point energy is also infinite. This infinite value of the zero-point energy is an undecided problem in the field quantization of the radiation field.

16.3 Quantization of a single mode

Now we will quantize a single-mode radiation field, the process being similar to the way we quantize a simple harmonic oscillator. We can write the Hamiltonian of the single-mode e-m field from equation (16.35) just dropping the suffix γ which indicates different modes.

$$\mathbf{H} = \hbar\omega(\hat{a}^{+}\hat{a} + \frac{1}{2}) \tag{16.36}$$

Suppose $|n>$ be the eigenstate of the number operator and the above Hamiltonian for a single-mode field. Then the corresponding eigenvalue equation of the single-mode can be written as,

$$\mathbf{H} \mid n > = \hbar\omega\left(\hat{a}^{+}\hat{a} + \frac{1}{2}\right) \mid n > = E_n \mid n> \tag{16.37}$$

Here E_n is the energy eigenvalue. Multiplying by the operator $\hat{a}^{+}$ from the left in equation (16.37) we get

$$\hbar\omega\left(\hat{a}^{+}\hat{a}^{+}\hat{a} + \frac{1}{2}\hat{a}^{+}\right) \mid n > = \hat{a}^{+}E_n \mid n> \tag{16.38}$$

By using the communicator relation $[\hat{a},\hat{a}^{+}] = 1$ we can simplify the equation (16.38) as

$$\hbar\omega\left(\hat{a}^{+}\hat{a} + \frac{1}{2}\right)(\hat{a}^{+} \mid n >) = (E_n + \hbar\omega)(\hat{a}^{+} \mid n >) \tag{16.39}$$

This corresponds to the eigenvalue equation for the state $(\hat{a}^{+}|n>)$ having energy eigenvalue $(E_n + \hbar\omega)$. So when we apply the creation operator $\hat{a}^{+}$ in the state $|n>$ the energy eigenvalue increases by $\hbar\omega$. In the same manner, if we act with destruction or annihilation operator a and apply the communication relation we will get.

$$\hbar\omega\left(\hat{a}^{+}\hat{a} + \frac{1}{2}\right)(\hat{a}|n>) = \mathbf{H}(\hat{a}|n>) = (E_n - \hbar\omega)(\hat{a} \mid n >) \tag{16.40}$$

Equation (16.40) tells us that the annihilation operator destroys the energy of one photon and the energy eigenvalue of the eigenstate $(a n>)$ becomes $(E_n - \hbar\omega)$. The energy eigenvalue of the lowest or ground state $(|0>)$ is $\mathbf{H}|0 > = \hbar\omega\left(\hat{a}^{+}\hat{a} + \frac{1}{2}\right)|0> = \frac{\hbar\omega}{2}$. So the lowest energy state has the eigenvalue equal to $\frac{\hbar\omega}{2}$ known as zero-point energy, as discussed in the last section. The number operator of the single-mode radiation field is $\hat{n} = \hat{a}^{+}\hat{a}$ and $\hat{n}|n > = n|n$. These number states satisfy the normalization condition, i.e. $<n|n> = 1$. We know from the standard quantum mechanics the property of annihilation $\hat{a}$ and creation $\hat{a}^{+}$ operators in the number state $|n>$ are,

$$\hat{a} \mid n > = \sqrt{n} \mid n - 1> \tag{16.41}$$

$$\hat{a}^{+} \mid n > = \sqrt{n + 1} \mid n + 1> \tag{16.42}$$

The number state ($|n>$) can be created upon repeated operation of the creation operator on the ground state ($|0>$), using the communication relation [H, $\hat{a}^+$]$=\hbar\omega\hat{a}^+$. We get $H\hat{a}^+ \mid 0 > = (\hat{a}^+H + \hbar\omega\hat{a}^+)\mid 0 > = \frac{3\hbar\omega}{2} \hat{a}^+ \mid 0>$. In the same manner, we can get

$$H(\hat{a}^+)^n|0 > = \hbar\omega\left(n + \frac{1}{2}\right)(\hat{a}^+)^n|0> \tag{16.43}$$

It is evident from equation (16.43) that the $(\hat{a}^+)^n \mid 0>$ is an eigenstate of the Hamiltonian with energy eigenvalue $\hbar\omega\left(n + \frac{1}{2}\right)$. So we can write $\mid n > = c_n(\hat{a}^+)^n \mid 0>$, here c_n is a normalization constant. Now we need to find the value of c_n. We can find the value of the c_n by repeated action of the $\hat{a}^+$ on the $|0>$.

$$\hat{a}^+ \mid 0 > = \mid 1> \tag{16.44}$$

$$\hat{a}^+ \mid 1 > = \sqrt{2} \mid 2> \tag{16.45}$$

and so on. By substituting this we can write

$$(\hat{a}^+)^n \mid 0 > = \sqrt{1.2.3.......(n + 1)} \mid n\rangle \tag{16.46}$$

Thus the normalized eigenvector for the photon number state or Fock state is

$$|n > = \frac{1}{\sqrt{n!}} (\hat{a}^+)^n \mid 0> \tag{16.47}$$

The Hamiltonian operator H and the number operator n both are Hermitian operators and so the $|n>$ create an orthonormal complete set of states,

$$<n|n' > = \delta_{nn'} \tag{16.48}$$

$$\sum_{n=0}^{\infty} <n \mid n > = \hat{I} \tag{16.49}$$

The matrix elements of the annihilation and creation operators which are existing can be written as,

$$<n - 1 \mid \hat{a} \mid n > = \sqrt{n} < n - 1 \mid n - 1 > = \sqrt{n} \tag{16.50}$$

$$<n + 1 \mid \hat{a}^+ \mid n > = \sqrt{n + 1} < n + 1 \mid n + 1 > = \sqrt{n + 1} \tag{16.51}$$

This entire discussion leads to the conclusion that a photon is a quantum of a single mode of the electromagnetic field and each mode has its own 'photon'. The energy of the field thus gets quantized. Each mode can have an integral number of photons and hence cannot have any arbitrary value of energy. The electromagnetic field that is prevailing in this world can be conceived of as a reservoir of an infinite number of photon modes. But it is difficult to visualize this in the real world since there we use the term photon to mean a localized packet of energy. However, our idea of Fourier analysis says that localization of any wave packet takes place when modes superpose.

16.4 Multimode radiation field

Eigen state of the total Hamiltonian ($H = \sum_\gamma H_\gamma$) for a multimode radiation field can be written as,

$$| n_1 > | n_2 > \dots | n_\gamma > \dots = | n_1, n_2 \dots n_\gamma, \dots > \qquad (16.52)$$

Here $n_1, n_2 \dots n_\gamma$ represents the number of photons in the modes characterized by 1, 2, $\dots \gamma$ states. The Schrödinger equation for the multimode field can be written as,

$$H \, | n_1, n_2 \dots n_\gamma, \dots > = \left[\sum_\gamma \left(n_\gamma + \frac{1}{2} \right) \hbar \omega_\gamma \right] \, | n_1, n_2 \dots n_\gamma, \dots > \qquad (16.53)$$

So the total eigenvalue is the sum of the eigenvalues i.e. $\left(n_1 + \frac{1}{2} \right) \hbar \omega_1$, $\left(n_2 + \frac{1}{2} \right) \hbar \omega_2$ ..... of different modes $n_1, n_2 \dots$, they are not interacting with each other. We can define the multimode vacuum state in which all the states are not occupied $| 0_1, 0_2 \dots 0_j, \dots >$. The action of annihilation and creation operator on the multimode state is also specific and acts on that particular mode only.

$$\hat{a}_\gamma \, | n_1, n_2 \dots n_\gamma, \dots > = \sqrt{n_\gamma} \, | n_1, n_2 \dots n_\gamma - 1, \dots > \qquad (16.54)$$

$$\hat{a}_\gamma^+ \, | n_1, n_2 \dots n_\gamma, \dots > = \sqrt{n_\gamma + 1} \, | n_1, n_2 \dots n_\gamma + 1, \dots > \qquad (16.55)$$

Equations (16.54) and (16.55) imply that the annihilation and creation operator acting on a particular mode γ destroys or creates one photon of that particular mode keeping other modes unaffected. Again, we can write the orthogonality relation of the different modes as,

$$< n_1', n_2', \dots n_\gamma', \dots | n_1, n_2, \dots n_\gamma, \dots > = \delta_{n_1 n_1'} \delta_{n_2 n_2'} \dots \delta_{n_\gamma n_\gamma'} \dots \qquad (16.56)$$

The state of the radiation field doesn't need to be an eigenstate of H. This state may be a superposition of the eigenstates and it can be written as

$$| \psi > = \sum_{n_1, n_2 \dots} C_{n_1, n_2 \dots n_\gamma} \, | n_1, n_2, \dots n_\gamma, \dots > \qquad (16.57)$$

Here $| C_{n_1} |^2$ is the probability of finding n_1 photons in the first mode, similarly $| C_{n_2} |^2$ is the probability of finding n_2 photons in the second mode and so on.

The multimode electric field operator is

$$\hat{\varepsilon}_\lambda(t) = i \sum_\lambda \sqrt{\frac{\hbar \omega_\lambda}{\epsilon_0 V}} (\hat{a}_\lambda(t) \exp(ik_\lambda . \, r) - \hat{a}_\lambda^+(t) \exp(-ik_\lambda . \, r)) \hat{e}_\lambda \qquad (16.58)$$

In equation (16.58) all the operators are in Heisenberg representation, in the Schrödinger representation it can be written as,

$$\hat{\varepsilon}_\lambda = i \sum_\lambda \sqrt{\frac{\hbar\omega_\lambda}{\epsilon_0 V}} (\hat{a}_\lambda \exp(i\boldsymbol{k}_\lambda.\ \boldsymbol{r}) - \hat{a}_\lambda^+ \exp(-i\boldsymbol{k}_\lambda.\ \boldsymbol{r}))\hat{e}_\lambda \qquad (16.59)$$

Now we would like to calculate the expectation value of this electric field operator $\hat{\varepsilon}_\lambda$, in the following way,

$$<n_1, n_2, \ldots\ldots n_\gamma, \ldots| \hat{\varepsilon}_\lambda | n_1, n_2, \ldots n_\gamma, \ldots> = <n_1 | n_1 > <n_2 | n_2 > \ldots <n_\gamma | \hat{\varepsilon}_\lambda | n_\gamma > \ldots$$

$$= 0 \qquad (16.60)$$

Since $<n_\gamma | \hat{a} | n_\gamma > = 0 = <n_\gamma | \hat{a}^+ | n_\gamma>$ we can calculate the expectation value of the $\hat{\varepsilon}_\lambda^2$ in the same way.

$$< n_1, n_2, \ldots\ldots n_\gamma, \ldots| \hat{\varepsilon}_\lambda^2 | n_1, n_2, \ldots n_\gamma, \ldots> = < n_1 | n_1 > <n_2 | n_2 > \ldots <n_\gamma | \hat{\varepsilon}_\lambda^2 | n_\gamma>$$
$$= \frac{\hbar\omega_\lambda}{\epsilon_0 V}\left(n_\gamma + \frac{1}{2}\right) \qquad (16.61)$$

In equation (16.60) we have used the expression of multimode electric field operator given in equation (16.59) and following mathematical relations,

$$< n_\gamma | \hat{a}_\lambda\hat{a}_\lambda^+ | n_\gamma> = \sqrt{(n_\gamma + 1)} < n_\gamma | \hat{a}_\lambda | n_\gamma + 1 > = (n_\gamma + 1),$$
$$< n_\gamma | \hat{a}_\lambda^+\hat{a}_\lambda | n_\gamma> = \sqrt{n_\gamma} < n_\gamma | \hat{a}_\lambda^+ | n_\gamma - 1 > = n_\gamma \text{ and} \qquad (16.62)$$
$$< n_\gamma | \hat{a}\hat{a} | n_\gamma> = 0 = <n_\gamma | \hat{a}^+\hat{a}^+ | n_\gamma > .$$

The uncertainty of the electric field operator $(\Delta\hat{\varepsilon}_\lambda)$ is denoted by the variance

$$\Delta\hat{\varepsilon}_\lambda = \sqrt{<\hat{\varepsilon}_\lambda^2 > - <\hat{\varepsilon}_\lambda>^2} = \sqrt{\frac{\hbar\omega_\lambda}{\epsilon_0 V}}\left(n_\gamma + \frac{1}{2}\right)^{\frac{1}{2}} \qquad (16.63)$$

Since from equation (16.60), we get $<\hat{\varepsilon}_\lambda>^2 = 0$ and $<\hat{\varepsilon}_\lambda^2 > = \frac{\hbar\omega_\lambda}{\epsilon_0 V}\left(n_\gamma + \frac{1}{2}\right)$, calculated in equation (16.61). Now the variance of the electric field operator in the vacuum state (i.e. $| 0 > = | 0_1, 0_2\ldots\ldots\ldots 0_\lambda, \ldots\ldots>$) is

$$\Delta\hat{\varepsilon}_\lambda = \sqrt{\frac{\hbar\omega_\lambda}{\epsilon_0 V}} \qquad (16.64)$$

Equation (16.63) tells us when none of the modes is occupied by the photons then also the uncertainty of the electric field is not zero but is of a finite value. This is due to vacuum field fluctuations and it is the reason for the many interesting effects like spontaneous emission, parametric down-conversion etc. We can introduce two field quadrature operators as,

$$\hat{X}_1 = \frac{(\hat{a} + \hat{a}^+)}{2} \text{ and } \hat{X}_2 = \frac{(\hat{a} - \hat{a}^+)}{2i} \qquad (16.65)$$

Here $\hat{X}_1$ and $\hat{X}_2$ are Hermitian operators though $\hat{a}$ and $\hat{a}^+$ are not. $\hat{X}_1$ and $\hat{X}_2$ are essentially the dimensionless position and momentum operators as shown in equations (16.32–16.33). It is possible to express the electric field operator (equation (16.58)) in terms of field quadratures.

$$\hat{\varepsilon}_\lambda(t) = \sqrt{\frac{\hbar\omega_\lambda}{\epsilon_0 V}}[\hat{X}_1 \sin(\omega t - \boldsymbol{k}.\boldsymbol{r}) - \hat{X}_2 \cos(\omega t - \boldsymbol{k}.\boldsymbol{r})]\hat{e}_\lambda \qquad (16.66)$$

It is seen from equation (16.66) that $\hat{X}_1$ multiply with sine and $\hat{X}_2$ by cosine terms. They are related to field amplitudes oscillating out of phase with each other by an angle of $90°$. $\hat{X}_1$ and $\hat{X}_2$ are in quadratures so they are called quadrature operators. These quadrature operators satisfy the following commutation relation,

$$[\hat{X}_1, \hat{X}_2] = \frac{i}{2} \qquad (16.67)$$

Now for the number state $|n\rangle$, we can show that $\langle n | \hat{X}_1 | n \rangle = 0 = \langle n | \hat{X}_2 | n\rangle$ but for the $\hat{X}_1^2$ and $\hat{X}_2^2$ it is not equal to zero.

$$\langle n|\hat{X}_1^2|n\rangle = \frac{1}{4}\langle n | (\hat{a} + \hat{a}^+)^2 | n \rangle = \frac{1}{4}(2n + 1) \qquad (16.68)$$

$$\langle n|\hat{X}_2^2|n\rangle = -\frac{1}{4}\langle n | (\hat{a} - \hat{a}^+)^2 | n \rangle = \frac{1}{4}(2n + 1) \qquad (16.69)$$

It is clear from equations (16.68) and (16.69) that the number state uncertainty in the $\hat{X}_1$ and $\hat{X}_2$ quadrature is the same and equal to $(\Delta X_1)^2 = (\Delta X_2)^2 = \frac{1}{4}(2n + 1)$. The product of the uncertainties in the two quadratures can be written as,

$$(\Delta X_1)(\Delta X_2) = \frac{1}{4}(2n + 1) \qquad (16.70)$$

For the vacuum state ($0\rangle$) this product is equal to $\frac{1}{4}$ and in this state number of the photon (n) is zero. A state where the product of the uncertainty is $\frac{1}{4}$ is known as *minimum uncertainty state* (MUS).

Let us elaborate on the implication of equation (16.70). This says that there is always uncertainty in the measurement of both the quadratures of the electromagnetic field and this equation puts a limit on the measurement accuracy. At the same time, an interesting feature comes out. The uncertainty in one of the quadratures may be less than $\frac{1}{2}$ but that in the other quadrature is greater so that their product becomes $\frac{1}{4}$. The states for which the uncertainty is below $\frac{1}{2}$ are termed as *squeezed states*.

16.5 Coherent states

The states which are the eigenstate of the annihilation operator and also minimize the Heisenberg uncertainty product ($\Delta p \Delta q$) are known as coherent states. Coherent states are denoted by Dirac notation as $| \alpha\rangle$ and it satisfies the following relation.

$$\hat{a} \mid \alpha > = \alpha \mid \alpha> \tag{16.71}$$

In equation (16.71) α is the eigenvalue of the operator a in general, it is a complex number otherwise arbitrary. A coherent state can be represented as

$$\mid \alpha > = e^{-\frac{|\alpha|^2}{2}} \sum_{n=0}^{\infty} \frac{\alpha^n}{\sqrt{n!}} \mid n> \tag{16.72}$$

Coherent states are not orthogonal to each other and they are not the eigenstates of the Hamiltonian operator. Whether coherent states are the eigenstate of the annihilation operator a or not can be clarified by calculating the relation $a|\alpha>$.

$$\hat{a} \mid \alpha> = e^{-\frac{|\alpha|^2}{2}} \sum_{n=0}^{\infty} \frac{\alpha^n}{\sqrt{n!}} \hat{a} \mid n>.$$

$$= e^{-\frac{|\alpha|^2}{2}} \sum_{n=0}^{\infty} \frac{\alpha^n}{\sqrt{n!}} \sqrt{n} \mid n-1> \tag{16.73}$$

$$= \alpha e^{-\frac{|\alpha|^2}{2}} \sum_{n=0}^{\infty} \frac{\alpha^{n-1}}{\sqrt{(n-1)!}} \mid n-1>$$

$$= \alpha \mid \alpha>$$

The Hermitian conjugate of the equation (16.73) tells us that the coherent states also remain eigenstates of the annihilation operator $\hat{a}^+$.

$$(\hat{a}|\alpha>)^+ = (\alpha|\alpha>)^+$$
$$< \alpha|\hat{a}^+ = < \alpha|\alpha^* \tag{16.74}$$

Now we can calculate the expectation value of the number operator n (single mode) in the state $|\alpha>$ by using the equation (16.73), and (16.74).

$$<\alpha \mid \hat{n} \mid \alpha > = \bar{n} = <\alpha \mid \hat{a}^+\hat{a} \mid \alpha>$$

$$= <\alpha \mid \alpha^*\alpha \mid \alpha>$$

$$= \alpha^*\alpha = \mid \alpha \mid^2 \tag{16.75}$$

Here $\mid \alpha \mid^2$ is the average photon number n of the field. To calculate the variance of the photon number (Δn), we need to calculate the expectation value of $\hat{n}^2$.

$$<\alpha \mid \hat{n}^2 \mid \alpha > = <\alpha \mid \hat{a}^+\hat{a}\hat{a}^+\hat{a} \mid \alpha > = \mid \alpha \mid^4 + \mid \alpha \mid^2 = \bar{n}^2 + \bar{n} \tag{16.76}$$

Thus the value of photon variance or fluctuation of photon number is

$$\Delta\bar{n} = \sqrt{<\hat{n}^2> - <\hat{n}>^2} = \sqrt{\bar{n}^2 + \bar{n} - \bar{n}^2} = \sqrt{\bar{n}}. \tag{16.77}$$

This is the property of a Poisson distribution, now we can calculate the probability of n photons in the coherent state. To calculate this probability P_n, we need to evaluate the value of $<n|\alpha>$.

$$\langle n \mid \alpha \rangle = e^{-\frac{|\alpha|^2}{2}} \sum_{m=0}^{\infty} \frac{\alpha^m}{\sqrt{m!}} \langle n \mid m \rangle$$

$$= e^{-\frac{|\alpha|^2}{2}} \sum_{m=0}^{\infty} \frac{\alpha^m}{\sqrt{m!}} \delta_{nm} \qquad (16.78)$$

$$= e^{-\frac{|\alpha|^2}{2}} \frac{\alpha^n}{\sqrt{n!}}$$

$\langle n \mid m \rangle = \delta_{nm}$ is the orthogonality relation of the number states, $\mid n \rangle$. The probability is $P_n = |\langle n \mid \alpha \rangle|^2$,

$$P_n = e^{-|\alpha|^2} \frac{|\alpha|^{2n}}{n!}$$

$$= e^{-\bar{n}} \frac{\bar{n}^n}{n!} \qquad (16.79)$$

In equation (16.79) we have used the relation $|\alpha|^2 = \bar{n}$. So P_n is a Poisson distribution with a mean of n. We conclude that coherent states obey Poisson photon statistics. The fractional uncertainty in the photon number is

$$\frac{\Delta \bar{n}}{\bar{n}} = \frac{1}{\sqrt{\bar{n}}} \qquad (16.80)$$

that decreases to zero as we increase the intensity, i.e. photon numbers. We can summarize the properties of the coherent states as follows:

   a) In the coherent states, the Heisenberg uncertainty product ($\Delta p \Delta q$) is minimum.
   b) Coherent states are an eigenstate of the annihilation operator.
   c) Coherent states satisfy the Glauber condition.

In figure 16.2 we have plotted P_n versus n for mean value (n) of 10 and 20. This shows that the size of the fluctuation decreases as we increase the value of n. From equation (16.80) we also get the same result, suppose for $n = 16$, $\Delta n = 4$ and $\frac{\Delta \bar{n}}{\bar{n}} = 0.25$. If we take $n = 64$ then $\Delta n = 8$ and $\frac{\Delta \bar{n}}{\bar{n}} = 0.125$.

Expression of the coherent state ($\mid \alpha \rangle$) given in the equation (16.72) can be rewritten by substituting $\mid n \rangle$ in terms of $\mid 0 \rangle$ as,

$$\mid \alpha \rangle = e^{-\frac{|\alpha|^2}{2}} e^{\alpha \hat{a}^+} \mid 0 \rangle \qquad (16.81)$$

Here we have used $\mid n \rangle = \frac{(a^+)^n}{\sqrt{n!}} \mid 0 \rangle$, again $e^{-\alpha^* \hat{a}} \mid 0 \rangle = \mid 0 \rangle$ so equation (16.81) can be modified in the following form

$$\mid \alpha \rangle = e^{-\frac{|\alpha|^2}{2}} e^{\alpha a^+} e^{-\alpha^* a} \mid 0 \rangle \qquad (16.82)$$

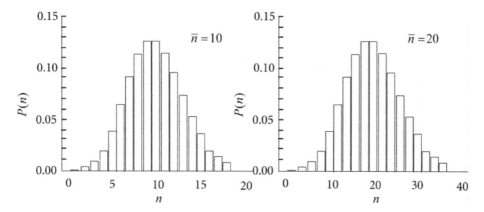

Figure 16.2. Plot of P_n versus n for mean value (n) of 10 and 20.

Now we can use the Baker–Hausdorff relation, which says if A and B are two operators such that $((A,B),A) = ((A,B),B) = 0$, then $e^{\hat{A}+\hat{B}} = e^{\hat{A}}e^{\hat{B}}e^{-\frac{[\hat{A},\hat{B}]}{2}}$. If we think $\alpha\hat{a}^{+} = \hat{A}$ and $-\alpha*\hat{a} = \hat{B}$ and $[\alpha\hat{a}^{+}, -\alpha*\hat{a}] = -|\alpha|^2$ then equation (16.82) becomes

$$|\alpha> = e^{\alpha\hat{a}^{+}-\alpha*\hat{a}} |0> = D(\alpha)|0> \tag{16.83}$$

Here $D(\alpha) = e^{\alpha\hat{a}^{+}-\alpha*\hat{a}}$, known as the *displacement operator* that generates the coherent state $|\alpha>$ by operating in the vacuum state $|0>$. The displacement operator has the following properties,

$$D^{+}(\alpha) = D^{-1}(\alpha) = D(-\alpha), D^{+}(\alpha)\hat{a}^{+}D(\alpha) = (\hat{a}^{+}+\alpha*) \text{ and } D^{+}(\alpha)\hat{a}D(\alpha)$$
$$= (\hat{a} + \alpha) \tag{16.84}$$

It can be verified that the coherent states are not orthogonal to each other by taking the scalar product of two coherent states

$$<\alpha | \beta> = <0 | D^{+}(\alpha)D(\beta) | 0> = e^{\left[-\frac{1}{2}(|\alpha|^2+|\beta|^2-2\alpha*\beta)\right]} \tag{16.85}$$

The absolute value of the above scalar product is $|<\alpha | \beta>|^2 = e^{-|\alpha-\beta|^2}$, so the coherent states with different eigenvalues are not orthogonal. Although these two coherent states become approximately orthogonal when $(\alpha-\beta) \gg 1$, i.e. α differs from β considerably.

In table 16.1 we have listed some standard definitions of the number states used in this section.

16.6 Squeezed states of light

In section 16.4, we have seen that the uncertainty in both quadratures for the coherent state is equal and minimum and so they can be written as,

$$(\Delta X_1) = (\Delta X_2) = \frac{1}{2} \tag{16.86}$$

Table 16.1. Some standard definitions of number state representation used in discussion.

	Symbol	Definition
Number state	$\lvert n \rangle$	$\hat{n} \lvert n \rangle = n \lvert n \rangle,\ \hat{H} \lvert n \rangle = E_n \lvert n \rangle,\ E_n = (n + \frac{1}{2})\hbar\omega$
Vacuum state	$\lvert 0 \rangle$	$\hat{a} \lvert 0 \rangle = 0$
Number operator	$\hat{n}$	$\hat{n} = \hat{a}^\dagger \hat{a}$
Creation operator	$\hat{a}^\dagger$	$\hat{a}^\dagger \lvert n \rangle = \sqrt{n+1} \lvert n+1 \rangle$
Annihilation operator	$\hat{a}$	$\omega \hat{a} \lvert n \rangle = \sqrt{n} \lvert n-1 \rangle$
Commutator	$[\hat{a}, \hat{a}^\dagger]$	$[\hat{a}, \hat{a}^\dagger] = \hat{a},\ \hat{a}^\dagger - \hat{a}^\dagger \hat{a} = 1$

In the case of squeezed states, the uncertainty of both quadratures is not the same but their product has a minimum value. Generally, a squeezed state has less noise in one quadrature than a coherent state. If we take two operators say $\hat{A}$ and $\hat{B}$ if they satisfy the commutation relation $(\hat{A}, \hat{B}) = i\,\hat{C}$ then

$$<(\Delta \hat{A})^2 > <(\Delta \hat{B})^2 > \geqslant \frac{1}{4} < (\Delta \hat{C})^2> \qquad (16.87)$$

ΔA or ΔB are said to be squeezed if they obey the following relations,

$$<(\Delta \hat{A})^2 > < \frac{1}{2} \lvert<\hat{C}>\rvert \text{ or } < (\Delta \hat{B})^2 > < \frac{1}{2} \lvert<\hat{C}>\rvert \qquad (16.88)$$

If we want to calculate the quadrature squeezing then we can take $\hat{A} = \hat{X}_1$ and $\hat{B} = \hat{X}_2$. Here $\hat{X}_1$ and $\hat{X}_2$ are the quadrature operators as discussed earlier and $\hat{C} = \frac{\hat{i}}{2}$. Squeezing in the $\hat{X}_1$ or $\hat{X}_2$ quadrature can be possible if $<(\Delta X_1)^2 > < \frac{1}{4}$ and $<(\Delta X_2)^2 > < \frac{1}{4}$. If we plot ΔX_1 versus ΔX_2 graph it will give a hyperbola, as shown in figure 16.3 and the minimum uncertainty states lie on point A. All the physical states lying on the right-hand side of this hyperbola and coherent states with $(\Delta X_1) = (\Delta X_2)$ is a special case of the state. In the case of a squeezed state, one quadrature reduces uncertainty while the other quadrature increases uncertainty in the same amount (i.e. $(\Delta X_1) < \frac{1}{2} < (\Delta X_2)$).

These squeezed states are shown by the shaded region in figure 16.3. A squeezed state can be generated mathematically by the action of the unitary 'squeeze operator'. Squeeze operator can be defined as,

$$S(\xi) = \exp\left[\frac{1}{2}(\xi_* \hat{a}^2 - \xi(\hat{a}^+)^2)\right] \qquad (16.89)$$

here $\xi = re^{2i\theta}$, is the squeezing parameter with amplitude r, $(0 \leqslant r < \infty)$ and phase θ, $(0 \leqslant \theta \leqslant 2\pi)$. The squeeze operator executes this relation $S^+(\xi) = S^-(\xi) = S(-\xi)$, which one can easily prove. The squeeze operator has the following transformation properties.

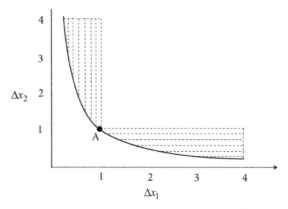

Figure 16.3. Minimum uncertainty state graph, 'A' point is the coherent state and the dotted area in the graph corresponds to the squeezed state.

$$S^+(\xi)\hat{a}S(\xi) = \hat{a}\cosh r - \hat{a}^+ e^{-2i\theta}\sinh r \qquad (16.90a)$$

$$S^+(\xi)\hat{a}^+ S(\xi) = \hat{a}^+ \cos h r - \hat{a}e^{-2i\theta} \sinh r \qquad (16.90b)$$

$$S^+(\xi)(Y_1 + iY_2)S(\xi) = Y_1 e^{-r} + iY_2 e^{r} \qquad (16.90c)$$

To derive equation (16.84a) we have used the formula $e^{\alpha}\beta e^{-\alpha} = \beta + [\alpha, \beta] + \frac{1}{2!}[\alpha, [\alpha, \beta]] + ...$, in equation (16.84c), $(Y_1 + iY_2) = (X_1 + iX_2)e^{-i\theta}$ a rotated complex amplitude at an angle θ. One component of the rotated complex amplitude gets squeezed, whereas another component is amplified on the application of the squeeze operator. We can define a factor by $r = (\xi)$ known as the *squeezing factor* which determines the degree of attenuation and amplification of a component. The coherent squeezed state $|\alpha, \xi\rangle$ is achieved by operating the displacement operator $D(\alpha)$ on vacuum state first then applying the squeezing operator $S(\xi)$.

$$|\alpha, \xi\rangle = S(\xi)D(\alpha)|0\rangle \qquad (16.91)$$

Here $\alpha = |\alpha|e^{i\theta}$, definitely, we get the coherent state by putting $\xi = 0$ and a squeezed vacuum state is obtained by putting $\alpha = 0$. From equation (16.83) we can see that the coherent state is produced by linear terms of annihilation and creation operators a and $\hat{a}^+$ in the exponent, whereas in the case of a squeezed coherent state quadratic terms of a and $\hat{a}^+$ are in the exponent.

Now we will calculate the quadrature variances of rotated amplitudes of Y_1 and Y_2. For that we start with calculating operator expectation values of the state $|\alpha, \xi\rangle$. The expectation value of $\langle\hat{a}\rangle$ can be calculated as follows,

$$\langle\hat{a}\rangle = \langle\alpha, \xi|\hat{a}|\alpha, \xi\rangle = \langle 0|D^+(\alpha)S^+(\xi)\hat{a}S(\xi)D(\alpha)|0\rangle$$
$$= \langle\alpha|(a\cos h r - a^+ e^{i\theta}\sin h r)|\alpha\rangle \qquad (16.92)$$
$$= \alpha\cos h r - \alpha^* e^{i\theta}\sin h r$$

$$\langle \hat{a}^2 \rangle = \langle (\hat{a}^+)^2 \rangle^*$$

$$= \langle 0 \left| D^+(\alpha) S^+(\xi) \hat{a}^2 S(\xi) D(\alpha) \right| 0 \rangle \tag{16.93}$$

$$= \langle \alpha \left| S^+(\xi) \hat{a} S(\xi) D(\alpha) \right| 0 \rangle$$

$$= \alpha^2 \cos h^2 r + \alpha^{*2} e^{2i\theta} \sin h^2 r - 2 \mid \alpha \mid^2 e^{i\theta} \sin hr \cos hr - e^{i\theta} \sin hr \cos hr$$

$$\langle a^+ a \rangle = \mid \alpha \mid^2 (\cos h^2 r + \sin h^2 r) - \alpha^{*2} e^{i\theta} \sin hr \cos hr - \alpha^2 e^{-i\theta} \sin hr$$
$$\cos hr + \sin h^2 r \tag{16.94}$$

By using the value of X_1 and X_2 i.e. $X_1 = \frac{(\hat{a} + \hat{a}^+)}{2}$ and $X_2 = \frac{(\hat{a} - \hat{a}^+)}{2i}$, we can get

$$(Y_1 + iY_2) = (X_1 + iX_2)e^{-i\theta} = \hat{a}e^{-i\theta} \tag{16.95}$$

Now the variances of Y_1 and Y_2 can be calculated as

$$\Delta Y_1 = (\langle Y_1^2 \rangle - \langle Y_1 \rangle^2)^{\frac{1}{2}} = \left[\frac{1}{4} \langle (\hat{a}e^{-i\theta} + \hat{a}^+ e^{i\theta})^2 \rangle - \frac{1}{4}(\langle \hat{a}e^{-i\theta} + \hat{a}^+ e^{i\theta} \rangle)^2 \right]^{\frac{1}{2}}$$

$$= [\frac{1}{4}\{\langle \hat{a}^2 e^{-2i\theta} + \hat{a}^{+2} e^{2i\theta} + \hat{a}\hat{a}^+ + \hat{a}^+ \hat{a} \rangle - (\langle \hat{a} \rangle e^{-i\theta} + \langle \hat{a}^+ \rangle e^{i\theta})^2\}]^{\frac{1}{2}} \tag{16.96}$$

$$= \frac{1}{2}e^{-r}$$

Similarly, we can calculate ΔY_2, which is

$$\Delta Y_2 = \frac{1}{2}e^{r} \tag{16.97}$$

In equations (16.88 and 16.89) we have used the expectation values of equations (16.86a)–(16.86c). The product of these variances is

$$\Delta Y_1 \Delta Y_2 = \frac{1}{4} \tag{16.98}$$

It is seen from equations (16.88) and (16.89) that ΔY_1 and ΔY_2 are not equal and so the *squeezed coherent state* is an excellent squeezed state. In figure 16.4 we have plotted the phase space representation of a coherent and squeezed coherent state. The squeezed state (figure 16.4(b)) has an error ellipse, whereas for a coherent state it is an error circle. The *squeezing parameter* is defined by $r = (\xi)$, which measures the degree of squeezing.

We shall introduce photon number states which are one of the most important types of squeezed states in chapter 18 during the discussion of sub-Poissonian photon distribution. These are states of perfectly defined photon number n, i.e. $\Delta n = 0$ and correspond to a completely undefined phase. This is in contrast to the coherent states having a better-defined phase which have larger photon number fluctuations $\Delta n = n$.

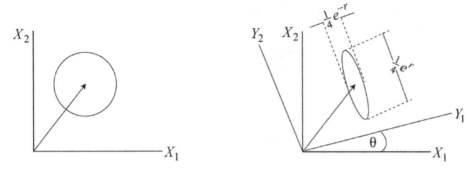

Figure 16.4. Phase space representation of error contour of a coherent state (a) and a squeezed coherent state (b).

16.7 Further reading

[1] Scully M O and Zubairy S 1997 *Quantum Optics* (Cambridge: Cambridge University Press)
[2] Fox M 2006 *Quantum Optics: An Introduction* (Oxford: Oxford University Press)
[3] Walls D F and Milburn G J 2008 *Quantum Optics* (Berlin: Springer)
[4] Loudon R 2001 *Quantum Theory of Light* 3rd edn (Oxford: Oxford University Press)
[5] Meystre P and Sergent III M 2009 *Elements of Quantum Optics* 3rd edn (Berlin: Springer)
[6] Thyagarajan K and Ghatak A 2010 *Lasers: Fundamentals and Applications* (Berlin: Springer)
[7] Glauber R J 1963 Coherent and incoherent states of the radiation field *Phys. Rev.* **131** 2766
[8] Scully M O and Sergent M 1973 The concept of the photons *Phys. Today* **25** 38
[9] Loudon R and Knight P L 1987 Squeezed light *J. Mod. Opt.* **34** 709

16.8 Problems

1. Show that the electric-field-per-photon for an electric field can be written as $\sqrt{\dfrac{\hbar\omega}{\epsilon_0 V}}$.

2. If two operators A and B do not commute then prove the following operator identity

$$e^{\beta A} B e^{-\beta A} = B + \beta[A, B] + \frac{1}{2!}\beta^2[A, [A, B]] + \cdots\cdots\frac{1}{n!}\beta^n[A, [A\cdots[A, B]\ldots]] + \cdots$$

3. In the case of a coherent state ($|\alpha>$) show that

$$\int d^2\alpha \quad |\alpha><\alpha| = \pi \sum_n |n><n| = \pi$$

4. Show that the displacement operator ($D(\alpha) = e^{\alpha a^+ - \alpha^* a}$) satisfies the following relations.

$$D^+(\alpha) = D^{-1}(\alpha) = D(-\alpha)$$

5. Show that,

$$D^+(\alpha)a^+D(\alpha) = (\hat{a}^+ + \alpha^*) \text{ and}$$

$$D^+(\alpha)aD(\alpha) = (\hat{a} + \alpha)$$

6. For an e-m field propagating along the z-direction and the electric field operator in the x-direction, then the single-mode electric field operator can be written as

$$\hat{\varepsilon}_x(z, t) = \varepsilon_0(\hat{a} + \hat{a}^+) \sin kz$$

Show that electric field operator does not commute with the number operator ($\hat{n} = \hat{a}^+\hat{a}$) and their commutation relation is

$$[\hat{n}, \hat{\varepsilon}_x] = \varepsilon_0(\hat{a}^+ - \hat{a}) \sin kz$$

Then find their uncertainty relation.

7. For a coherent state ($|\alpha\rangle$) with $|\alpha| = 10$, calculate the mean photon number and the standard deviation in photon number.

8. Show that in an e-m wave the time average of electric and magnetic fields energy are identical.

IOP Publishing

Quantum Optics and Quantum Computation
An introduction
Dipankar Bhattacharyya and Jyotirmoy Guha

Chapter 17

Interaction of an atom with a quantized field

In chapters 14 and 15, we employed semi-classical approximation to study the interactions of atoms with optical fields. The treatment gives results that corroborate well with many experimental findings. However, some significant experiments show strong disagreements with predictions made by the semi-classical theories. In this chapter, atom–field interaction will be looked upon fully based on quantum mechanics. This is sure to provide deeper insight and a basic understanding of spontaneous emission. Further, it would become easier to treat more advanced problems in quantum optics such as resonance fluorescence, squeezed states and laser linewidth etc. The foundation of the present treatment stems from a suitable combination of knowledge we have gained from the semi-classical theory of the atom–field interactions with the quantized field treatment of chapter 16.

17.1 Interaction Hamiltonian in terms of Pauli operators

We consider a two-level system as shown in figure 14.1, having an upper level $|b>$ and lower level $|a>$ having energy eigenvalues $E_b = \hbar\omega_b$ and $E_a = \hbar\omega_a$ and the transition frequency is $\omega = (\omega_b - \omega_a)$. If the eigenfunctions of the lower and upper levels are ψ_a and ψ_b then they may be expressed as the column vectors,

$$\psi_a = \begin{pmatrix} 0 \\ 1 \end{pmatrix} \text{ and } \psi_b = \begin{pmatrix} 1 \\ 0 \end{pmatrix} \tag{17.1}$$

The complete atomic wavefunction can be written as,

$$\phi = c_a\psi_a + c_b\psi_b \tag{17.2}$$

In the state vector representation, ϕ can be written $\phi = \begin{pmatrix} C_b \\ C_a \end{pmatrix}$.

We would like to express the energy and electric dipole operators in terms of Pauli spin matrices for convenience. Pauli spin matrices are.

$$\sigma_x = \begin{pmatrix} 0 & 1 \\ 1 & 0 \end{pmatrix}, \ \sigma_y = \begin{pmatrix} 0 & -i \\ i & 0 \end{pmatrix} \text{ and } \sigma_z = \begin{pmatrix} 1 & 0 \\ 0 & -1 \end{pmatrix} \tag{17.3}$$

We can define non-Hermitian spin–flip operators by using the Pauli spin matrices which are Hermitian matrices as seen from equation (17.3).

$$\sigma_+ = \frac{1}{2}(\sigma_x + i\sigma_y) = \begin{pmatrix} 0 & 1 \\ 0 & 0 \end{pmatrix} \tag{17.4}$$

$$\sigma_- = \frac{1}{2}(\sigma_x - i\sigma_y) = \begin{pmatrix} 0 & 0 \\ 1 & 0 \end{pmatrix} \tag{17.5}$$

The spin–flip operators flip the atoms from the lower level to the upper level or vice versa, transition from one level to another due to interaction of energy can easily be explained by these operations.

$$\sigma_+\psi_a = \psi_b \text{ and } \sigma_-\psi_b = \psi_a \tag{17.6}$$

These spin–flip operators are also known as rising (σ_+) and lowering (σ_-) operators. Now if we assume that the energy is zero at halfway between the upper and lower levels then we can say $E_a = -\frac{\hbar\omega}{2}$ and $E_b = \frac{\hbar\omega}{2}$. So the unperturbed Hamiltonian can be written as,

$$H_0 = \frac{\hbar\omega}{2}\begin{pmatrix} 1 & 0 \\ 0 & -1 \end{pmatrix} = \frac{\hbar\omega}{2}\sigma_z \tag{17.7}$$

In equation (16.59) we have presented the electric field operator for the multimode radiation field; now we can write the electric field operator for the single-mode electric field

$$\varepsilon = \varepsilon_0(a + a^+)\sin kz \tag{17.8}$$

Here $\varepsilon_0 = \sqrt{\frac{\hbar\nu}{\varepsilon_0 V}}$ is the amplitude of the electric field already discussed in the earlier chapter. Interaction Hamiltonian for the atom and quantized field can be written in the same fashion as we did in semi-classical theory. Applying rotating wave approximation (RWA) we can write H_I as,

$$H_I = \hbar g(\sigma_+ + \sigma_-)(a + a^+) \tag{17.9}$$

where $g = \frac{\mu\varepsilon_0}{2\hbar}\sin kz$, is the Rabi frequency as defined in the semi-classical section, here μ is the dipole moment. The total Hamiltonian of this two-level system is the sum of unperturbed Hamiltonian ($\frac{\hbar\omega}{2}\sigma_z$), field-free Hamiltonian ($\hbar\nu(a^+a + \frac{1}{2})$), and the interaction Hamiltonian ($\hbar g(\sigma_+ + \sigma_-)(a + a^+)$).

$$H = \frac{\hbar\omega}{2}\sigma_z + \hbar\nu\left(a^+a + \frac{1}{2}\right) + \hbar g(\sigma_+ + \sigma_-)(a + a^+) \tag{17.10}$$

Here we consider the two-level system is at rest and the quantization axis of the atom is chosen in such a way that g is real.

17.2 Absorption and emission phenomena

In equation (17.10) the term $\sigma_+ a$ describes the phenomena of absorption (figure 17.1). One photon is absorbed by an atom or a photon annihilated and the atom goes to the excited state ($|b>$) from the ground state ($|a>$). In this process energy is conserved. The term $\sigma_- a^+$ suggests the downward transition of an atom from the upper state to the ground state by emitting a photon. In this process

Energy is conserved and the process is known as stimulated emission (figure 17.2). Now let discuss the remaining two terms, i.e. $\sigma_+ a^+$ and $\sigma_- a$. The first term represents an upward transition of the atom with the stimulated emission of a photon (figure 17.3). In this process, conservation of energy is not possible so we can drop this term.

The term $\sigma_- a$ corresponds to the downward transition of an atom and absorption of a photon (figure 17.4). This process is also not allowed by the principle of conservation of energy.

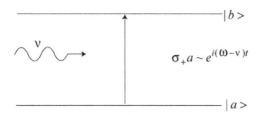

Figure 17.1. In the two-level atomic system atom in the ground state ($|a>$) absorbs a photon and moves to the excited state ($|b>$).

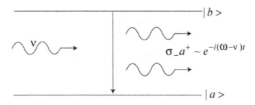

Figure 17.2. Stimulated emission process, an atom is going down to ground state by emission of a photon.

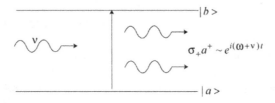

Figure 17.3. Emission of a photon and transition of an atom from ground to excited state which is an energy non-conserving process.

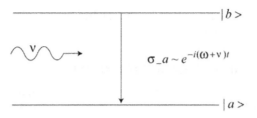

Figure 17.4. Transition of an atom from excited to the ground state by absorption of a photon is an energy non-conserving process.

Now we will show that to satisfy the rotating wave approximation (RWA) these two terms, $\sigma_+ a^+$ and $\sigma_- a$ do not take part in the interaction energy. In the Heisenberg picture both annihilation (a) and creation (a^+) operators have the time dependence. Now if we consider the time dependence of spin–flip operators then we can write the spin–flip operators as

$$\sigma_\pm(t) = \sigma_\pm(0)e^{\pm i\omega t} \tag{17.11}$$

In the previous chapter, we have shown the time dependence of annihilation (a) and creation (a^+) operators as

$$a(t) = a(0)e^{-i\nu t} \text{ and } a^+(t) = a^+(0)e^{i\nu t} \tag{17.12}$$

Hence, we can write the four terms of the interaction energy as follows

$$\sigma_+(t)a(t) = \sigma_+(0)a(0)e^{i(\omega-\nu)t} \tag{17.13}$$

$$\sigma_-(t)a^+(t) = \sigma_-(0)a^+(0)e^{-i(\omega-\nu)t} \tag{17.14}$$

$$\sigma_+(t)a^+(t) = \sigma_+(0)a^+(0)e^{i(\omega+\nu)t} \tag{17.15}$$

$$\sigma_-(t)a(t) = \sigma_-(0)a(0)e^{-i(\omega+\nu)t} \tag{17.16}$$

The terms of equations (17.13) and (17.14) vary with time at low frequency (i.e. $(\omega-\nu)$) near resonance, whereas the terms of equations (17.15) and (17.16) are the rapidly varying anti-resonance terms. These terms average to zero after a few optical cycles and so their contribution is nil in the interaction energy. In semi-classical treatment the terms containing far off-resonance frequency (i.e. $(\omega + \nu)$) go to zero. So this is related to the same physics as in the case of RWA. Omitting these two terms now we can write the interaction Hamiltonian as

$$H_{\text{Int}} = \hbar g(\sigma_+ a + \sigma_- a^+) \tag{17.17}$$

We shall begin with a relatively easy problem called the Jaynes–Cummings model. It is, however, ironic that Jaynes championed atom–field theories that use classical fields though this model considers atoms to be coupled to a single quantized mode of the field. We introduce the '*dressed-atom*' picture, which has the potential to explain the nature of elementary phenomena such as the 'light shift' and the occurrence of three peaks in strong field resonance fluorescence. Section 17.4 discusses the dynamics of the atom–field model for various states of the field. We obtain an elementary picture of spontaneous and stimulated emission and absorption and briefly discuss the 'Cummings collapse' and revivals due to the quantum granularity of the field.

We shall derive the spontaneous emission decay rate using the more generally accurate Weisskopf–Wigner theory at the end of this chapter.

17.3 Dressed states

The state vector of atom–field interaction corresponds to unperturbed Hamiltonian $H_0 = \frac{\hbar\omega}{2}\sigma_z + \hbar\nu\left(a^+a + \frac{1}{2}\right)$ (i.e. first two terms of equation (17.10)) is

$$| an > =| a > | n>$$ (17.18)

Here, an atom is in state $|a>$ and the field is containing n number of photons. The unperturbed Hamiltonian H_0 can satisfy the following equations.

$$H_0|an > =\left[-\frac{\hbar\omega}{2} + \hbar\nu\left(n + \frac{1}{2}\right)\right]|an>$$ (17.19)

$$H_0|bn > =\left[\frac{\hbar\omega}{2} + \hbar\nu\left(n + \frac{1}{2}\right)\right]|bn>$$ (17.20)

The first term of the interaction Hamiltonian (equation (17.17)) couples the states $|an + 1>$ and $|bn>$ as,

$$\sigma_+a \mid an + 1 > =\sqrt{n + 1} \mid bn>$$ (17.21)

In the same way,

$$\sigma_-a^+ \mid bn > =\sqrt{n + 1} \mid an + 1>$$ (17.22)

The lower level ($|a>$) has one excess photon over the upper level ($|b>$) to conserve the photon during interaction of a single-mode field with a closed two-level system. In our notation, we consider a lower level of the atomic state by 'a', it has nothing to do with the annihilation operator 'a'. Other interaction terms σ_+a^+ and σ_-a would not contribute since they are dropped by RWA. So we consider the manifold of states $|bn>$ and $|an + 1>$ as forming basis set and under this basis, the total Hamiltonian can be written as,

$$H = \hbar\nu\begin{pmatrix} n & 0 \\ 0 & n + 1 \end{pmatrix} + \frac{\hbar\omega}{2}\begin{pmatrix} 1 & 0 \\ 0 & -1 \end{pmatrix} + \hbar g\begin{pmatrix} 0 & \sqrt{n + 1} \\ \sqrt{n + 1} & 0 \end{pmatrix}$$ (17.23)

By introducing the detuning parameter term $\delta = (\omega - \nu)$ we can rewrite equation (17.23) as,

$$H = \hbar\nu\left(n + \frac{1}{2}\right)\begin{pmatrix} 1 & 0 \\ 0 & 1 \end{pmatrix} + \frac{\hbar}{2}\begin{pmatrix} \delta & 2g\sqrt{n+1} \\ 2g\sqrt{n+1} & -\delta \end{pmatrix} \tag{17.24}$$

In the first term of equation (17.24) the $\frac{1}{2}$ term arrives from the mathematical calculation and it is not related to zero-point energy since that is not included in the calculation. If $n = 0$ then transition occurs from the ground state $|a1\rangle$ and $|b0\rangle$ by absorbing a photon. In the same manner, a transition takes place from $|b0\rangle$ to $|a1\rangle$ by emitting a photon. So the transition occurs between the one manifold of states $\{|bn\rangle, |an+1\rangle\}$. The total Hamiltonian acts on all other similar manifolds with different photon numbers. In equation (17.24) the interaction matrix (i.e. second matrix of equation (17.24)) is the same form as in the case of the semi-classical approach. The only difference here is that the Rabi frequency is changed by $2g\sqrt{n+1}$. This $\sqrt{n+1}$ appears due to the quantum nature of the light. The energy eigenvalues and the energy eigenstates can be evaluated by diagonalizing the interaction matrix. The eigenvalue equation can be written as,

$$\begin{bmatrix} \delta - \lambda_n & 2g\sqrt{n+1} \\ 2g\sqrt{n+1} & -\delta - \lambda_n \end{bmatrix}\begin{bmatrix} |2n\rangle \\ |1n\rangle \end{bmatrix} = \begin{bmatrix} 0 \\ 0 \end{bmatrix} \tag{17.25}$$

We will get the values of λ_n as,

$$\lambda_n^2 = \delta^2 + 4g^2(n+1)$$
$$\lambda_n = \pm[\delta^2 + 4g^2(n+1)]^{\frac{1}{2}} = \pm\Omega_n \tag{17.26}$$

Here Ω_n is the generalized Rabi flopping frequency for a quantized field at resonance condition (i.e. $\delta = 0$). It becomes $2g\sqrt{n+1}$. Energy eigenvalues are,

$$E_{1n} = \hbar\nu\left(n + \frac{1}{2}\right) + \frac{\hbar\Omega_n}{2} = \hbar[-\frac{1}{2}\omega + (n+1)\nu + \frac{1}{2}(\Omega_n + \delta)] \tag{17.27a}$$

$$E_{2n} = \hbar\nu\left(n + \frac{1}{2}\right) - \frac{\hbar\Omega_n}{2} = \hbar[\frac{1}{2}\omega + n\nu - \frac{1}{2}(\Omega_n + \delta)] \tag{17.27b}$$

The corresponding energy eigen velctors can be written as,

$$|2n\rangle = \cos\theta_n | bn\rangle - \sin\theta_n | an+1\rangle \tag{17.28a}$$

$$|1n\rangle = \sin\theta_n | bn\rangle + \cos\theta_n | an+1\rangle \tag{17.28b}$$

The values of $\sin\theta_n$ and $\cos\theta_n$ are

$$\sin\theta_n = \frac{2g\sqrt{n+1}}{\sqrt{(\Omega_n - \delta)^2 + 4g^2(n+1)}} \quad \text{and}$$

$$\cos\theta_n = \frac{(\Omega_n - \delta)}{\sqrt{(\Omega_n - \delta)^2 + 4g^2(n+1)}} \tag{17.29}$$

at the-on resonance condition $\delta = 0$ and $\Omega_n = 2g\sqrt{n+1}$. So the value of $\sin\theta_n$ and $\cos\theta_n$ are the same and equal to $\frac{1}{\sqrt{2}}$ and $\theta_n = 45^0$. These two new states $|1n>$ and $|2n>$ are known as the *dressed state*. They can be obtained from the initial bare states (i.e. $|bn>$ and $|an+1>$) through a matrix rotation of an angle θ_n.

$$\begin{pmatrix} \cos\theta_n & -\sin\theta_n \\ \sin\theta_n & \cos\theta_n \end{pmatrix} \begin{pmatrix} |bn> \\ |an+1> \end{pmatrix} = \begin{pmatrix} |2n> \\ |1n> \end{pmatrix} \tag{17.30}$$

The first matrix of equation (17.30) is a transformation matrix (U_n) which transform the bare states into dressed states.

$$U_n\begin{pmatrix} |bn> \\ |an+1> \end{pmatrix} = \begin{pmatrix} |2n> \\ |1n> \end{pmatrix} \tag{17.31}$$

In the zero photon situation (i.e. $n = 0$) $\sin\theta_n \neq 0$ and $\cos\theta_n \neq 1$ and so dressing of the states is still possible though there are no photons present. The dressed states for no photon case are,

$$|20> = \cos\theta_0 |b0> - \sin\theta_0 |a1> \tag{17.32a}$$

$$|10> = \sin\theta_0 |b0> + \cos\theta_0 |a1> \tag{17.32b}$$

Here the $\cos\theta_0 = \dfrac{(\Omega_0 - \delta)}{\sqrt{(\Omega_0 - \delta)^2 + 4g^2}}$ and $\sin\theta_0 = \dfrac{2g}{\sqrt{(\Omega_0 - \delta)^2 + 4g^2}}$.

In figure 17.5, we have plotted the eigenvalues of the dressed states with the atomic transition frequency ω. The bare states or unperturbed states energy eigenvalues E_{bn+1} and E_{an} are drawn in dotted straight lines. These lines cross each other at on resonance position ($\omega = \nu$). Solid lines are representing the dressed state eigenvalues and at resonance, they repel each other. This repulsion is called anti-crossing, the minimum separation of energy that occurs at resonance. The value of this minimum separation is $E_{2n} - E_{1n} = 2g\hbar\sqrt{n+1}$. One can say that the bare states

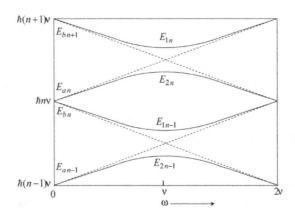

Figure 17.5. Energy level diagram of bare and dressed atomic states. The curved solid lines correspond to the energy of dressed state and straight dotted lines represent the energy level of bare states.

atomic levels $|an>$ and $|bn + 1>$ are 'dressed' by the radiation field. At on-resonance condition, the dressed states are represented as,

$$| 2n > = \frac{1}{\sqrt{2}}[| bn > - | an + 1 >] \qquad (17.33a)$$

$$| 1n > = \frac{1}{\sqrt{2}}[| bn > + | an + 1 >] \qquad (17.33b)$$

Corresponding energy eigenvalues are given below,

$$E_{2n} = \hbar v \left(n + \frac{1}{2} \right) - g\hbar\sqrt{n + 1} \qquad (17.34a)$$

$$E_{1n} = \hbar v \left(n + \frac{1}{2} \right) + g\hbar\sqrt{n + 1} \qquad (17.34b)$$

In the Schrödinger picture, a state vector of the atom can be written in terms of bare-atom state vectors as,

$$| \psi > = \sum_n [C_{an+1}(t) \mid an + 1 > + C_{bn}(t) \mid bn >] \qquad (17.35)$$

The above state vector in terms of the dressed state vectors as,

$$| \psi > = \sum_n [C_{2n}(t) \mid 2n > + C_{1n}(t) \mid 1n >] \qquad (17.36)$$

Probability amplitudes of bare-state atom and dressed state atoms are related through a transformation matrix (U_n) as

$$\begin{pmatrix} C_{2n}(t) \\ C_{1n}(t) \end{pmatrix} = U_n \begin{pmatrix} C_{bn}(t) \\ C_{an+1}(t) \end{pmatrix} \qquad (17.37)$$

17.4 Jaynes–Cummings model

We can elegantly describe the dynamics of a two-level atom interacting with a quantized field mode using the dressed-atom picture. In this section, we use this picture to treat Rabi flopping in a fully quantized way. We then compare the results with the corresponding semi-classical treatment adhered to in chapter 14. We describe an example of spontaneous emission, the phenomenon known as 'Cummings collapse' of Rabi flopping induced by a coherent state. Revivals of the Rabi flopping that result from the discrete nature of the photon field will also be addressed.

In 1963, Jaynes and Cummings introduced a model which describes the phenomena of the interaction of a two-level atom with a single-mode electro-magnetic field. Here we take a two-level atom having ground level $|a>$ and excited state $|b>$. A single-mode cavity with the electric field as given in equation (17.8) interacts with this two-level atom. The total Hamiltonian contains three terms we already discussed in the earlier section (i.e. equation (17.10)). In equation (17.10) the interaction Hamiltonian contains only two terms after adopting the rotating wave approximation, the form of the total Hamiltonian is given below.

$$H = \frac{\hbar\omega}{2}\sigma_z + \hbar\nu\left(a^+a + \frac{1}{2}\right) + \hbar g(\sigma_+a + \sigma_-a^+) \tag{17.38}$$

We consider that the atom is in the excited state $|b>$ and the interacting radiation field is initially in $|n>$ state. For simplicity, we are considering the on-resonant case (i.e. $\delta = 0$). The initial atom field state can be written as $|i> = |b> |n>$, The energy of this state ($|i>$) is $E_i = -\frac{\hbar\omega}{2} + n\hbar\omega$. The final state of this system is $|f> = |a>|n + 1>$ and the energy of this state is $E_f = \frac{\hbar\omega}{2} + (n + 1)\hbar\omega$. Hence $(E_f - E_i) = 0$. The state vector of the atom-field system is given by,

$$| \psi(t) > = c_i(t) | i > + c_f(t) |f> \tag{17.39}$$

By using the Hamiltonian of equation (17.38) in the time-dependent Schrödinger equation, i.e. $\frac{d|\psi>}{dt} = -\frac{i}{\hbar}H|\psi>$ we will get

$$\frac{dc_i}{dt} = -ig\sqrt{n + 1} \quad c_f \tag{17.40a}$$

$$\frac{dc_f}{dt} = -ig\sqrt{n + 1} \quad c_i \tag{17.40b}$$

Using the above two equations and by eliminating the c_f we can get,

$$\frac{d^2c_i}{dt^2} + g^2(n + 1)c_i = 0 \tag{17.41}$$

In our case, the atom is initially excited so the boundary condition is $c_i(0) = 1$ and $c_f(0) = 0$, the solution of equation (17.41) being

$$c_i(t) = \cos(gt\sqrt{n + 1}) \tag{17.42}$$

Similarly solution of c_f is

$$c_f(t) = -i\sin(gt\sqrt{n + 1}) \tag{17.43}$$

By putting the value of $c_i(t)$ and $c_f(t)$ in equation (17.39) the state vector of the atom-field system at a time t is

$$| \psi > = \cos(gt\sqrt{n + 1})| b > | n > -i\sin(gt\sqrt{n + 1})| a > | n + 1> \tag{17.44}$$

The probability that the system stays in the initial state $|i>$ is $| c_i(t) |^2 = \cos^2(gt\sqrt{n + 1})$ and the probability to make a transition in the state $|f>$

is $| c_f(t) |^2 = \sin^2(gt\sqrt{n+1})$. The population inversion $P_{in}(t)$ is the population difference between excited and ground state

$$P_{in}(t) = | c_i(t) |^2 - | c_f(t) |^2 = \cos(2gt\sqrt{n+1}) \tag{17.45}$$

The above result is similar to the semi-classical result, the only difference here being that the Rabi frequency is replaced by $\Omega_n = 2g\sqrt{n+1}$. So $P_{in}(t) = \cos(\Omega_n t)$. In figure 16.9 we have plotted population inversion with the time in which an atom interacts with a radiation field having initial photon number 10. It is seen from equation (17.45) that Rabi oscillation is still present even when the number of the photons is zero ($n = 0$), i.e. $P_{in}(t) = \cos(2gt)$. There is no classical analogue of this kind of Rabi oscillation. This is known as *vacuum field-induced Rabi oscillations.*

In the semi-classical approach, we have added the spontaneous decay rates of an atom phenomenologically. But in the field quantized approach spontaneous emission of a photon appeared automatically due to vacuum Rabi oscillations. So we can say that the vacuum Rabi oscillations are the cause of spontaneous emission. In a high Q cavity experiment, the vacuum Rabi oscillations are observed. Other than $n = 0$ photon state, the population inversion of an atom with a finite number of photon states is very similar to semi-classical Rabi oscillations.

Now we are considering a more general case, where the atom is initially ($t = 0$) present in a superposition state of $|a>$ and $|b>$.

$$| \psi(0)>_{atom} = c_a(0)| a > +c_b(0)| b> \tag{17.46}$$

The radiation field was initially in a state,

$$| \psi(0)>_{field} = c_0 | 0 > +c_1 | 1 > +c_2 | 2 > \ldots\ldots\ldots = \sum_{n=0}^{\infty} c_n | n> \tag{17.47}$$

The initial atom–field interaction state can be written as,

$$| \psi(0) > =| \psi(0)>_{atom} \otimes | \psi(0)>_{field} \tag{17.48}$$

Now the solution of the Schrödinger equation is

$$|\psi(t) > = \sum_{n=0}^{\infty} [\{c_n c_b \cos(gt\sqrt{n+1}) - i c_{n+1} c_a \sin(gt\sqrt{n+1})\} | b >$$

$$+\{-i c_{n-1} c_b \sin(gt\sqrt{n}) + c_n c_a \cos(gt\sqrt{n})\}]| a > | n> \tag{17.49}$$

In equation (17.49) we can put the initial boundary condition (i.e. $c_b = 1$ and $c_a = 0$) since the atoms are in the excited state. Then the above equation turns into

$$|\psi(t) > = \sum_{n=0}^{\infty} [\{c_n \cos(gt\sqrt{n+1})\} | b > +\{-i c_{n-1} \sin(gt\sqrt{n})\}]| a > | n> \tag{17.50}$$

Now in equation (17.50) as $\psi(t) > =| \psi_a(t) > |a > +\psi_b(t) > |b>$, we can think that $| \psi_a(t)>$ and $\psi_b(t)$ are the state vectors when the atoms are in the ground and excited state.

So

$$| \psi_a(t) > =-i\sum_{n=0}^{\infty} c_n\sin(gt\sqrt{n + 1}) | n + 1> \qquad (17.51a)$$

$$| \psi_b(t) > = \sum_{n=0}^{\infty} c_n\cos(gt\sqrt{n + 1}) | n> \qquad (17.51b)$$

Now the population inversion $P_{in}(t)$ in this case, is

$$P_{in}(t) = <\psi_b(t) | \psi_b(t) > -<\psi_a(t) | \psi_a(t)>$$

$$= \sum_{n=0}^{\infty}| c_n |^2 \cos(2gt\sqrt{n + 1}) \qquad (17.52)$$

If we compare the expression of the population inversion of equation (17.52) with equation (17.45) we will see that this is the sum of n-photons inversion of equation (17.45) weighted with the photon number distribution of the initial field state.

If we assume the initial field is a coherent state, then

$$c_n = e^{-\frac{|\alpha|^2}{2}} \frac{\alpha^n}{\sqrt{n!}} \qquad (17.53)$$

Substituting this value of c_n in equation (17.52) we will get

$$P_{in}(t) = \sum_{n=0}^{\infty} e^{-\bar{n}}\frac{\bar{n}^n}{n!}\cos(2gt\sqrt{n + 1}) \qquad (17.54)$$

Since in case of coherent state $| \alpha |^2 = \bar{n}$. In figure 17.7 we have plotted population inversion $(P_{in}(t))$ versus gt, it is seen that firstly Rabi oscillations die out or collapse as time passes and after a certain time Rabi oscillations start to revive. Though the revival of oscillations is not completely like the first oscillation, the amplitude of oscillation is less than the initial oscillation. We can see that for a larger time there are a series of collapses and revival of Rabi oscillations. Later on, the revival is less prominent than this collapse and revival of Rabi oscillation is a fully quantum mechanical phenomenon that is not observed in a semi-classical approach.

These collapses and revivals are due to the spread of probabilities about $\bar{n}$ for photon numbers in the range $\bar{n} \pm \Delta\bar{n}$ meaning that there will be many Rabi frequencies other than the dominant one, corresponding to the average photon number $\bar{n}$. One can make an approximate estimate of the collapse time τ_c from the time-frequency uncertainty relation and $\tau_c \simeq \frac{1}{2g}$.

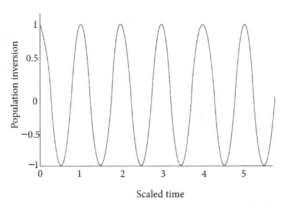

Figure 17.6. Population inversion versus scaled time (gt) graph. Here the initial state of the radiation field is Fock state 10 photons.

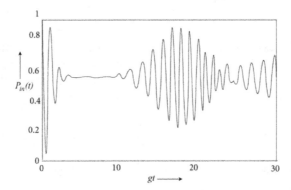

Figure 17.7. Population inversion $P_{in}(t)$ versus gt plot for a quantized single-mode field interactng with an atom.

17.5 Theory of spontaneous emission: Wigner–Weisskopf model

The total Hamiltonian of a two-level system interacting with a multimode field can be written as,

$$H = \sum_m \hbar\nu_m \left(a_m^+ a_m + \frac{1}{2}\right) + \hbar \begin{pmatrix} \omega_a & 0 \\ 0 & \omega_b \end{pmatrix} + \sum_m \hbar g_m (\sigma_+ a_m + \sigma_- a_m^+) \qquad (17.55)$$

The last term in equation (17.55) is the interaction Hamiltonian or interaction energy due to multimode after applying the rotating wave approximation (RWA).

$$H_I = \sum_m \hbar g_m (\sigma_+ a_m + \sigma_- a_m^+) = \sum_m H_m \qquad (17.56)$$

The unperturbed eigenvector of the system is

$$| \beta, n_1, n_2, n_3 \ldots n_m \ldots \ldots > here \quad \beta = a, b \qquad (17.57)$$

The state vector of the system can be represented as,

$$|\psi> = \sum_{\beta=a,b} \sum_{n_1} \sum_{n_2} \cdots\cdots \sum_{n_m} \cdots\cdots C_{\beta,n_1 n_2,n_3,\cdots\cdots n_m\cdots\cdots} |\beta, n_1 n_2, n_3, \cdots\cdots n_m\cdots\cdots> \quad (17.58)$$

The atom is either in the state $|a>$ or $|b>$ and the first mode contains n_1 number of photons, the second mode contains n_2 number of photons etc. The C-coefficients denote the probability amplitudes in such states. The summation of each n_m starts from zero and extends to infinity. The interaction Hamiltonian (equation (17.56)) only connects a state $|b,0>$ (where an atom is in the upper state with no photon) with the state $|a, 1_m>$. In this state ($|a, 1_m>$) atom is in the lower state with one photon in the mth mode and no photons in other modes. The state vector can be written by comparing equation (17.58) as,

$$|\psi> = C_{b,0} | b, 0 > + \sum_{a,1_m} | a, 1_m> \quad (17.59)$$

By putting equation (17.59) in the Schrödinger equation and taking projection on to the states $|b,0>$ and $|a, 1_m>$ we can find the equation of motion of the C-coefficients as,

$$\dot{C}_{b,0} = -i\sum_m g_m e^{i(\omega-\nu_m)t} C_{a,1_m} \quad (17.60)$$

$$\dot{C}_{a,1_m} = -i\sum_m g_m e^{-i(\omega-\nu_m)t} C_{b,0} \quad (17.61)$$

Integrating equation (17.61) over time and then substituting in equation (17.60) we get,

$$\dot{C}_{b,0} = -i\sum_m g_m^2 \int_0^t d\tau e^{i(\omega-\nu_m)(t-\tau)} C_{b,0}(\tau) \quad (17.62)$$

In the continuum of modes where the frequencies are very closely spaced, we can change the summation by integration of density of states $D(\nu)$.

$$\sum \rightarrow \int D(\nu)d\nu \quad (17.63)$$

The density of the state can be calculated by considering that the radiation is confined in a cubic cavity having length 'l' so the volume of the cavity $V = l^3$. $k_x = \frac{2\pi n_x}{l}$, $n_x = 1, 2, 3\ldots$ are the wavenumbers of the running modes in the cavity along the x-direction. Similarly, k_y and k_z are the wavenumbers along y and z directions. The number of modes between k_x and $k_x + dk_x$ can be calculated by differentiating the expression of k_x, which is $dn_x = \frac{dk_x l}{2\pi}$. Using the same calculation for other directions (i.e. y and z) we can evaluate the total number of modes in the elementary volume element $dk_x dk_y dk_z$.

$$dn = d^3k(\frac{l}{2\pi})^3 \tag{17.64}$$

If the cavity length l is large then we can replace the summation over k by an integration,

$$\frac{1}{V}\sum_k F(k) \to \frac{1}{V}\int dn \quad F(k) = \frac{1}{(2\pi)^3}\int d^3k \quad F(k) \tag{17.65}$$

The elementary volume element (d^3k) in the spherical polar coordinate is

$$d^3k = k^2\sin\theta \; dk \; d\theta \; d\phi \tag{17.66}$$

By using $k = c\nu$, in equation (17.65) we get here c is the velocity of light in free space, putting it.

$$\frac{1}{V}\sum_k F(k) \quad dk = \frac{1}{(2\pi)^3}\int d\nu \frac{\nu^2}{c^3}\int_0^\pi \sin\theta d\theta \int_0^{2\pi} F(\nu)d\phi \tag{17.67}$$

We need to take the sum over two polarization components of the electromagnetic field.

$$\frac{1}{V}\sum_k F(k) \quad dk = \int d\nu \quad D(\nu)F(\nu) \tag{17.68}$$

where the density of the state is,

$$D(\nu) = \frac{V\nu^2}{n^2c^3} \tag{17.69}$$

Now in equation (17.62), the interaction or coupling term g and density of states ($D(\nu)$) are slowly varying terms compared to the oscillatory term due to that we will take them out of the integration over ν.

$$\dot{C}_{b,0} = -\int D(\nu) \quad g^2(\nu)d\nu \quad \int_0^t d\tau e^{i(\omega-\nu)(t-\tau)}C_{b,0}(\tau) = -D(\nu) \quad g^2\int d\nu$$
$$\int_0^t d\tau e^{i(\omega-\nu)(t-\tau)}C_{b,0}(\tau) \tag{17.70}$$

We assume that $C_{b,0}(\tau)$ varies sufficiently slowly so we can evaluate it at $\tau = t$, the value of this integral is as follows,

$$\int_0^t d\tau e^{i(\omega-\nu)(t-\tau)} = \pi\delta(\omega - \nu) - \wp(\frac{i}{\omega - \nu}) \tag{17.71}$$

The second term $(\wp(\frac{i}{\omega-\nu}))$ of equation (17.71) is the principal part of the integral which leads to an energy level shift related to Lamb shift. The first term gives us,

$$\dot{C}_{b,0} = -\frac{2\pi D(\nu) \quad g^2(\nu)}{2}C_{b,0}(t) = -\frac{\Gamma}{2}C_{b,0}(t) \tag{17.72}$$

Here $\Gamma = 2\pi D(\nu)$ $g^2(\nu)$, known as *Wigner–Weisskopf* spontaneous decay rate.

$$| C_{b,0}(t) |^2 = e^{-\Gamma t} \tag{17.73}$$

An atom in the excited state can decay to the ground state without any interaction with the field, it decays exponentially with a decay rate Γ. This is known as the *Wigner–Weisskopf model* for spontaneous emission.

17.6 Further reading

[1] Scully M O and Zubairy S 1997 *Quantum Optics* (Cambridge: Cambridge University Press)

[2] Meystre P and Sergent III M 2009 *Elements of Quantum Optics* 3rd edn (Berlin: Springer)

[3] Gerry C C and P L 2005 *Knight Introductory Quantum Optics* (Cambridge: Cambridge University Press)

[4] Rand S C 2010 *Non-linear and Quantum Optics* (Oxford: Oxford University Press)

[5] Thyagarajan K and Ghatak A 2010 *Lasers: Fundamentals and Applications* (Berlin: Springer)

[6] Shore B W and Knight P L 1993 The Jaynes–Cummings model *J. Mod. Opt.* **40** 1195

17.7 Problems

1. Show that the Pauli spin operators obey the following communication relations.

$$[\sigma_x, \sigma_y] = 2i\sigma_z, [\sigma_y, \sigma_z] = 2i\sigma_x \text{ and } [\sigma_z, \sigma_x] = 2i\sigma_y$$

2. Show by mathematical induction that

$$[a, a^{+m}] = m(a^+)^{m-1} \text{ and } [a^+, a^m] = -m(a)^{m-1}$$

3. Show that the following relations hold

$$[\sigma_+, \sigma_-] = \sigma_z, [\sigma_-, \sigma_+] = 2\sigma_-$$

4. If two operators A and B do not commute then we have the following operator identity

$$e^{\beta A}Be^{-\beta A} = B + \beta[A, B] + \frac{1}{2!}\beta^2[A, [A, B]] + \cdots\cdots\frac{1}{n!}\beta^n[A, [A\cdots[A, B]......]] + \cdots$$

Using the above operator identity show that

$$\exp(i\nu a^+at)a \exp(-i\nu a^+at) = a \; exp(-i\nu t)$$

$$\exp\left(i\omega\sigma_z\frac{t}{2}\right)\sigma_+\exp\left(-i\omega\sigma_z\frac{t}{2}\right) = \sigma_+\exp(i\omega t)$$

5. Solve the dressed-atom eigenvalue problem and obtain the equations (17.28a)–(17.28b).
6. An interaction energy operator for all field modes other than the dressing mode can be written as

$$H_I = \hbar c(|\, a > <b\,| + |\, b> < a\,|)$$

Here c is some constant. H_I the operator allows the atom to change state keeping the dressing mode unaffected (some other mode or modes provide or get the atomic energy). Using (17.28a) and (17.28b), calculate the matrix elements of H_I between all combinations of the dressed states $|2n>$, $|1n>$, $|2n–1>$, $|1n–1>$.

IOP Publishing

Quantum Optics and Quantum Computation
An introduction
Dipankar Bhattacharyya and Jyotirmoy Guha

Chapter 18

Photon statistics

We start this chapter with the famous Young's double-slit experiment and explain the first-order correlation function and its physical significance. Later we discuss the classical description of the time-dependent intensity fluctuations in a light beam by using the Hanbury Brown and Twiss (HB–T) interferometer, which is an intensity interferometer and used to measure higher-order correlation function like second-order correlation function ($g^2(\tau)$). Different types of light sources can be specified based on their correlation function.

We shall end this chapter by viewing the subject from a different angle, namely from the perspective of the statistical properties of the photon stream. The three different types of photon statistics are Poissonian, super-Poissonian, and sub-Poissonian. In photodetection experiments, Poissonian and super-Poissonian statistics turn out to be consistent with the classical theory of light but not the sub-Poissonian statistics. Observation of sub-Poissonian photon statistics is a direct confirmation of the photon nature of light. With the development of high-efficiency detectors, this has been observed in spite of the fact that the sub-Poissonian light is very sensitive to optical losses.

18.1 Young's double-slit experiment

Young's double-slit experiment is shown in figure 18.1. In this figure, quasi-monochromatic light source is used. The light wave is divided into two parts as it passes through the two holes or slits P_1 and P_2. An electric field can be decomposed into two parts $E(r, t) = E^+(r, t) + E^-(r, t)$, $E^+(r, t)$ is the positive frequency part and $E^-(r, t)$ the negative frequency part. The electric field at point PD is the superposition of two fields coming from the slits P_1 and P_2.

$$E^+(r, t) = E^+(r_1, t_1) + E^+(r_2, t_2) \tag{18.1}$$

Here r_1 and r_2 are the position vectors of the two slits and t_1 and t_2 are the delayed times defined as,

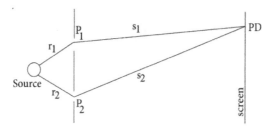

Figure 18.1. Young's double slit experiment set up P_1 and P_2 are the two slits and PD (photodetector) is present in the screen.

$$t_j = \left(t - \frac{s_j}{c}\right), j = 1, 2 \qquad (18.2)$$

$s_j(j = 1, 2)$ are the distances of slits (P_1 and P_2) to the photodetector. We would like to calculate the intensity at PD in screen S, where the photodetector is placed. By using equation (18.1) we can calculate the intensity,

$$I(r, t) = |E^+(r, t)|^2 = |E^+(r_1, t_1)|^2 + |E^+(r_2, t_2)|^2$$
$$+ 2Re\{E^-(r_1, t_1)E^+(r_2, t_2)\} \qquad (18.3)$$

Since in general light source contains noise one can use a statistical approach, repeat the measurements many times and average the results. Mathematically this looks like,

$$\langle I(r, t)\rangle = \langle |E^+(r, t)|^2\rangle$$

$$= \langle |E^+(r_1, t_1)|^2\rangle + \langle |E^+(r_2, t_2)|^2\rangle + 2Re\langle E^-(r_1, t_1)E^+(r_2, t_2)\rangle \qquad (18.4)$$

Here, (......) represents the ensemble average, we can write the first-order correlation function as,

$$G^1(r_1t_1, r_2t_2) = \langle E^-(r_1, t_1)E^+(r_2, t_2)\rangle \qquad (18.5)$$

Now we can rewrite equation (18.4) in terms of $G^1(r_1t_1, r_2t_2)$,

$$\langle I(r, t)\rangle = \langle |E^+(r, t)|^2\rangle = G^1(r_1t_1, r_1t_1) + G^1(r_2t_2, r_2t_2) + 2ReG^1(r_1t_1, r_2t_2) \quad (18.6)$$

The first two terms of the right hand of equation (18.6) are real and positive quantity, whereas $G^1(r_1t_1, r_2t_2)$ is in general complex. This complex correlation function $G^1(r_1t_1, r_2t_2)$ can be written as

$$G^1(r_1t_1, r_2t_2) = |G^1(r_1t_1, r_2t_2)| e^{i\theta(r_1t_1, r_2t_2)} \qquad (18.7)$$

Here, $\theta(r_1t_1, r_2t_2)$ is the phase factor, by putting the above value of $G^1(r_1t_1, r_2t_2)$ in equation (18.7) we get

$$\langle I(r, t)\rangle = G^1(r_1t_1, r_1t_1) + G^1(r_2t_2, r_2t_2) + 2|G^1(r_1t_1, r_2t_2)|\cos\theta \qquad (18.8)$$

In equation (18.8), the cosine term varies sinusoidally with time and it is responsible for the interference fringes in the photodetector. Now we will discuss the physical significance of $G^1(r_1t_1, r_2t_2)$, visibility (V) or contrast of the interference fringe is defined as

$$V = \frac{\langle I(r, t)\rangle_{\text{max}} - \langle I(r, t)\rangle_{\text{min}}}{\langle I(r, t)\rangle_{\text{max}} + \langle I(r, t)\rangle_{\text{min}}} \tag{18.9}$$

If $\cos\theta = 1$, then $\langle I \rangle = \langle I \rangle_{\text{max}}$ and for $\cos\theta = -1$, then $\langle I \rangle = \langle I \rangle_{\text{min}}$ by using the values of $\langle I \rangle_{\text{max}}$ and $\langle I \rangle_{\text{min}}$ in equation (18.9) we get visibility is equal to,

$$V = \frac{2 \mid G^1(r_1t_1, r_2t_2) \mid}{G^1(r_1t_1, r_1t_1) + G^1(r_2t_2, r_2t_2)} \tag{18.10}$$

$G^1(r_1t_1, r_1t_1)$ and $G^1(r_2t_2, r_2t_2)$ are the intensities at the photodetector due to the waves coming from the first and second slit of figure 18.1. So the visibility is proportional to $\mid G^1(r_1t_1, r_2t_2) \mid$, and if $\mid G^1(r_1t_1, r_2t_2) \mid$ is maximum the visibility is maximum. $G^1(r_1t_1, r_2t_2)$ is the electric field fluctuations and can be represented as

$$G^1(r_1t_1, r_2t_2) = \varepsilon^*(r_1, t_1)\, \varepsilon(r_2, t_2) \tag{18.11}$$

$\varepsilon(r_j, t_j), j = 1, 2$ are the electric field.

A monochromatic field is coherent to all orders but a first-order coherent field is not necessarily monochromatic. This is although we often deal with stationary light, such as that from stars and CW light sources, If a stationary field has two time-dependent properties then by definition it depends on the time difference. The corresponding first-order correlation function thus has the form,

$$G^1(t_1, t_2) = G^1(t_1 - t_2) = G^1(\tau) \tag{18.12}$$

So for a stationary first-order coherent field the correlation function shown in equation (18.11), can be written as

$$G^1(t_1 - t_2) = \varepsilon^*(t_1)\quad \varepsilon(t_2) \tag{18.13}$$

The above is true when $\varepsilon(t_1) = Ce^{-i\nu t_1}$, '$C$' is a constant. It implies that the stationary first-order coherent field is monochromatic.

18.2 Hanbury Brown–Twiss experiment

We now describe the celebrated HB–T experiment shown in figure 18.2. It probes the higher-order coherence properties of a field. A beam of light (taken from a star in the original experiment) is split into two beams, which are detected by detectors PD_1 and PD_2. Two amplifiers A_1 and A_2 amplify the signals of the two photodetectors. A correlator multiplies the signal and averages. Unlike in the Young double-slit experiment, here light intensities, rather than amplitudes, are compared. Two absorption measurements are performed on the same field at times t and $(t + \tau)$. It can be shown that these measure $\mid E^+(r, t)E^+(r, t + \tau)\mid^2$. Dropping the useless

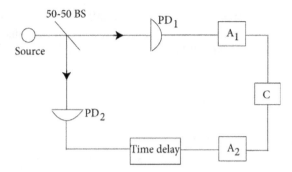

Figure 18.2. Schematic experimental set up of a HB–T interferometer. BS: beam splitter, PD_1, PD_2: photodetectors A_1 and A_2: amplifiers, C: correlator.

variable and averaging, we see that this is precisely the second-order correlation function.

Second-order correlation function at position r and time t and $(t + \tau)$ may be written as

$$G^2(r, \tau) = <E^-(r, t)E^-(r, t + \tau)E^+(r, t)E^+(r, t + \tau)> \qquad (18.14)$$

The normalized second-order correlation function $g^2(r, \tau)$ of light can be defined in the following way after dropping the position variable (r).

$$g^2(\tau) = \frac{<E^-(t)E^-(t + \tau)E^+(t)E^+(t + \tau)>}{<E^-(t)E^+(t) > <E^-(t + \tau)E^+(t + \tau)>}$$

$$= \frac{<I(t)I(t + \tau)>}{<I(t) > <I(t + \tau)>} = \frac{G^2(\tau)}{\mid G^1(0) \mid^2} \qquad (18.15)$$

Here $I(t)$ and $I(t + \tau)$ are the intensity of the light field at a time t and $(t + \tau)$, (...) implies the time average. If we consider $<I(t + \tau)> = <I(t)>$ it means that the light intensity is the same at different times, i.e. constant intensity. If we take a spatially coherent light arriving from a small area source then measurement of the $g^2(\tau)$ means we study the spatial coherence of this light source. If τ_c is the coherence time of the light source then at $\tau \gg \tau_c$, there is no correlation in the fluctuation of intensity at time t and $(t + \tau)$. Fluctuations $\Delta I(t)$ and $\Delta I(t + \tau)$ are uncorrelated with each other, their time average is also zero. We can write intensity at time t as

$$I(t) = <I > +\Delta I(t) \qquad (18.16)$$

$<I(t)I(t + \tau)>$ and at $\tau \gg \tau_c$ this can be written as

$$<I(t)I(t + \tau)>_{\tau \gg \tau_c} = <(<I > +\Delta I(t))(<I > +\Delta I(t + \tau)) > =<I>^2 \qquad (18.17)$$

since both $<\Delta I(t)>$ and $<\Delta I(t + \tau)>$ are equal to zero at $\tau \gg \tau_c$. Now from equation (18.15), we can write

$$g^2(\tau)_{\tau \gg \tau_c} = \frac{<I(t)I(t+\tau)>}{<I(t)><I(t)>} = \frac{<I(t)>^2}{<I(t)>^2} = 1 \tag{18.18}$$

In the time scale $\tau \ll \tau_c$, we will get correlations between the fluctuations at two times. The value of the second-order correlation function at $\tau = 0$ is

$$g^2(0) = \frac{<I(t)I(t)>}{<I(t)><I(t)>} = \frac{<I(t)^2>}{<I(t)>^2} \tag{18.19}$$

It can be shown by using equation (18.19) that for any plausible time dependence of $I(t)$ the value of $g^2(0)$ is always,

$$g^2(0) \geqslant 1 \tag{18.20}$$

Comparing equations (18.18) and (18.20) we can write

$$g^2(0) \geqslant g^2(\tau) \tag{18.21}$$

For a perfectly monochromatic light like laser source with constant intensity (I_0) it is easy to show the value of $g^2(\tau) = 1$ for all values of τ.

$$g^2(\tau) = \frac{<I(t)I(t+\tau)>}{<I(t)>^2} = \frac{I_0^2}{I_0^2} = 1 \tag{18.22}$$

All classical light $g^2(0) \geqslant 1$ for a source whose intensity is varying with time, we will see $g^2(\tau)$ increases with increasing time and for a large value of τ it becomes one. In the case of a perfectly coherent light where $I(t)$ is not varying with time we have seen $g^2(\tau) = 1$, So $g^2(\tau)$ does not depend on time delay τ. By using equation (18.15) we can write

$$g^2(0) = 1 + \frac{G^2(0)}{|G^1(0)|^2} - \frac{|G^1(0)|^2}{|G^1(0)|^2} = (1 + \frac{G^2(0) - |G^1(0)|^2}{|G^1(0)|^2}) \tag{18.23}$$

The above can be written by using the intensity of the field as

$$g^2(0) = 1 + \frac{<I^2> - <I>^2}{<I>^2} = 1 + \frac{<(I - <I>)^2>}{<I>^2} \tag{18.24}$$

The second term of equation (18.24) must be positive since $<(I - <I>)^2> \geqslant 0$, so $g^2(0) \geqslant 1$. In figure 18.3 we have plotted the second-order correlation function ($g^2(\tau)$) versus τ for a perfectly coherent light and a non-coherent light. We can see that the solid line indicates ($g^2(0) > 1$)meaning non-coherent light and in the dotted line $g^2(0) = 1$ so it is a perfectly coherent source.

We deal with the classical description of the time-dependent intensity fluctuations in a light beam, first investigated by R Hanbury Brown and R Q Twiss in the 1950s. Modern quantum optics revolves on their work significantly. The HB–T experiments naturally led to the concept of the second-order correlation function. We study the values that $g^2(\tau)$ can take for different types of classical light.

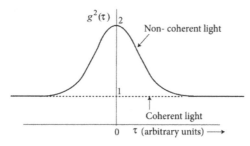

Figure 18.3. Plot of $g^2(\tau)$ versus τ for non-coherent (solid line) and perfectly coherent (dotted line) sources.

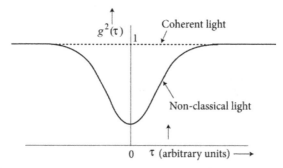

Figure 18.4. Plot of $g^2(\tau)$ versus τ for non-classical light (solid line).

Examples of classical lights are coherent light sources, non-coherent sources like thermal sources, chaotic sources etc. All classical lights obey the inequality given in equation (18.21). Violation of this inequality leads to *non-classical light*. It is possible to get a light source with $g^2(\tau) < 1$. Observation of such a result indicates the quantum nature of light, i.e. non-classical light. Later we will see the light sources that exhibit these non-classical results (i.e. $g^2(\tau) < 1$) are represented as *anti-bunched* light and it is particularly interesting in quantum optics. In figure 18.4 we have presented $g^2(\tau)$ versus τ plot for a non-classical light as indicated by the solid line for which $g^2(\tau) < 1$ and dotted lines show coherent light.

Till now we described the $g^2(\tau)$ function in terms of intensity correlations. One can write the equation (18.15) with photon number fluctuations. If $n_1(t)$ is number of photon count in the detector PD_1 at a time t and $n_2(t + \tau)$ is the same for the detector PD_2 at a time $(t + \tau)$ as shown in figure 18.2 of the HB–T interferometer. The $g^2(\tau)$ can be written as

$$g^2(\tau) = \frac{<n_1(t)n_2(t + \tau)>}{<n_1(t)> <n_2(t + \tau)>} \qquad (18.25)$$

Equation (18.25) says $g^2(\tau)$ depends on the probability of photon count by both photodetectors PD_1 and PD_2 at the time t and $(t + \tau)$.

18.3 Photon counter

We consider an experiment where a faint light source sends a beam that falls on a very sensitive photodetector like photomultiplier tube (PMT) or avalanche photodiode for a specific time interval T by having a shutter open in front of the detector for the time T. We assume that the light source is stationary, i.e., the long-time average of the intensity is fixed and independent of the period of measurement. The number of photoelectrons liberated is registered. After a certain time delay (that is longer than the coherence time of the source) the shutter is opened again for an equal time interval T and the number of photoelectrons counted during the interval is again registered in a time-correlated photon counter (TCPC). This experiment is repeated a large number of times ($\approx 10^5$) and the number of photoelectrons produced in equal intervals of time is counted. The results of such an experiment give the probability distribution $p(n,T)$ of counting n photons in a time T. The probability distribution so obtained contains information regarding the statistical properties of the source. Such studies have applications in spectroscopy, stellar interferometry etc.

A schematic diagram of the experimental setup is shown in figure 18.5, this experimental setup is known as a photon counter whose operating principle is similar to the Geiger counter used in a nuclear physics experiment. It is evident from the analogy between a photon counter and a Geiger counter that the number of expected counts in a given time interval would not be constant. The radioactive process is intrinsically random and so a Geiger counter, records a count rate that fluctuates about the average value. This is similar to what happens with a photon counter where the average count rate is determined by the intensity of the light beam, but the actual count rate fluctuates from measurement to measurement. We are concerned with these fluctuations in the count rate.

18.4 Outcome of the photon counter

Now we will discuss the outcome or the result coming from the photon counting experiment if we take a perfectly monochromatic light source. Then the intensity of the beam is constant (say I_0) and the linear frequency of the photon is ν. The photon flux (ϕ) of the beam can be written in terms of the power of the beam (P) divided by the energy of an individual photon.

$$\phi = \frac{P}{h\nu} = \frac{P}{\hbar\omega} \tag{18.26}$$

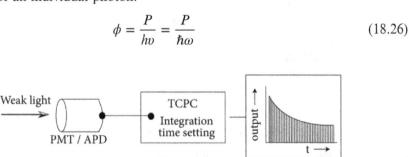

Figure 18.5. Schematic experimental diagram of photon counter. PMT: photomultiplier tube, APD: avalanche photodiode, TCPC: time correlated photon counter.

Here $\omega = 2\pi\upsilon$ is the circular frequency, power of the beam can be calculated as $P = I_0 A$, 'A' is area of the beam. The unit of ϕ is photons per second. As described in the last section, the detector is open for a time T, the number of photon count registered by the detector during this time is

$$n(T) = \eta\phi T \qquad (18.27)$$

where η is the quantum efficiency of the photodetector. The count rate is the number of counts registered per second, i.e.

$$\mathbb{C}_R = \frac{n(T)}{T} = \eta\frac{P}{\hbar\omega}. \qquad (18.28)$$

The count rate of a detector also depends on the dead time of the detector which is the recovery time of the detector (i.e. detectors need some time to recover after counting one event). If the dead time of a detector is about 1 μs, then this is the time gap between the two successive events. The maximum count rate ($\mathbb{C}_R$) will be 10^6 counts per second. The quantum efficiency of the detector is generally 10%. We can use equation (18.29) to get the power of the beam and it is in the order of 10^{-12} Watt or less. So it is clear that the photon counters can only be useful to measure properties of a very low power beam.

Now we want to calculate the photon flux by using equation (18.26) for a faint light beam having power $P = 2$ nW and photon energy is 4 eV. Then $\phi = \frac{2 \times 10^{-9}}{4 \times (1.6 \times 10^{-19})} = 3.1 \times 10^9$ photons s^{-1}. Since the velocity of light in free space is 3×10^8 m so we can say a beam of length 3×10^8 m can contain 3.1×10^9 photons on average.

Photons are discrete energy packets, and the actual number of photons in a beam segment (large or small) cannot be fractional and has to be an integer. In statistical physics, it is also assumed that the photons are equally likely to be at any point within the beam.

In an experiment with any number of beam segments (large or small), we find random fluctuations and even might get fractional values of mean and standard deviation.

Statistical fluctuations arise because we do not know whether the photons are really within the beam segment. Some beam segments may be empty and some may be populated with photons (that cannot be divided or split) but those segments cannot be identified. Also, the shorter the time interval, the more difficult it becomes to know where the photons are since then a larger number of beam segments fall empty and only a few get populated by photons. This is the cause of fractional values of mean or standard deviation.

18.5 Photon statistics of a perfectly coherent light

Now we will see the nature of the photon statistics of perfectly coherent light as we know a coherent light has constant intensity. If there are no intensity fluctuations then the power of the beam does not vary with time, then the average photon flux of the coherent light is constant with time. Since photon flux (ϕ) is defined by equation (18.26) if P is not varying with time then ϕ is constant.

We have noticed the statistical fluctuations on short time scales due to the discrete nature of the photons and so a beam of light with a time-invariant average photon flux would not consist of a stream of photons with regular time intervals between them. Perfectly coherent light with a constant intensity has *Poissonian* photon statistics which we will prove now.

If we consider the beam segment length l of a coherent light then the average number of photons within this length is

$$\bar{n} = \frac{l\phi}{c} \tag{18.29}$$

Here c is the velocity of the light in free space and ϕ is the photon flux as defined in equation (18.26). We chose the value of l to be large so that n is an integer value, if we sub-divide this length segment into N intervals. Then the length of each interval is $\frac{l}{N}$, the probability of finding one photon per interval is $p = \frac{\bar{n}}{N}$. Here we assume N is adequately large. The probability of finding no photons per interval is $(1 - p)$. The probability of finding the n photons in a beam length l with intervals N can be found by using binomial distribution

$$P(n) = \frac{N!}{n!(N-n)!}p^n(1-p)^{N-n} \tag{18.30}$$

Here $(1 - p)^{N-n}$ is the probability of finding no photons and $\frac{N!}{n!(N-n)!}p^n$ is the probability of finding one photon in n subsegments. Let us put $p = \frac{\bar{n}}{N}$ in equation (18.30) and under the assumption $N \to \infty$, we get

$$P(n) = \frac{N!}{n!(N-n)!}(\frac{\bar{n}}{N})^n \quad (1 - \frac{\bar{n}}{N})^{N-n}$$

$$P(n) = \frac{1}{n!}\left(\frac{N!}{N^n(N-n)!}\right)\bar{n}^n(1 - \frac{\bar{n}}{N})^{N-n} \tag{18.31}$$

At $N \to \infty$ we can show that $\left(\frac{N!}{N^n(N-n)!}\right) = 1$ and $(1 - \frac{\bar{n}}{N})^{N-n} = e^{-\bar{n}}$. From Stirling's formula, $\lim N \to \infty$, $\ln(N!) = N \ln N - N$ then we can write

$$\lim_{N\to\infty}\left\{\ln\left(\frac{N!}{N^n(N-n)!}\right)\right\} = 0$$

Then

$$\lim_{N\to\infty}\left\{\left(\frac{N!}{N^n(N-n)!}\right)\right\} = 1 \tag{18.32}$$

Now if we do the series expansion of the term $(1 - \frac{\bar{n}}{N})^{N-n}$ and under the assumption $N \to \infty$ it gives,

$$(1 - \frac{\bar{n}}{N})^{N-n} \to 1 - \bar{n} + \frac{\bar{n}^2}{2!} - \cdots\cdots$$

$$= e^{-\bar{n}} \tag{18.33}$$

Then equation (18.31) reduced to

$$\lim_{N \to \infty} \{P(n)\} = \left(\frac{1}{n!}\right) . 1 . \bar{n}^n . e^{-\bar{n}} \tag{18.34}$$

So the probability distribution of arriving photons from a perfectly coherent source is

$$P(n) = \frac{\bar{n}^n}{n!} e^{-\bar{n}} \tag{18.35}$$

Equation (18.35) is a Poisson distribution so we can say the photon statistics of coherent light (e.g. laser) is a *Poisson distribution*.

Random processes that can only return integer values can be described through Poissonian statistics in general, e.g. the number of counts detected by a Geiger tube studying a radioactive source where the number of counts is integral and the average count value n is decided by the half-life of a radioactive source, its amount, and the recording time interval. The probability for registering n counts is given by the Poissonian formula of equation (18.36), as because of the random nature of the radioactive source the actual count values fluctuate above and below the mean value. The count rate recorded by a photon-counting system observing individual photons from a light beam with constant intensity shows a similar trend, the randomness being due to disintegration of the continuous beam into discrete energy packets having an equal probability of being detected within any given time sub-interval.

In the earlier chapter, we have plotted the Poisson distribution ($P(n)$) with photon number (n) curve and also proved standard variation or variance is $\Delta n = \sqrt{\bar{n}}$ in the case of Poisson distribution.

18.6 Photon statistics of a thermal light

Thermal light is the electromagnetic wave coming from a hot body, also known as black-body radiation. It will follow Planck's distribution formula and the average or mean number of the photon in a single-mode thermal radiation field can be written as,

$$\bar{n} = \frac{1}{(e^{\hbar\omega/K_B T} - 1)} \tag{18.36}$$

Here ω is the circular frequency of the radiation field and T is the equilibrium temperature and K_B is the Boltzmann constant. The photon occupation probability or the probability of find n photons in a single mode (P_n) can easily be deduced from Boltzmann's law. The energy of the state with n photon is

$$E_n = (n + \frac{1}{2})\hbar\omega \tag{18.37}$$

Here we consider that electromagnetic radiation is generating due to vibration of the linear harmonic oscillator at a frequency ω and n is an integer and greater than zero.

$$P_n = \frac{e^{-\frac{E_n}{K_B T}}}{\sum\limits_{n=0}^{\infty} e^{-\frac{E_n}{K_B T}}}$$

$$= \frac{e^{-\frac{(n+\frac{1}{2})\hbar\omega}{K_B T}}}{\sum\limits_{n=0}^{\infty} e^{-\frac{(n+\frac{1}{2})\hbar\omega}{K_B T}}}$$

$$= \frac{e^{-\frac{n\hbar\omega}{K_B T}}}{\sum\limits_{n=0}^{\infty} e^{-\frac{n\hbar\omega}{K_B T}}}, \text{ putting } \frac{\hbar\omega}{K_B T} = x$$

$$= e^{-nx}(1 - e^{-x})(\because \sum\limits_{n=0}^{\infty} a^n = \frac{1}{(1-a)}, \text{ Summed geometric series})$$

$$= e^{-nx}(\frac{e^x - 1}{e^x})$$

$$= \frac{e^x - 1}{(e^x)^{n+1}}$$

$$= \frac{(e^x - 1)^{n+1}}{(e^x)^{n+1}} \frac{1}{(e^x - 1)^n}$$

$$= \frac{(\frac{1}{e^x - 1})^n}{(\frac{1}{e^x - 1} + 1)^{n+1}}$$

$$P_n = \frac{\bar{n}^n}{[\bar{n} + 1]^{n+1}} = \frac{1}{(1 + \bar{n})}(\frac{\bar{n}}{1 + \bar{n}})^n \tag{18.38}$$

In equation (18.38) we have used the expression of n taken from equation (18.36). The above distribution (P_n) is known as *Bose–Einstein* (B–E) distribution. It is seen from the equation (18.38) P_n is maximum at $n = 0$ and its value decreases exponentially with increasing n. If we put $y = e^{-\frac{\hbar\omega}{K_B T}}$, then the expression of thermal probability becomes $P_n = \dfrac{y^n}{\sum\limits_{n=0}^{\infty} y^n} = y^n(1-y)$, the mean or the average number of photon n excited in the field mode can be derived as,

$$\bar{n} = \sum_{n=0}^{\infty} nP_n = (1-y) \sum_{n=0}^{\infty} ny^n = (1-y)y\frac{\partial}{\partial y}\sum_{n=0}^{\infty} y^n$$

$$\bar{n} = (1-y)y\frac{\partial}{\partial y}\left(\frac{1}{1-y}\right) = \frac{y}{(1-y)} = \frac{e^{-\frac{\hbar\omega}{K_B T}}}{(1-e^{-\frac{\hbar\omega}{K_B T}})} = \frac{1}{(e^{\frac{\hbar\omega}{K_B T}}-1)} \qquad (18.39)$$

This is the derivation of equation (18.36). It is plotted in figure 18.6, n with photon frequency ω.

Now we will calculate the variance Δn of the B–E distribution. We know $\Delta n^2 = (\overline{n^2} - \bar{n}^2)$, let's calculate $\overline{n^2}$ following way,

$$\overline{n^2} = \sum_{n=0}^{\infty} n^2 P_n = \sum_{n=0}^{\infty} n^2 y^n(1-y) = y(1-y)\frac{\partial}{\partial y}\sum_{n=0}^{\infty}(ny^n)$$

$$= y(1-y)\sum_{n=0}^{\infty}\frac{\partial}{\partial y}(ny\,y^{n-1}) = y(1-y)\sum_{n=0}^{\infty}\frac{\partial}{\partial y}[y\frac{\partial}{\partial y}(y^n)]$$

$$= y(1-y)\frac{\partial}{\partial y}\left[y\frac{\partial}{\partial y}\sum_{n=0}^{\infty}(y^n)\right] = y(1-y)\frac{\partial}{\partial y}[y\frac{\partial}{\partial y}\left(\frac{1}{1-y}\right)]$$

$$= y(1-y)\left[\frac{(1-y)^2 + 2y(1-y)}{(1-y)^4}\right]$$

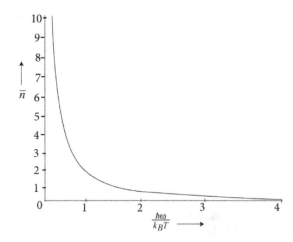

Figure 18.6. Average or mean photon number n versus photon frequency ω plot for thermal light.

$$=\frac{y}{(1-y)} + \frac{2y^2}{(1-y)^2} = (\bar{n} + 2\bar{n}^2) \tag{18.40}$$

The variance of B–E distribution is

$$\Delta n^2 = (\bar{n} + 2\bar{n}^2) - \bar{n}^2 = (\bar{n} + \bar{n}^2) \tag{18.41}$$

So we can see here $\Delta n > \sqrt{\bar{n}}$, in the case of thermal light which follows B–E distribution, the variance is always greater than the variance of the Poisson distribution obeyed by perfectly coherent light. We can conclude the B–E distribution is a *super-Poisson distribution*. The signal-to-noise ratio (SNR) is defined as SNR $= \frac{\bar{n}^2}{\Delta n^2}$, so in the case of thermal light SNR $= \frac{1}{(1+\bar{n})}$ which is always smaller than one. But if we take the coherent light SNR $= \frac{\bar{n}^2}{\bar{n}} = \bar{n}$, large photon numbers produces a high SNR. So if we use coherent light in an experiment we get better SNR in comparison to using thermal light.

Figure 18.6(a) and (b) shows the plot of the B–E distribution P_n given in equation (18.38) and Poisson distribution $P(n)$ shown in equation (18.35) for $n = 5$.

We have plotted both the distribution for $n = 5$, in the case of the Poisson distribution variance, it is $\Delta n = 2.23$ and for B–E distribution $\Delta n = 5.47$. So the photon coming from a thermal source is a more widely spread distribution than the photons arriving from a perfectly coherent source. It is also seen from figure 18.7(b) even for a beam of constant intensity light the fluctuations still exist.

18.7 Classification of light by second-order correlation function and photon statistics.

Now we will evaluate the second-order correlation ($g^2(\tau)$) function for thermal, coherent and non-classical fields. We can rewrite equation (18.24) in terms of variance Δn and photon number; here we will restrict ourselves to a single-mode field.

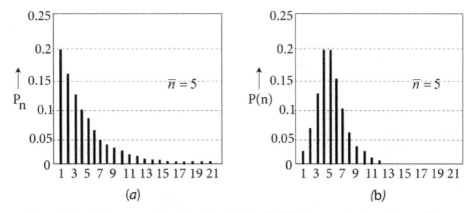

Figure 18.7. (a) P_n versus n plot for a thermal light, (b) $P(n)$ versus n plot for a coherent light.

$$g^2(0) = \frac{<a^+a^+aa>}{<a^+a>^2} = 1 + \frac{\Delta n^2 - \bar{n}}{\bar{n}^2} \tag{18.42}$$

where $\Delta n^2 = (<(a^+a)^2> -<a^+a>^2)$. $\Delta n^2 = (\bar{n} + \bar{n}^2)$for the thermal light which follows the B–E distribution if we substitute it in equation (18.42) we get

$$g^2(0) = 1 + \frac{(\bar{n} + \bar{n}^2) - \bar{n}}{\bar{n}^2} = 1 + 1 = 2 \tag{18.43}$$

So in the case of thermal light, the value of the second-order correlation function is $g^2(0) = 2$. Now we will check the value of $g^2(0)$ for perfectly coherent light and non-classical light. With $\Delta n^2 = \bar{n}$ for coherent light, we have from equation (18.42)

$$g^2(0) = 1 + \frac{\bar{n} - \bar{n}}{\bar{n}^2} = 1 \tag{18.44}$$

So we can conclude that for coherent light which has Poissonian photon distribution the value of $g^2(0) = 1$. A well-known example of non-classical light is Fock states or number states. Fock state is a state with a well-defined number of the photon in a mode they are eigenstate of the number operator it can be represented as,

$$| n > = \frac{(a^+)^n}{\sqrt{n!}} | 0> \tag{18.45}$$

Here $|0>$ is the vacuum state. They form a complete set of states, i.e. $\sum_{n=0}^{\infty} | n > <n | = \hat{I}$, here $\hat{I}$ is the identity operator in the Hilbert space of the single-mode system. The variance of a single-mode field in the Fock state is zero, one can calculate it as,

$$<n |\Delta n^2 | n > =<n | n^2 | n > -<n | n | n>^2 = 0 \tag{18.46}$$

This should vanish since the value of the photon number is 'sharp' in a Fock state and due to this there are no energy fluctuations in the Fock states. Now by using equation (18.42) and putting $\Delta n = 0$ we get

$$g^2(0) = 1 + \frac{0 - \bar{n}}{\bar{n}^2} = 1 - \frac{1}{\bar{n}} \tag{18.47}$$

So $g^2(0) < 1$ for the non-classical light. The probability distribution

$$P_N = |<n | n > |^2 = 1 \tag{18.48}$$

In table 18.1 we have classified the three types of light with their distribution function, variance and second-order correlation function.

Three types of photon statistics are possible in terms of the variance or photon fluctuations of the three types of lights (i.e. thermal, coherent and non-classical) they are as follows:

Thermal light: $\Delta n > \sqrt{\bar{n}}$ super-Poissonian photon statistics.
Coherent light: $\Delta n = \sqrt{\bar{n}}$ Poissonian photon statistics.

Table 18.1. Classification of lights according to photon statistics and second-order correlation function.

Types of light	Photon number distribution function		Variance Δn	Second order correlation function $g^2(\tau)$
Thermal light	$P_n = \frac{1}{1+\bar{n}}(\frac{\bar{n}}{1+\bar{n}})^n$	Bose Einstein distribution	$\Delta n > \sqrt{\bar{n}}$	$g^2(0) = 2\,g^2(0) \geqslant g^2(\tau)$
Coherent light	$P(n) = \frac{\bar{n}^n}{n!}e^{-\bar{n}}$	Poisson distribution	$\Delta n = \sqrt{\bar{n}}$	$g^2(0) = 1$ and $g^2(\tau) = 1$ for all values of τ
Non classical light	$P_N = 1$		$\Delta n = 0$	$g^2(0) < 1$, $g^2(0) < g^2(\tau)$

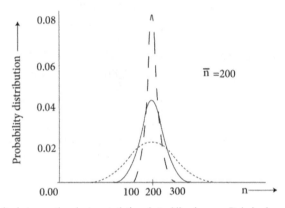

Figure 18.8. Disimilarity between the photon statistics: dotted line is super Poissionian, solid line is Poissionian and dashed line is sub Poissionain. All are for the same mean photon number $n = 200$.

Non-classical light: $\Delta n < \sqrt{\bar{n}}$ sub-Poissonian photon statistics.

Figure 18.8 highlights the difference between three different types of statistics where photon number distributions of super-Poissonian and sub-Poissonian light have been compared to that of a Poisson distribution with the same mean photon number $n = 200$. It is evident from the figure that distributions of super-Poissonian and sub-Poissonian lights are, respectively, broader or narrower than the Poisson distribution.

Let us now refer to the types of light having super-Poissonian statistics. Any classical fluctuations in the intensity mean that larger photon number fluctuations are expected compared to the constant intensity case. A perfectly stable intensity gives Poissonian statistics. It follows that all classical light beams with time-varying light intensities will have super-Poissonian photon number distributions, e.g. thermal light from a black-body source and the partially coherent light from a discharge lamp. They have larger variations in the intensity, i.e. they are 'noisier' than perfectly coherent light in the classical sense and have larger photon number fluctuations in the quantum sense.

On the other hand, sub-Poissonian light has a narrower distribution than the Poissonian one and is, therefore, 'quieter' than perfectly coherent light. A perfectly coherent beam is the most stable form of light possible in classical optics. Sub-Poissonian light has no classical counterpart. It is, therefore, the first example of *non-classical light*. Sub-Poissonian light is quite difficult and generally not dealt with in standard optics texts.

18.8 Photon bunching and anti-bunching

In the last section, we showed the difference between the three types of light based on which statistics they are following. Photons coming from black body radiation (i.e. thermal light, chaotic light etc) are non-coherent photons and they follow the super-Poissonian statistics. In the case of perfectly coherent light, these photons have the Poissonian statistics. Photons of non-classical light have photon distribution narrower than Poissonian which is sub-Poissonian statistics. We also divided these three classes of light in terms of the second-order correlation function $g^2(\tau)$ as shown in table 18.1. Now we can classify these lights based on $g^2(0)$ if $g^2(0) = 1$: *coherent light*, $g^2(0) > 1$: *bunched light* and $g^2(0) < 1$: *anti-bunched light*.

This classification is done depending upon how the photon streams are moving. Later we will discuss it in detail.

All classical light follows the Schwartz inequality ($<\hat{A}\hat{B}>^2 \leqslant <\hat{A}^2> <\hat{B}^2>$) but non-classical light violates this inequality and so it has no classical counterpart. Non-classical light can only be described by quantum mechanics. In table 18.2 we have tabulated the anti-bunched, bunched and coherent lights with their properties.

18.8.1 Coherent light

The schematic representation of the photon stream of coherent light is shown in figure 18.9. It is seen from this figure that the photons are coming randomly, with no time correlation between two photons.

Table 18.2. Anti-bunched, bunched and coherent lights with their property.

Types of light	Photon stream in the light beam	Photon statistics	$g^2(0)$
Chaotic light: pretty much everything other than lasers	Bunching	Super-Poissionian	$g^2(0) > 1$
Coherent light : the intensity of the beam is constant with time, e.g. laser	Random	Poissionian	$g^2(0) = 1$
No classical analogy	Anti-bunching	Sub-Poissionian	$g^2(0) < 1$

Figure 18.9. Stream of photons of perfectly coherent light.

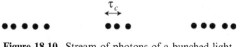

Figure 18.10. Stream of photons of a bunched light.

Figure 18.11. Photon stream of anti-bunched light.

In the earlier sections, we have discussed properties of perfectly coherent light and here we summarise those points. From a classical point of view, no intensity fluctuations in the coherent light in the quantum picture and no photon bunching is seen in figure 18.9. The source of the coherent light is laser emission.

1. It has Poissonian photon-number statistics, $P(n) = \frac{\bar{n}^n}{n!}e^{-\bar{n}}$.
2. Random photon arrival times, the time intervals between the photons are not constant, but rather, random.
3. The second-order correlation function $g^2(\tau) = 1$ for all values of τ.
4. Partially satisfies both Cauchy–Schwarz inequality and Heisenberg uncertainty.

18.8.2 Bunched light

An example of bunched light is thermal light, chaotic light etc, where the intensity of the beam is not constant in time. The photon stream of the bunched light is shown in figure 18.10.

Photons are bunches which are visible in figure 18.10, the time intervals of photon bunches are of the coherence time (say τ_c).

We list the main properties of bunched light below,

1. For bunched light second-order correlation function $g^2(0) > 1$, in case of thermal light $g^2(0) = 2$.
2. In a stream of photons of a bunched light, the photons arrive together as in a cluster.
3. In bunched light $g^2(0) > g^2(\tau = \infty)$ and so the probability of measuring a photon in a small value of τ is larger than in a high τ.
4. It follows super-Poissonian photon statistics, e.g. B–E distribution in the case of thermal light.

18.8.3 Anti-bunched light

Photons emerge out with regular gaps between them in anti-bunched light instead of maintaining a random spacing, as shown schematically in figure 18.11. For a regular photon flow, a long time is elapsed between two photon counts. In this case, the probability of getting a photon on PD_1 after detecting one on PD_2 (HB–T interferometer in figure 18.2) is small for small values of τ and then increases with τ. Hence anti-bunched light has

$$g^2(\tau) > g^2(0) \text{ and } g^2(0) < 1 \qquad (18.49)$$

Like coherent and bunched light here also we summarize the main features of anti-bunched light below.

1. It violates the Schwartz inequality which all classical lights follow. So photon anti-bunching phenomena have purely quantum behaviour of light without any classical analogy.
2. There is no probability to find photons with no time separation, the second-order correlation function increases with time, i.e. $g^2(\tau) > g^2(0)$.
3. Observation of $g^2(0) < 1$ is a clear signature of anti-bunched light.

It has been established from the study of sub–Poissonian light in earlier sections that, like anti-bunched light, sub–Poissonian light violates the classical inequality so it cannot be explained classically. It is only explained by the quantum nature of light. We now address whether photon anti-bunching and sub-Poissonian photon statistics are different manifestations of the same quantum optical phenomenon. According to X T Zou and P Mandel, they are not the same. However, a regular photon stream such as figure 18.11 has sub-Poissonian photon statistics. Thus, although these two phenomena are not exact, non-classical light will frequently possess both photon anti-bunching and sub-Poissonian photon statistics at the same time.

$g^2(\tau) \rightarrow 1$ at long times for all the fields (e.g. thermal, coherent and non-classical) which is clearly illustrated in figures 18.3 and 18.4. There are various experiments by which people observed the photon anti-bunching. Two types of non-classical light sources are very common: (a) single-photon sources; (b) photons pair/two-mode squeezing by using parametric down-conversion.

Significant progress has been made in atomic anti-bunching experiments of late. For example, it is possible to demonstrate anti-bunching from a one-atom laser in the strong coupling regime of cavity quantum electrodynamics. Further, many other types of light emitters including several solid-state sources such as organic molecules like terylene molecule, single CdSe/ZnS quantum dots, quantum dots in a micro-cavity, N-V centre in diamond, Fluorescent dye molecules doped in a crystal or glass show anti-bunching.

18.9 Further reading

[1] Fox M 2006 *Quantum Optics: An Introduction* (Oxford: Oxford University Press)
[2] Mandel L and Wolf E 1995 *Optical Coherence and Quantum Optics* (Cambridge: Cambridge University Press)
[3] Loudon R 2001 *Quantum Theory of Light* 3rd edn (Oxford: Oxford University Press)
[4] Walls D F and Milburn G J 2008 *Quantum Optics* (Berlin: Springer)
[5] Scully M O and Zubairy S 1997 *Quantum Optics* (Cambridge: Cambridge University Press)
[6] Zou X T and Mandel L 1990 Photon antibunching and sub-Poissonian photon statics *Phys. Rev. A* **41** 476

[7] Hanbury Brown R and Twiss R Q 1956 Correlation between photons in two coherent beams of light *Nature* **177** 27

[8] Kimble H K, Dagenais M and Mandel P 1977 Photon antibunching in resonance fluorescence *Phys. Rev. Lett.* **39** 691

[9] Dagenais M and Mandel P 1978 Investigation of two-time correlations in photon emissions from a single atom *Phys. Rev. A* **18** 2217

18.10 Problems

1. Photon number fluctuation is defined as $\Delta n^2 = (<(a^+a)^2> - <a^+a>^2)$, calculate Δn^2 for a coherent state $|\alpha>$.

2. Photon number distribution of a thermal light is described by equation (18.38), show that it exhibits bunching.

3. What is the probability of having n photons for (a) a Poisson distribution, and (b) a thermal distribution, considering one photon on the average?

4. Suppose we are investigating photon statistics of thermal light, the radiation from a blackbody source is filtered with an interference filter of bandwidth 0.2 nm centred at 550 nm. In the said experiment the radiation is allowed to fall on a photon-counting detector. What is the number of modes incident on the detector? Discuss the type of statistics that is expected.

5. Calculate the average and standard deviation of the photocount number if a beam with a photon flux of 500 photons s^{-1} is incident on a detector with a quantum efficiency of 15%, the time interval of the counter being set to 5 s in the following cases:

    (a) the light is in a photon number state.

    (b) the light has Poissonian statistics.

1. Evaluate the values of $g^2(0)$ in the case of a monochromatic light wave with a square wave intensity modulation of $\pm$ 10%.

CPSIA information can be obtained
at www.ICGtesting.com
Printed in the USA
BVHW010832010422
632297BV00011B/21

9 780750 327138